Probability and Mathematical Statistics (Continued)

PURI, VILAPLANA, and WERTZ • New Perspectives in Theoretical and Applied Statistics
RANDLES and WOLFE • Introduction to the Theory of Nonparametric Statistics
RAO • Linear Statistical Inference and Its Applications, *Second Edition*
RAO • Real and Stochastic Analysis
RAO and SEDRANSK • W.G. Cochran's Impact on Statistics
RAO • Asymptotic Theory of Statistical Inference
ROHATGI • An Introduction to Probability Theory and Mathematical Statistics
ROHATGI • Statistical Inference
ROSS • Stochastic Processes
RUBINSTEIN • Simulation and The Monte Carlo Method
SCHEFFE • The Analysis of Variance
SEBER • Linear Regression Analysis
SEBER • Multivariate Observations
SEN • Sequential Nonparametrics: Invariance Principles and Statistical Inference
SERFLING • Approximation Theorems of Mathematical Statistics
SHORACK and WELLNER • Empirical Processes with Applications to Statistics
TJUR • Probability Based on Radon Measures

Applied Probability and Statistics

ABRAHAM and LEDOLTER • Statistical Methods for Forecasting
AGRESTI • Analysis of Ordinal Categorical Data
AICKIN • Linear Statistical Analysis of Discrete Data
ANDERSON, AUQUIER, HAUCK, OAKES, VANDAELE, and WEISBERG • Statistical Methods for Comparative Studies
ARTHANARI and DODGE • Mathematical Programming in Statistics
ASMUSSEN • Applied Probability and Queues
BAILEY • The Elements of Stochastic Processes with Applications to the Natural Sciences
BAILEY • Mathematics, Statistics and Systems for Health
BARNETT • Interpreting Multivariate Data
BARNETT and LEWIS • Outliers in Statistical Data, *Second Edition*
BARTHOLOMEW • Stochastic Models for Social Processes, *Third Edition*
BARTHOLOMEW and FORBES • Statistical Techniques for Manpower Planning
BECK and ARNOLD • Parameter Estimation in Engineering and Science
BELSLEY, KUH, and WELSCH • Regression Diagnostics: Identifying Influential Data and Sources of Collinearity
BHAT • Elements of Applied Stochastic Processes, *Second Edition*
BLOOMFIELD • Fourier Analysis of Time Series: An Introduction
BOX • R. A. Fisher, The Life of a Scientist
BOX and DRAPER • Empirical Model-Building and Response Surfaces
BOX and DRAPER • Evolutionary Operation: A Statistical Method for Process Improvement
BOX, HUNTER, and HUNTER • Statistics for Experimenters: An Introduction to Design, Data Analysis, and Model Building
BROWN and HOLLANDER • Statistics: A Biomedical Introduction
BUNKE and BUNKE • Statistical Inference in Linear Models, Volume I
CHAMBERS • Computational Methods for Data Analysis
CHATTERJEE and PRICE • Regression Analysis by Example
CHOW • Econometric Analysis by Control Methods
CLARKE and DISNEY • Probability and Random Processes: A First Course with Applications, *Second Edition*
COCHRAN • Sampling Techniques, *Third Edition*
COCHRAN and COX • Experimental Designs, *Second Edition*
CONOVER • Practical Nonparametric Statistics, *Second Edition*
CONOVER and IMAN • Introduction to Modern Business Statistics
CORNELL • Experiments with Mixtures: Designs, Models and The Analysis of Mixture Data

Applied Probability and Statistics (Continued)

COX • Planning of Experiments
COX • A Handbook of Introductory Statistical Methods
DANIEL • Biostatistics: A Foundation for Analysis in the Health Sciences, *Fourth Edition*
DANIEL • Applications of Statistics to Industrial Experimentation
DANIEL and WOOD • Fitting Equations to Data: Computer Analysis of Multifactor Data, *Second Edition*
DAVID • Order Statistics, *Second Edition*
DAVISON • Multidimensional Scaling
DEGROOT, FIENBERG and KADANE • Statistics and the Law
DEMING • Sample Design in Business Research
DILLON and GOLDSTEIN • Multivariate Analysis: Methods and Applications
DODGE • Analysis of Experiments with Missing Data
DODGE and ROMIG • Sampling Inspection Tables, *Second Edition*
DOWDY and WEARDEN • Statistics for Research
DRAPER and SMITH • Applied Regression Analysis, *Second Edition*
DUNN • Basic Statistics: A Primer for the Biomedical Sciences, *Second Edition*
DUNN and CLARK • Applied Statistics: Analysis of Variance and Regression
ELANDT-JOHNSON and JOHNSON • Survival Models and Data Analysis
FLEISS • Statistical Methods for Rates and Proportions, *Second Edition*
FLEISS • The Design and Analysis of Clinical Experiments
FOX • Linear Statistical Models and Related Methods
FRANKEN, KÖNIG, ARNDT, and SCHMIDT • Queues and Point Processes
GALLANT • Nonlinear Statistical Models
GIBBONS, OLKIN, and SOBEL • Selecting and Ordering Populations: A New Statistical Methodology
GNANADESIKAN • Methods for Statistical Data Analysis of Multivariate Observations
GREENBERG and WEBSTER • Advanced Econometrics: A Bridge to the Literature
GROSS and HARRIS • Fundamentals of Queueing Theory, *Second Edition*
GUPTA and PANCHAPAKESAN • Multiple Decision Procedures: Theory and Methodology of Selecting and Ranking Populations
GUTTMAN, WILKS, and HUNTER • Introductory Engineering Statistics, *Third Edition*
HAHN and SHAPIRO • Statistical Models in Engineering
HALD • Statistical Tables and Formulas
HALD • Statistical Theory with Engineering Applications
HAND • Discrimination and Classification
HOAGLIN, MOSTELLER and TUKEY • Exploring Data Tables, Trends and Shapes
HOAGLIN, MOSTELLER, and TUKEY • Understanding Robust and Exploratory Data Analysis
HOEL • Elementary Statistics, *Fourth Edition*
HOEL and JESSEN • Basic Statistics for Business and Economics, *Third Edition*
HOGG and KLUGMAN • Loss Distributions
HOLLANDER and WOLFE • Nonparametric Statistical Methods
IMAN and CONOVER • Modern Business Statistics
JAGERS • Branching Processes with Biological Applications
JESSEN • Statistical Survey Techniques
JOHNSON • Multivariate Statistical Simulation: A Guide to Selecting and Generating Continuous Multivariate Distributions
JOHNSON and KOTZ • Distributions in Statistics
Discrete Distributions
Continuous Univariate Distributions—1
Continuous Univariate Distributions—2
Continuous Multivariate Distributions

(*continued on back*)

Applied Probability and Queues

Applied Probability and Queues

SØREN ASMUSSEN
Institute of Mathematical Statistics
University of Copenhagen
Denmark

JOHN WILEY & SONS

Chichester · New York · Brisbane · Toronto · Singapore

Library of Congress Cataloging-in-Publication Data:

Asmussen, Søren.
Applied probability and queues
(Wiley series in probability and mathematical statistics. Applied probability and statistics)
Bibliography: p.
Includes index.
1. Stochastic processes. 2. Queueing theory.
3. Markov processes. I. Title. II. Series.
QA274.A825 1987 519.2 86–13173

ISBN 0 471 91173 9

British Library Cataloguing in Publication Data:

Asmussen, Søren
Applied probability and queues
—(Wiley series in probability and mathematical statistics, applied)
1, Queueing theory
I. Title
519.8′2 T57.9

ISBN 0 471 91173 9

Printed and bound in Great Britain

Contents

Part B: Basic Mathematical Tools

Part C: Special Models and Methods

Preface

This book treats the mathematics of queueing theory and some related areas, as well as the basic mathematical tools relevant for the study of such models. It thus aims to serve as an introduction to queueing theory, to provide a thorough treatment of tools like Markov processes, renewal theory, regenerative processes and random walks, and to treat in some detail basic models like the $GI/G/1$ queue, risk processes, dams and so on. Particular attention has been given to modern probabilistic points of view, as alternatives to traditional analytic methods. Within this framework the choice of topics is, however, rather traditional. The aim has been to present what I consider the basic knowledge in the area, not to advocate special directions in which the area is at present developing.

The level is graduate or postgraduate, with the necessary background at the level of say parts of Breiman (1968) or Chung (1974) incorporating measure theory, conditioning, the law of large numbers, uniform integrability, basic facts about weak convergence and so on. Martingales are used occasionally, but are hardly crucial for the main stream of exposition. The same remark applies to an even higher degree to weak convergence in function spaces. A brief survey of some of the prerequisites is given in the Appendix. Problems, most of them hopefully rather easy, have been worked out for at least some central chapters. When lecturing on the first draft of the book in 1983–84, I covered most of the material in a two-term course of four hours per week (most of Chapters I–II could then be omitted). For shorter courses, Chapters III–V and VII–IX would seem to be in the core of the book.

The book being intended as a textbook and not a specialist's monograph, the text has, as far as possible, been kept free of references, which are instead collected in the Notes following the separate sections. These Notes are mainly intended as a first guidance for further reading, not as a bibliography (in many cases, encyclopaedic articles like those in Gnedenko and König, 1983/84, or Kotz *et al.*, 1982–, are excellent for both purposes but have not been listed separately in the Notes). Questions of priority or details of history are treated rather sporadically, and in most cases the reader must pursue these topics himself or consult other sources.

To work out this book has been an immense pleasure for me, due not least to the interest shared by a number of friends, colleagues and students. Comments, corrections and suggestions at all levels were provided by Aksel Bertelsen, Lennart Bondesson, Daryl Daley, Peter Dalgaard, Peter Glynn, Martin Jacobsen, Helle Johansen, Torgny Lindvall, Marcel Neuts, Jim Pitman, Holger Rootzén, Karl Sigman, Henrik Stryhn, Hermann Thorisson, an anonymous reviewer, Lindsay Prisgrove and Donald L. Iglehart. Thanks to all. It is a privilege to be part of a scientific community incorporating people of such dedication and competence. On the technical side, the drawing department at the H. C. Ørsted Institute at the University of Copenhagen did a splendid job in making the figures. Claus Holst Pedersen supplied computer data for some of the illustrations as well as for Problem V.3.3. In a difficult period with resources running short, my department did everything possible to ensure that the manuscript was typed without delay. Several typists were involved, but unquestionably the heaviest load was carried by Ursula Hansen. Her efforts are greatly appreciated.

Copenhagen, December 1985 Søren Asmussen

Part A: Simple Markovian Models

CHAPTER I

Markov Chains

1. PRELIMINARIES

We consider a Markov chain $X_0, X_1, \ldots$ with discrete (i.e. finite or countable) state space $E = \{i, j, k, \ldots\}$ and specified by the transition matrix $P = (p_{ij})_{i,j \in E}$. By this we mean that P is a given $E \times E$ matrix such that $p_{i\cdot} = (p_{ij})_{j \in E}$ is a probability (vector) for each i, and that we study $\{X_n\}$ subject to exactly those governing probability laws $\mathbb{P} = \mathbb{P}_\mu$ (*Markov probabilities*) for which

$$\mathbb{P}(X_0 = i_0, X_1 = i_1, \ldots, X_n = i_n) = \mu(i_0) p_{i_0 i_1} p_{i_1 i_2} \cdots p_{i_{n-1} i_n} \tag{1.1}$$

where $\mu(i) = \mathbb{P}(X_0 = i)$. The particular value of the initial distribution μ is unimportant in most cases and is therefore suppressed in the notation. An important exception is the case where X_0 is degenerate, say at i, and we write then $\mathbb{P}_i$ so that $\mathbb{P}_i(X_0 = i) = 1$. Given μ, it is readily checked that (1.1) uniquely determines a probability distribution on $\mathscr{F}_n = \sigma(X_0, \ldots, X_n)$. Appealing to basic facts from the foundational theory of stochastic processes (to be discussed in Section 6), this set of probabilities on $\mathscr{F}_n$ can be uniquely extended to a probability law $\mathbb{P}_\mu$ governing the whole chain. Thus since the transition matrix P is fixed here and in the following, the Markov probabilities are in one-to-one correspondence with the set of initial distributions.

If $\mathbb{P}$ is a Markov probability, then (with the usual a.s. interpretation of conditional probabilities and expectations)

$$p_{ij} = \mathbb{P}_i(X_1 = j) = \mathbb{P}(X_{n+1} = j \mid X_n = i), \tag{1.2}$$

$$\mathbb{P}(X_{n+1} = j \mid \mathscr{F}_n) = p_{X_n j} = \mathbb{P}_{X_n}(X_1 = j), \tag{1.3}$$

$$\mathbb{E}[h(X_n, X_{n+1}, \ldots) \mid \mathscr{F}_n] = \mathbb{E}_{X_n} h(X_0, X_1, \ldots), \qquad h: E \times E \times \cdots \to \mathbb{R} \tag{1.4}$$

(here e.g. $\mathbb{P}_{X_n}(X_1 = j)$ means $g(x) = \mathbb{P}_x(X_1 = j)$ evaluated at $x = X_n$). Conversely, either of (1.3) or (1.4) are sufficient for p to be a Markov probability. The formal proof of these facts is an elementary (though in part lengthy) exercise in conditioning arguments and will not be given here. However, equations (1.2), (1.3), (1.4) have important intuitive contents. Thus (1.4) means that at time n, the chain is restarted with the new initial value X_n. Equivalently, the post-n-chain $X_n, X_{n+1}, \ldots$ evolves as the Markov chain itself, started at X_n but otherwise independently of the past. Similarly, in simulation terminology (1.3) means that the chain can be stepwise constructed by, at step n, drawing X_{n+1} according to $p_{X_n \cdot}$ (to get started, draw X_0 according to μ).

Recall from Appendix A8 that a *stopping time* σ is a random variable with values in $\mathbb{N} \cup \{\infty\}$ and satisfying $\{\sigma = n\} \in \mathscr{F}_n$ for all n, that $\mathscr{F}_\sigma$ denotes the σ-algebra which consists of all disjoint unions of the form $\bigcup_0^\infty A_n$ with $A_n \in \mathscr{F}_n$, $A_n \subseteq \{\sigma = n\}$ (here $n = \infty$ is included with the convention $\mathscr{F}_\infty = \sigma(X_0, X_1, \ldots)$), and that σ and X_σ are measurable with respect to $\mathscr{F}_\sigma$. The important *strong Markov property* states that for the sake of predicting the future development of the chain a stopping time may be treated as a fixed deterministic point of time. For example, we have the following extension of (1.4):

Theorem 1.1 (STRONG MARKOV PROPERTY). *Let σ be a stopping time. Then for any* (say) *bounded $h: E \times E \times \cdots \to \mathbb{R}$ it a.s. holds on $\{\sigma < \infty\}$ that*

$$\mathbb{E}[h(X_\sigma, X_{\sigma+1}, \ldots) | \mathscr{F}_\sigma] = \mathbb{E}_{X_\sigma} h(X_0, X_1, \ldots). \tag{1.5}$$

Proof. We must show that for any $A \in \mathscr{F}_\sigma$ with $A \subseteq \{\sigma < \infty\}$ it holds that

$$\mathbb{E}[h(X_\sigma, X_{\sigma+1}, \ldots); A] = \mathbb{E}[\mathbb{E}_{X_\sigma} h(X_0, X_1, \ldots); A].$$

However, if $A \in \mathscr{F}_n$, $\sigma = n$ on A this is immediate from (1.4). Replace A by $A \cap \{\sigma = n\}$ and sum over n. □

The mth power (iterate) of the transition matrix is denoted by $P^m = (p_{ij}^m)$. An easy calculation (e.g. let $n = nm$ in (1.1) and sum over the i_k with $k \notin \{0, m, \ldots, nm\}$ to evaluate the joint distribution of $X_0, X_m, \ldots, X_{nm}$) shows that $X_0, X_m, X_{2m}, \ldots$ is a Markov chain and that its transition matrix is simply P^m.

Associated with each state is the hitting time

$$\tau(i) = \inf\{n \geqslant 1 : X_n = i\}$$

(with the usual convention $\tau(i) = \infty$ if no such n exists) and the number of visits $N_i = \sum_1^\infty I(X_n = i)$ to i. Clearly, $\{\tau(i) < \infty\} = \{N_i > 0\}$ and we call i *recurrent* if the recurrence-time distribution $\mathbb{P}_i(\tau(i) = k)$ is proper, i.e. if $\mathbb{P}_i(\tau(i) < \infty) = 1$, and *transient* otherwise. The chain itself is recurrent (transient) if all states are so.

Proposition 1.2. *Let i be some fixed state. Then:*

(i) *the following assertions* (a), (b), (c) *are equivalent*: (a) *i is recurrent*; (b) $N_i = \infty$ $\mathbb{P}_i$ *a.s.*; (c) $\mathbb{E}_i N_i = \sum_1^\infty p_{ii}^m = \infty$;

(ii) *the following assertions* (a′), (b′), (c′) *are equivalent as well*: (a′) *i is transient*; (b′) $N_i < \infty$ $\mathbb{P}_i$ *a.s.*; (c′) $\mathbb{E}_i N_i = \sum_1^\infty p_{ii}^m < \infty$.

Proof. Define $\tau(i; 1) = \tau(i)$,

$$\tau(i; k+1) = \inf\{n > \tau(i; k) : X_n = i\} \qquad \theta = \mathbb{P}_i(\tau(i; 1) < \infty).$$

Then N_i is simply the number of k with $\tau(i; k) < \infty$, and by the strong Markov property and $X_{\tau(i;k)} = i$,

$$\begin{aligned}\mathbb{P}_i(\tau(i;k+1)<\infty) &= \mathbb{E}_i\mathbb{P}(\tau(i;k+1)<\infty,\quad \tau(i;k)<\infty\,|\,\mathscr{F}_{\tau(i;k)})\\ &= \mathbb{E}_i[\mathbb{P}(\tau(i;k+1)<\infty\,|\,\mathscr{F}_{\tau(i;k)});\quad \tau(i;k)<\infty]\\ &= \mathbb{E}_i[\mathbb{P}_{X_{\tau(i;k)}}(\tau(i;1)<\infty);\quad \tau(i;k)<\infty] = \theta\mathbb{P}_i(\tau(i;k)<\infty) = \cdots\\ &= \theta^{k+1}. \end{aligned} \tag{1.6}$$

If (a) holds, then $\theta = 1$ so that it follows that all $\tau(i;k)<\infty$ $\mathbb{P}_i$ a.s., and (b) also holds. Clearly, (b)$\Rightarrow$(c) so that for part (i) it remains to prove (c)$\Rightarrow$(a) or equivalently (a′)$\Rightarrow$(c′). But if $\theta < 1$, then

$$\begin{aligned}\mathbb{E}_i N_i = \sum_{k=0}^{\infty}\mathbb{P}_i(N_i > k) = \sum_{k=1}^{\infty}\mathbb{P}_i(\tau(i;k)<\infty) = \sum_{k=1}^{\infty}\theta^k\\ = \theta(1-\theta)^{-1} < \infty\end{aligned}$$

Part (ii) follows by negation. □

It should be noted that though Proposition 1.2 gives necessary and sufficient conditions for recurrence/transience, the criteria are almost always difficult to check: even for extremely simple transition matrices P, it is usually impossible to find closed expressions for the p_{ii}^m. Some alternative general approaches are discussed in Section 5, but in many cases the recurrence/transience classification leads into arguments particular for the specific model.

Our emphasis in the following is on the recurrent case and we shall briefly discuss some aspects of the set-up. Two states i, j are said to communicate, written $i\leftrightarrow j$, if i can be reached from j (i.e. $p_{ji}^m > 0$ for some m) and vice versa. Clearly, this relation is transitive and symmetric. Now suppose i is recurrent and that j can be reached from i. Then also i can be reached from j. In fact even $\tau(i)<\infty$ $\mathbb{P}_j$ a.s. since otherwise $\mathbb{P}_i(\tau(i)=\infty)>0$. Furthermore, j is recurrent since

$$\sum_{m=1}^{\infty} p_{jj}^m \geqslant \sum_{m=1}^{\infty} p_{ji}^{m(1)} p_{ii}^m p_{ij}^{m(2)} = \infty$$

if $m(1), m(2)$ are chosen such that the relevant factors do not vanish. Obviously $i\leftrightarrow i$ and it follows that $\leftrightarrow$ is an equivalence relation on the recurrent states so that we may write

$$E = T\cup R_1\cup R_2\cup\cdots, \tag{1.7}$$

where $R_1, R_2, \ldots$ are the equivalence classes (*recurrent classes*) and T the set of transient states. It is basic to note that the recurrent classes are *closed* (or *absorbing*), i.e.

$$\mathbb{P}_i(X_n\in R_k \quad\text{for all}\quad n) = 1 \quad\text{when}\quad i\in R_k$$

(this follows from the above characterization of R_k as the set of states which can be reached from i). When started at $i\in R_k$ the chain therefore evolves within R_k only, and the state space may be reduced to R_k. If, on the other hand, $X_0 = i$ is transient, two types of paths may occur: either $X_n\in T$ for all n (with any $j\in T$ visited at most finitely often) or at some stage the chain enters a recurrent class R_k and is *absorbed*, i.e. evolves from then on in R_k.

Most often one can restrict attention to *irreducible* chains, defined by the requirement that all states in E communicate. Such a chain is either transient or E consists of exactly one recurrent class. In fact, if a recurrent state, say i, exists at all, it follows from the above that any other state j is in the same recurrent class as i.

A recurrent state is called *positive recurrent* if the mean recurrence time $\mathbb{E}_i\tau(i)$ is finite. Otherwise i is *null recurrent*. The *period* $d = d(i)$ is the period of the recurrence-time distribution, i.e. the greatest integer d such that $\mathbb{P}_i(\tau(i) \in L_d) = 1$ where $L_d = \{d, 2d, 3d, \ldots\}$. If $d = 1$, i is *aperiodic*.

Proposition 1.3. *Let R be a recurrent class. Then the states in R*

(i) *are either all positive recurrent or all null recurrent;*
(ii) *have all the same period.*

Proof. (i) is deferred to Section 3. Let $i, j \in R$ and choose r, s with $p_{ij}^r > 0, p_{ji}^s > 0$. Then $p_{ii}^{r+s} > 0$, i.e. $r + s \in L_{d(i)}$, and whenever $p_{jj}^n > 0$, $p_{ii}^{r+s+n} > 0$ also, i.e. $r + s + n \in L_{d(i)}$ so that $n \in L_{d(i)}$ also. It follows that $\mathbb{P}_j(\tau(j) \in L_{d(i)}) = 1$, i.e. $d(j) \geqslant d(i)$. By symmetry, $d(j) \leqslant d(i)$. □

Problems

1.1. Explain that $\mathbb{P}_\mu = \sum_{i \in E} \mu_i \mathbb{P}_i$.

1.2. Show that if $\theta = p_{ii} > 0$, then the exit time $\eta(i) = \inf\{n \geqslant 1 : X_n \neq i\}$ has a geometric distribution

$$\mathbb{P}_i(\eta(i) = n) = (1 - \theta)\theta^{n-1} \qquad n = 1, 2, \ldots.$$

1.3. In a number of population processes one encounters Markov chains with $E = \mathbb{N}$, X_n representing the population size at time n, state 0 absorbing and $\mathbb{P}_i(\tau(0) < \infty) > 0$ for all i. Explain why it is reasonable to denote $\{\tau(0) < \infty\}$ as the event of *extinction*. Show that any state $i \geqslant 1$ is transient and that $X_n \to \infty$ a.s. in the event $\{\tau(0) = \infty\}$ of non-extinction.

Notes

Markov chains with a discrete state space (and their continuous-time generalizations) form in many ways a natural starting-point of applied probability: when considering a specific phenomenon, the first attempt to formulate and solve a stochastic model is usually performed within the Markovian set-up, and also the mathematical questions arising in connection with Markov chains are to a large extent the same as for more general models (in particular this is so in queueing theory).

The present text therefore starts with a treatment of the relevant features of Markov chains and (in Chapter II) Markov jump processes. The expósition is in principle self-contained, but might for many readers rather serve as a refresher and/or reference section. Of the many textbooks containing introductory chapters on Markov chains and Markov jump processes we mention for example Breiman (1968), Çinlar (1975), Feller (1966), Grimmett and Stirzaker (1982) and Karlin and Taylor (1975). More advanced treatments of the discrete state space case are in Chung (1967), Freedman (1971), Kemeny *et al.* (1976) and Orey (1971). Virtually all the material of this and the next chapter is standard.

2. ASPECTS OF RENEWAL THEORY IN DISCRETE TIME

Let $f_1, f_2, \ldots$ be the point probabilities of a distribution on $\{1, 2, \ldots\}$. Then by a (discrete time) *renewal process* governed by $\{f_n\}$ we understand a point process (see Appendix A4 for the terminology) on $\mathbb{N}$ with epochs $S_0 = 0$, $S_n = Y_1 + \cdots + Y_n$, where the Y_i are independent identically distributed with common distribution $\{f_n\}$. Instead of epochs we usually speak of *renewals*. The associated *renewal sequence* $u_0, u_1, \ldots$ is defined by $u_k = \mathbb{P}(\text{some } S_n = k)$, that is, the probability of a renewal at k.

A renewal occurs at k if either $Y_1 = k$ which happens with probability $f_k = f_k u_0$, or if $Y_1 = l < k$ and $Y_2 + \cdots + Y_n = k - l$ for some n. The probability of this being $f_l u_{k-l}$, we obtain

$$u_k = f_k u_0 + f_{k-1} u_1 + \cdots + f_1 u_{k-1}, \qquad k \geq 1, \tag{2.1}$$

i.e. in convolution notation $u = \delta_0 + u * f$, where $\delta_{0i} = I(i = 0)$. In conjunction with $u_0 = 1$, (2.1) clearly uniquely determines $\{u_n\}$.

These concepts are intimately related to Markov chains. Consider some fixed recurrent state i, let $Y_1 = \tau(i)$ and more generally let Y_k be the inter-occurrence time between the $(k-1)$th and the kth visit to i. Then $Y_1, Y_2, \ldots$ are independent identically distributed with respect to $\mathbb{P}_i$ according to the strong Markov property, the common distribution $\{f_n\}$ is the recurrence-time distribution of i, and the renewals are the times of visits to i so that $u_n = \mathbb{P}_i(X_n = i) = p_{ii}^n$. Conversely, *any* renewal process can be constructed in this way from a Markov chain which we shall denote by $\{A_n\}$. Indeed, define $A_n = n - \sup\{S_k : S_k \leq n\}$ as the *backwards recurrence time at n*, i.e. the time passed since the last renewal, see Fig. 2.1. Then the paths of $\{A_n\}$ are at 0 exactly at the renewals, i.e. the renewals

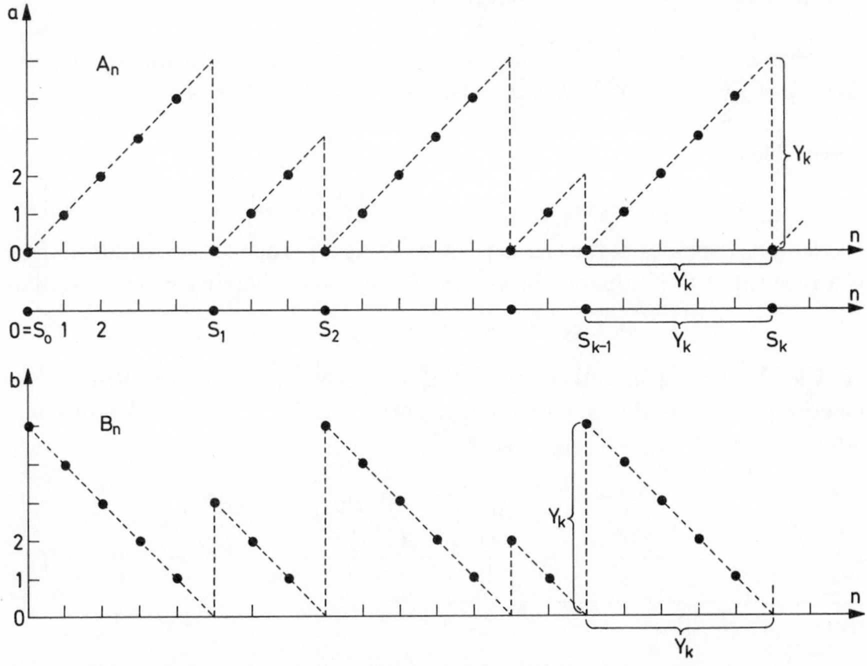

Fig. 2.1 The recurrence-time chains of a renewal process in discrete time

are the recurrence times of 0, and the Markov property follows by noting that $\{A_n\}$ moves from i to either $i+1$ or 0, the probability of $i+1$ being $\mathbb{P}(Y_k > i+1 \mid Y_k > i)$ independently of $A_0, \ldots, A_{n-1}$. The state space E is $\mathbb{N}$ if $f_k > 0$ infinitely often, and $\{0, 1, \ldots, K-1\}$ with $K = \inf\{k: f_1 + \cdots + f_k = 1\}$ otherwise. A closely related important Markov chain is the *forwards recurrence-time chain* $\{B_n\}$, i.e. B_n is the waiting time until the next renewal after n, again see Fig. 2.1. The Markov property is even more immediate since the paths decrease deterministically from i to $i-1$ if $i > 1$, whereas the value of B_{n+1} following $B_n = 1$ is chosen according to $\{f_n\}$ independently of the past. The state space is $\{1, 2, \ldots\}$ if $f_k > 0$ infinitely often and $\{1, \ldots, K\}$ otherwise, and a renewal occurs at n if and only if $B_{n-1} = 1$.

Lemma 2.1. *$\{u_n\}$ and $\{f_n\}$ have the same period d.*

Proof. Since $u_n \geqslant f_n$, it is clear that the period d_f of $\{f_n\}$ is at least that d_u of $\{u_n\}$. Conversely, it is only possible that $\mathbb{P}(S_k = n) > 0$ and hence $u_n > 0$ if n is a multiple of d_f. Hence $d_u \geqslant d_f$. □

If $d = 1$ in Lemma 2.1, we call the renewal sequence (process) *aperiodic*.

Renewal processes with the Y_k having a possibly continuous distribution will play a major role in later parts of the book. We shall here exploit the connection between (discrete) renewal processes and Markov chains in the limit theory. Within the framework of renewal processes, the main result is as follows (to be translated to Markov chains in Section 4):

Theorem 2.2. *Let $\{u_n\}$ be an aperiodic renewal sequence governed by $\{f_n\}$ and define $\mu = \sum_1^\infty n f_n = \mathbb{E}Y_1$. Then $u_n \to 1/\mu$ as $n \to \infty$* (here $1/\infty = 0$).

Proof. Define

$$r_n = f_{n+1} + f_{n+2} + \cdots = \mathbb{P}(Y_1 > n) \tag{2.2}$$

and let L be the index of the last renewal in $\{0, \ldots, n\}$. Then $\{L = l\}$ occurs if there is a renewal at l and the next Y is $> n - l$, i.e. the probability is $u_l r_{n-l}$ so that

$$1 = \mathbb{P}(0 \leqslant L \leqslant n) = r_0 u_n + r_1 u_{n-1} + \cdots + r_n u_0. \tag{2.3}$$

Now let $\lambda = \limsup u_n$ and choose $\{n(k)\}$ such that $u_{n(k)} \to \lambda$. Let i satisfy $f_i > 0$. Choosing N such that $r_N < \varepsilon$, we obtain from (2.1) and $u_n \leqslant 1$ that for k sufficiently large

$$\begin{aligned} \lambda - \varepsilon \leqslant u_{n(k)} &\leqslant r_N + \sum_{j=1}^{N} f_j u_{n(k)-j} \\ &\leqslant \varepsilon + (1 - f_i)(\lambda + \varepsilon) + f_i u_{n(k)-i} \end{aligned} \tag{2.4}$$

so that

$$\lambda \leqslant (1 - f_i)\lambda + f_i \liminf_{k \to \infty} u_{n(k)-i} \leqslant (1 - f_i)\lambda + f_i \lambda = \lambda \tag{2.5}$$

which is only possible if $u_{n(k)-i} \to \lambda$. Repeating the argument we see that this also holds for any i of the form

$$i = x_1 a_1 + \cdots + x_t a_t \quad \text{where} \quad x_k \in \mathbb{N}, \qquad f_{a_k} > 0.$$

But since $\{f_j\}$ is aperiodic, we can choose t and the a_k such that g.c.d. of the a_k is 1, and it then follows by elementary number theory (e.g. Feller, 1966, p. 366) that any sufficiently large i, say $i \geqslant a$, can be represented in this form. Thus letting $n = n(k) - a$ in (2.3) we obtain for any N

$$1 \geqslant \sum_{j=0}^{N} r_j u_{n(k)-a-j} \to \lambda \sum_{j=0}^{N} r_j. \tag{2.6}$$

Since $r_0 + r_1 + \cdots = \mu$, this proves $1 \geqslant \lambda\mu$. It then remains only to show that $v = \liminf u_n \geqslant \mu^{-1}$. This is clear if $\mu = \infty$ and can be proved similarly as above if $\mu < \infty$. In fact, if $\{m(k)\}$ is chosen such that $u_{m(k)} \to v$, we obtain, instead of (2.4),

$$\begin{aligned} v + \varepsilon \geqslant u_{m(k)} &\geqslant \sum_{j=1}^{N} f_j u_{m(k)-j} \geqslant (v - \varepsilon) \sum_{j=1, j \neq i}^{N} f_j + f_i u_{m(k)-i} \\ &= (1 - f_i)(v - \varepsilon) - r_N(v - \varepsilon) + f_i u_{m(k)-i}, \\ v &\geqslant (1 - f_i)v + f_i \limsup u_{m(k)-i} \geqslant (1 - f_i)v + f_i v = v. \end{aligned}$$

Thus $u_{m(k)-i} \to v$ and hence for fixed N

$$1 \leqslant \sum_{j=0}^{N} r_j u_{m(k)-a-j} + \sum_{j=N+1}^{\infty} r_j \to v \sum_{j=0}^{N} r_j + \sum_{j=N+1}^{\infty} r_j$$

which tends to $v\mu + 0$ as $N \to \infty$. □

Corollary 2.3. *Let* $\{u_n\}$, $\{f_n\}$ *have period* $d > 1$. *Then:*

(i) $\{u_{nd}\}_{n=1}^{\infty}$ *is an aperiodic renewal sequence governed by* $\{f_{nd}\}_{n=1}^{\infty}$;
(ii) $u_m = 0$ *whenever* m *is not of the form* $m = nd$;
(iii) $u_{nd} \to d/\mathbb{E}Y$ *as* $n \to \infty$.

Proof. Here (i) and (ii) are obvious, and from Theorem 2.2 and (i) we get

$$u_{nd} \to 1 \Big/ \sum_{n=1}^{\infty} n f_{nd} = d/\mathbb{E}Y.$$ □

Sometimes one also encounters defective governing distributions $\{f_n\}$, i.e. $f_\infty = 1 - f_1 - f_2 - \cdots > 0$. The corresponding renewal sequence is still uniquely determined by $u_0 = 1$ and (2.1), and can be interpreted in terms of a *terminating* or *transient* renewal process. This is defined simply by attaching the Y_k mass f_∞ at ∞. If $f_\infty > 0$, then $\sigma = \inf\{n \geqslant 1 : Y_n = \infty\}$ is finite a.s., and $S_n < \infty$ for $n = 0, \ldots, \sigma - 1$, $= \infty$ for $n \geqslant \sigma$. In particular, the number σ of renewals is finite a.s., and hence the probability u_n of a renewal at n tends to zero as $n \to \infty$. More precisely:

Proposition 2.4. *If* $f_\infty > 0$, *then the expected number of renewals is given by* $\sum_{n=0}^{\infty} u_n = 1/f_\infty$.

Proof. Since u_n is the probability of a renewal at n, the expected number of renewals is indeed $\sum_0^\infty u_n$. But it is also

$$\mathbb{E}\sigma = \sum_{n=1}^{\infty} \mathbb{P}(\sigma \geqslant n) = \sum_{n=1}^{\infty} \mathbb{P}(Y_k < \infty\ k = 1, \ldots, n-1)$$

$$= \sum_{n=0}^{\infty} (1 - f_\infty)^n = 1/f_\infty. \qquad \square$$

Problems

2.1. Define the generating function of $\{f_n\}$ by $\hat{f}(s) = \sum_0^\infty s^n f_n$ $(f_0 = 0)$. Show that $\hat{u}(s) = \sum_0^\infty s^n u_n = (1 - \hat{f}(s))^{-1}$.
2.2. Consider the geometric case $f_n = (1-\theta)\theta^{n-1}$. Show that u_n is constant for $n > 0$, $u_n = 1 - \theta$.
2.3. Show that $\{u_n v_n\}$ is a renewal sequence if $\{u_n\}$, $\{v_n\}$ are so.
2.4. Let $\{u_n\}$ be a renewal sequence with $\sum_1^\infty f_n \neq 1$. Assume that $\sum_1^\infty \rho^n f_n = 1$ for some ρ. Show that $\{\rho^n u_n\}$ is a renewal sequence and that $u_n \cong c\rho^{-n}$ for some c (provided $\{f_n\}$ is aperiodic).

Notes

The proof of Theorem 2.2. is a classical argument due to Erdös, *et al.* (1949). Additional material on renewal sequences and related topics can be found in Kingman (1972).

3. STATIONARITY

Let $\nu = (\nu_i)_{i \in E}$ be any non-negative measure on E (it is not assumed that ν is a distribution, $|\nu| = \sum \nu_i = 1$, neither that ν is finite, $|\nu| < \infty$, nor even that all $\nu_i < \infty$). We can then define a new measure νP by usual matrix multiplication, i.e. νP attaches mass $\sum_{i \in E} \nu_i p_{ij}$ to j. We call $\nu \neq 0$ *stationary* if all $\nu_i < \infty$ and $\nu P = \nu$, i.e. if in algebraic terms ν is left eigenvector of the transition matrix P corresponding to the eigenvalue 1.

Of particular importance is the case where ν is a distribution. Irrespective of whether or not ν is stationary, we have

$$\mathbb{P}_\nu(X_1 = j) = \sum_{i \in E} \mathbb{P}_\nu(X_0 = i) p_{ij} = \sum_{i \in E} \nu_i p_{ij} = (\nu P)_j.$$

Thus νP can be interpreted as the $\mathbb{P}_\nu$-distribution of X_1, and in a similar manner the $\mathbb{P}_\nu$-distribution of X_m is νP^m. In particular, if ν is stationary, then $\nu P^m = \nu$ for all m so that the distribution of X_m is independent of m. More generally:

Theorem 3.1. *Suppose that ν is a stationary distribution. Then*:

(i) *the chain is strictly stationary with respect to $\mathbb{P}_\nu$, i.e. the $\mathbb{P}_\nu$-distribution of $\{X_n, X_{n+1}, \ldots\}$ does not depend on n;*
(ii) *there exists a strictly stationary version $\{X_n\}_{n \in \mathbb{Z}}$ of the chain with doubly infinite time and $\mathbb{P}(X_n = i) = \nu_i$ for all $n \in \mathbb{Z}$.*

Proof. (i) Clearly $\{X_n, X_{n+1}, \ldots\}$ is a Markov chain with transition matrix P with respect to $\mathbb{P}_\nu$. Thus the distribution is uniquely given by the initial probabilities $\mathbb{P}_\nu(X_n = i) = \nu_i$ and hence independent of n.

(ii) This requires some background in the foundations of the general theory of stochastic processes so we shall sketch only two different proofs. The first applies Kolmogorov's consistency theorem to a family $(\mathbb{P}^{n(1);\ldots;n(k)})$ with $n(i)\in\mathbb{Z}$ and $n(1) < n(2) < \cdots < n(k)$. Here $\mathbb{P}^{n(1);\ldots;n(k)}$ is taken as the $\mathbb{P}_\nu$-distribution of $(X_0, X_{n(2)-n(1)}, \ldots, X_{n(k)-n(1)})$ and consistency becomes an immediate consequence of the stationarity of $\{X_0, X_1 \cdots\}$ with respect to $\mathbb{P}_\nu$. In the second proof, one considers a sequence $\mathbb{P}^{(n)}$ of probabilities on $E^{\mathbb{Z}}$ such that $\{X_{-n}, X_{-n+1}, \ldots\}$ has the $\mathbb{P}_\nu$-distribution of $\{X_0, X_1, \ldots\}$ and the distribution of $\{X_k\}_{k<-n}$ is arbitrary, say X_k is degenerate at i_0. Then $\{\mathbb{P}^{(n)}\}$ is tight and any weak limit of the sequence serves as the desired probability on $E^{\mathbb{Z}}$. □

Questions of existence and uniqueness of stationary measures is one of the main topics of Markov chain theory. We start by an explicit construction (a generalization of which will also turn out to be basic for non-Markovian processes, cf. Ch. V):

Theorem 3.2. *Let i be a fixed recurrent state. Then a stationary measure ν can be defined by letting ν_j be the expected number of visits to j in between two consecutive visits to i,*

$$\nu_j = \mathbb{E}_i \sum_{n=0}^{\tau(i)-1} I(X_n = j) = \sum_{n=0}^{\infty} \mathbb{P}_i(X_n = j, \tau(i) > n). \tag{3.1}$$

Proof. Since $X_0 = X_{\tau(i)} = i$ $\mathbb{P}_i$ a.s. and $\{\tau(i) > n-1\}\in\mathscr{F}_{n-1}$, we get

$$\begin{aligned}
\nu_j &= \mathbb{E}_i \sum_{n=0}^{\tau(i)-1} I(X_n = j) = \mathbb{E}_i \sum_{n=1}^{\tau(i)} I(X_n = j) \\
&= \sum_{n=1}^{\infty} \mathbb{P}_i(X_n = j, \tau(i) \geqslant n) = \sum_{n=1}^{\infty} \mathbb{E}_i \mathbb{P}_i(X_n = j, \tau(i) > n-1 \,|\, \mathscr{F}_{n-1}) \\
&= \sum_{n=1}^{\infty} \mathbb{E}_i[p_{X_{n-1}j}; \tau(i) > n-1] = \sum_{k\in E} p_{kj} \sum_{n=1}^{\infty} \mathbb{P}_i(X_{n-1} = k, \tau(i) > n-1) \\
&= \sum_{k\in E} p_{kj}\nu_k,
\end{aligned}$$

proving $\nu P = \nu$ so that we must only check that all $\nu_j < \infty$. Clearly, $\nu_j = 0$ if j is not in the same recurrent class as i. Otherwise $p_{ji}^m > 0$ for some m and since clearly $\nu_i = 1 < \infty$, $\nu_j < \infty$ follows from

$$\nu_i = \sum_{k\in E} \nu_k p_{ki}^m \geqslant \nu_j p_{ji}^m. \tag{3.2}$$

□

Theorem 3.3. *If the chain is irreducible and recurrent, then a stationary measure ν exists, satisfies $0 < \nu_j < \infty$ for all j and is unique up to a multiplicative factor. That is, if ν, ν^* are stationary, then $\nu = c\nu^*$ for some $c\in(0,\infty)$.*

Here existence is immediate from Theorem 3.2 (we denote in the following by $\nu^{(i)}$ the measure given by (3.1)). Also, $\nu_i > 0$ for any $i \in E$ and any stationary measure ν is clear from (3.2) since we may choose j with $\nu_j > 0$. The key step in the proof of uniqueness is the following:

Lemma 3.4. *Let i be some fixed state and let ν be superstationary* (i.e. $\nu P \leqslant \nu$) *with $\nu_i \geqslant 1$. Then $\nu_j \geqslant \nu_j^{(i)}$ for all $j \in E$.*

Proof. With $\tilde{P}$ the matrix obtained from P by replacing the ith column by zeros, it is easily seen by induction that $\tilde{p}_{kj}^n$ is the *taboo probability* $\mathbb{P}_k(X_n = j, \tau(i) > n)$. In particular, if we let $k = i$ and sum over n, we get $\nu^{(i)} = \delta^{(i)} \sum_0^\infty \tilde{P}^n$, where $\delta^{(i)}$ is the distribution degenerate at i. We next claim that $\nu_j \geqslant \delta_j^{(i)} + (\nu\tilde{P})_j$. Indeed, for $j = i$ this follows from $\nu_i \geqslant \delta_i^{(i)} = 1$ and for $j \neq i$ we have $(\nu\tilde{P})_j = (\nu P)_j \leqslant \nu_j$. Hence

$$\nu \geqslant \delta^{(i)} + \nu\tilde{P} \geqslant \delta^{(i)}(I + \tilde{P}) + \nu\tilde{P}^2 \geqslant \cdots$$
$$\geqslant \delta^{(i)} \sum_{n=0}^{N} \tilde{P}^n + \nu\tilde{P}^{N+1} \geqslant \delta^{(i)} \sum_{n=0}^{N} \tilde{P}^n,$$

and letting $N \to \infty$, $\nu \geqslant \nu^{(i)}$ follows. □

Proof of Theorem 3.3. If ν is stationary, then $\nu_i > 0$ as observed above. Thus we may assume $\nu_i = 1$ and the proof will then be complete if we can show $\nu = \nu^{(i)}$. But according to the lemma, we have $\nu \geqslant \nu^{(i)}$. Hence $\mu = \nu - \nu^{(i)}$ is non-negative with $\mu P = \mu$. As noted above $\mu_i = 0$ then implies $\mu = 0$ and $\nu = \nu^{(i)}$. □

Clearly, the total mass of the stationary measure $\nu^{(i)}$ given by (3.1) is

$$|\nu^{(i)}| = \sum_{j \in E} \nu_j^{(i)} = \mathbb{E}_i \sum_{n=0}^{\tau(i)-1} 1 = \mathbb{E}_i \tau(i). \tag{3.3}$$

Now if the chain is irreducible and recurrent, it follows by uniqueness that the $|\nu^{(j)}|, j \in E$, are either all finite or all infinite, i.e. that the states are either all positive recurrent or all null recurrent, proving the remaining part of Proposition 1.3. In the first case, ν can hence be normalized to a stationary distribution $\pi = \nu/|\nu|$ which is unique. In particular, for each j we have $\pi_j = \nu_j^{(j)}/|\nu^{(j)}| = 1/\mathbb{E}_j\tau(j)$ which yields an expression for π independent of a reference state i. In summary:

Corollary 3.5. *If the chain is irreducible and positive recurrent, there exists a unique stationary distribution π given by*

$$\pi_j = \frac{1}{\mathbb{E}_i\tau(i)} \mathbb{E}_i \sum_{n=0}^{\tau(i)-1} I(X_n = j) = \frac{1}{\mathbb{E}_j\tau(j)}. \tag{3.4}$$

Corollary 3.6. *Any irreducible Markov chain with a finite state space is positive recurrent.*

Proof. With $S_i = \sum_0^\infty I(X_n = i)$, we have $\sum_{i \in E} S_i = \infty$ so that by finiteness $S_i = \infty$ for at least one i. But then i is recurrent, and therefore by irreducibility the chain is recurrent. Since obviously the stationary measure cannot have infinite mass in the finite case, we have positive recurrence. □

Example 3.7. Consider the backwards and forwards recurrence time chains $\{A_n\}$, $\{B_n\}$ of a renewal process governed by $\{f_n\}$. From the discussion in Section 2 it is immediately apparent that both chains are irreducible on the appropriate state spaces. It is also clear that 0 is recurrent for A_n and 1 for B_n with $\{f_n\}$ as recurrence-time distribution in both cases. In particular, positive recurrence is equivalent to $\mu = \sum nf_n < \infty$. For $\{A_n\}$, the stationary measure (3.1) with $i = 0$ becomes $\nu_n = r_n = f_{n+1} + f_{n+2} + \cdots$. Indeed, n is visited once in between two consecutive visits to 0 if the recurrence time is $\geqslant n+1$. This occurs with probability r_n and otherwise n is not visited. In particular, if $\mu < \infty$ then the stationary distribution is $\pi_n = r_n/\mu$. In an entirely similar manner it is seen that the stationary measure for $\{B_n\}$ is $\nu_n = r_{n-1}$ and if $\mu < \infty$ then $\pi_n = r_{n-1}/\mu$ defines the stationary distribution. □

The above assumption of irreducibility and recurrence (i.e. one recurrent class) can easily be weakened by invoking the decomposition (1.7) of the state space. For example, if $\nu^{(r)}$ is a stationary measure on the rth recurrent class, it is easy to see that $\nu = \sum_r \nu^{(r)}$ is stationary for the whole chain. Conversely, the restriction of a stationary ν to R_r is stationary (for the chain restricted to R_r). Also, some *transient* chains have a stationary measure. The theory is more difficult than for the recurrent case and will not be discussed or used here. We remark only that a stationary *distribution* always attaches mass zero to the transient states because $\mathbb{P}(X_n = i) \to 0$ when i is transient. Combining (1.7) and Corollary 3.1, it is then easy to see that the most general form of a stationary distribution is a convex combination of the unique stationary distributions on the positive recurrent classes.

An alternative proof of the uniqueness of the stationary measure will be given in VI.3. It relies on restricting the Markov chain to a subset of F of the state space, a procedure which has also other applications and which we now take the opportunity to discuss briefly. Let $\tau(F;k)$ be the time of the kth visit of $\{X_n\}$ to F, and define $\tau(F) = \tau(F;1)$, $X_k^F = X_{\tau(F;k)}$. In the recurrent case, $\tau(F;k) < \infty$ for all k, and because of the strong Markov property it can be easily seen that $\{X_k^F\}$ is a Markov chain. The transition matrix has elements $p_{kl}^F = \mathbb{P}_k(X_{\tau(F)} = l)$, $k, l \in F$, but these cannot in general be found explicitly in terms of the p_{ij}. Nevertheless, we have the following result:

Proposition 3.8. *If $\{X_n\}$ is irreducible and recurrent with stationary measure ν, then $\{X_k^F\}$ is also irreducible and recurrent, and the stationary measure $\nu^F = (\nu_l^F)_{l \in F}$ can be obtained by restricting ν to F, i.e. up to a multiplicative constant $\nu_l^F = \nu_l$, $l \in F$. In particular, if $\{X_k^F\}$ is positive recurrent, then the stationary distribution is given by*

$$\pi_l^F = \frac{\nu_l}{\sum_{k \in F} \nu_k}.$$

Proof. The first assertion is obvious. If we choose the state i in (3.1) in F, then both $\{X_n\}$ and $\{X_k^F\}$ visits $l \in F$ the same number of times in between visits to i. Hence, also constructing v^F according to (3.1) yields $v_l^F = v_l$. □

The formula, which conversely expresses v in terms of v^F (and P), is given later in X.3.

Occasionally the following criterion is useful:

Lemma 3.9. *Let $\{X_n\}$ be irreducible and F a finite subset of the state space. Then the chain is positive recurrent if $\mathbb{E}_i\tau(F) < \infty$ for all $i \in F$.*

Proof. Define

$$\sigma(i) = \inf\{k \geq 1 : X_k^F = i\}, \qquad Y_k = \tau(F;k) - \tau(F;k-1), \qquad (\tau(F;0) = 0).$$

Then with $m = \max_{j \in F} \mathbb{E}_j\tau(F)$ we have for $i \in F$ that

$$\mathbb{E}_i\tau(i) = \mathbb{E}_i \sum_{k=1}^{\sigma(i)} Y_k = \sum_{k=1}^{\infty} \mathbb{E}_i[\mathbb{E}(Y_k | \mathscr{F}_{\tau(F;k-1)}); k \leq \sigma(i)]$$

$$\leq m \sum_{k=1}^{\infty} \mathbb{P}_i(k \leq \sigma(i)) = m\mathbb{E}_i\sigma(i).$$

Since E is finite, $\{X_k^F\}$ is positive recurrent. Thus $\mathbb{E}_i\sigma(i) < \infty$, and $\mathbb{E}_i\tau(i) < \infty$ and positive recurrence follows. □

Problems

3.1. Compute a stationary measure of a *doubly stochastic* matrix P, i.e. where both the rows and columns sum to 1.

3.2. Show that a *Bernoulli random walk*, $E = \mathbb{Z}$, $p_{n(n+1)} = 1 - p_{n(n-1)} = p$, is doubly stochastic and, if in addition $p \neq \frac{1}{2}$, transient. Show also that both $v_n = 1$ and $\mu_n = p^n/(1-p)^n$ are stationary.

3.3. (continuation of Problem 2.1). Show that the generating function $\hat{v}(s)$ of the stationary measure of the backwards recurrence-time chain of a renewal process is given by $\hat{v}(s) = (\hat{f}(s) - 1)/(s - 1)$.

3.4. A set A of states is called an *atom* if $p_{i\cdot}$ is the same for all $i \in A$. Show that $\tau(A)$ is finite $\mathbb{P}_i$ a.s., either for all $i \in A$ or for no $i \in A$, and that in the first case a stationary measure can be defined by

$$v_j = \mathbb{E}_i \sum_{n=1}^{\tau(A)} I(X_n = j) \quad \text{with} \quad i \in A \quad \text{arbitrary.}$$

3.5. Consider the recurrence times A_n, B_n of a renewal process. Show that $\{(A_n, B_n)\}$ is Markov with the set of states of the form $(i, 1)$ being an atom, and that the stationary measure is given by $v_{ij} = f_{i+j}$.

3.6. Show that $\{(X_n, X_{n+1})\}$ is a Markov chain, and compute the stationary measure in terms of that of $\{X_n\}$.

3.7. Let $\{X_n\}$ be stationary w.r.t. π and $\tau = \inf\{n \geq 1 : X_n = X_0\}$ the time of return to the initial state. Evaluate $\mathbb{E}_\pi\tau$.

4. ERGODIC THEORY

The aim is to obtain the limiting behaviour of the p_{ij}^n. We start by noting that this is non-trivial only in the positive recurrent case:

Proposition 4.1. *If the chain is irreducible and either transient or null recurrent, then $p_{ij}^n \to 0$ as $n \to \infty$ for any $i, j \in E$.*

Proof. In the transient case, $\{X_n = j\}$ occurs only finitely often so that its $\mathbb{P}_i$-expectation must tend to zero. In the null recurrent case, write

$$p_{ij}^n = \sum_{k=1}^{\infty} \mathbb{P}_i(\tau(j) = k) u_{n-k}, \quad \text{where} \quad u_n = p_{jj}^n. \tag{4.1}$$

Now $\{u_n\}$ is a renewal sequence governed by a distribution with infinite mean, and therefore by Corollary 2.3 $u_n \to 0$. Letting $n \to \infty$ in (4.1) and appealing to dominated convergence yields $p_{ij}^n \to 0$. □

In the aperiodic positive recurrent case it follows from Theorem 2.2 that $u_n \to \mu^{-1}$, where μ is the mean recurrence time $\mathbb{E}_j \tau(j) = \pi_j^{-1}$ with π the stationary distribution, cf. Corollary 3.5. Exactly as above, we may then conclude from (4.1) that $p_{ij}^n \to \pi_j$, and we have proved the following main result:

Theorem 4.2. (ERGODIC THEOREM FOR MARKOV CHAINS). *Suppose that the chain is irreducible and aperiodic positive recurrent, and let $\pi = \{\pi_j\}_{j \in E}$ be the stationary distribution. Then $p_{ij}^n \to \pi_j$ for all i, j.*

That is, the limiting distribution of X_n is π, irrespective of the initial state i. Replacing $\mathbb{P}_i$ by $\mathbb{P}_\nu$ in (4.1) shows that the same conclusion more generally holds for *any* initial distribution ν.

The case $d > 1$ can be quite easily reduced to the case $d = 1$. To this end, we need the concept of *cyclic classes*, i.e. a partitioning $E = \bigcup_0^{d-1} E_r$ of the state space with the property that the only possible transitions are of the form $E_r \to E_{r+1}$, where we identify E_d with E_0, E_{d+1} with E_1 and so on.

Proposition 4.3. *Consider an irreducible chain with period d, let i be some arbitrary but fixed state and define for $r = 0, \ldots, d-1$*

$$E_r = \{j \in E : p_{ij}^{nd+r} > 0 \text{ for some } n\}.$$

Then $E_0, \ldots, E_{d-1}$, partition E into non-empty disjoint sets and if $j \in E_r$ then $\mathbb{P}_j(X_1 \in E_{r+1}) = 1$, and more generally $\mathbb{P}_j(X_m \in E_{m+r}) = 1$. Furthermore, these properties determine the E_r uniquely up to cyclic rotations.

Proof. It is obvious that $E_r \neq \varnothing$ (take $n = 0$). By irreducibility, each j is in some E_r, i.e. $\bigcup_0^{d-1} E_r = E$. Suppose that p_{ij}^{nd+r} and p_{ij}^{md+s} are both > 0, and choose t with $p_{ji}^t > 0$. Then $nd + r + t$ and $md + s + t$ must both be multiples of d, i.e. $r - s = 0 \pmod{d}$ so that the E_r are disjoint. Clearly, $j \in E_r$ and $p_{jk}^m > 0$ implies $k \in E_{r+m}$. Summing over all k yields $\mathbb{P}_j(X_m \in E_{m+r}) = 1$. Uniqueness is easy and is omitted. □

It follows that if $d > 1$, then the chain $X_0, X_d, X_{2d}, \ldots$ has $E_0, \ldots, E_{d-1}$ as disjoint closed sets. In the irreducible positive recurrent case, it is furthermore clear that $\{X_{nd}\}$ is aperiodic positive recurrent on each E_r, i.e. admits a unique stationary distribution $\pi^{(r)}$ concentrated on E_r. Now if π is stationary for $\{X_n\}$, it is also stationary for $\{X_{nd}\}$, and thus by uniqueness π is a convex combination $\sum_0^{d-1}\alpha_r\pi^{(r)}$ of the $\pi^{(r)}$. Since

$$\alpha_{r+1} = \mathbb{P}_\pi(X_{n+1}\in E_{r+1}) = \mathbb{P}_\pi(X_n\in E_r) = \alpha_r,$$

we must even have $\alpha_r = d^{-1}$. Also, the limiting behaviour of p^n_{jk} can be easily seen from $p^{nd}_{jl}\to\pi^{(r)}_l$ if $j,l\in E_r$. Indeed, supposing that $j\in E_r$, then $p^{nd+s}_{jk}=0$ for all n if $k\notin E_{r+s}$, whereas if $k\in E_{r+s}$, then by dominated convergence

$$p^{nd+s}_{jk} = \sum_{l\in E_{r+s}} p^s_{jl}p^{nd}_{lk} \to \sum_{l\in E_{r+s}} p^s_{jl}\pi^{(r+s)}_k = \pi^{(r+s)}_k. \tag{4.2}$$

A further noteworthy property of the stationary distribution is, as the limit of time averages,

$$n^{-1}\sum_{k=0}^{n} f(X_k)\to\mathbb{E}_\pi f(X_k) = \sum_{i\in E} f(i)\pi_i \quad \text{a.s.} \tag{4.3}$$

which holds under suitable conditions on f, e.g. for all bounded f. The proof is carried out in a more general setting in V.3.

In view of this discussion one can assume in most cases that the period is unity. An irreducible aperiodic positive recurrent chain is frequently simply called *ergodic*.

Also in the null recurrent case it is sometimes possible in various ways to obtain limit statements in terms of the stationary measure which are more refined than just $p^n_{ij}\to 0$. For example:

Proposition 4.4. *If the chain is irreducible recurrent with stationary measure* ν, *then for all* $i,j,k,l\in E$

$$\frac{\sum_{n=0}^{m} p^n_{ij}}{\sum_{n=0}^{m} p^n_{lk}} \to \frac{\nu_j}{\nu_k}, \qquad m\to\infty. \tag{4.4}$$

For the proof, we need two lemmas (the proof of the first is a straightforward verification and is omitted; generalizations are in Problem 5.1 and X.1):

Lemma 4.5. *The matrix* $\tilde{P}$ *with elements* $\tilde{p}_{ij} = \nu_j p_{ji}/\nu_i$ *is a transition matrix. Furthermore, the* ijth *element* $\tilde{p}^m_{ij}$ *of the* mth *power* $\tilde{P}^m$ *is given by* $\tilde{p}^m_{ij} = \nu_j p^m_{ji}/\nu_i$.

Lemma 4.6. *Define* $N^m_i = \sum_{n=0}^m I(X_n = i)$ *as the number of visits to i before time m. Then if the chain is irreducible and recurrent,* $\lim_{m\to\infty}\mathbb{E}_j N^m_i/\mathbb{E}_k N^m_i = 1$ *for any* $j,k\in E$.

Proof. It may be assumed that $k=i$. Since $\mathbb{E}_i N^m_i\to\mathbb{E}_i N_i=\infty$ and $N^{m-n}_i = N^m_i + O(1)$, we get by dominated convergence

$$\frac{\mathbb{E}_j N^m_i}{\mathbb{E}_i N^m_i} = \sum_{n=0}^{m}\mathbb{P}_j(\tau(i)=n)\frac{\mathbb{E}_i N^{m-n}_i}{\mathbb{E}_i N^m_i} \to \sum_{n=0}^{\infty}\mathbb{P}_j(\tau(i)=n) = 1. \qquad \square$$

Proof of Proposition 4.4. Consider a Markov chain with transition matrix $\tilde{P}$ given by Lemma 4.1. The expression for $\tilde{p}_{ij}^n$ shows that P and $\tilde{P}$ are irreducible at the same time, and also that recurrence holds at the same time in view of the criterion $\sum_1^\infty p_{ii}^m = \infty$ for all (some) i, cf. Prop. 1.2. Hence $\tilde{X}_0, \tilde{X}_1, \ldots$ satisfies the assumptions of Lemma 4.6 and we obtain

$$1 = \lim_{m\to\infty} \frac{\tilde{\mathbb{E}}_j N_i^m}{\tilde{\mathbb{E}}_k N_i^m} = \lim_{m\to\infty} \frac{\sum_{n=0}^m \tilde{p}_{ji}^n}{\sum_{n=0}^m \tilde{p}_{ki}^n} = \frac{v_k}{v_j} \lim_{m\to\infty} \frac{\sum_{n=0}^m p_{ij}^n}{\sum_{n=0}^m p_{ik}^n}$$

$$= \frac{v_k}{v_j} \lim_{m\to\infty} \frac{\sum_{n=0}^m p_{ij}^n}{\sum_{n=0}^m p_{lk}^n},$$

using Lemma 4.6 once more in the last step. □

Notes

The terminology 'ergodic' is in agreement with probabilistic ergodic theory (e.g. Breiman, 1968, Ch. 6) in the sense that the Markov chains which are ergodic in this second meaning are precisely the stationary versions of positive recurrent aperiodic chains. The key link is the fact that in general the invariant σ-field of $\{X_n\}$ is spanned exactly by the events $\{X_0 \in E_r\}$ with $E_0, \ldots, E_{d-1}$ the cyclic classes, see for example Orey (1971, Ch. 1).

One might expect from Proposition 4.4 and the ergodic theorem for Markov chains that if the chain is also aperiodic, then the *strong ratio property* $p_{ij}^n / p_{lk}^n \to v_j / v_k$ holds. This is, however, not true for all null recurrent chains and presents in fact difficult and far from completely solved problems, see again Orey (1971, Ch. 3).

5. HARMONIC FUNCTIONS, MARTINGALES AND RECURRENCE/TRANSIENCE CRITERIA

There is a concept dual to that of a stationary measure, namely that of a *harmonic function* defined as a right eigenvector h of P corresponding to the eigenvalue 1. The requirement $Ph = h$ means

$$h(i) = \sum_{j\in E} p_{ij} h(j) = \mathbb{E}_i h(X_1) = \mathbb{E}[h(X_{n+1}) | X_n = i],$$

i.e. that $\{h(X_n)\}$ is a martingale. Similarly, one defines h to be *subharmonic* if $Ph \geqslant h$, i.e. $\{h(X_n)\}$ is a submartingale, and *superharmonic* or *excessive* if $Ph \leqslant h$, i.e. $\{h(X_n)\}$ is a supermartingale.

Proposition 5.1. *If the chain is irreducible and recurrent, then any non-negative superharmonic function h is necessarily constant. Similarly, any bounded subharmonic function is constant.*

Proof. We must show that $h(i)=h(j)$ for $i \neq j$. Now from the convergence of any non-negative supermartingale we have that $Z=\lim h(X_n)$ exists $\mathbb{P}_i$ a.s. Since $\mathbb{P}_i(X_n=i$ infinitely often$)=1$, it follows that $Z=h(i)$ $\mathbb{P}_i$ a.s. Similarly $\mathbb{P}_i(X_n=j$ infinitely often$)=1$ implies that $Z=h(j)$ $\mathbb{P}_i$ a.s. and hence $h(i)=h(j)$. The subharmonic case is treated in a similar manner using the a.s. convergence of any bounded submartingale. □

In view of this result, (super or sub) harmonic functions do not play a very prominent role here since we are mainly concerned with the recurrent case. It turns out, however, that a number of useful recurrence/transience criteria can be stated in terms of functions h with properties looking rather similar and allowing arguments along the lines of the proof of Proposition 5.1. We start in Proposition 5.2 by a general standard result, and proceed next in Propositions 5.3 and 5.4 to *Foster's criteria*, which are tailored to the type of Markov chains occurring in many contexts in queueing theory.

Proposition 5.2. *Suppose the chain is irreducible and let i be some fixed state. Then the chain is transient if and only if there is a bounded non-zero function* $h: E\backslash\{i\} \to \mathbb{R}$ *satisfying*

$$h(j)=\sum_{k \neq i} p_{jk} h(k), \qquad j \neq i. \tag{5.1}$$

Proof. Obviously $h(j)=\mathbb{P}_j(\tau(i)=\infty)$ is bounded and satisfies (5.1). If the chain is transient, then furthermore $h \neq 0$. Suppose, conversely, there is an h as stated and define $\tilde{h}(j)=h(j)$, $j \neq i$, $\tilde{h}(i)=0$, $\alpha=P\tilde{h}(i)$. By changing the sign if necessary, we may assume that $\alpha \geqslant 0$ so that $P\tilde{h}(i) \geqslant \tilde{h}(i)$. Since $P\tilde{h}(j)=\tilde{h}(j)$ for $j \neq i$, h is thus subharmonic. Hence if the chain is recurrent, we have by Proposition 5.1 for all j that $h(j)=\tilde{h}(j)=\tilde{h}(i)=0$, contradicting $h \neq 0$. Hence the chain is transient. □

Proposition 5.3. *Suppose that the chain is irreducible and let* E_0 *be a finite subset of the state space E. Then:*

(i) *The chain is recurrent if there exists a function* $h: E \to \mathbb{R}$ *such that* $\{i: h(i)<K\}$ *is finite for each K and*

$$\sum_{k \in E} p_{jk} h(k) \leqslant h(j), \qquad j \notin E_0 \tag{5.2}$$

(ii) *The chain is positive recurrent if for some* $h: E \to \mathbb{R}$ *and some* $\varepsilon>0$ *we have* $\inf_j h(j)>-\infty$ *and*

$$\sum_{k \in E} p_{jk} h(k)<\infty, \qquad j \in E_0,$$

$$\sum_{k \in E} p_{jk} h(k) \leqslant h(j)-\varepsilon, \qquad j \notin E_0. \tag{5.3}$$

Proof. By adding a constant if necessary, we may assume that $h \geqslant 0$. Write $T=\tau(E_0)$ and define $Y_n=h(X_n) I(T>n)$.

(i) Note first that (5.2) may be rewritten $\mathbb{E}(h(X_{n+1}) \mid X_n=j) \leqslant h(j)$ for $j \notin E_0$. Let

$X_0 = i \notin E_0$. Then on $\{T > n\}$, $X_n \notin E_0$ (this fails for $n = 0$ if $X_n \in E_0$) and hence

$$\begin{aligned}\mathbb{E}_i(Y_{n+1}|\mathscr{F}_n) &\leqslant \mathbb{E}_i[h(X_{n+1}); T > n|\mathscr{F}_n] \\ &= I(T > n)\mathbb{E}_i[h(X_{n+1})|\mathscr{F}_n] \leqslant I(T > n)h(X_n) = Y_n. \qquad (5.4)\end{aligned}$$

If $T \leqslant n$, then $Y_n = Y_{n+1} = 0$ and thus $\mathbb{E}_i(Y_{n+1}|\mathscr{F}_n) \leqslant Y_n$, i.e. $\{Y_n\}$ is a non-negative supermartingale and hence converges a.s., $Y_n \to Y_\infty$. Suppose the chain is transient. Then $h(X_n) < K$ only finitely often, i.e. $h(X_n) \to \infty$ $\mathbb{P}_i$ a.s. and since $Y_\infty < \infty$ we must have $\mathbb{P}_i(T = \infty) = 0$ for all $i \notin E_0$. But $\mathbb{P}_i(T < \infty) = 1$ for all $i \notin E_0$ implies that some $j \in E_0$ is recurrent, a contradiction.

(ii) Again let $X_0 = i \notin E_0$. Then, as in (5.4),

$$\mathbb{E}_i(Y_{n+1}|\mathscr{F}_n) \leqslant I(T > n)\mathbb{E}_i[h(X_{n+1})|\mathscr{F}_n] \leqslant Y_n - \varepsilon I(T > n)$$

on $\{T > n\}$. Again the same is obvious on $\{T \leqslant n\}$ and hence

$$0 \leqslant \mathbb{E}_i Y_{n+1} \leqslant \mathbb{E}_i Y_n - \varepsilon \mathbb{P}_i(T > n) \leqslant \cdots \leqslant \mathbb{E}_i Y_0 - \varepsilon \sum_{k=0}^{n} \mathbb{P}_i(T > k).$$

Letting $n \to \infty$ and using $Y_0 = h(i)$ yields $\mathbb{E}_i T \leqslant \varepsilon^{-1} h(i)$. Thus for $j \in E_0$

$$\mathbb{E}_j T = \sum_{i \in E_0} p_{ji} + \sum_{i \notin E_0} p_{ji} \mathbb{E}_i(T + 1) \leqslant 1 + \varepsilon^{-1} \sum_{i \notin E_0} p_{ji} h(i)$$

which is finite by (5.3). Positive recurrence now follows by Lemma 3.9. □

Proposition 5.4. *Suppose that the chain is irreducible and let E_0 be a finite subset of E such that* (5.2) *holds for some bounded h, satisfying $h(i) < h(j)$ for some $i \notin E_0$ and all $j \in E_0$. Then the chain is transient.*

Proof. Define T as above but now let $Y_n = h(X_{n \wedge T})$. It is then readily verified that Y_n is again a supermartingale with respect to $\mathbb{P}_i$ so that $Y_n \to Y_\infty$, where $\mathbb{E}_i Y_\infty \leqslant \mathbb{E}_i Y_0 = h(i)$. But $Y_\infty > h(i)$ on $\{T < \infty\}$, so that $\mathbb{P}_i(T < \infty) < 1$ and transience follows. □

The conditions of Propositions 5.3(i) and 5.4 are also known to be necessary. The proof is not unreasonably complicated but will be omitted since our main application here is of Proposition 5.3(ii) with $E = \mathbb{N}$, $h(i) = i$ as a sufficient criterion for positive recurrence. To illustrate the usefulness of the approach, we consider one of the very simplest queueing situations:

Example 5.5. Consider a queue where service takes place at a discrete sequence of instants $n = 0, 1, 2, \ldots$, let X_n be the queue length at time n, B_n the number of customers arriving between n and $n + 1$ and A_n the maximal number of customers which can be served at the $(n + 1)$th service event. Thus with $Y_n = B_n - A_n$,

$$X_{n+1} = (X_n + Y_n)^+, \qquad (5.4)$$

a recurrence relation also typical for many other queueing situations and discussed in detail in III.7. Assume further that (A_0, B_0), $(A_1, B_1), \ldots$ are independent identically distributed. For example, this could describe the queue at the stop of a bus with a regular schedule, with A_n the number of available seats in the $(n+1)$th bus. Suppose $\mathbb{E}A_n > \mathbb{E}B_n$ and let $\varepsilon_i = -\mathbb{E}Y_n I(Y_n > -i)$, $\varepsilon = -\mathbb{E}Y_n$. Then $\varepsilon_i \to \varepsilon > 0$, $i \to \infty$, and hence for i so large that $\varepsilon_i \geqslant \varepsilon/2$,

$$\begin{aligned}\mathbb{E}(X_{n+1} \mid X_n = i) = \mathbb{E}(i + Y_n)^+ &= \mathbb{E}(i + Y_n)I(Y_n > -i) \leqslant i - \varepsilon_i \\ &\leqslant i - \varepsilon/2.\end{aligned}$$

Thus Proposition 5.3(i) applies to ensure that $\{X_n\}$ is positive recurrent.

Problems

5.1. Suppose the chain is irreducible and that $h \geqslant 0$ is harmonic. Show that $h(i) > 0$ for all i and that the matrix $\tilde{p}$ with elements $\tilde{p}_{ij} = h(j)p_{ij}/h(i)$ is a transition matrix.

5.2. Consider a population process satisfying the assumptions of Problem 1.3 and with all states $i, j \geqslant 1$ communicating. Show that the extinction probability $q_i = \mathbb{P}_i(\tau(0) < \infty)$ is either $= 1$ for all $i \geqslant 1$ or < 1 for all $i \geqslant 1$. Let E_0 be finite with $0 \in E_0$ and suppose (5.2) holds. Show that $q_i = 1$ if $h(n) \to \infty$ and that $q_i < 1$ if h is bounded with $h(i) < h(j)$ for some $i \notin E_0$ and all $j \in E_0$ [hint: consider $\{\tilde{X}_n\}$ evolving as $\{X_n\}$ except that $\tilde{p}_{01} = 1$ rather than $p_{00} = 1$, and apply Foster's criteria to $\{\tilde{X}_n\}$]. Show in particular that if $\mathbb{E}_i(X_{n+1} \mid X_n) \leqslant X_n$ (i.e. the expected number of children per individual does not exceed 1), then extinction occurs a.s.

Notes

The present proof of Propositions 5.3(i) and 5.4 is from Mertens *et al.* (1978), whereas that of Proposition 5.3(ii) is a modification of an argument communicated by Tony Pakes. Further developments of Foster's criteria can be found in Tweedie (1983).

6. FOUNDATIONS OF THE GENERAL THEORY OF MARKOV PROCESSES

We shall consider two generalizations, first that of the state space E being general (i.e. possibly continuous or non-discrete), and next that of a continuous time parameter $t \in [0, \infty)$.

If E is not necessarily discrete, we need to be given a measurable structure on E, i.e. a σ- algebra $\mathscr{E}$ to which all subsets of E are assumed to belong in the following. Instead of the transition matrix we have a *transition* (or *Markov*) *kernel* P, i.e. a function $P(x, A)$ of $x \in E$ and $A \in \mathscr{E}$ such that $P(x, \cdot)$ is a probability on $(E, \mathscr{E})$ for each x and $P(\cdot, A)$ is a measurable function for each A.

Markov chains $\{X_n\}$ with transition kernel P and the corresponding Markov probabilities $\mathbb{P}_\mu$ are defined by the requirements $\mathbb{P}_\mu(X_0 \in A) = \mu(A)$,

$$\mathbb{P}_\mu(X_{n+1} \in A \mid \mathscr{F}_n) = P(X_n, A), \tag{6.1}$$

where $\mathscr{F}_n = \sigma(X_0, \ldots, X_n)$. With the usual a.s. interpretation of conditional probabilities, it follows from (6.1) that

$$\mathbb{P}_\mu(X_{n+1} \in A \mid X_n = x) = P(x, A). \tag{6.2}$$

Also, say by induction, the following is easily obtained:

$$\mathbb{P}_\mu(X_0 \in A_0, X_1 \in A_1, \ldots, X_n \in A_n)$$
$$= \int_{A_0} \mu(\mathrm{d}x_0) \int_{A_1} P(x_0, \mathrm{d}x_1) \cdots$$
$$\int_{A_{n-1}} P(x_{n-2}, \mathrm{d}x_{n-1}) P(x_{n-1}, A_n). \tag{6.3}$$

This formula also immediately suggests how to define the Markov probabilities and the Markov chain: take $X_0, X_1, \ldots$ as the projections $E^{\mathbb{N}} \to E$ and let

$$\mathscr{E}_n = \sigma(X_0, \ldots, X_n), \qquad \mathscr{E}_\infty = \sigma(X_0, X_1, \ldots) = \mathscr{E}^{\mathbb{N}}.$$

Then by standard arguments from measure theory it can be seen that the r.h.s. of (6.3) can be uniquely extended to a probability $\mathbb{P}^n_\mu$ on $(E^{\mathbb{N}}, \mathscr{E}_n)$. The $\mathbb{P}^n_\mu$ have the consistency property $\mathbb{P}^n_\mu(A) = \mathbb{P}^m_\mu(A)$, $m \leqslant n$, $A \in \mathscr{E}_m$, and hence define a finitely additive probability on the algebra $\bigcup_0^\infty \mathscr{E}_n$. The desired $\mathbb{P}_\mu$ is the (necessarily unique) extension to $\mathscr{E}_\infty = \sigma(\bigcup_0^\infty \mathscr{E}_n)$. The existence, i.e. the σ-additivity on $\bigcup_0^\infty \mathscr{E}_n$, may be seen either from Kolmogorov's consistency theorem which requires some topological assumptions like E being Polish and $\mathscr{E}$ the Borel σ-algebra, or by a measure-theoretic result of Ionesca Tulcea (see e.g. Neveu, 1965).

The continuous-time case is substantially more involved. What will be needed in later chapters is, however, only a few basic facts and we shall therefore just outline a theory which needs several amendments when pursuing Markov process theory in its full generality.

One does not get very far without topology, so we assume right from the start that E is Polish with $\mathscr{E}$ the Borel σ-algebra. That a process $\{X_t\}_{t \geqslant 0}$ with state space E is Markovian means intuitively just the same as in discrete time: given the history $\mathscr{F}_t = \sigma(X_s; s \leqslant t)$ of the process up to time t, the process evolves from then on as restarted at time 0 in state X_t, and depending on $\mathscr{F}_t$ through X_t only. Formally, this may be expressed by the existence of a family of probability measures $\mathbb{P}_\mu$ with the property $\mathbb{P}_\mu(X_0 \in A) = \mu(A)$,

$$\mathbb{E}_\mu[h(X_{s+t}; t \geqslant 0) | \mathscr{F}_s] = \mathbb{E}_{X_s} h(X_t; t \geqslant 0) \tag{6.4}$$

where $\mathbb{P}_x$, $\mathbb{E}_x$ refers to $X_0 = x$ and (6.4) should hold for a class of functions h of the process sufficiently rich to determine the distribution of $\{X_t\}_{t \geqslant 0}$. For example, it would suffice to consider the class H of all h of the form

$$h(x_t; t \geqslant 0) = \prod_{i=0}^{n} I(x_{t_i} \in A(i)). \tag{6.5}$$

If $\{X_t\}_{t \geqslant 0}$ has paths say in $D = D([0, \infty), E)$, then (6.4) for all $h \in H$ will be equivalent to (6.4) to hold for all bounded measurable $h: D \to \mathbb{R}$. In fact, an easy induction argument shows that it is even sufficient to let $n = 0$ in (6.5), and the Markov property in this equivalent formulation then becomes

$$\mathbb{P}(X_{s+t} \in A | \mathscr{F}_s) = P^t(X_s, A) \quad \text{where} \quad P^t(x, A) = \mathbb{P}_x(X_t \in A). \tag{6.6}$$

Given a Markov process, it is clear that $P^t(x, A)$ as defined by (6.6) is a transition kernel. Using the Markov property we get

$$P^{t+s}(x, A) = \mathbb{E}_x \mathbb{P}(X_{t+s} \in A \mid \mathscr{F}_s) = \mathbb{E}_x P^t(X_s, A)$$

$$= \int P^t(y, A) P^s(x, \mathrm{d}y),$$

which in operator notation is written $P^{t+s} = P^t P^s$ and referred to as the *Chapman–Kolmogorov equations* (or the *semi-group property*). Conversely, given a family $\{P^t\}_{t \geqslant 0}$ of transition kernels satisfying the Chapman–Kolmogorov equations, it is possible to construct a corresponding Markov process. To this end, we proceed as in discrete time: let $X_t : E^{[0,\infty)} \to E$ be the projection and define for $0 = t_0 < t_1 < \cdots < t_n$ a probability on the sub-σ-algebra $\sigma(X_{t_i}; i = 0, \ldots, n)$ of $\mathscr{E}^{[0,\infty)}$ by

$$\mathbb{P}_\mu^{t_0, \ldots, t_n}(X_{t_0} \in A_0, \ldots, X_{t_n} \in A_n) = \int_{A_0} \mu(\mathrm{d}x_0) \int_{A_1} P^{t_1 - t_0}(x_0, \mathrm{d}x_1) \cdots$$

$$\int_{A_{n-1}} P^{t_{n-1} - t_{n-2}}(x_{n-2}, \mathrm{d}x_{n-1}) P^{t_n - t_{n-1}}(x_{n-1}, A_n). \tag{6.7}$$

That this defines a consistent family is readily apparent from the Chapman–Kolmogorov equations. As E is Polish, there thus exists a unique extension to $\mathscr{E}^{[0,\infty)}$, and the Markov property (6.4) with $h \in H$ is inherent in the definition (6.7).

There are, however, severe difficulties associated with this approach. First, the intuitive description of a particular model is seldom in terms of the P^t. Next, the construction makes $\mathscr{E}^{[0,\infty)}$ the collection of measurable sets, i.e. when $A \notin \mathscr{E}^{[0,\infty)}$ one cannot make sense of $\mathbb{P}_\mu(A)$. But $\mathscr{E}^{[0,\infty)}$ is not very rich since one can easily see that $A \in \mathscr{E}^{[0,\infty)}$ implies that A depends on the X_t for t in a countable collection $T_A \subseteq [0, \infty)$ of time points. Thus for example sets like

$$\{\omega : X_t(\omega) \text{ is a continuous function of } t\}$$

are not in $\mathscr{E}^{[0,\infty)}$, and when $E = \mathbb{R}$ similarly $\max_{0 \leqslant t \leqslant T} X_t$ and $\inf\{t : X_t = 0\}$ are not measurable. Hence it is necessary to construct versions of the process with sample paths say in D. This requires further explicit knowledge of P^t, typically continuity requirements. We shall not go into this since the explicit examples that we shall encounter will always a priori satisfy such path regularity conditions. For example, queues are constructed by simple transformations of sequences of service times and inter-arrival times, and not starting from semigroups, consistent families and so on.

Now let σ be a stopping time with respect to $\{\mathscr{F}_t\}_{t \geqslant 0}$ and let $\mathscr{F}_\sigma$ be the corresponding stopping time σ-field, cf. App. A8. We say that $\{X_t\}_{t \geqslant 0}$ has the *strong Markov property with respect to* σ if a.s. on $\{\sigma < \infty\}$

$$\mathbb{P}_\mu(X_{\sigma+t} \in A \mid \mathscr{F}_\sigma) = P^t(X_\sigma, A). \tag{6.8}$$

(again, this implies a functional form

$$\mathbb{E}_\mu[h(X_{\sigma+t}; t\in T)|\mathscr{F}_\sigma] = \mathbb{E}_{X_\sigma} h(X_t; t\in T)).$$

The process is *strong Markov* if it has the strong Markov property with respect to any stopping time σ.

Proposition 6.1. *A Markov process $\{X_t\}_{t\in T}$ has the strong Markov property with respect to any stopping time σ which assumes only a countable number of finite values, $\sigma\in\{\infty, s_1, s_2, \ldots\}$.*

Proof. We must show that for $A\in\mathscr{E}$, $F\in\mathscr{F}_\sigma$

$$\mathbb{P}_\mu(X_{\sigma+t}\in A; F) = \mathbb{E}_\mu[P^t(X_\sigma, A); F].$$

However, if $F\subseteq\{\sigma = s_k\}$ this is immediate from the Markov property (6.4). In the general case, decompose F as the disjoint union of the sets $F\cap\{\sigma = s_k\}$ and sum over k. □

As an immediate consequence, we have:

Corollary 6.2. *Any discrete time Markov chain* (with discrete or general state space) *has the strong Markov property.*

Also for $T = [0, \infty)$ Proposition 6.1 is greatly helpful in establishing the strong Markov property. A typical example is the following:

Corollary 6.3. *Suppose that $T = [0, \infty)$, that $\{X_t\}_{t\geqslant 0}$ has right-continuous paths and that for any bounded continuous $f: E\to\mathbb{R}$ it holds that $\mathbb{E}_x f(X_s)$ is a continuous function of x or, more generally, that the paths of $\mathbb{E}_{X_t} f(X_s)$ are right-continuous functions of t. Then the strong Markov property holds.*

Proof. Let σ be a given stopping time and define $\sigma(k) = n2^{-k}$ on $\{(n-1)2^{-k} < \sigma \leqslant n2^{-k}\}$. Then the $\sigma(k)$ are stopping times and $\sigma(k)\downarrow\sigma$ as $k\to\infty$. By Proposition 6.1 we have furthermore

$$\mathbb{E}_\mu[f(X_{\sigma(k)+s})|\mathscr{F}_{\sigma(k)}] = \mathbb{E}_{X_{\sigma(k)}} f(X_s). \tag{6.9}$$

If $F\in\mathscr{F}_\sigma$, then $F\in\mathscr{F}_{\sigma(k)}$ as well, and hence (6.9) implies

$$\mathbb{E}_\mu[f(X_{\sigma(k)+s}); F] = \mathbb{E}_\mu[\mathbb{E}_{X_{\sigma(k)}} f(X_s); F].$$

A check of the assumptions shows that the integrands converge pointwise. Thus by dominated convergence

$$\mathbb{E}_\mu[f(X_{\sigma+s}); F] = \mathbb{E}_\mu[\mathbb{E}_{X_\sigma} f(X_s); F].$$

The truth of this for all bounded continuous f and all $F\in\mathscr{F}_\sigma$ implies (6.8). □

Define now the *hitting time* of a Borel subset A of E as $\tau(A) = \inf\{t > 0; X_t\in A\}$. That $\tau(A)$ is a stopping time is a triviality in discrete time since then obviously

$$\{\tau(A)\leqslant n\} = \bigcup_{k=1}^{n} \{X_k\in A\}.$$

However, in continuous time some (perhaps unexpected) difficulties arise even for elementary sets like closed and open ones, and this is in fact one of the reasons that one needs to amend and extend the theory which has been discussed so far and which may still appear reasonably simple and intuitive. We discuss these points briefly below, but first state and prove a more elementary result which is sufficient to deal with virtually all the processes to be met and all the questions to be asked in this book.

A function is called piecewise continuous if the jump times $0 < \iota(1) < \iota(2) < \cdots$ do not accumulate, $\iota(k) \uparrow \infty$.

Proposition 6.4. *Suppose that the paths of X_t are right-continuous and piecewise continuous. Then:*

(a) *the jump times $0 < \iota(1) < \iota(2) < \cdots$ are stopping times with respect to $\{\mathscr{F}_t\}$;*
(b) *if A is closed, then $\tau(A)$ is a stopping time with respect to $\{\mathscr{F}_t\}$.*

Proof. Let $\mathbb{Q}(t)$ be the set of numbers of the form qt with q rational and $0 \leqslant q \leqslant 1$, and let d be some metric on E. Then the sets $A = \{\iota(1) \leqslant t\}$ and

$$B = \bigcup_{m=1}^{\infty} \bigcap_{n=1}^{\infty} \bigcup_{u,\, s \in \mathbb{Q}(t)} \left\{ |u - s| \leqslant \frac{1}{n}, d(X_u, X_s) > \frac{1}{m} \right\}$$

coincide. In fact, on A we have for some m a jump of size at least m^{-1}, and this easily yields $A \subseteq B$. Conversely, the uniform continuity of $\{X_s\}_{s \leqslant t}$ on A^c easily shows that $A^c \subseteq B^c$. Since $d(X_u, X_s)$ is $\mathscr{F}_t$-measurable for $u, s \leqslant t$, we have $A = B \in \mathscr{F}_t$ and thus $\iota(1)$ is a stopping time. For $\iota(2)$, just add the requirement $u, s \geqslant \iota(1)$ in the definition of B, and so on. To prove (b), define $m(S) = \inf\{d(X_u, A): u \in S\}, S \subseteq [0, \infty)$. If A is closed, we have $X_{\tau(A)} \in A$ and hence in the special case of continuous paths

$$\{\tau(A) \leqslant t\} = \{m([0, t]) = 0\} = \{m(\mathbb{Q}(t)) = 0\} \in \mathscr{F}_t. \tag{6.10}$$

The problem in the general case is to deal with the case

$$\liminf_{u \uparrow \iota(k)} d(X_u, A) = 0, \; X_{\iota(k)} \notin A.$$

But let

$$I_{k,n} = \left\{ u : \iota(k) - \frac{1}{n} \leqslant u < \iota(k) \leqslant t \right\}$$

so that

$$\{u \in I_{k,n}\} = \left\{ u < \iota(k) \leqslant t \wedge \left(u + \frac{1}{n} \right) \right\} \in \mathscr{F}_t.$$

Thus, as in (6.10),

$$\{\tau(A) \leqslant t\} = \lim_{n \to \infty} \left\{ \tau(A) \in [0, t] \Big\backslash \bigcup_{k=1}^{\infty} I_{k,n} \right\}$$

$$= \lim_{n \to \infty} \left\{ m\left(\mathbb{Q}(t) \Big\backslash \bigcup_{k=1}^{\infty} I_{k,n} \right) = 0 \right\} \in \mathscr{F}_t. \qquad \square$$

We conclude with a brief discussion of some of the difficulties in the general case (this is not essential for the rest of the book). Define $\mathscr{F}_{t+} = \bigcap_{s>t}\mathscr{F}_s$ and let $\mathscr{G}^{(\mu)}$ denote the $\mathbb{P}_\mu$-completion of $\mathscr{G}$ (some arbitrary σ-field), i.e. the smallest σ-field containing $\mathscr{G}$ and all $\mathbb{P}_\mu$-null sets. Then:

Proposition 6.5. *Suppose that $\{X_t\}$ has right-continuous paths. Then:*

(a) *if A is open, then $\tau(A)$ is a stopping time with respect to $\{\mathscr{F}_{t+}\}$;*
(b) *for any Borel set A, $\tau(A)$ is a stopping time with respect to $\{\mathscr{F}_{t+}^{(\mu)}\}$.*

Proof of (a). If A is open and $X_u \in A$, then $X_{u+v} \in A$ when v is also sufficiently small and positive. Hence the event $\{\tau(A) \leqslant t\}$ may be written as

$$\bigcap_{n=1}^{\infty} \bigcup_{s\leqslant t+1/n} \{X_s \in A\} = \bigcap_{n=1}^{\infty} \bigcup_{s\in\mathbb{Q}(t+1/n)} \{X_s \in A\}$$

and here the event on the r.h.s. is clearly in $\mathscr{F}_{t+}$. □

The proof of (b) is far beyond the present scope (and need!), and we refer, e.g. to Chung (1982).

One may now define a *history* of the process as any increasing family $\{\mathscr{G}_t\}_{t\geqslant 0}$ of σ-fields with $\mathscr{F}_t \subseteq \mathscr{G}_t$ (or equivalently X_t $\mathscr{G}_t$-measurable) and say that $\{X_t\}$ is Markov with transition semigroup $\{P^t\}$ with respect to $\{\mathscr{G}_t\}$ and some fixed governing probability measure if

$$\mathbb{P}(X_{t+s} \in A \mid \mathscr{G}_s) = P^t(X_s, A). \tag{6.11}$$

Apart from $\{\mathscr{F}_t\}$, some main candidates for the history are $\{\mathscr{F}_{t+}\}$ and $\{\mathscr{F}_{t+}^{(\mu)}\}$. It follows immediately from properties of conditional expectations that if $\{X_t\}$ is Markov with respect to some history, then $\{X_t\}$ is Markov with respect to $\{\mathscr{F}_t\}$ as well. Conversely:

Proposition 6.6. *Let $\{X_t\}$ be Markov with respect to $\{\mathscr{F}_t\}$ and satisfy the regularity assumptions of Corollary 6.3. Then:*

(a) *for each μ and each measurable h we have $\mathbb{P}_\mu$ a.s.*

$$\begin{aligned}\mathbb{E}_\mu[h(X_{s+t}; t \geqslant 0) \mid \mathscr{F}_s] &= \mathbb{E}_\mu[h(X_{s+t}; t \geqslant 0) \mid \mathscr{F}_{s+}] \\ &= \mathbb{E}_\mu[h(X_{s+t}; t \geqslant 0) \mid \mathscr{F}_{s+}^{(\mu)}].\end{aligned}$$

(b) (BLUMENTHAL'S 0–1 LAW) *if $A \in \mathscr{F}_{0+}$ and $x \in E$, then either $\mathbb{P}_x A = 0$ or $\mathbb{P}_x A = 1$.*
(c) *$\{X_t\}$ is Markov with respect to $\{\mathscr{F}_{t+}\}$ and $\{\mathscr{F}_{t+}^{(\mu)}\}$ as well.*

Proof. (a) The second identity is just a general property of the completion operator. For the first, arguments similar to those used many times above show that it suffices to take h of the form $h(X_t)$ with $t > 0$ and h continuous and bounded. Since $h(X_{s+t+1/n})$ then converges to $h(X_{s+t})$, it follows from a continuity result for conditional expectations (Chung, 1974, p. 340) that indeed

$$\mathbb{E}_\mu[h(X_{s+t})|\mathscr{F}_{s+}] = \lim_{n\to\infty} \mathbb{E}_\mu[h(X_{s+t+1/n})|\mathscr{F}_{s+1/n}]$$
$$= \lim_{n\to\infty} \mathbb{E}_{X_{s+1/n}} h(X_t) = \mathbb{E}_{X_s} h(X_t)$$
$$= \mathbb{E}_\mu[h(X_{s+t})|\mathscr{F}_s]$$

and the proof of (a) is complete. For (b), let $t = 0$ and $h = I(A)$ in (a) to obtain $\mathbb{P}_x(A|\mathscr{F}_{0+}) = \mathbb{P}_x(A|\mathscr{F}_0)$ a.s. Here the l.h.s. is just $I(A)$ and since $\mathscr{F}_0$ is $\mathbb{P}_x$-trivial, the r.h.s. is constant a.s. Hence $I(A)$ is constant a.s., which is only possible if the probability is either zero or 1. Finally (c) is an immediate consequence of (a). □

We stop the discussion of the foundations of the general theory of Markov processes at this point. As for the topics discussed in Sections 2–5, state space properties and ergodic theory will be discussed in Chapter II for $T = [0, \infty)$ and a discrete state space E. The case of a general E is much more complicated even for $T = \mathbb{N}$. For example, it is not clear what recurrence should mean, since even in simple-minded continuous state space models $\mathbb{P}_x(X_n = x \text{ for some } n \geqslant 1)$ will in general be zero. Some results (in fact the best ones known) are given in VI.3 and can, somewhat surprisingly, be derived as simple consequences of the ergodic theorem for discrete Markov chains. For $T = [0, \infty)$ a satisfying general theory hardly exists at all, but a number of special cases will be encountered. For example the main problem within the whole area of renewal theory (Ch. IV) will be seen to be equivalent to the ergodicity question for the continuous-time and -state version of the recurrence-time chains in Section 2.

Notes

General Markov chains (in discrete time) are treated, e.g. in Neveu (1965) and Revuz (1975). For up-to-date and readable accounts of the continuous-time case, see Williams (1979) and Chung (1982).

CHAPTER II

Markov Jump Processes

1. BASIC STRUCTURE

Let E be a discrete (finite or countable) state space and $\{X_t\}_{t\geqslant 0}$ a Markov process with state space E as defined in I.6. A measure μ on E being defined by its point probabilities μ_i, we write $p_{ij}^t = P^t(i, \{j\}) = \mathbb{P}_i(X_t = j)$ and we may identify the transition semigroup by the family $\{P^t\}t \geqslant 0$ of transition matrices. The Chapman–Kolmogorov equations $P^{t+s} = P^t P^s$ may then be interpreted in the sense of usual matrix multiplication.

Problems arising when pursuing the theory without further regularity conditions have already been discussed in I.6. As a further unpleasant possibility, we mention here that some (or even all) states i may be *instantaneous*, i.e. the process jumps out of i immediately after i has been entered. We shall avoid these problems by imposing upon the process a further regularity property, which is inherent in the intuitive picture of any of the models we are concerned with, and which turns out to be sufficient for developing the theory quite smoothly.

The feature that we concentrate on is that of a *pure jump* structure illustrated in Fig. 1.1: the amount of time spent in each state is positive so that the sample paths are piecewise constant. For a pure jump process, we denote the times of jumps by $S_0 = 0 < S_1 < S_2 < \cdots$, the sojourn times (or holding times) by $T_n = S_{n+1} - S_n$ and the sequence of states visited by $Y_0, Y_1, \ldots$. Thus the sample paths are constant between consecutive S_n and we define the value at S_n by right-continuity, i.e. $X_{S_n} = Y_n$. Two possible phenomena require some further comment. The process may be absorbed, say at i. In that case there is a last finite S_n (the time to absorption) and we use the convention that $T_n = T_{n+1} = \cdots \infty$, $Y_n = Y_{n+1} = \cdots = i$. This still yields a very simple structure of paths. More troublesome from that point of view is the possibility of the jumps to accumulate, i.e. of the *explosion time* $\omega(\Delta) = \sup_n S_n$ to be finite (in that case the Y_n and T_n determine the process only up to $\omega(\Delta)$). This seems contrary to intuition in most cases, but is perfectly feasible from the point of view of general theory. We discuss the point in more detail later in Sections II.2 and II.3, and proceed here to discuss some fundamental properties of a Markov jump process.

Theorem 1.1. *Any Markov jump process has the strong Markov property.*

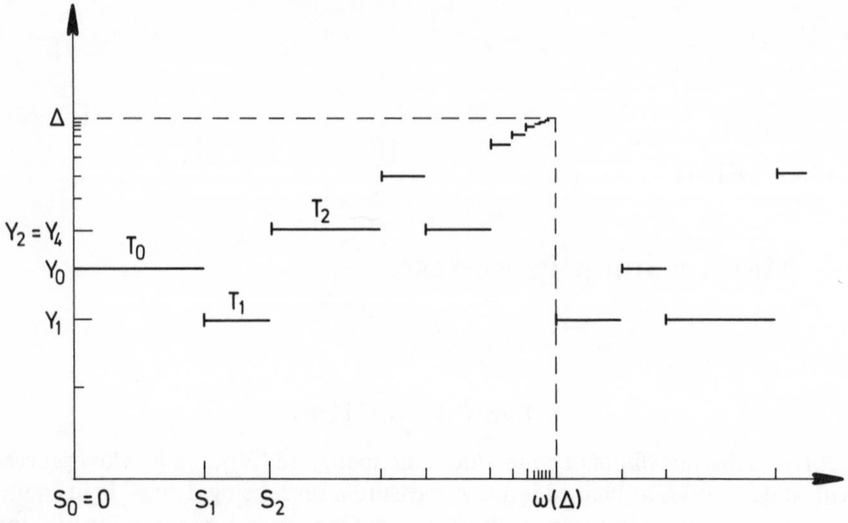

Fig. 1.1 Sample path of a pure jump process. The scale of the state space is chosen so as to illustrate explosion within finite time

Proof. This is a trivial consequence of the first part of 1.6.3 since when E is discrete, then *any* function g on E (in particular $g(x) = \mathbb{E}_x f(X_s)$) is continuous. □

The next result describes the basic structure of a Markov jump process up to the time of explosion. Consider the exponential distribution with density $\lambda e^{-\lambda x}, x > 0$, and denote by the *intensity* (or sometimes *rate*) the parameter λ (by the exponential distribution with intensity $\lambda = 0$ we understand the distribution degenerate at ∞).

Theorem 1.2. *Consider a Markov jump process. Then the joint distribution of the sequences* $\{Y_n\}_{n\in\mathbb{N}}$ *of states visited* (before explosion) *and* $\{T_n\}_{n\in\mathbb{N}}$ *of holding times is given by*: (i) $\{Y_n\}_{n\in\mathbb{N}}$ *is a Markov chain*; (ii) *there exist* $\lambda(i) \geqslant 0$ *such that given* $\{Y_n\}_{n\in\mathbb{N}}$ *the* T_l *are independent, with* T_k *being exponentially distributed with intensity* $\lambda(Y_k)$.

Proof. The joint distribution of the Y_n, T_n is completely specified by probabilities of the form

$$p_n = \mathbb{P}_i(Y_k = i(k), T_{k-1} > t(k) \quad k = 1, \ldots, n).$$

Letting $i(0) = i$, the assertion of the theorem is equivalent to

$$p_n = \prod_{k=1}^{n} q_{i(k-1)i(k)} \exp\{-\lambda(i(k-1))t(k)\} \tag{1.1}$$

for some transition matrix Q and suitable intensities $\lambda(i)$. It is clear that the only possible candidate for Q is $q_{ij} = \mathbb{P}_i(Y_1 = j)$. To determine $\lambda(i)$, we let $z(t) =$

$\mathbb{P}_i(T_0 > t)$. Since $X_t = i$ on $\{T_0 > t\}$, the Markov property yields

$$z(t+s) = \mathbb{E}_i\mathbb{P}_i(T_0 > t+s \mid \mathscr{F}_t) = \mathbb{E}_i[\mathbb{P}_i(T_0 > s); T_0 > t] = z(s)z(t). \tag{1.2}$$

Since z is non-increasing, elementary facts on functional equations yield $z(t) = \mathrm{e}^{-\lambda(i)t}$ for some $\lambda(i) \geqslant 0$ (the pure jump property implies $z(t)\uparrow 1$ as $t\downarrow 0$ so that $z(t) = I(t=0)$, i.e. $\lambda(i) = \infty$, is excluded).

Applying the Markov property once more, we get similarly

$$\begin{aligned}\mathbb{P}_i(Y_1 = j, T_0 > t) &= \mathbb{E}_i\mathbb{P}_i(Y_1 = j, T_0 > t \mid \mathscr{F}_t)\\ &= \mathbb{E}_i[\mathbb{P}_i(Y_1 = j); T_0 > t] = q_{ij}\mathrm{e}^{-\lambda(i)t}\end{aligned}$$

which is (1.1) for $n = 1$. The case $n > 1$ now follows easily by the strong Markov property and induction. Indeed, evaluating p_n upon conditioning upon $\mathscr{F}_{S_{n-1}}$ we obtain from $X_{S_{n-1}} = Y_{n-1}$ that

$$\begin{aligned}p_n &= \mathbb{E}_i[\mathbb{P}_{X_{S_{n-1}}}(Y_1 = i(n), T_0 > t(n)); \quad Y_k = i(k), T_{k-1} > t(k) \quad k = 1,\ldots,n-1]\\ &= \mathbb{P}_{i(n-1)}(Y_1 = i(n), T_0 > t(n))p_{n-1}\\ &= q_{i(n-1)i(n)}\exp\{-\lambda(i(n-1))t(n)\}p_{n-1}.\end{aligned}$$

□

Problems

1.1. Show that the explosion time is a stopping time with respect to $\{\mathscr{F}_t\}_{t\geqslant 0}$.

2. THE MINIMAL CONSTRUCTION

The intuitive description of a practical model is usually given in terms of the intensities $\lambda(i)$ and the jump probabilities q_{ij} rather than in terms of the transition matrices P^t which are difficult to evaluate even in extremely simple cases. The question therefore arises whether *any* set of $\lambda(i)$, q_{ij} leads to a Markov jump process. The construction (given below) is immediately suggested by Theorem 1.2 and the problem becomes to check whether indeed a Markov process comes out. As will be seen, the answer is affirmative.

We assume therefore that we are given a set $\lambda(i) \geqslant 0$, $i \in E$, and a transition matrix Q on E (the Q of Theorem 1.2 has the property $q_{ii} = 0$ if and only if $\lambda(i) > 0$, but this need not be assumed here). Let $\Delta \notin E$ be some extra state (needed to describe the process after a possible explosion), write $E_\Delta = E \cup \{\Delta\}$ and define $\lambda(\Delta) = 0$, $q_{\Delta\Delta} = 1$. We consider the sample space

$$\Omega = (0,\infty]^{\mathbb{N}} \times E_\Delta^{\mathbb{N}} = \{(t_0, t_1, \ldots, y_0, y_1, \ldots) : 0 < t_k \leqslant \infty, y_k \in E_\Delta\}$$

and let $T_0, T_1, \ldots, Y_0, Y_1, \ldots$ be the obvious coordinate functions (projections) on Ω. It is then a matter of routine to construct probabilities $\mathbb{P}_i$, $i \in E_\Delta$, on Ω with the following properties:

(a) $\{Y_n\}_0^\infty$ is a Markov chain with transition matrix Q;
(b) given $\{Y_n\}_0^\infty$, the T_l are independent, with T_k being exponentially distributed with intensity $\lambda(i)$ on $\{Y_k = i\}$.

We construct $\{X_t\}_{t\geqslant 0}$ up to the time of explosion simply by reversing the construction of the Y_k, T_k following Fig. 1.1 (and if needed letting X_t remain in Δ after explosion). That is, we let $S_0 = 0$,

$$S_n = T_0 + \cdots + T_{n-1}, \qquad \omega(\Delta) = \sup S_n = T_0 + T_1 + \cdots,$$

$$X_t = \begin{cases} Y_k & \text{if } S_k \leqslant t < S_{k+1} \\ \Delta & \text{if } t \geqslant \omega(\Delta) \end{cases}$$

We shall prove the following main result:

Theorem 2.1. *$\{X_t\}_{t\geqslant 0}$ is a Markov jump process on E_Δ.*
In the proof, we need to study the residual sojourn time R_t at time t, i.e. $R_t = S_{n(t)} - t$, where $n(t) = \inf\{n : S_n > t\}$

Lemma 2.2. *Given $\mathscr{F}_t = \sigma(X_s; s \leqslant t)$, the conditional distribution of R_t is exponential with intensity $\lambda(X_t)$.*

Proof. The intuitive argument is just that given $\mathscr{F}_t$, the distribution of $T_{n(t)-1}$ is that of T given $T > u$, where $u = t - S_{n(t)-1}$ and T is exponential with intensity $\lambda(Y_{n(t)-1}) = \lambda(X_t)$. To spell out a formal proof we must show that

$$\mathbb{P}_i(R_t > r, A) = \mathbb{E}_i[\exp\{-\lambda(X_t)r\}; A] \tag{2.1}$$

for all $r < \infty$ and all $A \in \mathscr{F}_t$. If $A \subseteq \{\omega(\Delta) \leqslant t\}$, then both sides are just $\mathbb{P}_i A$ so we may assume $A \subseteq \{\omega(\Delta) > t\}$ and it then suffices to consider A of the form

$$\begin{aligned}&\{n(t) - 1 = n, Y_k = i(k), T_{k-1} > t(k), k = 1, \ldots, n\} \\ &\quad = \{Y_k = i(k), T_{k-1} > t(k), k = 1, \ldots, n, S_n \leqslant t, S_n + T_n > t\}.\end{aligned}$$

Thus if we condition upon the $Y_k, T_{k-1}, k = 1, \ldots, n$ and use the formula $\mathbb{P}(T > t + r) = e^{-\lambda r}\mathbb{P}(T > t)$ for the exponential distribution, we may evaluate the l.h.s. of (2.1) as

$$\begin{aligned}&\mathbb{P}_i(Y_k = i(k), T_{k-1} > t(k), k = 1, \ldots, n, S_n \leqslant t, S_n + T_n > t + r\} \\ &\quad = \exp\{-\lambda(i(n))r\}\mathbb{P}(Y_k = i(k), T_{k-1} > t(k), k = 1, \ldots, n, S_n \leqslant t, S_n + T_n > t)\end{aligned}$$

which is the same as the r.h.s. □

Proof of Theorem 2.1. $\{X_s\}_{s\geqslant 0}$ is clearly pure jump, so it suffices to show that on $\{X_t = i\}$ the conditional distribution of $\{X_{t+s}\}_{s\geqslant 0}$ given $\mathscr{F}_t$ is just the $\mathbb{P}_i$-distribution of $\{X_s\}_{s\geqslant 0}$. Define

$$M_t = (T_{n(t)}, T_{n(t)+1}, \ldots, Y_{n(t)-1}, Y_{n(t)}, \ldots).$$

Then $\{X_{t+s}\}_{s\geqslant 0}$ is constructed from (R_t, M_t) in just the same way as $\{X_s\}_{s\geqslant 0}$ is constructed from

$$(R_0, M_0) = (T_0, M_0) = (T_0, T_1, \ldots, Y_0, Y_1, \ldots).$$

Hence we must show that on $\{X_t = i\}$, the conditional distribution of (R_t, M_t) given $\mathscr{F}_t$ is the P_i-distribution of (T_0, M_0), i.e. that in the conditional distribution (i) R_t, M_t are independent, (ii) R_t has the $\mathbb{P}_i$-distribution of R_0, (iii) M_t has the $\mathbb{P}_i$-distribution of M_0. Now clearly $\{(Y_n, T_n)\}$ is a Markov chain with state space $E_\Delta \times (0, \infty]$ and transition kernel given by

$$\mathbb{P}(Y_{n+1} = j, T_{n+1} > t \mid \mathscr{H}_n) = q_{Y_n j} \exp\{-\lambda(j)t\} \tag{2.2}$$

where $\mathscr{H}_n = \sigma(Y_k, T_k : k \leqslant n)$. Also $n(t) - 1$ is a stopping time with respect to this chain and we shall evaluate the distribution of (M_t, R_t) conditionally upon $\mathscr{F}_t$ by first conditioning upon the larger σ-algebra $\mathscr{H}_{n(t)-1}$. Since $Y_{n(t)-1} = X_t$, the strong Markov property I.6.2 and (2.2) implies that given $\mathscr{H}_{n(t)-1}$, M_t has the $\mathbb{P}_{X_t}$-distribution of M_0, whereas R_t (being $\mathscr{H}_{n(t)-1}$-measurable) is degenerate. These facts and the $\mathscr{F}_t$-measurability of X_t imply (i) and (iii), whereas (ii) is the statement of Lemma 2.2. □

It should be noted, that if the process is explosive (i.e. $\mathbb{P}_i(\omega(\Delta) < \infty) > 0$ for some $i \in E$), then (see Problems) there are in general several ways of continuing the process after $\omega(\Delta)$ which will lead to a Markov jump process (to use a common phrase, the process 'runs out of instructions' at the explosion time). Among such processes all behaving in the same way up to the explosion time, the one in Theorem 2.1 obviously minimizes $\mathbb{P}_i(X_t = j)$ for any $i, j \in E$, and for this reason we talk about the *minimal construction.*

Some further discussion of the basic structure of a Markov jump process will be given in Sections 3a, 3b (though essentially this is only a reformulation of what has been shown so far), and we return here to the explosion problem. In most cases this presents an unwanted technicality, and one wants to assert as quickly as possible that a given Markov jump process is non-explosive (e.g., in the minimal construction one can then restrict the state space to E). Necessary and sufficient conditions are given in the following Proposition 2.3, and in Proposition 3.3 of the next section, whereas Proposition 2.4 gives some sufficient conditions which are easier to work with in many cases.

Proposition 2.3. *Define* $R = \sum_0^\infty \lambda(Y_n)^{-1}$. *Then for any* $i \in E$, *the sets* $\{\omega(\Delta) < \infty\}$ *and* $\{R < \infty\}$ *coincide* $\mathbb{P}_i$ a.s.

Proof. Conditionally upon $\{Y_n\}$, $\omega(\Delta) = \sup_n S_n = \sum_0^\infty T_n$ is distributed as $\sum_0^\infty \lambda(Y_n)^{-1} V_n$, where the V_n are independent identically distributed and exponential with intensity 1. The result therefore comes out by standard facts on weighted sums of independent identically distributed random variables. Thus $R < \infty$ implies $\omega(\Delta) < \infty$ because of $R = \mathbb{E}(\omega(\Delta) \mid Y_0, Y_1, \ldots)$, and the converse may be seen, for example by an application of the three-series criterion. □

Proposition 2.4. *Sufficient criteria for* $\mathbb{P}_i(\omega(\Delta) < \infty) = 0$ *for all* $i \in E$ *are*: (i) $\sup_{i \in E} \lambda(i) < \infty$; (ii) *$E$ is finite*; (iii) *$\{Y_n\}$ is recurrent.*

Proof. It follows from Proposition 2.3 that $\lambda(Y_n) \to \infty$ on $\{\omega(\Delta) < \infty\}$. Hence the sufficiency of (i) is clear, and (ii) is a consequence of (i). If $\{Y_n\}$ is recurrent, and $X_0 = Y_0 = i$, then $\lambda(i)$ is a limit point of $\{\lambda(Y_n)\}$. Thus $\lambda(Y_n) \to \infty$ cannot hold and $\mathbb{P}_i(\omega(\Delta) < \infty) = 0$. □

Problems

2.1. Let $\{X_t\}$ be explosive and modify the process so as to restart in some fixed state i after each explosion. Show that we obtain a Markov jump process.

2.2. Let $E = \mathbb{Z}\backslash\{0\}$ and $\lambda(k) = k^2$, $q_{n(-n-1)} = q_{(-n)(n+1)} = 1/n^2$, $q_{n(n+1)} = q_{(-n)(-n-1)} = 1 - 1/n^2$, $n > 0$. Show that the process is explosive and that $0 < \mathbb{P}F_+ < 1$, $\mathbb{P}F_+ + \mathbb{P}F_- = 1$, where $F_\pm = \{\lim_{t\uparrow\omega(\Delta)} X_t = \pm\infty\}$. Show that we get a Markov process by letting $X_{\omega(\Delta)} = 1$ on F_+, $X_{\omega(\Delta)} = -1$ on F_- (and similarly for explosions $2, 3, \ldots$).

2.3. Let $E = \mathbb{Z} \cup \{\Delta\}$ and $\lambda(k) = (k+1)^2$, $q_{k(k+1)} = 1$ for all $k \in \mathbb{Z}$. Show that the process is explosive and (at least heuristically) that there exists version with Δ as instantaneous state and $X_t \to -\infty$, $t \downarrow \omega(\Delta)$ [such a version cannot be pure jump in the present strict sense, of course].

3. THE INTENSITY MATRIX

3a. Definition and uniqueness

Assume from now on $q_{ii} = 0$ when $\lambda(i) > 0$ and define the *intensity matrix* $\Lambda = (\lambda(i,j))_{i,j \in E}$ of the process by

$$\lambda(i,j) = \lambda(i) q_{ij}, \quad j \neq i, \quad \lambda(i,i) = -\lambda(i). \tag{3.1}$$

Proposition 3.1. *An $E \times E$ matrix Λ is the intensity matrix of a Markov jump process $\{X_t\}_{t \geq 0}$ if and only if*

$$\lambda(i,i) \leq 0, \quad \lambda(i,j) \geq 0, \quad i \neq j, \quad \sum_{j \in E} \lambda(i,j) = 0. \tag{3.2}$$

Furthermore, Λ is in one-to-one correspondence with the distribution of the minimal process.

Proof. If Λ is an intensity matrix, it follows from (3.1) by considering the cases $\lambda(i) = 0$ and $\lambda(i) > 0$ separately that $\sum_{j \neq i} \lambda(i,j) = \lambda(i)$ and therefore that (3.2) holds. Conversely, if (3.2) is satisfied, then we let $\lambda(i) = -\lambda(i,i)$ and define q_{ij} by (3.1) and $q_{ij} = 0$ if $\lambda(i) > 0$, and let $q_{ij} = \delta_{ij}$ otherwise. It is then a matter of routine to check from (3.2) that Q is a transition matrix, and clearly the Markov jump process determined by Q and the $\lambda(i)$ has intensity matrix Λ. The stated one-to-one correspondence is obvious from Sections 1 and 2. □

Notes

In much of the literature, it is standard to denote the intensity matrix by Q. instead as here by Λ. Thus one even sometimes refers to the intensity matrix as 'the Q-matrix'.

3b. Reformulations and examples

It is now possible to give a reinterpretation of the intuitive picture of the evolvement of a Markov jump process which has been developed in Sections 1, 2. So far, by the well-known interpretation of the intensity parameter of the exponential distribution this picture has been that the process, when in state i at time t, exits from i before $t + \mathrm{d}t$ with probability (risk) $\lambda(i)\mathrm{d}t$. The next value j is selected independently of the time of exit from i and according to q_{ij}. We can now, however, instead consider the process as subject to (with a terminology used in survival analysis) *competing risks* with intensities $\lambda(i,j)$, $j \neq i$. That is, after entrance to i the jth type of event has an exponential waiting time Z_{ij} and the Z_{ij} are independent. Physically only the first (say $J = j$) of the events occur at time $Z_i = \inf_j Z_{ij}$ and the process then jumps to j. That this yields the given transition mechanism is checked as follows:

$$\begin{aligned}
\mathbb{P}(Z_i > z, J = j) &= \mathbb{P}(Z_{ik} > Z_{ij} > z, k \neq j) \\
&= \lambda(i,j)\int_z^\infty \mathbb{P}(Z_{ik} > y, k \neq j | Z_{ij} = y)\exp\{-\lambda(i,j)y\}\,\mathrm{d}y \\
&= \lambda(i,j)\int_z^\infty \prod_{k \neq j}\exp\{-\lambda(i,k)y\}\exp\{-\lambda(i,j)y\}\,\mathrm{d}y \\
&= \lambda(i,j)\int_z^\infty \exp\{-\lambda(i)y\}\,\mathrm{d}y = \frac{\lambda(i,j)}{\lambda(i)}\exp\{-\lambda(i)z\} \\
&= q_{ij}\exp\{-\lambda(i)z\}
\end{aligned}$$

In infinitesimal terms, the probability of a transition to j before $t + \mathrm{d}t$ is $\lambda(i,j)\,\mathrm{d}t$.

A description along these lines is usually the most natural in a given practical situation, and the intensity matrix is therefore the parameter in terms of which the process is usually specified. An obvious example is a queue where arrivals occur at rate β and service is completed at rate δ (the $M/M/1$ queue, cf. III. 1b). Here $E = \mathbb{N}$ and

$$\Lambda = \begin{bmatrix} -\beta & \beta & 0 & 0 & 0 & \cdots \\ \delta & -\beta-\delta & \beta & 0 & 0 & \cdots \\ 0 & \delta & -\beta-\delta & \beta & 0 & \cdots \\ \vdots & & & & & \ddots \end{bmatrix}.$$

When started at state $i > 0$ at time $t = 0$, we may think of $Z_{i(i-1)}$ as the service time of the customer being presently handled by the server, and of $Z_{i(i+1)}$ as the waiting time until the next arrival. In contrast, the holding time $T_0 = Z_i$ is the waiting time until either an arrival occurs or service is completed and is not quite as intuitive as the Z_{ij}.

In some situations it may also be convenient to extend the sample space of the minimal construction in order that certain random variables naturally associated with the process are well defined. An example is a linear birth–death process, i.e. $E = \mathbb{N}$,

$$\Lambda = \begin{bmatrix} 0 & 0 & 0 & 0 & 0 & \cdots \\ \delta & -\beta-\delta & \beta & 0 & 0 & \\ 0 & 2\delta & -2\beta-2\delta & 2\beta & 0 & \cdots \\ \vdots & & & & & \ddots \end{bmatrix},$$

where one may think of X_t as the total size at time t of a population with individuals who (independently of each other) terminate their lives with intensity δ and give birth with intensity β. Here quantities like the individual lifetimes or the number of children of an individual are not recognizable from the minimal construction and a more natural construction proceeds as follows: Represent each individual by its *life*, i.e. its lifetime Z (exponential with intensity δ) and an independent Poisson process with intensity β whose events in $[0, Z)$ correspond to the birth times. Construct the process started from say $X_0 = 1$, from a sequence of independent identically distributed lives by letting the first correspond to the ancestor, the second to his first child, ..., the nth to the nth individual being born, see Fig. 3.1. Such variants of the minimal construction will sometimes be used without further notice.

As a by-product and further illustration of the above discussion, we shall also show an important property of the exponential distribution (which is also easily proved by a direct analytical argument, cf. Problem 3.1):

Lemma 3.2. *Let $T_0, T_1, \ldots$ be independent identically distributed and exponential with intensity δ and let N be independent of the T_n and geometric, $\mathbb{P}(N = n) = (1-\rho)\rho^{n-1}$ $n = 1, 2, \ldots$. Then $S = T_0 + \cdots + T_{N-1}$ is exponential with intensity $\eta = \delta(1-\rho)$.*

Proof. Let the intensities of a three-state process be specified by Fig. 3.2 where $\beta = \rho\delta$. Then if we start the process in 1, the sojourn times $T_0, T_1, \ldots$ and

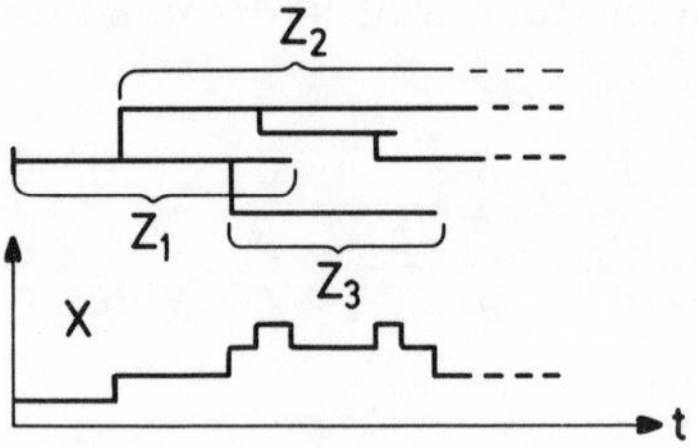

Fig. 3.1

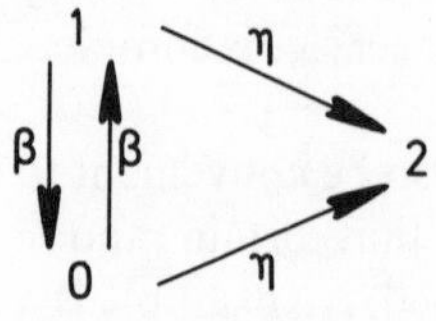

Fig. 3.2

$N=\inf\{n\geqslant 1: Y_n=2\}$ satisfies the given assumptions because of $\beta/(\beta+\eta)=\rho$ and $\beta+\eta=\delta$, and S is just the entrance time $\omega(2)=\inf\{t: X_t=2\}$ of 2. On the other hand, the symmetry between 0 and 1 ensures that the distribution of $\omega(2)$ is left unchanged if we collapse 0, 1 into the single state 1 according to $1\xrightarrow{\eta}2$. This makes it clear that $\mathbb{P}_1(\omega(2)>s)=e^{-\eta s}$. □

Problems

3.1. Show Lemma 3.2 using Laplace transforms.

3c. Reuter's explosion condition

The following result is of a similar form as the transience criterion I.5.2 for Markov chains and gives a necessary and sufficient condition (known as *Reuter's condition*) for a Markov jump process to be explosive:

Proposition 3.3. *A Markov jump process is non-explosive if and only if the only non-negative bounded solution $k=(k_i)_{i\in E}$ to the set of equations $\Lambda k=k$ is $k\equiv 0$.*

Proof. Suppose first that the process is explosive and define $k_i=\mathbb{E}_i e^{-\omega(\Delta)}$. Then $k_i>0$ at least for one i, and conditional upon the time $T_0=y$ of the first jump we get

$$k_i=\int_0^\infty \sum_{j\neq i}\lambda(i,j)\mathbb{E}_j e^{-y-\omega(\Delta)}\exp\{-\lambda(i)y\}\,dy=\sum_{j\neq i}\frac{\lambda(i,j)k_j}{1+\lambda(i)}.$$

Using $\lambda(i)=-\lambda(i,i)$, this implies $k_i=\sum_{j\in E}\lambda(i,j)k_j$ and $k=\Lambda k$. Suppose, conversely, the process is non-explosive, and define $h_i^{(n)}=\mathbb{E}_i\exp\{-T_0-\cdots-T_{n-1}\}$, $h_i^{(0)}=1$. Then just as above

$$h_i^{(n+1)}=\int_0^\infty \sum_{j\neq i}\lambda(i,j)e^{-y}h_j^{(n)}\exp\{-\lambda(i)y\}\,dy=\sum_{j\neq i}\frac{\lambda(i,j)h_j^{(n)}}{1+\lambda(i)}. \tag{3.3}$$

Now let $k\geqslant 0$ be bounded (without loss of generality $k_i\leqslant 1$) with $\Lambda k=k$. Then $k_i=\sum_{j\neq i}\lambda(i,j)k_j/(1+\lambda(i))$, and since $1=h_i^{(0)}\geqslant k_i$, it follows by induction from (3.3) that $h_i^{(n)}\geqslant k_i$ for all n. But $T_0+\cdots+T_n\uparrow\omega(\Delta)=\infty$ implies $h_i^{(n)}\to 0$. Hence $k_i=0$ for all i. □

Problems

3.2. Consider a pure birth process ($E=\mathbb{N}$, $\lambda(i,i+1)=\lambda(i)=\beta_i$). Show that the process is non-explosive if and only if $\sum_0^\infty\beta_n^{-1}<\infty$, and check that Proposition 2.3 and Proposition 3.3 yield the same result.

3d. The forwards and backwards equations

We now turn to one of the most celebrated classical topics in Markov process theory:

Theorem 3.4. *Let* $\Lambda = (\lambda(i,j))_{i,j\in E}$ *be an intensity matrix on E and* $\{X_t\}_{t\geqslant 0}$ *the corresponding minimal Markov jump process on E constructed in Theorem 3.1,* $p_{ij}^t = \mathbb{P}_i(X_t = j)$. *Then the* $E \times E$*-matrices* P^t *satisfy the backwards equation* $(\mathrm{d}/\mathrm{d}t)P^t = \Lambda P^t$, *i.e.*

$$\frac{\mathrm{d}p_{ij}^t}{\mathrm{d}t} = \sum_{k\in E} \lambda(i,k)p_{kj}^t \tag{3.4}$$

and the forwards equation $(\mathrm{d}/\mathrm{d}t)P^t = P^t\Lambda$, *i.e.*

$$\frac{\mathrm{d}p_{ij}^t}{\mathrm{d}t} = \sum_{k\in E} p_{ik}^t\lambda(k,j). \tag{3.5}$$

Proof. Conditioning upon $T_0 = s$ yields

$$\begin{aligned} p_{ij}^t &= \mathbb{P}_i(T_0 > t)\delta_{ij} + \int_0^t \lambda(i)\exp\{-\lambda(i)s\}\left\{\sum_{k\neq i} q_{ik}p_{kj}^{t-s}\right\}\mathrm{d}s \\ &= \exp\{-\lambda(i)t\}\left[\delta_{ij} + \int_0^t \sum_{k\neq i}\lambda(i,k)\exp\{-\lambda(i)s\}p_{kj}^s\,\mathrm{d}s\right]. \end{aligned}$$

The integrand $f(s) = \sum_{k\neq i}\cdots$ is well defined and bounded since $\sum|\lambda(i,k)| = 2\lambda(i) < \infty$. This shows first that p_{ij}^t (and similarly all p_{kj}^t) is continuous and thereafter that $f(s)$ is continuous and bounded. Therefore p_{ij}^t is differentiable with derivative

$$\begin{aligned} &-\lambda(i)\exp\{-\lambda(i)t\}\left[\delta_{ij} + \int_0^t f(s)\,\mathrm{d}s\right] + \exp\{-\lambda(i)t\}f(t) \\ &= -\lambda(i)p_{ij}^t + \sum_{k\neq i}\lambda(i,k)p_{kj}^t = \sum_{k\in E}\lambda(i,k)p_{kj}^t. \end{aligned}$$

The proof of the forwards equation is more involved and *will only be given subject to the assumption*

$$\sup_{i\in E}\lambda(i) < \infty \tag{3.6}$$

which will be used to infer that the $(p_{kj}^s - \delta_{kj})/s$ are bounded uniformly in s, k, j. This follows since

$$0 \leqslant p_{kj}^s \leqslant \lambda(k)\int_0^s \exp\{-\lambda(k)u\}\,\mathrm{d}u,\ k\neq j,$$

$$0 \leqslant 1 - p_{kk}^s \leqslant \lambda(k)\int_0^s \exp\{-\lambda(k)u\}\,\mathrm{d}u$$

and (3.5) comes out by dominated convergence (using $\sum p_{ik}^t < \infty$) from

$$\frac{p_{ij}^{t+s} - p_{ij}^t}{s} = \sum_{k\in E} p_{ik}^t\frac{p_{kj}^s - \delta_{kj}}{s} \to \sum_{k\in E} p_{ik}^t\lambda(k,j). \qquad \square$$

In the case of a finite E, standard results on existence and uniqueness of systems of differential equations yield

Corollary 3.5. *If E is finite, then for all $t \geqslant 0$*

$$P^t = e^{\Lambda t} = \sum_{n=0}^{\infty} \frac{t^n}{n!} \Lambda^n. \tag{3.7}$$

Example 3.6. Suppose that E has just $p = 2$ states 1, 2 and, to avoid trivialities, that $\lambda(1)$ and $\lambda(2)$ are not both zero. Then Λ has eigenvalues 0 and $\lambda = -\lambda(1) - \lambda(2) \neq 0$ with corresponding right eigenvectors (1, 1), $(\lambda(1), -\lambda(2))$. Hence

$$\Lambda = \begin{pmatrix} -\lambda(1) & \lambda(1) \\ \lambda(2) & -\lambda(2) \end{pmatrix} = B \begin{pmatrix} 0 & 0 \\ 0 & \lambda \end{pmatrix} B^{-1},$$

where

$$B = \begin{pmatrix} 1 & \lambda(1) \\ 1 & -\lambda(2) \end{pmatrix}, \qquad B^{-1} = \frac{1}{\lambda(1)+\lambda(2)} \begin{pmatrix} \lambda(2) & \lambda(1) \\ 1 & -1 \end{pmatrix},$$

$$P^t = e^{\Lambda t} = \sum_{n=0}^{\infty} \frac{t^n}{n!} B \begin{pmatrix} 0^n & 0 \\ 0 & \lambda^n \end{pmatrix} B^{-1} = B \begin{pmatrix} 1 & 0 \\ 0 & e^{\lambda t} \end{pmatrix} B^{-1}$$

$$= \frac{1}{\lambda(1)+\lambda(2)} \begin{pmatrix} \lambda(2) + \lambda(1)e^{\lambda t} & \lambda(1) - \lambda(1)e^{\lambda t} \\ \lambda(2) - \lambda(2)e^{\lambda t} & \lambda(1) + \lambda(2)e^{\lambda t} \end{pmatrix}. \qquad \square$$

For purposes like those of the present book, the backwards and forwards equations are of quite limited utility. This is so in particular for an infinite state space, but even for $p < \infty$ states the Jordan canonical form and hence the algebra corresponding to Example 3.1 becomes much more cumbersome for $p > 2$ due to the possibility of eigenvalues which are complex or of multiplicity > 1. One common application is to look for a stationary probability distribution ($\pi P^t = \pi$) by means of $D(\pi P^t)|_{t=0} = 0$, i.e. $\pi\Lambda = 0$. This equation comes out, however, quite easily by a direct argument in the next section. Also the time-dependent solution (i.e. the p_{ij}^t for $t < \infty$) can be found explicitly only in very special cases with E infinite and it is even then frequently easier to obtain by different means. Examples are the linear birth–death process (see e.g. Harris, 1963) and the $M/M/1$ queue to be discussed in III.9.

4. ERGODIC THEORY

4a. State space properties
4b. Stationary measures
4c. Ergodicity criteria and limit results

4a. State space properties

When defining concepts like irreducibility, recurrence or transience in continuous time, one may either mimic the discrete time definition or refer to the jump chain $\{Y_n\}$. We consider throughout in the following a minimal process and look first at irreducibility. Then:

Proposition 4.1. *The following properties are equivalent:* (a) $\{Y_n\}$ *is irreducible;* (b) *for any* $i, j \in E$ *we have* $p_{ij}^t > 0$ *for some* $t > 0$; (c) *for any* $i, j \in E$ *we have* $p_{ij}^t > 0$ *for all* $t > 0$.

Proof. Denote here and in the following

$$\omega(i) = \inf\left\{t > 0 : X_t = i, X_{t-} = \lim_{s \uparrow t} X_s \neq i\right\} \tag{4.1}$$

($\omega(i) = \infty$ if no such t exists) so that $\omega(i)$ is the hitting time of i if $X_0 \neq i$ and the recurrence time of i if $X_0 = i$. Since j has an exponential holding time, it is clear that $p_{ij}^t > 0$ if and only if $\mathbb{P}_i(\omega(j) \leqslant t) > 0$ (similarly we always have $p_{ii}^t > 0$). Now $\mathbb{P}_i(\omega(j) < \infty) > 0$ if and only if some path $ii_1 \cdots i_n j$ from i to j is possible for $\{Y_n\}$, and in that case we may evaluate the conditional distribution F of $\omega(j)$ given $\{\omega(j) < \infty\}$ by conditioning on the various paths. Thus F is mixture of convolutions of exponential distributions with intensities $\lambda(i, i_1), \lambda(i_1, i_2) \cdots$ and hence has a density > 0 on $(0, \infty)$. Thus $\mathbb{P}_i(\omega(j) \leqslant t) > 0$ if and only if $\{Y_n\}$ can reach j from i, proving the proposition. □

Accordingly we define $\{X_t\}_{t \geqslant 0}$ to be irreducible if one of properties (a), (b), (c) hold. Similarly (but easier), it is seen that we can define i to be transient (recurrent) for $\{X_t\}$ if either (a) the set $\{t : X_t = i\}$ is bounded (unbounded) $\mathbb{P}_i$ a.s., (b) i is transient (recurrent) for $\{Y_n\}$ or (c) $\mathbb{P}_i(\omega(i) < \infty) < 1 (= 1)$. As will be seen in the following, the distinction between null recurrence and positive recurrence can not, however, be related to $\{Y_n\}$ alone. Note also that we do not pay attention to periodicity. This is due to the fact that even though $\{Y_n\}$ may be periodic, the exponential holding times smooth away any such behaviour in continuous time.

4b. Stationary measures

A measure $\nu \neq 0$ is *stationary* if $0 \leqslant \nu_j < \infty$, $\nu P^t = \nu$ for all t.

Theorem 4.2. *Suppose that* $\{X_t\}_{t \geqslant 0}$ *is irreducible recurrent on E. Then there exists one, and up to a multiplicative factor only one, stationary measure* ν. *This* ν *has the property* $0 < \nu_j < \infty$ *for all* j *and can be found in either of the following ways:*

(i) *For some fixed but arbitrary state* i, ν_j *is the expected time spent in* j *between successive entrances to* i;

$$\nu_j = \mathbb{E}_i \int_0^{\omega(i)} I(X_t = j)\, \mathrm{d}t \tag{4.2}$$

(*with* $\omega(i)$ *given by* (4.1));

(ii) $\nu_j = \mu_j / \lambda(j)$, *where* μ *is stationary for* $\{Y_n\}$;

(iii) *as solution of* $\nu\Lambda = 0$.

Proof. We first prove uniqueness by considering the Markov chain $X_0, X_1, \ldots$.

This is irreducible since all $p_{ij}^t > 0$, and any ν stationary for $\{X_t\}$ is also stationary for $\{X_n\}$, so in order to apply I.3.3 we just have to show that $\{X_n\}$ is recurrent. But for any i, the sequence $U_1, U_2, \dots$ of holding times of i is non-terminating since i is recurrent. The U_k being independent identically distributed with $\mathbb{P}(U_k > 1) > 0$, we have $U_k > 1$ infinitely often and therefore also $X_n = i$ infinitely often. For (i) we show stationarity of (4.2) by evaluating the jth component of νP^t in a similar manner as in the proof of I.3.2:

$$\sum_{k \in E} \nu_k p_{kj}^t = \mathbb{E}_i \int_0^{\omega(i)} \sum_{k \in E} p_{kj}^t I(X_s = k)\, \mathrm{d}s$$

$$= \mathbb{E}_i \int_0^{\omega(i)} p_{X_s, j}^t\, \mathrm{d}s = \mathbb{E}_i \int_0^{\infty} p_{X_s, j}^t I(\omega(i) > s)\, \mathrm{d}s$$

$$= \mathbb{E}_i \int_0^{\infty} P(X_{t+s} = j, \omega(i) > s \mid \mathscr{F}_s)\, \mathrm{d}s$$

$$= \mathbb{E}_i \int_0^{\infty} I(X_{t+s} = j, \omega(i) > s)\, \mathrm{d}s$$

$$= \mathbb{E}_i \int_0^{\omega(i)} I(X_{t+s} = j)\, \mathrm{d}s = \mathbb{E}_i \int_t^{\omega(i)+t} I(X_u = j)\, \mathrm{d}u$$

$$= \mathbb{E}_i \int_t^{\omega(i)} I(X_u = j)\, \mathrm{d}u + \mathbb{E}_i \int_{\omega(i)}^{\omega(i)+t} I(X_u = j)\, \mathrm{d}u \tag{4.3}$$

$$= \mathbb{E}_i \int_t^{\omega(i)} I(X_u = j)\, \mathrm{d}u + \mathbb{E}_i \int_0^{t} I(X_u = j)\, \mathrm{d}u \tag{4.4}$$

$$= \mathbb{E}_i \int_0^{\omega(i)} I(X_u = j)\, \mathrm{d}u = \nu_j. \tag{4.5}$$

Here (4.3) and (4.5) follow just by standard integral formulae (irrespective of whether or not $t \leqslant \omega(i)$), whereas for (4.4) we use the strong Markov property and $X_{\omega(i)} = i$.

Next (ii) follows upon computation of (4.2) by using the interpretation in I.3.2 of μ. With $\tau(i) = \inf\{n : Y_n = i\}$, we get

$$\mathbb{E}_i \int_0^{\omega(i)} I(X_t = j)\, \mathrm{d}t = \mathbb{E}_i \sum_{n=0}^{\tau(i)-1} T_n I(Y_n = j)$$

$$= \sum_{n=0}^{\infty} \mathbb{E}_i \mathbb{E}_i [T_n;\ Y_n = j,\ \tau(i) > n \mid \{Y_n\}_0^{\infty}]$$

$$= \frac{1}{\lambda(j)} \mathbb{E}_i \sum_{n=0}^{\infty} I(Y_n = j, \tau(i) > n) = \frac{1}{\lambda(j)} \frac{\mu_j}{\mu_i}.$$

That is, ν_j is proportional to $\mu_j / \lambda(j)$.

For (iii) we note that according to (ii) ν is stationary for $\{X_t\}$ if and only if

$(v_j\lambda(j))_{j\in E}$ is stationary for $\{Y_n\}$, that is, if and only if

$$\sum_{i\in E} v_i\lambda(i)q_{ij} = v_j\lambda(j)$$

for all $j\in E$, or since $q_{ii}=0$ if and only if

$$0 = -v_j\lambda(j) + \sum_{i\neq j} v_i\lambda(i,j) = \sum_{i\in E} v_i\lambda(i,j).$$

Finally, $0 < v_j < \infty$ follows easily say by (ii), since in the recurrent case $0 < \lambda(j) < \infty$. □

4c. Ergodicity criteria and limit results

An irreducible recurrent process with the stationary measure having finite mass is called *ergodic*, and we have:

Theorem 4.3. *An irreducible non-explosive Markov jump process is ergodic if and only if one can find a probability solution* π $(|\pi|=1,\ 0\leqslant \pi_i \leqslant 1)$ *to* $\pi\Lambda = 0$. *In that case* π *is the stationary distribution.*

Proof. That π exist and is stationary in the ergodic case follows immediately from Theorem 4.2. Suppose conversely that π exists and define

$$p_{ij}^{t;n} = \mathbb{P}_i(X_t = j;\ T_0 + \cdots + T_n > t), \quad n = 0, 1, 2, \ldots$$

Now a path starting from i contributes to $p_{ij}^{t;n}$ either if it has no jumps before t (and $i=j$) or it has a last jump, say from k to j at time $s\leqslant t$, and at most $n-1$ jumps before s. Thus collecting terms we get

$$p_{ij}^{t;n} = \delta_{ij}\exp\{-\lambda(i)t\} + \int_0^t \sum_{k\neq j} p_{ik}^{s;(n-1)}\lambda(k,j)\exp\{-\lambda(j)(t-s)\}\,\mathrm{d}s,$$

$$\sum_{i\in E}\pi_i p_{ij}^{t;n} = \pi_j\exp\{-\lambda(j)t\} + \int_0^t \exp\{-\lambda(j)(t-s)\}\sum_{k\neq j}\lambda(k,j)\sum_{i\in E}\pi_i p_{ik}^{s;(n-1)}\,\mathrm{d}s \tag{4.6}$$

Obviously

$$\sum_{i\in E}\pi_i p_{ik}^{s;0} = \pi_k\exp\{-\lambda(k)s\} \leqslant \pi_k, \quad \text{i.e. } \pi P^{s;0} \leqslant \pi.$$

It thus follows by induction from (4.6) that $\pi P^{t;n} \leqslant \pi$ since then

$$\sum_{i\in E}\pi_i p_{ij}^{t;(n+1)} \leqslant \pi_j\exp\{-\lambda(j)t\} + \int_0^t \exp\{-\lambda(j)(t-s)\}\sum_{k\neq j}\lambda(k,j)\pi_k\,\mathrm{d}s$$

$$= \pi_j\exp\{-\lambda(j)t\} + \pi_j\lambda(j)\int_0^t \exp\{-\lambda(j)s\}\,\mathrm{d}s = \pi_j$$

But since the process is non-explosive, we have $p_{ij}^{t;n}\to p_{ij}^t$ and $\sum_{j\in E}p_{ij}^t = 1$. Hence

$$\sum_{i\in E} \pi_i p_{ij}^t \leqslant \pi_j$$

and since summing both sides over j yields 1, equality must hold so that $\pi P^t = \pi$. Thus π is a stationary distribution. This implies recurrence (since in the transient case $\mathbb{P}_\pi(X_t = j) \to 0$) and $|\pi| = 1$ then finally shows ergodicity. □

As noted in Section 3, the equation $\pi\Lambda = 0$ is the same as that which comes out by formal manipulations with the differential equations. In the literature one occasionally proves ergodicity by checking irreducibility and finding a probability solution to $\pi\Lambda = 0$. *This procedure is, however, not valid without having excluded explosion.* To see this, consider for example a *transient* $\{Y_n\}$ with a stationary measure μ (for an example, see Problem I.3.2) and choose the $\lambda(j)$ such that $\pi_j = \mu_j/\lambda(j)$ has mass 1. Then as in the proof of (iii), it holds that $\pi\Lambda = 0$, and clearly the transience of $\{Y_n\}$ excludes recurrence of $\{X_t\}$ (it follows from Theorem 4.3 that $\{X_t\}$ must even be explosive). However:

Corollary 4.4. *A sufficient condition for ergodicity of an irreducible process is the existence of a probability π which solves $\pi\Lambda = 0$ and has the additional property $\sum \pi_j \lambda(j) < \infty$ (which is automatic if $\sup_{i\in E}\lambda(i) < \infty$).*

Proof. Letting $\mu_j = \pi_j\lambda(j)$, it follows as in the proof of part (iii) of Theorem 4.2 that μ is stationary for $\{Y_n\}$. Since μ has finite mass, $\{Y_n\}$ is positive recurrent, in particular recurrent, hence $\{X_t\}$ is non-explosive and Theorem 4.3 applies. □

Exactly as in I.3.1(ii) one also has:

Proposition 4.5. *If the process is ergodic, then there exists a strictly stationary version $\{X_t\}_{-\infty<t<\infty}$ with doubly infinite time.*

We next turn to the limiting behaviour of the p_{ij}^t and have as expected:

Theorem 4.6. *If $\{X_t\}$ is ergodic and π the stationary distribution, then for all i, j $p_{ij}^t \to \pi_j$ as $t \to \infty$.*

Proof. As noted above in the case $\delta = 1$, $\{X_{n\delta}\}$ is an irreducible recurrent Markov chain for each $\delta > 0$. It is ergodic since π is stationary, and hence $p_{ij}^{n\delta} \to \pi_j$ as $n \to \infty$. The continuity of the p_{ij}^t being straightforward to verify, the assertion thus follows by the method of discrete skeletons, Appendix A 9.2. Alternatively, we may apply the more elementary Appendix A9.1. The uniform continuity follows say from the backwards equation (3.4) which in conjunction with $\sum|\lambda(i, k)| < \infty$ shows that dp_{ij}^t/dt exists and is bounded in t. □

As in discrete time, I.(4.3), time-average properties like

$$\frac{1}{T}\int_0^T f(X_t) \to \mathbb{E}_\pi f(X_t) = \sum_{i\in E} f(i)\pi_i$$

a.s. hold under suitable conditions on f, see V.3.

Exactly the same argument as for Theorem 4.6 yields

Corollary 4.7. *If* $\{X_t\}$ *is irreducible recurrent but not ergodic* (i.e. $|\nu| = \infty$), then $p^t_{ij} \to 0$ *for all* $i, j \in E$.

Corollary 4.8. *For any minimal Markov jump process* (irreducible or not), *the limits* $\lim_{t\to\infty} p^t_{ij}$ *exist* (recall that in discrete time periodicity might cause an exception to the analogue result).

Proof. Clearly $p^t_{ij} \to 0$ if j is transient. If j is in a recurrent class C, let $\nu^{(C)}$ be stationary for the process restricted to C. Then by Theorem 4.6 and Corollary 4.7

$$p^t_{ij} \to \mathbb{P}_i\,(\text{some } X_t \in C)\frac{\nu^{(C)}_j}{|\nu^{(C)}|}. \qquad \square$$

Problems

4.1. Consider a Markov jump process $\{X_t\}$ with bounded intensities, say $\lambda(i) \leqslant \lambda < \infty$. Show that $\tilde{Q} = \lambda^{-1}\Lambda + I$ is a transition matrix. Now consider a Poisson process $\{N_t\}$ with intensity λ and a process $\{\tilde{X}_t\}$ which jumps according to $\tilde{Q}$ at the epochs of $\{N_t\}$, say $\tilde{X}_t = \tilde{Y}_n$ on $\{N_t = n\}$, where $\{\tilde{Y}_n\}$ is a Markov chain governed by $\tilde{Q}$ and independent of $\{N_t\}$. Show that $\{\tilde{X}_t\}$ is a version of $\{X_t\}$, that $\{X_t\}$ is ergodic if and only if $\{\tilde{Y}_n\}$ is positive recurrent and that then the stationary distributions are the same (this procedure is known as *uniformization* or *randomization*, see e.g. Keilson, 1979. An application is in X.5)

Notes

Further material related to the equation $\pi\Lambda = 0$ in the explosive case can be found in Kelly (1983).

5. TIME REVERSIBILITY

Time reversibility (or just *reversibility*) of a process means loosely that the process evolves in just the same way irrespective of whether time is read forwards (as usual) or backwards. The concept is studied here mainly for the purpose of certain queueing applications in III.4, but its scope is in fact rather more general. For example it could be mentioned that time reversibility of processes occurring in physics is considered a property of intrinsic physical interest.

Our main interest is in Markov jump processes, but we start by the Markov chain case in order to motivate the definition to follow and to make some simple observations.

Proposition 5.1. *Let* $X_0, \ldots, X_N$ *be a time-homogeneous Markov chain with transition matrix* p *and define the time-reversed chain* $\tilde{X}_0, \ldots, \tilde{X}_N$ by $\tilde{X}_n = X_{N-n}$. *Consider some fixed Markov probability* $\mathbb{P}$, *define* $\pi^n_i = \mathbb{P}(X_n = i)$ *and assume that* $\pi^n_i > 0$ *for all* $i \in E, n = 0, \ldots, N$. *Then:*

(a) $\tilde{X}_0, \ldots, \tilde{X}_N$ *is a time-inhomogeneous Markov chain with transition matrices* $\tilde{P}(n) = (\tilde{p}_{ij}(n))$ *given by*

$$\tilde{p}_{ij}(n) = \mathbb{P}(\tilde{X}_{n+1} = j \,|\, \tilde{X}_n = i) = \frac{\pi^{N-n-1}_j p_{ji}}{\pi^{N-n}_i}; \tag{5.1}$$

If furthermore all $p_{ij} > 0$, *then*

(b) $\tilde{X}_0, \ldots, \tilde{X}_N$ *is time-homogeneous, i.e.* $\tilde{P}(n)$ *independent of* n, *if and only if* $X_0, \ldots, X_N$ *is stationary, i.e.* $\pi_i^n = \pi_i$ *independent of* n;

(c) $\tilde{X}_0, \ldots, \tilde{X}_N$ *has the same distribution as* $X_0, \ldots, X_N$ *if and only if* $X_0, \ldots, X_N$ *is stationary and* $\pi_j p_{ji} = \pi_i p_{ij}$.

Proof. (a) Letting $\tilde{\pi}_i^n = \mathbb{P}(\tilde{X}_n = i) = \pi_i^{N-n}$, we must show that

$$\mathbb{P}(\tilde{X}_0 = i_0, \ldots, \tilde{X}_N = i_N) = \tilde{\pi}_{i_0}^0 \tilde{p}_{i_0 i_1}(0) \tilde{p}_{i_1 i_2}(1) \cdots \tilde{p}_{i_{N-1} i_N}(N-1). \tag{5.2}$$

But the l.h.s. of (5.2) is

$$\mathbb{P}(X_0 = i_N, \ldots, X_N = i_0) = \pi_{i_N}^0 p_{i_N i_{N-1}} p_{i_{N-1} i_{N-2}} \cdots p_{i_1 i_0}. \tag{5.3}$$

Inserting the definition of $\tilde{P}(n)$ in the r.h.s. of (5.2), the π_j^n telescope and the r.h.s. of (5.3) comes out.

(b) It is clear that $\pi_i^n = \pi_i$ implies that $\tilde{P}(n)$ is independent of i. For the converse, first let $i = j$ in (5.1). It then follows that $\pi_i^n = \pi_i \rho_i^n$ for suitable π_i, ρ_i. Since $p_{ji} > 0$, the independence of $\tilde{p}_{ij}(n)$ of n then yields $\rho_i = \rho_j$. Hence all ρ_i are equal, say $\rho_i = \rho$. But then $\sum_E \pi_i^n = \rho^n \sum_E \pi_i = 1$ implies $\rho = 1$, i.e. $\pi_i^n = \pi_i$, and stationarity. Finally in (c) stationarity is necessary by (b), and $\pi_j p_{ji} = \pi_i p_{ij}$ is then equivalent to $\tilde{p}_{ij} = p_{ij}$ by (5.1). □

This result does not cover the Markov chain case in full generality since all π_i^n and all p_{ij} being non-zero is a restriction. However, if $X_0, \ldots, X_N$ are obtained by observing an irreducible Markov jump process at times $\delta, 2\delta, \ldots, (N+1)\delta$, this assumption is automatic. Since time reversibility in continuous time should imply reversibility of such discrete skeletons, it follows by (b) that we can safely restrict attention to stationary versions of ergodic processes.

Consider thus as in Proposition 4.5 a stationary version $\{X_t\}_{t \in \mathbb{R}}$ with doubly infinite time. We define the time-reversed process $\{\tilde{X}_t\}_{t \in \mathbb{R}}$ by $\tilde{X}_t = X_{-t-} = \lim_{s \uparrow -t} X_s$ and call $\{X_t\}$ *time-reversible* if $\{\tilde{X}_t\}$ has the same distribution as $\{X_t\}$. The reason for not simply letting $\tilde{X}_t = X_{-t}$ is to obtain right-continuous paths. Of course, this is immaterial for the distribution of finite-dimensional sets (since the probability of a jump at t is always zero) and therefore the whole process.

Theorem 5.2. *Let* π *be the ergodic distribution. Then a necessary and sufficient condition for time reversibility is* $\pi_i \lambda(i,j) = \pi_j \lambda(j,i)$ *for all* $i \neq j$.

Proof. Time reversibility is equivalent to $\{X_{n\delta}\}_{n \in \mathbb{Z}}$ and $\{\tilde{X}_{n\delta}\}_{n \in \mathbb{Z}}$ having the same distribution for all δ, i.e. by stationarity and Proposition 5.1(c) to $\pi_i p_{ij}^\delta = \pi_j p_{ji}^\delta$. Letting $\delta \downarrow 0$ and taking derivatives shows that the stated condition is necessary. Suppose, conversely, that $\pi_i \lambda(i,j) = \pi_j \lambda(j,i)$ for all i, j. Since $\{X_t\}$ is stationary and time-homogeneous, $\{\tilde{X}_t\}$ must be so according to Proposition 5.1(b). Thus, as an ergodic Markov jump process, $\{\tilde{X}_t\}$ is uniquely determined by its intensities $\tilde{\lambda}(i,j)$ and we have to show $\tilde{\lambda}(i,j) = \lambda(i,j)$ for $i \neq j$. But

$$
\begin{aligned}
\tilde{\lambda}(i,j) &= \lim_{t\downarrow 0}\frac{\tilde{p}^t_{ij}}{t} = \lim_{t\downarrow 0}\frac{\mathbb{P}(\tilde{X}_0=j,\tilde{X}_{-t}=i)}{t\mathbb{P}(\tilde{X}_{-t}=i)}\\
&= \lim_{t\downarrow 0}\frac{\mathbb{P}(X_0=j,X_t=i)}{t\mathbb{P}(X_t=i)} = \lim_{t\downarrow 0}\frac{\pi_j p^t_{ji}}{t\pi_i}\\
&= \frac{\pi_j\lambda(j,i)}{\pi_i} = \lambda(i,j). \qquad \square
\end{aligned}
$$

The term $\pi_i\lambda(i,j)$ is the rate at which transitions $i\to j$ occur in stationarity and is often denoted as the *probability flux* (or *flow*) from i to j. Thus the reversibility conditions means that the flux from i to j is the same as the flux from j to i, and for this reason is called the conditions of *detailed balance* in contrast to the equilibrium equation $\pi\Lambda=0$ which is the condition of *full balance*. More precisely, rewriting $\pi\Lambda=0$ in the form

$$\pi_i\lambda(i) = \sum_{j\neq i}\pi_j\lambda(j,i),$$

the l.h.s. is the total flow out of state i and the r.h.s. the total flow into state i.

Notes

Time reversibility is studied for example in Kelly (1979) and Keilson (1979).

CHAPTER III

Queueing Theory at the Markovian Level

1. GENERALITIES

1a. Queueing theory and some of its daily life motivations

Though the general field of applied probability has by now developed into a diversity of subareas, queueing theory is not only one of the oldest, but also by far the most notable and prominent. Queueing problems come up in a variety of situations in the real world and have stimulated an enormous literature which, though in part quite mathematical and abstract, is not of a purely academic nature. In fact, there has been a considerable interaction between the developments at the various degrees of abstraction in the field. Thus, though the more theoretical-orientated part of the literature (incorporating this book) tends to deal with models and problems too simplified to be of any great direct practical applicability, the notions and techniques that are studied are also important for the practical worker in the field. Conversely, the call for solution to particular problems has of course stimulated not only the theory of queueing but also that of probability as a whole, fields like Markov processes, renewal theory and random walks owing their present state and importance to a large extent from the impact from queueing theory. Queueing problems present a great challenge to the probabilist and a *memento mori* to probability theory as a whole: The development of abstract probability theory may be of great beauty, but seldom sheds much light on how to come up with the numbers the practical worker asks for. The crux is more often a thorough understanding of the particular features of the model combined with a few basic mathematical techniques, and it is a feeling for this that the present treatment aims at bringing the reader.

Queueing situations from daily life are almost too obvious, but we shall list a few anyway: customers queueing up before the m cashiers in a supermarket;

telephone callers waiting for one of the lines of an exchange to become available; aircraft circling over the airport before a runway becomes free; machines under care of a repairman who can handle only one at a time; and so on. Of more recent date than these classical examples are a number of problems connected with computer organization or networks in teletraffic theory or data transmission: in a time-sharing computer, we may think of the jobs as customers which are served by the central processor unit (CPU) and possibly input/output facilities. At each of these units queues may form, and in particular the queue at the CPU has some rather specific features (feedback, simultaneous service). In telephone networks there is a hierarchy of exchanges, so that, for example, local calls need only to pass an exchange at the lowest level, whereas long-distance calls may be directed among one of several possible paths connecting exchanges at various levels. Queues may form at the exchanges and are highly interactive.

We finally mention that a number of other situations may either directly be formulated in queueing terms or at least are closely related. Examples occur in storage processes and insurance risk. For example, in a store with items placed from time to time and taken out as demand arises, we may think of the items as customers and of the removals as service events.

1b. Classification of simple queues

The great diversity of queueing problems gives rise to an enormous variety of models each with their specific features. Incorporating more than one or two such features usually makes the model not only complicated but also analytically intractable. Therefore a substantial part of the literature deals with models of a very simple structure.

Without attempting anything near a classification of all queueing situations, one might tentatively single out the following relevant features for the description of a queue of reasonably simple structure: (a) the *input* or *arrival process*, i.e. the way in which the customers arrive to the queue; (b) the *service facilities*, i.e. the way in which the systems handles a given input stream. Logically incorporated in (b) but treated separately in Section 1c is (c) the *queue discipline*, i.e. the algorithm determining the order in which the customers are served. The descriptions of these features may be quite complicated and are, at least in their verbal form, always lengthy. A convenient shorthand notation system was suggested by D. G. Kendall (1953) and has to a large extent become standard since then. It enables one to replace phrases like 'the single-server queue with completely random arrivals and general service times' with symbolic notation like '$M/G/1$'. The notation covers some simple and basic queueing systems (but by no means all important ones) which have the following characteristics:

(a) Customers arrive one at a time according to a renewal process in discrete or continuous time. That is, the intervals between successive arrivals of customers are independent identically distributed (i.i.d.) and governed by a distribution A on $\mathbb{N}$ or $(0, \infty)$. We number the customers $0, 1, 2, \ldots$ and assume most often that customer 0 arrives at time 0. Thus, if T_n denotes the interval between the arrival of customers n and $n+1$, the T_n are i.i.d. governed by A and the arrival instants are $0, T_0, T_0 + T_1, \ldots$.

(b) The service times of different customers are i.i.d. and independent of the arrival process. We denote the governing distribution (concentrated on $(0, \infty)$) by B and the service time of customer n by U_n. Thus $U_0, U_1, \ldots$ are i.i.d. governed by B and independent of the T_n.

In Kendall's notation, a queueing system of this type is denoted by a string of the type $\alpha/\beta/m$, where α refers to the form of the interarrival time distribution, β to the form of the service time distribution and m is the number of servers. The most common values of α, β are as follows:

M The exponential distribution. (M = Markovian. Other terms are 'completely random' and 'Poissonian'.)

D The distribution degenerate at some point $d \in (0, \infty)$, frequently $d = 1$. (D = deterministic. Also, the term 'regular' is used.)

E_k The Erlang distribution with k stages, see Section 6.

H_k The hyperexponential distribution with k parallel channels, see Section 6.

PH A more general phase-type distribution, see Section 6.

GI or G No restrictions on the form of the distribution. (GI = General Independent, G = General; we shall here follow the tradition to use GI when referring to the interarrival distribution and G for the service time distribution.)

Thus examples of the particular queueing models become $M/D/1$, $GI/G/\infty$, $E_k/M/1$, $M/H_k/m$, etc. with, for example, $M/D/1$ denoting the single-server queue with Poisson arrivals and deterministic service times.

The notation is widely accepted, but notice should be taken that variants and extensions abound in the literature. One variant has a different distinction between GI and G than the (usual) one given here. This is motivated from the considerable attention which has in recent years been given to queues where the independence assumptions are replaced by the sequences $\{T_n\}$ and $\{U_n\}$ being only strictly stationary. One then writes $G/G/1$ etc. and uses GI to denote the classical independent case (e.g. in $G/GI/1$ service times will then be independent but interarrival times not). Other extensions (that will not be used in the present book) are for example $M^{(X)}/D/m$ and $M/M/m/n$. The first case refers to customers arriving in groups (bulks) distributed as the random variable X at the epochs of a Poisson process. The second may be used for a $M/M/m$ queue with finite waiting room of size n, a finite population of n customers or even other models.

1c. The queue discipline

We start by a list of some of the main types of queue disciplines.

FIFO First In, First Out. Also denoted FCFS = First Come, First Served. The customers are served in the order of arrival. Apparently this is the usual procedure at an ordered queue and therefore the predominant assumption in the literature. *Unless otherwise stated, this is the queue discipline in force throughout this book.*

LIFO Last In, First Out. Also denoted LCFS = Last Come, First Served. After having completed a service the server turns to the latest arrived customer. This would occur, for example, in storage situations where the items (customers) are stacked and all in-out operations occur at the top of the pile.

SIRO Service In Random Order. After having served a customer, the server picks the next at random among the remaining ones. This would occur, for example, in technical systems like telephone exchanges where the system does not remember when the customers arrived.

PS Processor Sharing. The customers share the server, i.e. when n customers are present, the server devotes $1/n$ of his capacity to each. Equivalently, the customers attain service at rate $1/n$ and leave the system once the attained service reaches the service time. The situation is illustrated in Fig. 1.1. The main example is a computer with several jobs running simultaneously. Here PS is really only an approximation to what physically occurs, namely the next discipline in the list.

RR Round Robin. Here the server works on the customers one at a time in a fixed time quantum δ. A customer not having completed service within this time is put back in the queue, and before he can retain service the other customers are each allowed their quantum of δ (or less, if service is completed). The situation is illustrated in Figs. 1.2, 1.3. As δ becomes infinitely small, PS is obtained as a limiting case of RR.

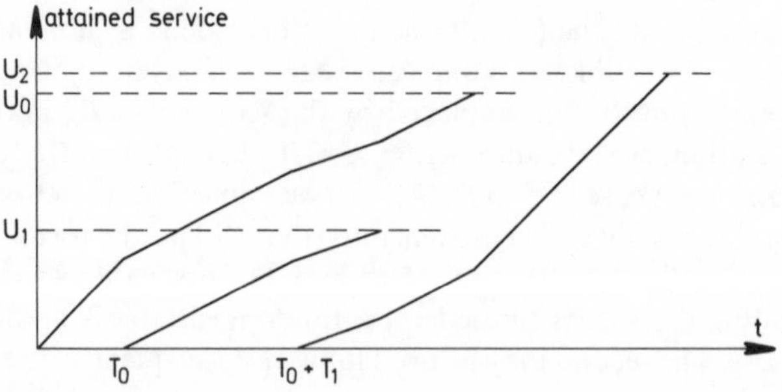

Fig. 1.1

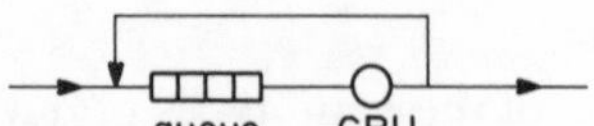

Fig. 1.2

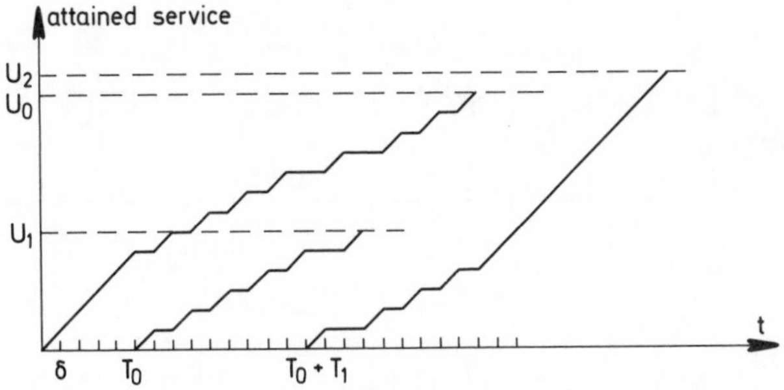

Fig. 1.3

This list is by no means complete and does not either cover all aspects. For example, it is not quite clear what is meant by a FIFO $GI/G/m$ queue since the customers may either queue up in one line (what we shall assume in the following) or in some way form m separate waiting lines. Further examples of queue disciplines are found above all in the area of *priority queueing*. Here the customers are divided into priority classes $1, 2, \ldots, K$, a customer from a lower class having priority before one from a higher class. The system may be *pre-emptive* or *non-pre-emptive*. In the first case the customer being served is interrupted in his service if a new one of higher priority arrives, in the second case not. In the pre-emptive case an interrupted customer may then either have attained some service or not, and so on, a great number of variations being possible.

1d. Queue lengths, waiting times and other functionals

In connection with a given queueing system, a great variety of stochastic processes and functionals arise. The main ones that we shall study are the following three (defined for $GI/G/m$, but with obvious generalizations to many other models):

- Q The *queue length* at time t (denoted X_t in the present chapter where $\{X_t\}_{t \geqslant 0}$ is a Markov jump process). Also denoted as the *number in system* to stress that the customer being presently handled by the server is included.
- W_n The *actual waiting time* (or just *waiting time*) of customer n, that is, the time from when he arrives at the system until service starts.
- V_t The *residual work* in the system at time t, that is, the total time the m servers have to work to clear the system. Thus V_t is the sum of the residual service times of customers being presently served and the customers awaiting service. In the case $m = 1$ of a single server, this is simply the time needed for the server to clear the system provided that no new customers arrive, that is, the waiting time of a hypothetical customer arriving just after t. For this reason V_t is denoted the *virtual waiting time* at time t for $m = 1$.

The connection is illustrated in Fig. 1.4. It is simplest to visualize for $m = 1$, where

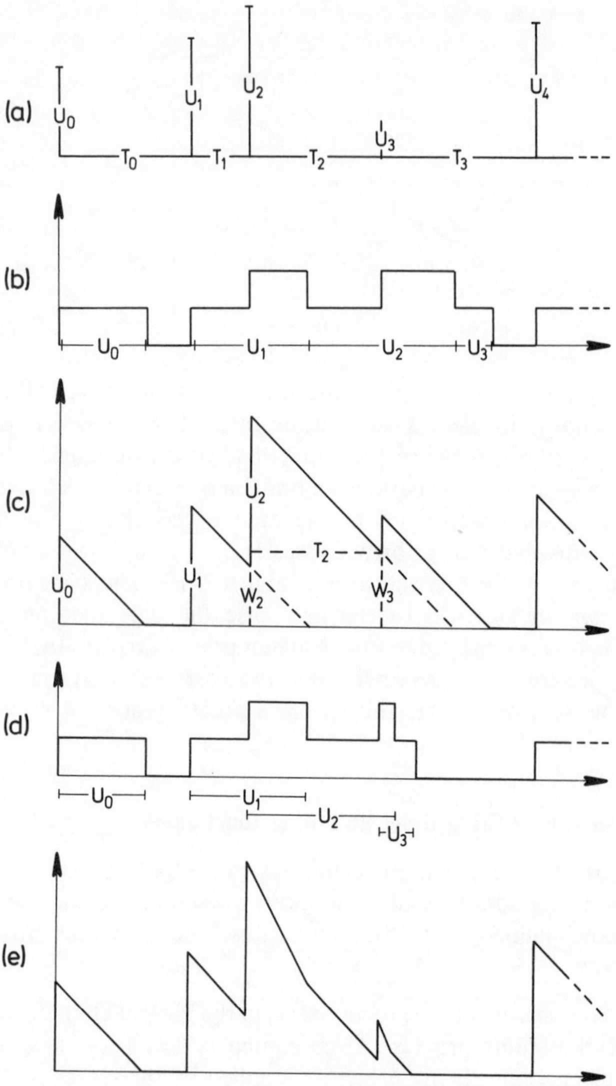

Fig. 1.4. (a) Input of service times and inter-arrival times; (b) the corresponding single-server queue length process; (c) the single-server residual work or virtual waiting-time process; (d) the two-server queue length process; (e) the two-server residual work process

the actual waiting time of customer $n+1$ is the virtual waiting time $V_{\sigma(n)-}$ just before the time $\sigma(n) = T_0 + \cdots + T_n$ of his arrival. Thus on the figure, $W_0 = W_1 = W_4 = 0$ and $W_2 > 0$, $W_3 > 0$. For other aspects, see the Problems.

There are two points worth noting when concentrating interest around these processes: (a) the processes or functionals of interest are not always of one of these three types, but have frequently very close relations; (b) it depends very much on the practical situation whether it is the queue length, the actual or the virtual waiting time or some other functionals that are of interest. An obvious example of (a) is the *sojourn time* of customer n, i.e. the total time he spends in the system. This is the waiting time followed by the service time, i.e. $W_n + U_n$, and since W_n, U_n are clearly independent, the sojourn-time distribution is a simple functional of the waiting-time distribution, namely the convolution with the service-time distribution B. Another example is busy and idle times, which in $GI/G/1$ can be described by the time intervals where $Q_t > 0$ (or equivalently $V_t > 0$) and $Q_t = V_t = 0$ respectively. We also mention that the interest in $\{V_t\}_{t \geqslant 0}$ is due to a large extent to the reinterpretations of this process within the areas of storage, dams and insurance risk, see Chapter XIII. For (b), note that, for example, sometimes the interest centres around the workload put on the system itself, in other cases around the inconvenience caused to the customers by exceedingly long waiting times (more typically, the aim is to balance these points of view). A typical example would be the design problem for the cash system in a supermarket: say for simplicity that we have m identical servers and want to choose the best value of m. If m is large, we expect the system to be idle for a considerable amount of time and thereby being insufficiently utilized compared to the cost of running. If, on the other hand, m is small, then we expect long waiting times for the customers which will encourage them to use instead a less congested competing shop near by. The quantitative evaluation of this effect of discouragement is of course a matter of management judgement and not mathematics. However, once this has been settled we need to say something about both idle times and waiting times for a given arrival rate. Possibly the discouragement could be an effect of the visible queue length and not the related but unobservable waiting time. Therefore, the queue length is also of potential interest here. It is certainly so in other situations like telephone exchanges with a limited number K of lines, where queue lengths $\geqslant K$ mean the possibility of calls being lost.

1e. Measures of performance. The traffic intensity

Seen from a practical point of view, the purpose of theoretical analysis is to shed some light on the queueing situation in question. The meaning of this may be rather vague and, for example, it may be argued that just formulating a simplified mathematical model is helpful since it necessitates thinking through and properly clarifying which features of the system are the basic ones. Having passed this point interest centres, however, on evaluating the performance of a given system (and possibly some related ones, for the purpose of assessing the effect of a change). That is, the first step is to define some appropriate measures of performance.

In rather general terms, we want to describe the properties of the basic processes of queue lengths and waiting times. A main step in that direction is a study of one-dimensional distributions, say for example F_n given by $F_n(t) = \mathbb{P}(W_n \leqslant t)$. Now this is difficult to compute in most situations and the dependence on t is a complicating factor for the sake of comparisons (so is the dependence on n, but we defer the discussion of this to Subsection 1f). Therefore, it becomes appropriate to consider some simple characteristics of a distribution F on $[0, \infty)$, and some of the main ones that are usually considered relevant are the following:

(a) The *mean* $\mu = \int_0^\infty x\,\mathrm{d}F(x)$, measuring the average values;
(b) Possibly some of the *higher order moments* $\mu_k = \int_0^\infty x^k\,\mathrm{d}F(x)$;
(c) The *variance* $\sigma^2 = \mu_2 - \mu^2$, measuring the dispersion around the mean; possibly also some higher order *cumulants*;
(d) The squared *coefficient of variation* σ^2/μ^2, giving a scale invariant measure of dispersion;
(e) The *tail characteristics* describing the asymptotic behaviour of the tail $1 - F(x)$. For example the relation $1 - F(x) \approx C\,\mathrm{e}^{-\eta x}$ holds for many distributions in queueing theory, cf. XII.5, and appropriate tail characteristics are then C and (in particular) η.

Which of these characteristics that are appropriate depends on the situation.

There is, however, one measure of performance of a queueing system which is of quite universal interest. This is the so-called *traffic intensity* ρ, which we define here for $GI/G/m$ by

$$\rho = \mathbb{E}U_k/m\mathbb{E}T_k = \int_0^\infty x\,\mathrm{d}B(x)/m\int_0^\infty x\,\mathrm{d}A(x) \tag{1.1}$$

(there are appropriate generalizations for most other queueing systems) and the interpretation is as follows. Suppose that for a very large amount of time t the system is working at full capacity, i.e. that all servers are busy. Then by the LLN there will be about $t/\mathbb{E}T_k$ arrivals and a total of about $mt/\mathbb{E}U_k$ services ($t/\mathbb{E}U_k$ for each server). Thus ρ is about the ratio, i.e. when $\rho > 1$ the number of arrivals exceeds the number of services so that we expect the queue to grow indefinitely. In contrast, when $\rho < 1$ then eventually even a very long initial queue will be cleared (in the sense that not all servers are busy; after that the queue may build up again, but will again be cleared up for the same reason, and so on, the system evolving in cycles). Thus the behaviour should be like transience when $\rho > 1$ and like recurrence when $\rho < 1$. This will be made more precise later in the various models and also results will be shown stating that the behaviour for $\rho = 1$ is like null recurrence.

1f. Equilibrium theory versus time dependence

The notion of *equilibrium* is within the setting of Markov processes just what so far has been called stationarity: a Markov chain or Markov jump process is in equilibrium if it is ergodic and stationary. The results developed in Chapters I–II state that after a long period of time an ergodic process settles in equilibrium or, with a different term, attains the *steady state.* A similar behaviour is on intuitive grounds to be expected far beyond the Markovian setting: if the capacity of the queueing system is sufficient to deal with the arriving workload, say the traffic intensity is less than 1, one expects the system to alternate between being busy and idle, and that the initial conditions will be smoothed away by the stochastic variation in the length of the cycles. Thus, under appropriate conditions there should exist limiting distributions of Q_t, V_t, W_n as $t \to \infty$ and $n \to \infty$ respectively, and there is then an apparent possibility of studying the characteristics of the queueing system by means of these limiting distributions. More generally, when studying functionals of the whole process like departure processes, one could restrict attention to a stationary or equilibrium version. In the following this will be represented by a governing probability distribution $\mathbb{P}_e$ (in some later chapters, we also use notation like W to denote a random variable having the limiting equilibrium waiting-time distribution). Thus, for an ergodic Markov chain $\{X_n\}$ with stationary distribution π, we have

$$\pi_j = \mathbb{P}_e(X_n = j) = \mathbb{P}(X = j) = \lim \mathbb{P}_i(X_n = j),$$

and $\mathbb{P}_e$ is the same as the $\mathbb{P}_\pi$ of Chapter I.

The idea of passing from the study of say $\mathbb{P}(W_n \leqslant t), n = 0, 1, 2, \ldots,$ to $\mathbb{P}_e(W_n \leqslant t)$ is clearly convenient, if nothing else, by eliminating the dependence on n. The motivations are in fact deeper than just this, with the two following points as the corner-stones: (1) a queueing system will frequently be operating for such long periods of time that equilibrium is entered rather quickly; (2) in addition to its limiting interpretation, $\mathbb{P}_e$ also describes the long-term behaviour in terms of time averages. For example, one has for a Markovian queue in continuous time that subject to suitable conditions $T^{-1}\int_0^T Q_t\,dt \approx \mathbb{E}_e Q_t$, cf. II.4, and this average is frequently an appropriate characteristic of the whole segment $\{Q_t\}_{0 \leqslant t \leqslant T}$.

The overwhelming majority of queueing theory (and also the material presented in this book) is concerned with the equilibrium or steady-state properties of the systems rather than finding *time-dependent* quantities like p_{ij}^n in a Markov chain (instead of time-dependent, frequently the somewhat unfortunate term 'transient' is used). The reasons for this are most often motivated by (1), (2) above. However, without any doubt the fact that time-dependent solutions are exceedingly more difficult to come by than the equilibrium ones also plays an important role in practice. Thus, it seems clear that in many situations it is not clear a priori what a long time period in (1) means. Hence it is necessary to have at least some estimate on the rate of convergence to equilibrium, that is, some ideas on the time-dependent behaviour. Also, it is clear that in other situations like the presence of a rush-hour where the queue suddenly builds up after having behaved stably, the equilibrium point of view is not adequate at all.

1g. Queueing theory in this book

A particular practical problem will usually exhibit a considerable number of the great variety of aspects presented so far, and most likely some further specific ones. Comprehensive mathematical models will therefore tend to be complicated and usually intractable: possibly the existence of a limiting steady state can be proved, but the derivation of its properties in a form suitable for numerical calculations is usually out of the question. Therefore the practitioner may have to use either empirical or semi-empirical methods like simulation, approximations or bounds, or to trust that solutions of greatly simplified models have something to say about his problem as well.

It is not our aim here to present queueing theory in a form ready for practical implementation, but rather to study some of the basic mathematical problems and techniques. In the present chapter a rather broad class of problems are studied within the Markovian setting, and after having developed the necessary mathematical tools in Part B, we then deal with a more narrow class of problems associated with general distributions of interarrival times and service times in Part C.

The Markovian assumptions greatly simplify the modelling and solution. They are therefore also frequently the first step when faced with a new type of problem, and they will be used here to look into phenomena requiring considerable effort in more general settings. Examples are queueing networks, time dependence, the busy-period distribution, the effect of queue disciplines other than the FIFO one and also some finite models (clearly, many important models are not touched upon at all). The set-up has its drawbacks, however. One is that queue lengths as discrete variables are more naturally incorporated than the continuous waiting and sojourn times. For example, in a network we can study the length of the various waiting lines, but not the presumably more interesting total sojourn time of a customer. Another deficit is the reliance on assumptions like Poisson arrivals and (probably more seriously) exponential service times. The phase method (to be developed in Section 6) presents a partial solution by extending the Markovian set-up to a class of models which is in a certain sense dense, but it is not entirely satisfying neither from the theoretical nor the practical point of view.

Finally, we mention that one of the classical topics in Markovian queueing theory, imbedded Markov chains, have been deferred to Chapter IX. A Markov chain is imbedded in a (typically non-Markovian) queue if it is obtained by observing the queue length at certain random times. Main examples are $M/G/1$ just after departure times and $GI/M/1$ just before arrival times. However, in particular in $M/G/1$, the imbedded Markov chain is only of limited intrinsic interest, and it requires the more advanced tools of Part B to relate it to the queue length in continuous time and the waiting times.

Problems

1.1. Consider the LIFO single-server queue. Show that the waiting times corresponding to the input in Fig. 1.4(a) are as for the FIFO case, and draw a different figure where this is not the case.

1.2. Draw a figure of the PS single-server queue length process corresponding to the input in Fig. 1.4(a). Find the sojourn times of the customers.

1.3. Compute the actual waiting times of the customers in the $GI/G/2$ case of Fig. 1.4.

1.4. Consider $GI/G/1$ with traffic intensity $\rho < 1$. Show heuristically that the server is busy (idle) in an average proportion $\rho\,(1-\rho)$ of the time.

1.5. Consider $GI/G/1$ with $\rho < 1$. Show by heuristical time-average considerations *Little's formula* $l = \lambda w$. Here l, w denote the equilibrium mean of the queue length and the sojourn time respectively, and $\lambda = 1/\mathbb{E}\,T_k$ the average arrival rate [hint: evaluate $\int_0^T Q_t\,dt$ in terms of the sojourn times of customers having arrived in $[0, T]$, neglecting boundary effects. A formal proof is in VIII.4]. Is the FIFO assumption essential?

Notes

Queueing theory as a whole is an enormous area. Most of the standard textbooks are listed in the references (special directions are only covered to the extent to which explicit references are made in the text or in these Notes). Cooper (1981) has an annotated brief bibliography and Cohen (1982) a survey of recent work and references.

For the current development of research in the area, some of the main journals to consult are the *Journal of Applied Probability*, *Advances in Applied Probability*, *Stochastic Processes and their Applications*, *Operations Research*, *Mathematics of Operations Research*, *Stochastic Models*, as well as the standard probability journals like the *Annals of Probability*, *Zeitschrift für Wahrscheinlichkeitstheorie und verwandte Gebiete*, *Theory of Probability and its Applications*, a number of further operations research journals and also some computer journals like the *Journal of the Association of Computing Machinery*. For more thorough treatments of aspects of Markov process methods, we mention in particular Kelly (1979) and Neuts (1981). The relaxation of the assumptions of independence and stationarity is an area still in active development. It is not treated in the present book, but we refer to Borovkov (1976), Franken *et al.* (1981), Bremaud (1981) and Rolski (1981).

2. GENERAL BIRTH–DEATH PROCESSES

By a *birth–death process* we understand a Markov jump process $\{X_t\}_{t\geqslant 0}$ on $E=\mathbb{N}$ which is *skip-free*, i.e. from state n it can only move to $n-1$ or $n+1$ (from 0 even only to 1). That is, the intensity matrix is of the form

$$\Lambda=\begin{pmatrix} -\beta_0 & \beta_0 & 0 & 0 & \cdots \\ \delta_1 & -\beta_1-\delta_1 & \beta_1 & 0 & \cdots \\ 0 & \delta_2 & -\beta_2-\delta_2 & \beta_2 & \cdots \\ \cdot & & & & \cdot \\ \cdot & & & & \cdot \\ \cdot & & & & \cdot \end{pmatrix}.$$

We denote β_n as *birth intensities* and δ_n as *death intensities*. In this terminology, one thinks of the process as the total size of a population and the most well-known example is the *linear birth–death process* $\beta_n=n\beta$, $\delta_n=n\delta$ which corresponds to the individuals giving birth and dying independently of each other, and in the same way for all population sizes. In our applications we interpret instead X_t as the number of customers in a queue at time t: a jump upwards corresponds to a customer arriving at the queue and a jump downwards to a customer having completed service and leaving the system. Thus in this generality the arrival rate β_n and the service rate δ_n depend in an unspecified manner on the number n of customers present. For example β_n could be decreasing in n, corresponding to customers being discouraged by long queues, and δ_n increasing, corresponding to the server working more rapidly when faced with a long queue. However, the main interest in birth–death processes is due to the more concrete interpretation of the models associated with the specific choices of β_n, δ_n to be presented in Sections 3a–3g. We proceed here to develop the general theory.

The jump chain $\{Y_n\}$ is clearly skip-free as well and may be viewed as a state-dependent Bernoulli random walk (i.e. the increments are ± 1), with reflection at zero. The transition matrix is

$$Q = \begin{pmatrix} 0 & 1 & 0 & 0 & \cdots \\ q_1 & 0 & p_1 & 0 & \cdots \\ 0 & q_2 & 0 & p_2 & \\ \cdot & & & & \cdot \\ \cdot & & & & \cdot \\ \cdot & & & & \cdot \end{pmatrix},$$

where $p_n = \beta_n/(\beta_n + \delta_n)$, $q_n = 1 - p_n = \delta_n/(\beta_n + \delta_n)$. We assume for a while that no p_n, $n \geqslant 1$, can take the values 0 or 1. This obviously implies irreducibility.

Proposition 2.1. *Recurrence of* $\{X_t\}_{t\geqslant 0}$ *or equivalently* $\{Y_n\}$ *is equivalent to*

$$\sum_{n=1}^{\infty} \frac{\delta_1 \cdots \delta_n}{\beta_1 \cdots \beta_n} = \sum_{n=1}^{\infty} \frac{q_1 \cdots q_n}{p_1 \cdots p_n} = \infty. \tag{2.1}$$

Proof. We apply the transience criterion I.5.2 with $i = 0$ to $\{Y_n\}$ and have to look for $h(k)$, $k \geqslant 1$, satisfying $h(j) = \sum_{k \neq 0} q_{jk} h(k)$, $j \neq 0$, i.e.

$$\begin{aligned} h(1) &= p_1 h(2), \\ h(2) &= q_2 h(1) + p_2 h(3), \\ &\vdots \\ h(n) &= q_n h(n-1) + p_n h(n+1). \\ &\vdots \end{aligned}$$

If on the l.h.s. we write $h(n) = (p_n + q_n)h(n)$ and solve for $h(n) - h(n-1)$, we get

$$h(2) - h(1) = q_1 h(1)/p_1,$$

$$h(n+1) - h(n) = \frac{q_n}{p_n}(h(n) - h(n-1)) = \cdots = \frac{q_n q_{n-1} \cdots q_2}{p_n p_{n-1} \cdots p_2}(h(2) - h(1))$$

$$= \frac{q_n \cdots q_1}{p_n \cdots p_1} h(1)$$

and it is clear that there is one, and up to proportionality only one, non-zero solution which is bounded if and only if

$$\sup_n h(n) = h(1) + \sum_{n=1}^{\infty} \{h(n+1) - h(n)\} = h(1)\left\{1 + \sum_{n=1}^{\infty} \frac{q_n \cdots q_1}{p_n \cdots p_1}\right\}$$

is finite. Thus transience is equivalent to (2.1) to fail. □

The criterion (2.1) states loosely that the q_n in some average sense should be as large as the p_n, i.e. that there is no drift to infinity. Assume, for example, some smooth behaviour like the existence of $\sigma = \lim p_n/q_n$. Then if $\sigma < 1$, (2.1) is infinite and we have recurrence, whereas (2.1) is finite for $\sigma > 1$ and we have transience (for $\sigma = 1$ both possibilities may occur, cf. Problem 2.2).

Lemma 2.2. *Irrespective of recurrence or transience, there is one, and up to proportionality only one, solution v to $v\Lambda = 0$, given by*

$$v_n = \frac{\beta_0 \cdots \beta_{n-1}}{\delta_1 \cdots \delta_n} v_0, \qquad n = 1, 2, \ldots. \tag{2.2}$$

Proof. The condition $v\Lambda = 0$ means

$$\beta_0 v_0 = \delta_1 v_1, \qquad (\beta_n + \delta_n) v_n = \beta_{n-1} v_{n-1} + \delta_{n+1} v_{n+1}, \qquad n \geqslant 1.$$

It is clear that given v_0, these equations uniquely determine v, and insertion shows that (2.2) is indeed a solution. □

Corollary 2.3. *In the recurrent case, the stationary measure μ for $\{Y_n\}$ is given by*

$$\mu_n = \frac{p_1 \cdots p_{n-1}}{q_1 \cdots q_n} \mu_0, \qquad n = 1, 2, \ldots. \tag{2.3}$$

Proof. Take μ as in II.4.2(ii)–(iii), $\mu_n = v_n \lambda(n)$. Then $\mu_0 = v_0 \beta_0$ and for $n = 1, 2, \ldots$

$$\mu_n = v_n \lambda(n) = v_n(\beta_n + \delta_n) = \frac{\beta_0 \cdots \beta_{n-1}}{\delta_1 \cdots \delta_n} v_0(\beta_n + \delta_n)$$

$$= \frac{p_1 \cdots p_{n-1}}{q_1 \cdots q_n} \frac{\beta_0 q_n}{\delta_n} (\beta_n + \delta_n) v_0 = \frac{p_1 \cdots p_{n-1}}{q_1 \cdots q_n} \mu_0$$ □

Now define

$$S = 1 + \sum_{n=1}^{\infty} \frac{\beta_0 \cdots \beta_{n-1}}{\delta_1 \cdots \delta_n}.$$

Corollary 2.4. *$\{X_t\}_{t \geqslant 0}$ is ergodic if and only if (2.1) holds and $S < \infty$, in which case the ergodic distribution π is given by*

$$\pi_0 = \frac{1}{S}, \qquad \pi_n = \frac{1}{S} \frac{\beta_0 \cdots \beta_{n-1}}{\delta_1 \cdots \delta_n}, \qquad n = 1, 2, \ldots. \tag{2.4}$$

Proof. Recurrence is equivalent to (2.1), and in that case the total mass of (2.2) is $|v| = S v_0$ so that according to II.4.3 ergodicity is equivalent to $S < \infty$. In that case, $\pi = v/|v|$. □

We conclude with some formulae for the case of a finite state space $\{0, \ldots, K\}$. This occurs if $\beta_K = 0$ since then $\{0, \ldots, K\}$ is a closed set. Irreducibility and hence ergodicity will hold if

$$\beta_k > 0, \qquad k = 0, \ldots, K-1, \qquad \beta_K = 0, \qquad \delta_k > 0, \qquad k = 1, \ldots, K \tag{2.5}$$

and in just the same manner as in Lemma 2.2 and Corollary 2.4 one obtains the stationary distribution as

$$\pi_0 = \frac{1}{S}, \qquad \pi_n = \frac{1}{S} \frac{\beta_0 \cdots \beta_{n-1}}{\delta_1 \cdots \delta_n}, \qquad n = 1, \ldots, K, \tag{2.6}$$

where

$$S = 1 + \sum_{n=1}^{K} \frac{\beta_0 \cdots \beta_{n-1}}{\delta_1 \cdots \delta_n}.$$

Remark 2.5. In many examples, the finite case arises as a modification of an infinite model by letting some $\beta_K = 0$. If the stationary distributions are π and $\pi^{(K)}$ respectively, it is seen from (2.6) that $\pi^{(K)}$ is simply obtained by conditioning (or truncation) of π on $\{0, \ldots, K\}$, $\pi_n^{(K)} = \pi_n/(\pi_0 + \cdots + \pi_K)$. Compare also I.3.8 (or rather the continuous-time analogue).

Problems

2.1. Show that recurrence holds if $\beta_n \leqslant \delta_n$ for all sufficiently large n.
2.2. Suppose $\beta_n = 1$, $\delta_n = (1 - 1/2n)^\gamma$, $\gamma \geqslant 0$. Show that we have transience for $\gamma > 2$ and null recurrence for $\gamma \leqslant 2$.
2.3. Suppose $\beta_n = 1$, $\delta_n = (1 + 1/n)^\gamma$, $\gamma \geqslant 0$. Show that we have ergodicity for $\gamma > 1$ and null recurrence for $\gamma \leqslant 1$.
2.4. Show that we have transience if $\delta_n = 1$ for all n, $\beta_{2^k} = k$, all other $\beta_n = 1$.
2.5. Consider for $k = 0, 1, 2, \ldots$ birth–death processes with $\beta_n^{(0)} = 1$, $\delta_n^{(0)} = 2$ and, for $k \geqslant 1$, $\delta_n^{(k)} = 2$, $\beta_k^{(k)} = 2^k$, all other $\beta_n^{(k)} = 1$. Show that we have ergodicity for all $k = 0, 1, 2, \ldots$, that $\beta_n^{(k)} \to \beta_n^{(0)}$, $\delta_n^{(k)} \to \delta_n^{(0)}$ as $k \to \infty$, but that $\pi_n^{(k)} \to \pi_n^{(0)}$ fails.
2.6. Let π be a distribution on $\mathbb{N}$ satisfying $\pi_n > 0$ for all n. Show that there exists an ergodic birth–death process with π as stationary distribution. Are the β_n, δ_n unique? Are they unique up to proportionality?

Notes

The more refined theory of birth–death processes owes much to a series of papers by Karlin and McGregor in the 1950s. Results beyond the present (standard) ones can be found in Keilson (1979) and van Doorn (1980).

3. BIRTH–DEATH PROCESSES AS QUEUEING MODELS

3a. The $M/M/1$ queue

The $M/M/1$ queue length process as defined in Section 1d clearly corresponds to a birth–death process with $\beta_n = \beta$ and $\delta_n = \delta$ independent of n. This is by far the conceptually most simple queueing system, the one of the greatest analytical tractability (at least for an infinite state space) and therefore it plays a prominent role in the literature.

The traffic intensity as defined in Section 1e is $\rho = \beta/\delta$. Thus the recurrence criterion (2.1) becomes $1 + \sum_1^\infty \rho^{-n} = \infty$ and we have at once

Proposition 3.1. *The $M/M/1$ queue with traffic intensity ρ is recurrent if and only if $\rho \leqslant 1$.*

This is intuitively reasonable at least if $\rho \neq 1$ by recalling the interpretation of ρ as the ratio (1.1) and will be seen to hold for more general queues (e.g. $GI/G/m$). Similarly, the ergodicity conditions derive immediately from Corollary 2.4. We get $S = 1 + \sum_1^\infty \rho^n = (1-\rho)^{-1}$ for $\rho \leqslant 1$ and thus:

Proposition 3.2. *The $M/M/1$ queue with traffic intensity ρ is ergodic if and only if $\rho < 1$. In that case, the equilibrium distribution π of the queue length is geometric, $\pi_n = (1-\rho)\rho^n$ $n = 0, 1, 2, \ldots$.*

This permits us immediately to calculate a number of interesting quantities. For example, the probability that the server is idle (busy), in equilibrium is

$$\mathbb{P}_e(X_t = 0) = \pi_0 = 1 - \rho, \qquad (\mathbb{P}_e(X_t > 0) = 1 - \pi_0 = \rho) \tag{3.1}$$

whereas by standard formulae for the geometric distribution we have

$$\mathbb{E}_e X_t = \frac{\rho}{1-\rho}, \qquad \mathbb{V}\mathrm{ar}_e X_t = \frac{\rho}{(1-\rho)^2}, \qquad \mathbb{P}_e(X_t \geqslant N) = \rho^N. \tag{3.2}$$

These formulae show among other things that as $\rho \uparrow 1$, then (not unexpectedly) with high probability ρ the server is busy and the mean queue length $\rho/(1-\rho)$ is large. Again, these properties are qualitatively (but not quantitatively) typical of more general queues.

3b. The $M/M/\infty$ queue

This corresponds clearly to the case $\beta_n = \beta$, $\delta_n = n\delta$. We may think of each customer being handled by his own server so that his sojourn time in the system is exponential with intensity δ and independent of all other customers. A different interpretation is therefore an *immigration–death process* with immigration according to a Poisson process and each individual dying after an exponential time.

The definition (1.1) of the traffic intensity yields $\rho = 0$. Instead, the interesting parameter is $\eta = \beta/\delta$ and we get

$$\sum_{n=1}^{\infty} \frac{\delta_1 \cdots \delta_n}{\beta_1 \cdots \beta_n} = \sum_{n=1}^{\infty} n!\,\eta^{-n} = \infty, \qquad S = 1 + \sum_{n=1}^{\infty} \frac{\eta^n}{n!} = \mathrm{e}^{\eta}$$

Thus Corollary 2.4 yields:

Proposition 3.3. *The $M/M/\infty$ queue is ergodic for all values of η. The equilibrium distribution π is Poisson with mean η, $\pi_n = \mathrm{e}^{-\eta}\eta^n/n!$*

For a different case of a Poisson π, see Problem 3.1.

3c. The $M/M/m$ queue

Here $\beta_n = \beta$ and $\delta_n = m(n)\delta$, where $m(n)$ is the number of busy servers in state n, i.e. $m(n) = m$, $n \geqslant m$, $m(n) = n$, $1 \leqslant n \leqslant m$. The traffic intensity is $\rho = \beta/m\delta$ and we have

$\beta_n/\delta_n = \rho, n \geqslant m$. Thus, as in the case $m = 1$, (2.1) and recurrence hold if and only if $\sum \rho^{-n} = \infty$, i.e. $\rho \leqslant 1$. Similarly, with $\eta = \beta/\delta$

$$S = 1 + \sum_{n=1}^{\infty} \frac{\beta_0 \cdots \beta_{n-1}}{\delta_1 \cdots \delta_n} = \sum_{n=0}^{m-1} \frac{\eta^n}{n!} + \frac{\eta^m}{m!} \sum_{n=0}^{\infty} \rho^n = \sum_{n=0}^{m-1} \frac{\eta^n}{n!} + \frac{\eta^m}{m!}(1-\rho)^{-1}$$

is finite if and only if $\rho < 1$, and we get

Proposition 3.4. *The $M/M/m$ queue with traffic intensity ρ is ergodic if and only if $\rho < 1$. In that case the ergodic distribution π is given by*

$$\pi_n = \begin{cases} \dfrac{1}{S} \dfrac{\eta^n}{n!} & 0 \leqslant n \leqslant m \\ \dfrac{1}{S} \dfrac{\eta^m}{m!} \rho^{n-m} & m \leqslant n < \infty \end{cases}$$

This solution is analytically more complicated than those encountered so far since the functional form of π_n is not the same for $n < m$ and $n \geqslant m$, and also S is more complicated. The probabilistic interpretation is, however, quite interesting: π is a combination of the $M/M/\infty$ solution and the $M/M/1$ solution, with the $M/M/\infty$ solution on the states $\{0, 1, \ldots, m\}$ with full server availability (no customers awaiting service) and the $M/M/1$ solution on the states $\{m+1, m+2, \ldots\}$ where some customers must await service.

Again it is straightforward to evaluate functionals. For example, the probability that all servers are busy and the mean queue lengths are

$$\pi_m + \pi_{m+1} + \cdots = \frac{1}{S} \frac{\eta^m}{m!} \frac{1}{1-\rho}$$

and

$$\mathbb{E}_e X_t = \sum_{n=0}^{\infty} n\pi_n = \frac{1}{S} \left\{ \sum_{n=1}^{m-1} \frac{\eta^n}{(n-1)!} + \frac{\eta^m}{m!} \left[\frac{\rho}{(1-\rho)^2} + \frac{m}{1-\rho} \right] \right\}.$$

3d. The $M/M/1$ queue with limited waiting room

So far we have had the infinite state space $\{0, 1, 2, \ldots\}$ in all examples. However, clearly in many practical situations there is a limited capacity of the system so that the queue may not be arbitrarily long, and examples of this will now be given here and in the following subsections.

A simple basic case is the $M/M/1$ queue with waiting room of size K. That is, at most K customers at a time can be present in the system (including the one being served) and customers arriving to a full system are lost. Thus $\beta_n = \beta$, $n = 0, \ldots, K-1$, $\delta_n = \delta$, $n = 1, \ldots, K$. Referring to Remark 2.5, we get for $\rho = \beta/\delta < 1$ the stationary distribution by conditioning (or *truncation*) of the geometric $M/M/1$ solution,

$$\pi_n = \frac{\rho^n}{1 + \rho + \cdots + \rho^K} = \frac{1-\rho}{1-\rho^{K+1}} \rho^n, \qquad n = 0, \ldots, K.$$

It can be immediately checked from (2.6) that this also holds for $\rho > 1$, whereas for $\rho = 1$ all $\pi_n = (1+K)^{-1}$.

3e. Erlang's loss system

A well-known and historically important example was considered by Erlang in connection with design problems for telephone exchanges. Suppose we have an exchange of K lines, that calls arrive at rate β and have exponential durations with intensity δ, and that calls arriving while all lines are busy are lost. Let $\eta = \beta/\delta$. What is (in equilibrium) $E_K(\eta)$, the fraction of calls that are lost?

To solve this problem, we may model the number of busy lines as a birth–death process on $\{0,\ldots,K\}$ with $\beta_k = \beta, k = 0,\ldots,K-1, \delta_k = k\delta, k = 1,\ldots,K$. This set is obtained by letting $\beta_K = 0$ in a $M/M/\infty$ (or $M/M/K$) queue so that by Remark 2.5 the stationary distribution is conditional Poisson,

$$\pi_n = \frac{\eta^n/n!}{1+\eta+\cdots+\eta^K/K!} \qquad n = 0,\ldots,K.$$

The probability of a particular call being lost in equilibrium is now simply the probability π_K of arriving at a full system so that

$$E_K(\eta) = \frac{\eta^K/K!}{1+\eta+\cdots+\eta^K/K!} \tag{3.3}$$

This is the well-known *Erlang's loss formula* (also denoted as Erlang's 1. formula or Erlang's B-formula) and of considerable interest in teletraffic theory. It can be proved to hold under substantially more general conditions, for example arbitrary distribution of the duration of calls.

3f. Engseth's loss system

All examples considered so far have Poisson arrivals, i.e. $\beta_n = \beta$. This is adequate if we have a finite but large population of customers. Here 'large' also means large compared to the sizes of the queues building up, so that even with queues of rather unlikely lengths the proportion of customers in the system is vanishing, i.e. the intensity of the source does not decrease significantly. Clearly this is not the case in all practical situations, and here and in Section 3g we shall consider two of the models which have been suggested in specific situations.

The first example is a teletraffic model considered by Engseth, essentially by just modifying Erlang's loss system to a finite population of N subscribers. Let K be the total number of lines and X_t the number of busy lines at time t. Any call is assumed to involve only one subscriber. Assuming that $N > K$, we then have a birth–death process with $\beta_n = (N-n)\beta$, $\delta_n = n\delta$. Thus

$$\beta_0 \cdots \beta_{n-1} = N^{(n)}\beta^n, \qquad \delta_1 \cdots \delta_n = n!\,\delta^n$$

and letting $\eta = \beta/\delta$ we obtain the stationary distribution as

$$\pi_n = \frac{\binom{N}{n}\eta^n}{1 + N\eta + \cdots + \binom{N}{K}\eta^K}, \qquad n = 0, 1, \ldots, K.$$

Choosing $p \in (0, 1)$ such that $p/(1-p) = \eta$ (i.e. $p = \beta/(\beta + \delta) = \rho/(1+\rho)$), we see that this is the binomial distribution with parameters (N, p) conditioned to be in $\{0, \ldots, K\}$, i.e. a *truncated binomial* or *Engseth distribution.*

3g. Palm's machine repair problem

Consider a population of K machines which each break down with intensity β and is immediately taken care of by one of N repairmen working at rate δ, as soon as one becomes available. Thus if X_t is the number of machines under repair or awaiting repair, we have

$$\beta_n = (K-n)\beta, \qquad n = 0, \ldots, K-1, \quad \text{and} \quad \delta_n = (N \wedge n)\delta, \qquad n = 1, \ldots, K.$$

We might wish to study the way in which the production loss due to stoppages and repairs depends on N, for thereby allocating the optimal number N of servers. To this end we need the wage expenses per unit time which are N times a known constant, and the average number of stopped machines per unit time, i.e. the equilibrium mean $\sum_0^K n\pi_n$. Obviously π can be immediately computed by means of (2.6). We shall not spell out the formulae, but mention only an important reinterpretation in the case $N = 1$. Considering the number $\tilde{X}_t = K - X_t$ of working machines instead of X_t, the intensities here are

$$\tilde{\beta}_n = \delta_{K-n} = \delta, \qquad n = 0, \ldots, K-1, \qquad \tilde{\delta}_n = \beta_{K-n} = n\beta, \qquad n = 1, \ldots, K.$$

This is of the same form as for Erlang's loss system, and we conclude immediately that $\tilde{X}_t$ is truncated Poisson in equilibrium.

This model has recently received renewed attention due to a computer system interpretation, where one thinks of the customers as terminals and the repairman as the computer handling calls from the terminals (without processor-sharing). Thus the terminals call the computer each with intensity β, and X_t is the number of calls being presently handled by the computer.

Problems

3.1. Consider the case $\beta_n = \beta/(n+1)$, $\delta_n = \delta$ of customers being discouraged by long queues. Show that the ergodic distribution π exists and is Poisson.

3.2. Show the recursion formula $E_{K+1}(\eta) = \eta E_K(\eta)/[K + 1 + \eta E_K(\eta)]$, cf. (3.3).

3.3. Let $\beta_n = \beta(N+n)$, $\delta_n = n\delta$. Show that we have ergodicity for $\rho = \beta/\delta < 1$ and that π is negative binomial,

$$\pi_n = \binom{-N}{n}(-\rho)^n(1-\rho)^N.$$

Give a demographic interpretation of the model.

3.4. Consider Erlang's loss system and let $H(k)$ denote the equilibrium probability that k fixed lines are busy. Show the *Palm–Jacobæus formula* $H(k) = E_K(\eta)/E_{K-k}(\eta)$.
3.5. Consider the same model as in Engseth's loss system except that now $N \leqslant K$. Show that π is binomial (N, p), where $p = \beta/(\beta + \delta)$.
3.6. Consider the $M/M/2$ queue with *heterogeneous servers*. That is, the servers 1, 2 have intensities $\delta^{(1)} > \delta^{(2)}$ and if server 1 becomes idle, the customer being served by 2 switches to 1. Explain that this corresponds to $\beta_n = \beta$, $\delta_1 = \delta^{(1)}$, $\delta_n = \delta^{(1)} + \delta^{(2)}$, $n = 2, 3, \ldots$. Show that we have ergodicity if and only if $\beta < \delta^{(1)} + \delta^{(2)}$ and find the stationary distribution.
3.7. Same questions as in Problem *3.6*, but the model is now modified such that the customers cannot switch to server 1 (a customer arriving to an empty system then always joins server 1) [hint: look first at the process restricted to $\{2, 3, \ldots\}$ as a birth–death process, and split next state 1 into two states indicating which server is busy].

Notes

Birth–death queueing models is a favourite textbook topic, see for example Cox and Smith (1961, II.4), Gross and Harris (1974), Kleinrock (1975, Ch. 3), Allen (1978, 5.2) and Iversen (1981).

4. POISSON DEPARTURE PROCESSES AND SERIES OF QUEUES

We start by noting

Proposition 4.1. *Any ergodic birth–death process is time reversible.*

Proof. We must check the conditions $\pi_i\lambda(i,j) = \pi_j\lambda(j,i)$ of detailed balance in II.5.2. If $|i-j| > 1$, then by the skip-free property both sides are zero so we can suppose $|i-j| = 1$, say $i = n, j = n+1$, where the condition reduces to $\pi_n\beta_n = \pi_{n+1}\delta_{n+1}$ which is clear from (2.4). It is instructive to indicate how alternatively the proof can be carried out without first computing π: in equilibrium, the flux from $\{0, 1, \ldots, n\}$ to $\{n+1, n+2, \ldots\}$ must balance the flux the other way. But the only possible transition the first way is from n to $n+1$ so the flux is $\pi_n\lambda(n, n+1)$. Similarly, the flux the other way is $\pi_{n+1}\lambda(n+1, n)$. □

Consider now a doubly infinite stationary version $\{X_t\}_{t\in\mathbb{R}}$ and define $\tilde{X}_t = X_{-t-}$. Then a departure for $\{X_t\}$ at time s corresponds to an arrival for $\{\tilde{X}_t\}$ at time $-s$. Considering the case of Poisson arrivals, $\beta_n = \beta$, the instants $-s$ form a Poisson process with intensity β by reversibility. Hence the instants s do so too, and we have proved

Corollary 4.2. *The departure process of an ergodic birth–death queue with Poisson arrivals with intensity β is itself a Poisson process with intensity β.*
This is the first in one of a series of results in this and the next section, which may be argued to be contrary to intuition or at least surprising. A discussion of this is deferred to Section 5d.

A main application of the Poisson departure property is to series of K queues, where customers enter queue 1 according to a Poisson process with intensity β, after being served proceed to queue 2, from there to queue 3 and so on (the system with $K = 2$ is called a *tandem queue*). Suppose that queue k has exponential services with intensities $\delta_n^{(k)}$. Then queue 1 is a birth–death queue with Poisson arrivals, hence has a Poisson departure process in equilibrium. But this process is just the arrival process to queue 2 so that this is a birth–death queue with Poisson arrivals which hence delivers Poisson input to queue 3 and so on. It follows that in equilibrium the number of customers at queue k and their waiting time have the same characteristics as if the queue was considered alone subject to Poisson arrivals. We generalize below this reasoning to the simultaneous behaviour of the queues, but first we shall give one more of the classical examples of a Poisson departure process.

Theorem 4.3. *The stationary $M/G/\infty$ queue $\{X_t\}_{-\infty<t<\infty}$ with doubly infinite time has a Poisson departure process.*

Proof. We shall show (slightly more generally) that if $\{\sigma(n)\}_{n\in\mathbb{Z}}$ are the epochs of a Poisson process N on $\mathbb{R}$ with intensity β (namely the arrival times) and$\ldots, U_{-1}, U_0, U_1,\ldots$ (the service times) are independent identically distributed and independent of N, then the point process M with epochs $\{\sigma(n)+U_n\}_{n\in\mathbb{Z}}$ (the departure process) is again Poisson with intensity β. The idea is to observe that this is trivial if the U_n are discrete, and next to apply a discrete approximation. Suppose first that the U_n can assume only the values $0, \pm\delta$, $\pm 2\delta,\ldots$ and let $p_k = \mathbb{P}(U_n = k\delta)$, $k\in\mathbb{Z}$. Then clearly the $\sigma(n)$ with $U_n = k\delta$ form a Poisson process $N^{(k)}$ with intensity βp_k, and the $N^{(k)}$ are independent. Letting $M^{(k)}$ denote $N^{(k)}$ translated by $k\delta$, $M^{(k)}$ is Poisson with intensity βp_k and the $M^{(k)}$ are independent. Hence $M = \sum_{k\in\mathbb{Z}} M^{(k)}$ is Poisson with intensity $\sum \beta p_k = \beta$ as asserted.

To deal with the general case, let $U_n^{(\delta)} = k\delta$, $k\delta \leqslant U_n < (k+1)\delta$ and let M_δ have epochs $\{\sigma(n)+U_n^{(\delta)}\}_{n\in\mathbb{N}}$. If I_r, $r = 1,\ldots,R$, are disjoint intervals, then $M_\delta(I_r) \to M(I_r)$ a.s. as $\delta\downarrow 0$ since $U_n^{(\delta)} \to U_n$ and the probability of an epoch of M at a boundary point of I_r is zero. Hence $\{M_\delta(I_r)\}_1^R \to \{M(I_r)\}_1^R$ in the sense of weak convergence. But by what has just been proved, M_δ is Poisson with intensity β. Hence the joint distribution of the $M(I_r)$ is the common joint distribution of the $M_\delta(I_r)$ so that $M \overset{\mathcal{D}}{=} M_\delta$. □

Preparing for a more thorough study of series of queues, we start by noting:

Proposition 4.4. *Consider the equilibrium version $\{X_t\}_{t\in\mathbb{R}}$ of an ergodic birth–death process with Poisson arrivals at rate β. Then the departure process prior to t is Poisson at rate β and independent of X_t.*

Proof. It only remains to check the independence. The argument is a slight variant of the proof of Corollary 4.2: the departure process prior to t is the arrival process after $-t$ of the time-reversed process and hence independent of its queue length $\tilde{X}_{-t}$ which a.s. coincides with X_t. □

Corollary 4.5. *Consider a series of K queues where the arrivals to the first are*

Poisson at rate β and the K servers work independently, with rate $\delta_n^{(k)}$ for server k at queue length n. Suppose that for each k the birth–death queue with $\beta_n = \beta$, $\delta_n = \delta_n^{(k)}$ is ergodic and let $\pi^{(k)}$ denote its stationary distribution, $\pi_n^{(k)} = S_k^{-1}\beta^n/\delta_1^{(k)}\cdots\delta_n^{(k)}$. Then the system of K queues is also ergodic and the steady state is described by the queue lengths $X_t^{(1)},\ldots,X_t^{(K)}$ being independent with $X_t^{(k)}$ governed by $\pi^{(k)}$,

$$\mathbb{P}_e(X_t^{(1)} = n(1),\ldots, X_t^{(K)} = n(K)) = \pi_{n(1)}^{(1)}\cdots\pi_{n(K)}^{(K)}. \tag{4.1}$$

Proof. Letting $\pi = \pi^{(1)}\otimes\cdots\otimes\pi^{(K)}$ be the distribution (4.1), we proceed by showing $\pi P^t = \pi$ for any fixed t (an alternative proof involving $\pi\Lambda = 0$ is in Problem 4.2). Suppose thus that the $X_0^{(k)}$ has been assigned initial joint distribution π and let $N^{(k)}$ be the departure process from queue k in $[0, t]$. The conclusion will follow if we can show that for each k $X_t^{(0)},\ldots, X_t^{(k)}, N^{(k)}$ are independent, governed by $\pi^{(0)},\ldots,\pi^{(k)}$ and the distribution of the Poisson process respectively. The case $k = 1$ is just Proposition 4.4. Suppose the assertion holds for k. Both $X_t^{(k+1)}$ and $N^{(k+1)}$ depend on $N^{(k)}$, $X_0^{(k+1)}$ and the action of server $k+1$ only, hence the set $(X_t^{(k+1)}, N^{(k+1)})$ is independent of $X_t^{(0)},\ldots, X_t^{(k)}$. But since $N^{(k)}$ is Poisson and $X_0^{(k)}$ governed by $\pi^{(k)}$, it follows by applying Proposition 4.4 once more that the joint distribution of $X_t^{(k+1)}, N^{(k+1)}$ is as asserted. □

The simplest case is of course that of $M/M/1$ queues in series, $\delta_n^{(k)} = \delta^{(k)}$. Then $\rho_k = \beta/\delta^{(k)}$ is the traffic intensity at queue k and

$$\pi_{n(1)}^{(1)}\cdots\pi_{n(K)}^{(K)} = \prod_{k=1}^{K}(1-\rho_k)\rho_k^{n(k)}.$$

Problems

4.1. Consider the case of $K = 3$ queues, but assume that customers leaving 1 do not necessarily go to 2, but choose between 2 and 3 with probabilities p, $q = 1 - p$. Show that if queue 1 has Poisson arrivals and is ergodic, then in equilibrium the input processes to 2, 3 are independent Poisson processes. Formulate the criterion for ergodicity of the whole system and show that the stationary distribution is of product form as in (4.1).
4.2. Check that (4.1) satisfies $\pi\Lambda = 0$ and that $\{X_t^{(1)},\ldots, X_t^{(K)}\}$ is non-explosive.

Notes

The results of this section are standard. Further similar applications of time reversibility are in Kelly (1979), and queueing output processes are surveyed by Daley (1976).

5. QUEUEING NETWORKS

5a. Models and examples

One of the simplest examples of a queueing network, queues in series, has already been encountered in the preceding section. It imposes, however, the restriction that customers can only move along one possible path.

We define now more generally a *queueing network* as follows. There is a finite number K of individual queues, the *nodes* of the networks, at which customers arrive from external sources according to independent Poisson processes with intensities $\alpha_1, \ldots, \alpha_K$. A customer having completed service at node k goes to node l with probability γ_{kl} and leaves the system with probability $\gamma_{ko} = 1 - \sum_1^K \gamma_{kl}$ (external drain). A graphical illustration is given in Fig. 5.1(a), an arrow from k to l denoting $\gamma_{kl} > 0$ and arrows to and from the external world denoting $\gamma_{ko} > 0$ and $\alpha_k > 0$ respectively.

For simplicity, we consider only the case of a single exponential server at each node (one then talks frequently about *Jackson networks*), and we denote the corresponding service rates by $\delta_1, \ldots, \delta_K$. Two main types are considered, *open* networks where external input is received and external output is delivered by customers entering and leaving the system, and *closed* networks where the customers can only move internally in the system. Thus a closed network has all $\alpha_k = 0$ and all $\gamma_{ko} = 0$, and the total number of customers in the system does not vary with time. In an open network some $\alpha_k > 0$ and some $\gamma_{ko} > 0$, and the number in system is a non-degenerate stochastic process.

Networks of this type come up in a great variety of problems, the most important of which are of rather recent date and associated with data communication systems and the internal organization of time-sharing computers. Also, colonies of biological individuals with migration between colonies have been modelled in this way, and as an example of a complicated system with a great number of nodes one could mention the waiting lines at the various ski-lifts at one or more winter resorts. We shall give two examples, one of an open and one of a closed network, with the reservation that it should be stressed that the above model does not pretend to be anything but a crude first approximation. For example, there would frequently be a strong positive correlation between the service times of a customer at the different nodes. In other cases the choice of the customer of the path along the nodes could depend on the length of the waiting lines.

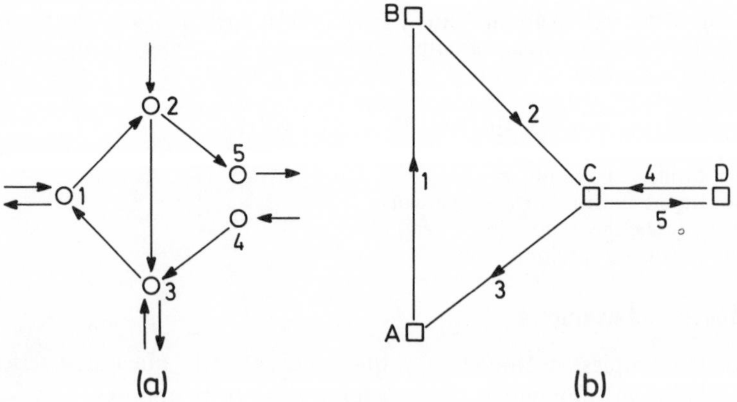

Fig. 5.1

Figure 5.1(b) depicts a communication network, say connecting three branches A, B, D, of a bank, or three computers via a transmission station C. Messages are sent by the directed channels 1, 2, 3, 4, 5 and if a channel is busy, a queue may be formed. Thus if we reinterpret channels as nodes and messages as customers, we arrive at the network in Fig. 5.1(a). This is open since new messages are created currently and a message leaves the system after having reached its destination.

Consider next jobs (customers) circulating in a time-sharing computer as in Fig. 5.2. The nodes are the CPU and input/output facilities, the allowance of feedback at the CPU corresponding to some sort of PS or RR. At first glance this looks like an open network. However, the number of steps taken by each job is typically very large and thus within time intervals of moderate length, the number of jobs is fixed and a description by a closed network may be more appropriate.

Now let $X_t^{(k)}$ denote the number of customers at node k at time t and let $X_t = (X_t^{(1)}, \ldots, X_t^{(K)})$. The state space is $\mathbb{N}^K$ and the states are denoted $n = n(1) \cdots n(K)$, $n(k) \in \mathbb{N}$. Introduce operators describing an arrival at k, a customer leaving the system at k and a customer moving from k to l by

$$\begin{cases} T(0,k)n = n(1) \cdots n(k) + 1 \cdots n(K), \\ T(k,0)n = n(1) \cdots n(k) - 1 \cdots n(K), & n(k) > 0, \\ \quad T(k,l) = T(k,0)T(0,l) = T(0,l)T(k,0), & n(k) > 0. \end{cases}$$

Excluding the case $\gamma_{kk} = 1$, we may assume $\gamma_{kk} = 0$ (otherwise just change δ_k to $\delta_k(1-\gamma_{kk})$ and γ_{kl} to $\gamma_{kl}/(1-\gamma_{kk})$) and the possible transitions and corresponding intensities become

$$\begin{cases} n \to T(0,k)n & \alpha_k & \\ n \to T(k,0)n & \delta_k \gamma_{k0} & (n(k) > 0) \\ n \to T(k,l)n & \delta_k \gamma_{kl} & (n(k) > 0) \end{cases} \tag{5.1}$$

(here the two first types do not occur in a closed network).

We first need to make an appropriate definition of the throughput rate β_k at node k, that is, the common rate of the input and output process (these need not be Poisson but the rate should exist in terms of long-term averages). The input rate is the sum of the rate α_k of external arrivals and the rates of internal arrivals from nodes $l \neq k$. But customers leave node l at rate β_l and go then to

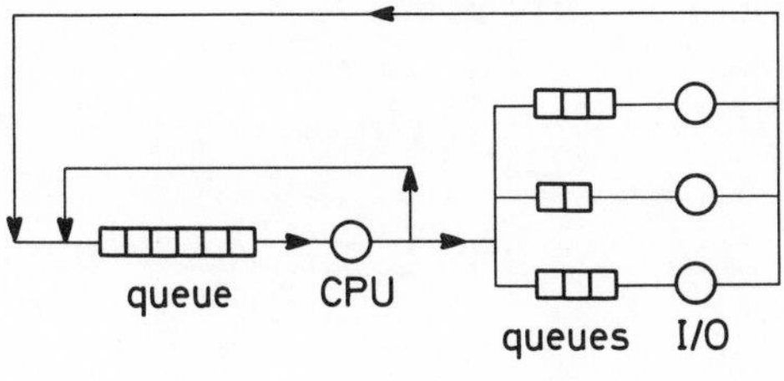

Fig. 5.2

k with probability γ_{lk}, so that we should have

$$\beta_k = \alpha_k + \sum_{l=1}^{K} \beta_l \gamma_{lk} \qquad k = 1, \ldots, K \tag{5.2}$$

5b. Ergodic theory for open networks

We shall assume that each node K may both receive external input and deliver external output (possibly via other nodes), that is, for each k (i) either $\alpha_k > 0$ or some $\alpha_{l(1)}\gamma_{l(1)l(2)} \cdots \gamma_{l(n)k} > 0$, and (ii) either $\gamma_{k0} > 0$ or some $\gamma_{kk(1)}\gamma_{k(1)k(2)} \cdots \gamma_{k(n)0} > 0$. This is easily seen to imply irreducibility of $\{X_t\}_{t \geqslant 0}$.

Proposition 5.1. *The set of equations* (5.2) *have a unique non-negative solution* $(\beta_1, \ldots, \beta_K)$. *It satisfies* $0 < \beta_k < \infty$.

Proof. Consider a Markov jump process on $\{0, 1, \ldots, K\}$ with off-diagonal intensities $\lambda(k, l) = \gamma_{kl}$, $k \neq 0$, $\lambda(0, k) = \alpha_k$. Our assumptions imply irreducibility and hence the existence of a stationary distribution π, uniquely given by $\pi\Lambda = 0$ which is easily seen to amount to

$$\pi_k = \pi_0 \alpha_k + \sum_{l=1}^{K} \pi_l \gamma_{lk} \tag{5.3}$$

$$\pi_0 \sum_{k=1}^{K} \alpha_k = \sum_{l=1}^{K} \pi_l \gamma_{l0} \tag{5.4}$$

It is therefore immediately apparent from (5.3) that $\beta_k = \pi_k/\pi_0$ solves (5.2). Suppose conversely $\beta_1, \ldots, \beta_K$ is a solution and define $\beta_0 = 1$, $\pi_k^* = \beta_k / \sum_0^K \beta_l$. Then (5.3) holds for π^* and (5.4) is a consequence of (5.3), as is seen by summing over $k = 1, \ldots, K$ and performing some algebra. Thus $\pi_k^* = \pi_k$, i.e. $\sum_0^K \beta_l = 1/\pi_0$, $\beta_k = \pi_k/\pi_0$, proving uniqueness. □

Theorem 5.2. *If* $\rho_k = \beta_k/\delta_k < 1$, $k = 1, \ldots, K$, *then* $\{X_t\}_{t \geqslant 0}$ *is ergodic with stationary distribution*

$$\pi_n = \pi_{n(1)\cdots n(K)} = \prod_{k=1}^{K} (1 - \rho_k)\rho_k^{n(k)}.$$

Proof. The intensities are bounded, cf. (5.1), and hence it suffices to show that $\pi\Lambda = 0$, cf. II.4.4, which by (5.1) amounts to

$$\pi_n \sum_{k=1}^{K} \{\alpha_k + \delta_k I(n(k) > 0)\} = \sum_{k=1}^{K} \{\pi_{T(k,0)n}\alpha_k I(n(k) > 0) + \pi_{T(0,k)n}\delta_k \gamma_{k0}\}$$
$$+ \sum_{k,l=1}^{K} \pi_{T(k,l)n}\delta_l \gamma_{lk} I(n(k) > 0). \tag{5.5}$$

Now

$$\pi_{T(0,k)n} = \rho_k \pi_n, \qquad \pi_{T(k,l)n} = \rho_k^{-1} \rho_l \pi_n.$$

Hence using (5.2) we get

$$\sum_{k=1}^{K} \pi_{T(0,k)n}\delta_k\gamma_{k0} = \pi_n \sum_{k=1}^{K} \beta_k\gamma_{k0} = \pi_n \sum_{k=1}^{K} \beta_k\left(1 - \sum_{l=1}^{K} \gamma_{kl}\right)$$

$$= \pi_n \sum_{l=1}^{K}\left\{\beta_l - \sum_{k=1}^{K} \beta_k\gamma_{kl}\right\} = \pi_n \sum_{l=1}^{K} \alpha_l, \tag{5.6}$$

$$\pi_{T(k,0)n}\alpha_k + \sum_{l=1}^{K} \pi_{T(k,l)n}\delta_l\gamma_{lk} = \pi_n\rho_k^{-1}\left\{\alpha_k + \sum_{l=1}^{K} \beta_l\gamma_{lk}\right\}$$

$$= \pi_n\rho_k^{-1}\beta_k = \pi_n\delta_k, \qquad n(k) > 0, \tag{5.7}$$

and (5.5) follows. □

5c. Ergodic theory for closed networks

Clearly, $E_N = \{n : \sum_1^K n(k) = N\}$ is closed and to discuss irreducibility and ergodicity, we therefore have to restrict the state space of $\{X_t\}_{t\geqslant 0}$ to E_N. Equation (5.2) reduces in matrix notation to $\beta = \beta\Gamma$, $\Gamma = (\gamma_{ij})_{i,j=1,\ldots,K}$ and since $\gamma_{k0} = 0$, Γ is a transition matrix. We shall assume that Γ is irreducible on $\{1,\ldots,K\}$. This implies the existence of a β (unique up to a constant) which satisfies $\beta = \beta\Gamma$, and also, as is readily seen, that $\{X_t\}_{t\geqslant 0}$ is irreducible on E_N and hence ergodic since E_N is finite.

Theorem 5.3. *Under the above assumptions, $\{X_t\}_{t\geqslant 0}$ is ergodic on E_N with invariant distribution*

$$\pi_n = C^{-1}\prod_{k=1}^{K} \rho_k^{n(k)},$$

where $\rho_k = \beta_k/\delta_k$ and C is a normalization constant ensuring $|\pi| = 1$.

Proof. In the same way as for (5.5), we have to check that

$$\pi_n \sum_{k=1}^{K} \delta_k I(n(k) > 0) = \sum_{k,l=1}^{K} \pi_{T(k,l)n}\delta_l\gamma_{lk} I(n(k) > 0). \tag{5.8}$$

But since $\pi_{T(k,l)n} = \rho_k^{-1}\rho_l\pi_n$, we get for $n(k) > 0$ that

$$\sum_{l=1}^{K} \pi_{T(k,l)n}\delta_l\gamma_{lk} = \pi_n\rho_k^{-1}\sum_{l=1}^{k} \beta_l\gamma_{lk} = \pi_n\rho_k^{-1}\beta_k = \pi_n\delta_k$$

which implies the truth of (5.8). □

Theorem 5.3 will now be shown to have as a consequence a *bottleneck* type of system behaviour: if N is large, then with high probability most of the N customers will be in the waiting line at the node with the highest ρ_k. Such a knowledge could be useful say for design purposes, since it would in some situations suggest an allocation of the total service capacity $\delta_1 + \cdots + \delta_k$ such that $\max \rho_k$ is minimized.

To illustrate this effect, we shall assume that one ρ_k, say ρ_1, is effectively largest and we may then choose the scale of β such that

$$\rho_1 = 1, \qquad \rho_2 < 1, \ldots, \rho_K < 1. \tag{5.9}$$

Consider the marginal equilibrium distribution of $X_t^{(1)}$ and $(X_t^{(2)}, \ldots, X_t^{(K)})$ respectively,

$$\eta_n^{(N)} = \mathbb{P}_e(X_t^{(1)} = n) = \sum_{n(2)+\cdots+n(K)=N-n} \pi_{nn(2)\cdots n(K)},$$

$$\theta_{n(2)\cdots n(K)}^{(N)} = \mathbb{P}_e(X_t^{(2)} = n(2), \ldots, X_t^{(K)} = n(K)) = \pi_{(1)\cdots n(K)}^{(N)},$$
$$n(1) = N - n(2) - \cdots - n(K).$$

Then in the limit $N \to \infty$, $\eta^{(N)}$ becomes degenerate at ∞, whereas $\theta^{(N)}$ has a proper limit of the same form as the equilibrium solution of an open network:

Corollary 5.4. *As $N \to \infty$, $\eta_n^{(N)} \to 0$ for all n and*

$$\theta_{n(2)\cdots n(K)}^{(N)} \to \prod_{k=2}^{K} (1 - \rho_k)\rho_k^{n(k)}.$$

Proof. Writing $C = C_N$, we have

$$C_N = \sum_{n(1)+\cdots+n(K)=N} \prod_{k=1}^{K} \rho_k^{n(k)} = \sum_{n(2)+\cdots+n(K) \leqslant N} \prod_{k=2}^{K} \rho_k^{n(k)}$$

$$\to \sum_{n(2)=0}^{\infty} \cdots \sum_{n(k)=0}^{\infty} \prod_{k=2}^{K} \rho_k^{n(k)} = \prod_{k=2}^{K} (1 - \rho_k)^{-1} = C_\infty \quad \text{(say)}.$$

Here $0 < C_\infty < \infty$ and hence if $\rho_k < \delta < 1$, $k = 2, \ldots, K$, we get

$$\eta_n^{(N)} = C_N^{-1} \sum_{n(2)+\cdots+n(K)=N-n} \prod_{k=2}^{K} \rho_k^{n(k)} < C_N^{-1} \binom{K+N-n-2}{N-n} \delta^{N-n} \to 0,$$

$$\theta_{n(2)\cdots n(K)}^{(N)} = C_N^{-1} \prod_{k=2}^{K} \rho_k^{n(k)} \to \prod_{k=2}^{K} (1 - \rho_k)\rho_k^{n(k)}. \qquad \square$$

5d. Pitfalls for intuition

We have now given rigorous mathematical proofs of a number of results on queues delivering Poisson output, and queues in series or networks which behave in a certain sense as if they were totally independent and each subject to Poisson arrivals. Many of these results may be difficult to understand on more intuitive grounds. For example one may ask:

(a) How can even such a simple queue as $M/M/1$ deliver Poisson output at rate β? The server has idle periods with no output and busy periods where departures are Poisson at rate δ. Also, observing the output alone, we can tell the value of β but not δ.

(b) How can the departure process $N^{(t)}$ prior to t be independent of the departures $M^{(t)}$ after t? Observation of $N^{(t)}$ should tell us something on $(X_s)_{s \leqslant t}$ (e.g. if there are few departures just before t, we expect $X_t = 0$ with greater probability than the average $1 - \rho$), and conditionally upon X_t, $M^{(t)}$ is certainly not Poisson.

(c) In a network rather than a series, the arrivals to node K are not in general Poisson. Why then is this not reflected in the behaviour of $X_t^{(k)}$?

(d) How can $X_t^{(k)}$ and $X_t^{(l)}$ in a network be independent? If $X_t^{(k)}$ is large and $\gamma_{kl} > 0$, then node l should have received more input prior to t than on the average, hence also $X_t^{(l)}$ should be large.

Of course, the mathematical proofs tell that such reasoning has to have gaps or errors, and it is not difficult either to elaborate further on the intuitive reasoning to deduce that important aspects have been ignored (for example, the independence in a network refers only to a fixed instant of time and not the time evolution). Our point here is merely to stress that intuitive reasoning, though an indispensable part of applied probability, has its pitfalls and that care has to be taken that it can be followed up with a more rigorous proof.

Notes

The results of this section are standard. Main references for queueing networks in general are Kleinrock (1976) and Kelly (1979). The area is rapidly developing, and a survey of recent work as well as an extensive bibliography are given in Disney and König (1985). See also Cohen (1982, IV.2.2).

6. THE PHASE METHOD

The amenability of Markovian models to analysis should by now have become apparent and further examples are given in the following sections. However, the Markovian set-up puts some restriction on the modelling and one of the most serious ones seems to be that whereas it will frequently be very reasonable to assume that interarrival times are exponential (i.e. we have Poisson arrivals), then this is not the case for service times.

The first idea on how to overcome this apparent difficulty was the so-called *method of stages* due to Erlang. The idea is to think of the customer as being composed of K stages each having an exponential service time, say with intensity δ. The stages are then served one at a time, and the service is completed when all stages are served. That is, the service time of the customer himself is the *Erlang distribution* E_k with k stages, namely a convolution of k exponentials with the same intensity and having density

$$\delta^k \frac{x^{k-1}}{(k-1)!} e^{-\delta x}, x > 0. \tag{6.1}$$

The point is now that if we count stages instead of customers and the arrival process is say Poisson, then we get a Markov process $\{X_t\}_{t \geqslant 0}$. Indeed, since the arrival of a customer corresponds to the arrival of k stages, the non-zero off-diagonal intensities are $\lambda(n, n+k) = \beta$, $\lambda(n, n-1) = \delta \; n \geqslant 1$. The queue length

$\{Q_t\}_{t\geqslant 0}$ is then obtained simply by summing the stages out. For example, if we want to determine $\mathbb{P}_e(Q_t = n)$, we solve for the stationary distribution π for X_t and have

$$\mathbb{P}_e(Q_t = n) = \pi_{(n-1)k+1} + \pi_{(n-1)k+2} + \cdots + \pi_{nk}.$$

Apparently what we just have described is, in the Kendall notation, the queueing system $M/E_k/1$. It should be stressed that this way of imbedding an apparently non-Markovian queue into a Markovian set-up is essentially an artifice: the stages themselves usually can be given no physical interpretation. The gain is the greater flexibility in the choice of the service-time distribution, as will be explained in more detail below.

This idea can be considerably generalized. One step would be to allow the stages different intensities $\delta_1, \ldots, \delta_k$. This scheme is for apparent reasons denoted as *exponential distributions in series*, cf. Fig. 6.1, and leads to $\lambda(n, n+k) = \beta$, $\lambda(nk+i+1, nk+i) = \delta_{k-i}$, $i = 0, \ldots, k-1$. But instead we might consider *exponential distributions in parallel channels*, that is, the service time is chosen at random between k exponential distributions with different intensities $\delta_1, \ldots, \delta_k$ (this corresponds to k types of customers). The corresponding service-time distribution

$$\mathbb{P}(U \leqslant u) = \sum_{i=1}^{k} \alpha_i(1 - \mathrm{e}^{-\delta_i u})$$

($\sum_1^k \alpha_i = 1, 0 \leqslant \alpha_i \leqslant 1$) is thus a convex combination of exponential distributions and denoted as H_k, the *hyperexponential distribution with k parallel channels.* The queueing system just described is thus $M/H_k/1$ and to fit it into a Markovian set-up, we need to let the state space be $\mathbb{N} \times \{1, \ldots, k\}$, the first component indicating the queue length, and the second the channel in which the server is currently operating or the type of the customer being served. The non-trivial intensities are

$$\lambda((n, i), (n+1, i)) = \beta, \qquad \lambda((n, i), (n-1, j)) = \delta_i \alpha_j \qquad n \geqslant 1.$$

In conclusion, we have now extended within the Markovian setting the class of possible service-time distributions (and quite similarly of interarrival distributions) from the exponential distribution M to the Erlangian E_k or hyperexponential H_k (both reducing to M for $k = 1$). The resulting gain in flexibility is illustrated in Fig. 6.2 where we have drawn some examples of the densities.

A further noteworthy property is the behaviour of the squared coefficient η of variation (s.c.v.). This is $1/k$ for E_k, i.e. $\eta \in (0, 1]$ with 1 attained for $k = 1$ (the exponential case) and 0 in the limit $k \to \infty$ (i.e. E_k approaches D). For H_k, the s.c.v. is always > 1 (provided at least two δ_i are different) and ranges in fact all over $(1, \infty)$. To derive these properties, use for E_k either standard moment

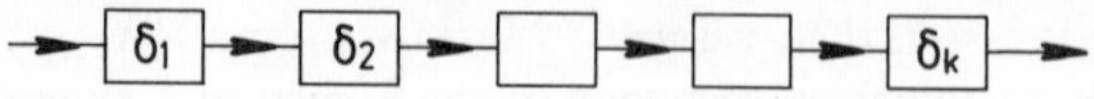

Fig. 6.1

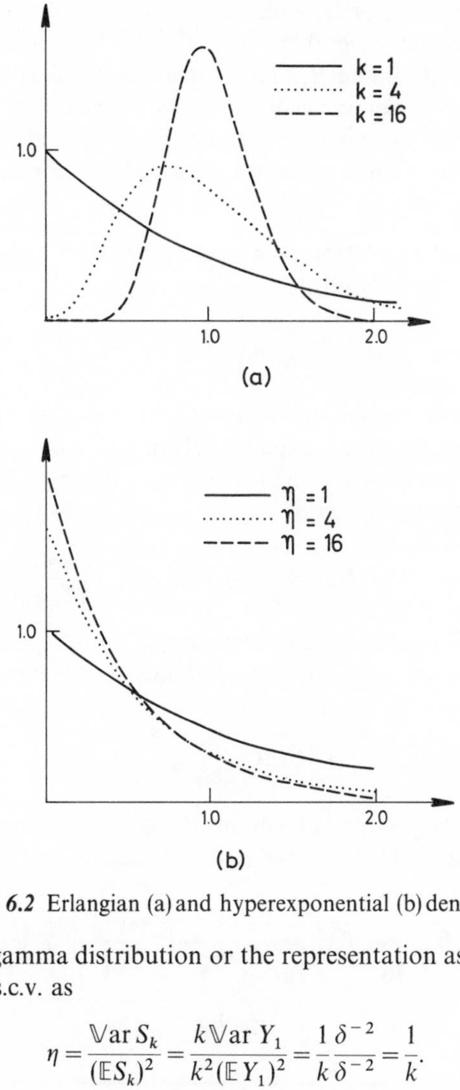

Fig. 6.2 Erlangian (a) and hyperexponential (b) densities

formulae for the gamma distribution or the representation as $S_k = Y_1 + \cdots + Y_k$ which yields the s.c.v. as

$$\eta = \frac{\mathbb{V}\mathrm{ar}\, S_k}{(\mathbb{E}S_k)^2} = \frac{k\mathbb{V}\mathrm{ar}\, Y_1}{k^2(\mathbb{E}\, Y_1)^2} = \frac{1}{k}\frac{\delta^{-2}}{\delta^{-2}} = \frac{1}{k}.$$

For H_k, we may use the representation Y_τ where $\mathbb{P}(\tau = i) = \alpha_i$, Y_i has density $\delta_i e^{-\delta_i y}$ (i.e. mean $\mu_i = \delta_i^{-1}$ and variance $\sigma_i^2 = \mu_i^2$) and is independent of τ. Then conditioning upon τ yields the s.c.v. as

$$\frac{\mathbb{V}\mathrm{ar}\, Y_\tau}{(\mathbb{E}Y_\tau)^2} = \frac{\mathbb{V}\mathrm{ar}\, \mu_\tau + \mathbb{E}\sigma_\tau^2}{(\mathbb{E}\mu_\tau)^2} = \frac{2\mathbb{E}\mu_\tau^2}{(\mathbb{E}\mu_\tau)^2} - 1$$

which is > 1 provided $\mathbb{V}\mathrm{ar}\, \mu_\tau > 0$, i.e. not all μ_i are equal. Further inspection clearly shows that the range of the s.c.v. is $(1, \infty)$ already for H_2.

In early literature, the discussion of phase-type distribution sometimes stopped at this point, the argument being that for a given distribution we may now choose an appropriate E_k or H_k which fits the s.c.v. reasonably well. Of course, this point of view is very rigid: there are many distributions with s.c.v. (say) > 1 which do not look at all like H_k. To resolve this problem we need to go one step further and define suitable general classes of phase-type distributions, and we shall briefly review some of the suggestions in the literature on how to define such classes (the notation is intended as a support for memory but is not standard in the literature):

$\mathscr{PH}_{\text{ME}}$ The class of all *mixtures of Erlang distributions with the same intensity*, i.e. densities of the form

$$\sum_{i=1}^{k} \alpha_i \delta^{n(i)} \frac{x^{n(i)-1}}{(n(i)-1)!} \mathrm{e}^{-\delta x}, \qquad x > 0. \tag{6.2}$$

$\mathscr{PH}_{\text{C}}$ The class of *Coxian distributions* with representation as in Fig. 6.3. That is, we consider n exponentials in series, but having passed stage k, we choose to leave with probability p_k (thus $p_n = 1$). With $q_k = 1 - p_k$, the Laplace transform is

$$\hat{G}(\beta) = \sum_{k=1}^{n} q_1 \cdots q_{k-1} p_k \prod_{i=1}^{k} \frac{\delta_i}{\delta_i + \beta} \tag{6.3}$$

$\mathscr{PH}_{\text{S/P}}$ The class of *exponential distributions in series and/or parallel*, i.e. with representation as in Fig. 6.4 or equivalently with Laplace transform of the form

$$\hat{G}(\beta) = \sum_{i=1}^{k} \alpha_i \prod_{j=1}^{n(i)} \frac{\delta_{ij}}{\delta_{ij} + \beta} \tag{6.4}$$

$\mathscr{PH}_{\text{AT}}$ The class of *absorption times for Markov* (jump) *processes* with a finite state space. That is, $G \in \mathscr{PH}_{\text{AT}}$ if one can find a Markov jump process

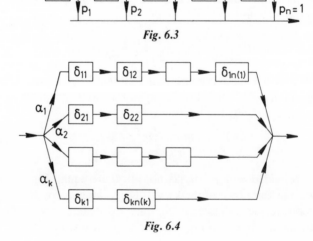

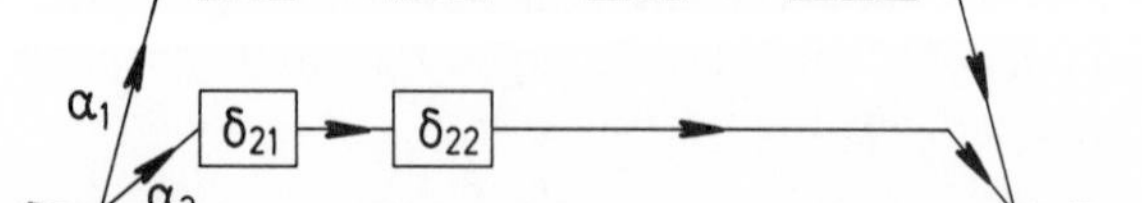

Fig. 6.3

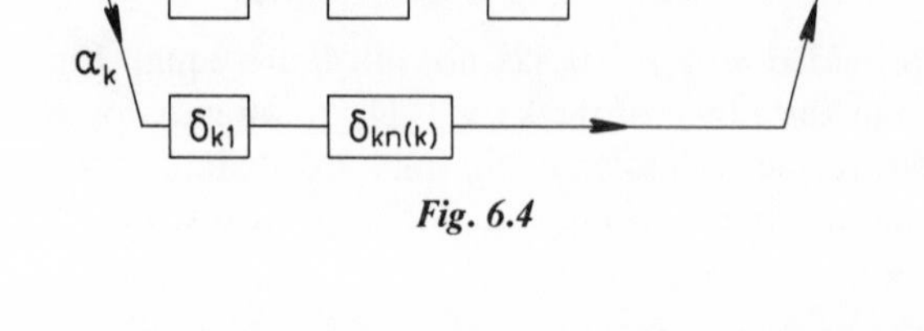

Fig. 6.4

$\{X_t\}_{t\geqslant 0}$, with finite state space $E\cup\{\Delta\}$ such that Δ is absorbing and the states in E transient, and an initial distribution π such that G is the $\mathbb{P}_\pi$-distribution of the time to absorption in Δ. The intensity matrix is easily seen to be given by its restriction Q to $E\times E$, and we then call (E, Q, π) a *representation* of G.

The explicit form of a $\mathscr{PH}_{\mathrm{C}}$-, $\mathscr{PH}_{\mathrm{S/P}}$- or $\mathscr{PH}_{\mathrm{AT}}$-distribution is in general more involved than for $\mathscr{PH}_{\mathrm{ME}}$, see Proposition 6.3 below.

$\mathscr{PH}_{\mathrm{RLT}}$ The class of distributons with rational Laplace transform, $\hat{G}(\beta)=Q(\beta)/R(\beta)$, where Q, R are polynomials.

Theorem 6.1. $\mathscr{PH}_{\mathrm{ME}}\subset\mathscr{PH}_{\mathrm{C}}=\mathscr{PH}_{\mathrm{S/P}}\subset\mathscr{PH}_{\mathrm{AT}}\subset\mathscr{PH}_{\mathrm{RLT}}$.

Proof, Assume without loss of generality that all $n(i)=i$ in (6.2). Then letting $n=n(k)$, $\delta_i=\delta$, $p_1=\alpha_1$, $p_2=\alpha_2/(1-\alpha_1)$, $p_3=\alpha_3/(1-\alpha_1-\alpha_2),\ldots$ in Fig. 6.3 shows that $\mathscr{PH}_{\mathrm{ME}}\subseteqq\mathscr{PH}_{\mathrm{C}}$ (that the inclusion is strict is obvious, e.g. $\mathscr{PH}_{\mathrm{C}}$ contains the convolution of two exponential distributions with different intensities which is clearly not in $\mathscr{PH}_{\mathrm{ME}}$). That $\mathscr{PH}_{\mathrm{C}}\subseteqq\mathscr{PH}_{\mathrm{S/P}}$ follows (say) from the expressions for the Laplace transforms. For the converse, consider $G\in\mathscr{PH}_{\mathrm{S/P}}$ represented as in Fig. 6.4. We may clearly assume that $\delta_{i1}\geqslant\delta_{i2}\geqslant\cdots$ for all i, and define λ_1 as the largest δ_{i1}. Now for any $\beta<\lambda_1$ a simple calculation shows that the distribution in Fig. 6.5(a) is simply the exponential distribution with intensity β. Applying this to a channel with $\beta=\delta_{i1}<\lambda_1$ yields the representation in Fig. 6.5(b), and altogether we may represent G as in Fig. 6.6(a) where G_1 has the same intensities as G (except that λ_1 may have been removed). In any case, the maximal number of occurrences of λ_1 in any channel has been decreased by one, and continuing in this manner we end up with the situation in Fig. 6.6(b) where G_r has a representation with a great number of channels but only one intensity. That is, G_r is an exponential distribution so that indeed $G\in\mathscr{PH}_{\mathrm{C}}$. For $\mathscr{PH}_{\mathrm{S/P}}\subseteqq\mathscr{PH}_{\mathrm{AT}}$, assume that G has representation as in Fig. 6.4 or (6.4), and

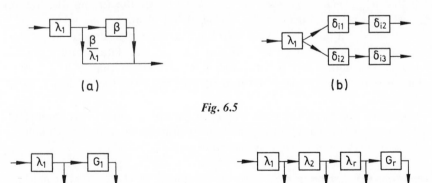

Fig. 6.5

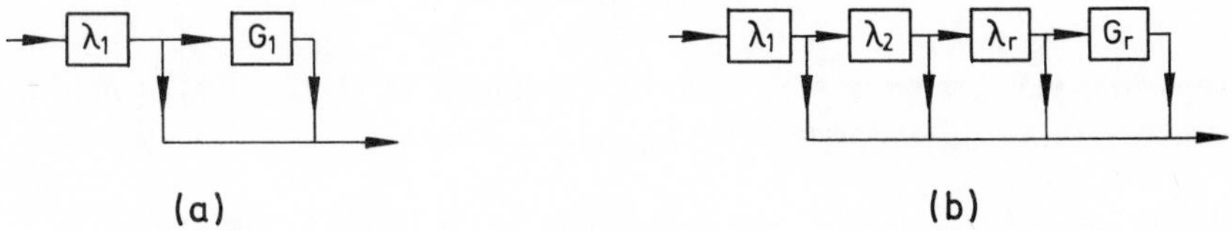

Fig. 6.6

define $E = \{ij: 1 \leqslant i \leqslant k,\ 1 \leqslant j \leqslant n(i)\}$,

$$\pi_{i1} = \alpha_i, \qquad \pi_{ij} = 0 \quad j > 1, \qquad \pi_\Delta = 0,$$
$$\lambda(ij, i(j+1)) = \delta_{ij}, j < n(i), \qquad \lambda(in(i), \Delta) = \delta_{in(i)},$$

all other off-diagonal intensities $= 0$. Then it is immediately clear that $G(t) = \mathbb{P}_\pi(\omega(\Delta) \leqslant t)$. That the inclusion is strict is shown in Problem 6.4. Finally $\mathscr{PH}_{\mathrm{AT}} \subseteqq \mathscr{PH}_{\mathrm{RLT}}$ follows from Proposition 6.3 below, and the strict inclusion from Problem 6.5. □

Theorem 6.2. *The class $\mathscr{PH}_{\mathrm{ME}}$ is dense in the set $\mathscr{P}$ of probability distributions on $(0, \infty)$. More generally, to any $F \in \mathscr{P}$ with $\mu(F; p) = \int_0^\infty x^p \, \mathrm{d}F(x) < \infty$ for some $p \geqslant 0$, there are $F_k \in \mathscr{PH}_{\mathrm{ME}}$ with $F_k \xrightarrow{w} F$ and $\mu(F_k; q) \to \mu(F; q)$ for $q \leqslant p$. The same conclusions hold true if $\mathscr{PH}_{\mathrm{ME}}$ is replaced by any of the classes $\mathscr{PH}_{\mathrm{C}}, \mathscr{PH}_{\mathrm{AT}}, \mathscr{PH}_{\mathrm{RLT}}$.*

Proof. In view of Theorem 6.1, it is obviously sufficient to consider $\mathscr{PH}_{\mathrm{ME}}$. Let d be a metric for weak convergence in $\mathscr{P}$ and define

$$d_p(F, G) = d(F, G) + |\mu(F; p) - \mu(G; p)|, \qquad p \geqslant 0$$

(so that $d_0 = d$). Since $F_k \to F$ and $\mu(F_k; p) \to \mu(F; p)$ implies $\mu(F_k; q) \to \mu(F; q)$ for $q \leqslant p$ (uniform integrability!), we thus have to show that $\mathscr{PH}_{\mathrm{ME}}$ is dense with respect to d_p in $\mathscr{P}_p = \{F \in \mathscr{P}: \mu(F; p) < \infty\}$. Letting $F^{(A)}$ be F truncated at A, $F^{(A)}(x) = F(x \wedge A)/F(A)$, it is easily seen that $d_p(F, F^{(A)}) \to 0$ as $A \to \infty$. Hence if $\mathscr{P}^{(A)}$ is the set of distributions supported by $[0, A]$, $\bigcup_{A > 0} \mathscr{P}^{(A)}$ is dense in $\mathscr{P}_p$. Further, it is a standard fact that the subset $\tilde{\mathscr{P}}^{(A)}$ of $\mathscr{P}^{(A)}$ consisting of discrete distributions is dense with respect to d, and hence to d_p since d and d_p are equivalent on $\mathscr{P}^{(A)}$. Now let $G \in \tilde{\mathscr{P}}^{(A)}$ be a discrete distribution with atoms $t_1, \ldots, t_k$ of weights $\alpha_1, \ldots, \alpha_k$ ($\sum_1^k \alpha_k = 1$). For each m and i choose integers $n_{m,i}$ such that $n_{m,i}/m \to t_i$ as $m \to \infty$, $i = 1, \ldots, k$. Consider the Erlang distribution $G_{m,i}$ with $n_{m,i}$ stages and intensity $\delta_m = m$. It has mean $n_{m,i}/m$ and s.c.v. $n_{m,i}^{-1}$, hence as $m \to \infty$ it converges weakly to the distribution degenerate at t_i. Thus with $G_m = \sum_1^k \alpha_i G_{m,i}$ we have $d(G_m, G) \to 0$. An easy calculation shows that all moments also converge. In particular, $d_p(G_m, G) \to 0$ so that $\overline{\mathscr{PH}_{\mathrm{ME}}} \supseteqq \tilde{\mathscr{P}}^{(A)}$. Taking first the union over A and next the closure shows that $\mathscr{PH}_{\mathrm{ME}}$ is dense in $\mathscr{P}_p$. □

In modern literature, the class $\mathscr{PH}_{\mathrm{AT}}$ is becoming increasingly important. One reason (among many) is the general formalism made available through the Markov process point of view. A typical example is as follows:

Proposition 6.3. *Let $G \in \mathscr{PH}_{\mathrm{AT}}$ have representation (E, Q, π), let π_E denote the restriction of π to E and write $|v| = \sum_{j \in E} v_j$ etc. Then:*

(i) *For $t \geqslant 0$, the cumulative distribution function is $G(t) = 1 - |\pi_E \mathrm{e}^{Qt}|$.*
(ii) *The Laplace transform is $\hat{G}(\beta) = 1 - |\pi_E(I - Q/\beta)^{-1}|$, $\beta \geqslant 0$, and is rational.*

Proof. If we think of Δ as the last state in $E \cup \{\Delta\}$, then the full intensity matrix $\tilde{Q}$ satisfies

$$\tilde{Q}^n = \begin{pmatrix} Q^n & R_n \\ 0 & 0 \end{pmatrix}, \qquad \text{hence } e^{\tilde{Q}x} = \begin{pmatrix} e^{Qx} & T \\ 0 & 0 \end{pmatrix} \tag{6.5}$$

for suitable vectors R_n, T (this is clear for $n = 1$ since Δ is absorbing and shown by induction for $n > 1$). Since the P_π-distribution of X_t is $\pi e^{\tilde{Q}t}$, we thus get

$$1 - G(t) = \mathbb{P}_\pi(X_t \in E) = |\pi_E e^{Qt}|$$

and (i) follows. The key for (ii) is the fact that all eigenvalues λ of Q satisfy $\operatorname{Re}\lambda < 0$. Indeed, if $Qh = \lambda h$ with $\operatorname{Re}\lambda \geqslant 0$, then $\mathbb{E}_\nu h(X_t) = e^{\lambda t}\nu h$ does not tend to zero, contradicting E being a transient set. Thus $I - Q/\beta$ has only eigenvalues with positive real part and in particular is invertible. Since

$$\hat{G}(\beta) = 1 - \beta \int_0^\infty e^{-\beta t}(1 - G(t))\mathrm{d}t, \tag{6.6}$$

(ii) then follows by (i) and the same formal rules for integration of matrix exponentials as in the univariate case. Alternatively, if β is so large that $\sum Q^n/\beta^n$ converges absolutely, then (6.6) becomes

$$\begin{aligned} 1 - \beta \sum_{i,j \in E} \sum_{n=0}^{\infty} \pi_i \frac{Q^n(i,j)}{n!} \int_0^\infty e^{-\beta t} t^n \,\mathrm{d}t &= 1 - \beta \sum_{i,j \in E} \sum_{n=0}^{\infty} \pi_i Q^n(i,j)/\beta^{n+1} \\ &= 1 - |\pi_E(I - Q/\beta)^{-1}| \end{aligned}$$

and by analytic continuation, this then holds for all $\beta > 0$. Finally, all elements of $(I - Q/\beta)^{-1}$ are rational functions of β since the determinant and all subdeterminants are also rational functions. □

When faced with a queueing problem say on the $M/G/1$ queue, we may now approximate the service-time distribution B with a distribution $\tilde{B}$ of phase type, for example by first considering a discrete approximation and next proceeding as in the proof of Theorem 6.2. We then model the approximating queue with service-time distribution $\tilde{B}$ by representing $\tilde{B}$ as in the definition of $\mathscr{PH}_{\mathrm{AT}}$ and thinking of the server as moving in E in the same way as $\{X_t\}$. The server is restarted in π each time he starts serving a new customer, i.e. when a customer arrives to an empty queue or $\{X_t\}$ hits Δ. Without spelling out the state space and the intensities, this is readily seen to lead to a Markov jump process, the solution of which may then be used as an approximation to the given $M/G/1$ queue.

It would be reasonable to admit that the picture of the role of phase-type distributions given by the above discussion is somewhat idealized and simplified for a number of reasons:

1. From the theoretical point of view, it must be proved that approximation of B with $\tilde{B}$ also implies approximation of the corresponding steady-state characteristics of the queue. This is a question of so-called *robustness* and far from trivial or without pitfalls, see VIII.5.

2. From the practical point of view, a good approximation of B will usually require a very large E giving rise to computational problems. Even for small E, it is not usually possible to find the solution in explicit form. We return to the problem in X.5.
3. The possibility of Markovian modelling is not the only way in which phase-type distributions arise in queueing theory. One alternative important property is the possibility of explicit solutions of fundamental problems in renewal theory and random walks, see IX.6. Also, this is subject to some of the same criticism as in (1) and (2). Still another point of view is taken in the literature based on transform methods, where the class $\mathscr{PH}_{\mathrm{RLT}}$ is exploited in a purely analytical way.

Problems

6.1. Write up the appropriate state space and intensities for some queueing systems like $H_k/E_l/1$, $M/H_k/m$, etc.
6.2. Show that the s.c.v. η of H_2 can have any value in $(1, \infty)$.
6.3. Let $G_1, G_2 \in \mathscr{PH}_{\mathrm{AT}}$. Show that the convolution $G_1 * G_2$ and a convex combination $\theta G_1 + (1-\theta)G_2$ are again in $\mathscr{PH}_{\mathrm{AT}}$.
6.4. Show that (a) if $G \in \mathscr{PH}_{\mathrm{S/P}}$ and $\hat{G} = Q/R$ with Q/R polynomials without common factors, then R cannot have complex roots; (b) if $G = (1-\theta)\sum_1^\infty \theta^{n-1} F^{*n}$ with $F \in \mathscr{PH}_{\mathrm{AT}}$, then $G \in \mathscr{PH}_{\mathrm{AT}}$; (c) if $F = E_3$ in (b), then $G \notin \mathscr{PH}_{\mathrm{S/P}}$.
6.5. Show that (a) the density of $G \in \mathscr{PH}_{\mathrm{AT}}$ cannot be zero [hint: G contains a component of exponentials in series]; (b) the distribution with density proportional to $(1 + \sin x)e^{-x}$ is in $\mathscr{PH}_{\mathrm{RLT}} \backslash \mathscr{PH}_{\mathrm{AT}}$.
6.6. Let $e = (1 \cdots 1)'$. Show that $R_1 = -Qe$ in (6.5), and hereby that an alternative expression for $\hat{G}(\beta)$ is $\pi_\Delta + \pi_E(\beta I - Q)^{-1} R_1$.
6.7. Show that the kth moment of $G \in \mathscr{PH}_{\mathrm{AT}}$ is $(-1)^k k! |\pi_E Q^{-k}|$.

Notes

The modern revival of phase-type distributions and the class $\mathscr{PH}_{\mathrm{AT}}$ in particular is due largely to M. F. Neuts, a main reference being his 1981 book. The literature relating to Theorem 6.2 is confusing and contains a variety of erroneous statements of the type '$\mathscr{PH}_{\mathrm{C}} = \mathscr{PH}_{\mathrm{RLT}}$'. The present proof of $\mathscr{PH}_{\mathrm{S/P}} \subseteq \mathscr{PH}_{\mathrm{C}}$ was communicated to the author by Peter Dalgaard and is essentially the same as that of Cumani (1982).

7. LINDLEY PROCESSES IN DISCRETE TIME

By a *Lindley process*, in discrete time and starting from w, we understand a process of the form

$$W_0 = w, \qquad W_{n+1} = (W_n + X_n)^+ \qquad n = 0, 1, \ldots, \tag{7.1}$$

where $x^+ = \max(x, 0)$ and $X_0, X_1, \ldots$ are independent identically distributed (in general assuming both positive and negative values) with common distribution F. Equivalently, the process may be described as a Markov chain with state space $E = [0, \infty)$ and transition kernel given by

$$\begin{aligned} P(w, [0, m]) &= \mathbb{P}(W_1 \leqslant m \mid W_0 = w) = \mathbb{P}(w + X_0)^+ \leqslant m \\ &= \mathbb{P}(w + X_0 \leqslant m) = F(m - w), \qquad w, m \geqslant 0. \end{aligned} \tag{7.2}$$

If F is lattice concentrated on $\{0, \pm\delta, \pm 2\delta, \ldots\}$ and we consider only initial values of the form $w = k\delta$, then the state space may be reduced to $E = \{0, \delta, 2\delta, \ldots\}$.

The interest in the Lindley process stems classically from the way it comes up in the $GI/G/1$ queue (see the basic Example 7.1 below), but it is quite common that in a particular queueing model one or more of the processes of interest may be related to a process which is Lindley or at least of a somewhat similar structure. One example has already been given in I.5.5 and one more is given here in Example 7.2. Further examples are given in the Problems and (in continuous time) Section 8.

Example 7.1. Consider the $GI/G/1$ queue and, as in Section 1(d), let W_n be the actual waiting time of customer n. What is the relation between W_{n+1} and W_n? Say that customer n arrives at time t and $n+1$ at $t + T_n$. The residual work in the queue is W_n just before t, $W_n + U_n$ just after t and W_{n+1} just before $t + T_n$. Since the residual work decreases at a linear rate in between arrivals when positive, W_{n+1} will be $W_n + U_n - T_n$ if $W_n + U_n - T_n \geqslant 0$, whereas otherwise $W_{n+1} = 0$ (a graphical illustration of the argument is given in Fig. 1.4(c) with $n = 2$). Hence (7.1) holds with $X_n = U_n - T_n$ and clearly the X_n are independent identically distributed. For example in the $M/M/1$ case $\mathbb{P}(U_n > u) = \mathrm{e}^{-\delta u}$, $\mathbb{P}(T_n > t) = \mathrm{e}^{-\beta t}$, it is readily seen that F is the *doubly exponential* or *Laplacian* distribution with density

$$\beta\delta/(\beta+\delta)\mathrm{e}^{-\delta x}, x \geqslant 0, \qquad \beta\delta/(\beta+\delta)\mathrm{e}^{\beta x}, x \leqslant 0. \qquad \square$$

Example 7.2. This is the imbedded Markov chain in $GI/M/1$ mentioned in Section 1(g), that is, W_n denotes the number of customers just before the arrival of customer n. Let A denote the interarrival distribution and δ the service intensity. We may think of services being described in terms of a Poisson process with intensity δ, such that an event in the Poisson process implies a customer being served if the queue is non-empty and is just redundant otherwise. To describe the relation between W_n and W_{n+1}, let K_n be the number of Poisson events in the interval between the arrivals of customers n and $n+1$ and

$$q_k = P(K_n = k) = \int_0^\infty \mathrm{e}^{-\delta t}\frac{(\delta t)^k}{k!}\mathrm{d}A(t). \tag{7.3}$$

Then clearly $W_n + 1$ customers are present just after the arrival of n and $W_{n+1} = (W_n + 1 - K_n)^+$ just before the arrival of $n+1$. Thus (7.1) holds with $X_n = 1 - K_n$ and the independence of the X_n is also immediate. We have here $E = \mathbb{N}$ and letting $r_n = q_{n+1} + q_{n+2} + \ldots$, the transition matrix of $\{W_n\}$ is immediately seen to be

$$\begin{pmatrix} r_0 & q_0 & 0 & 0 & \\ r_1 & q_1 & q_0 & 0 & \cdots \\ r_2 & q_2 & q_1 & q_0 & \\ & \vdots & & & \end{pmatrix}. \tag{7.4}$$

$\square$

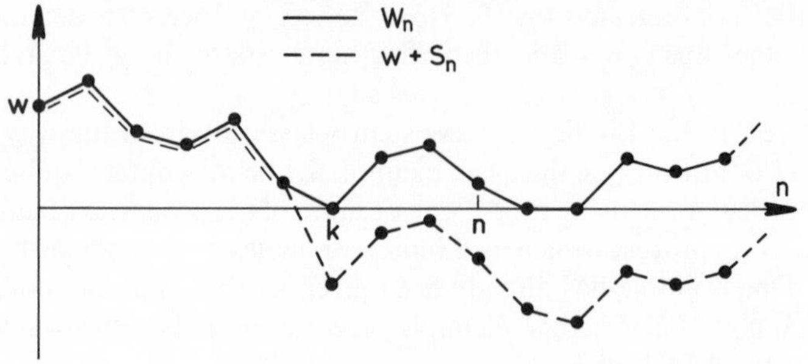

Fig. 7.1

Now define $S_0 = 0$, $S_n = X_0 + \cdots + X_{n-1}$. Then (7.1) reflects that the Lindley process has the same transition mechanism as the random walk $\{S_n\}$ except when the random walk crosses from positive to negative values (the Lindley process then stays at zero). The relation is illustrated in Fig. 7.1. It is, by the way, typical for many queueing processes that they are non-negative, but can be described by modification at (or near) zero of a process on the whole line with some basic simple structure (such modifications may, however, be more complex than in the present case).

Exploiting the relation between the paths of $\{W_n\}$ and $\{S_n\}$ even further yields:

Proposition 7.3. $W_n = \max(W_0 + S_n, S_n - S_1, \ldots, S_n - S_{n-1}, 0)$.

Proof. By (7.1), the increments of $\{W_n\}$ are at least those of $\{S_n\}$ so that

$$W_n - W_{n-k} \geqslant S_n - S_{n-k} \qquad k = 0, \ldots, n. \tag{7.5}$$

Letting $k = n$ yields $W_n \geqslant W_0 + S_n$, and using $W_{n-k} \geqslant 0$ we get $W_n \geqslant S_n - S_{n-k}$, proving '$\geqslant$'. For the converse, we shall show that either $W_n = W_0 + S_n$ or $W_n = S_n - S_{n-k}$ for some k. The first case occurs apparently if $W_0 + S_k \geqslant 0$, $k \leqslant n$. Otherwise $W_l = 0$ for some $l \leqslant n$ and letting k be the last such l, (7.1) yields $W_n = S_n - S_{n-k}$, see Fig. 7.1. □

Now define $M_n = \max_{0 \leqslant k \leqslant n} S_k$, $M = \max_{0 \leqslant k < \infty} S_k$. Since the distribution of $(S_n, S_n - S_1, \ldots, S_n - S_{n-1}, 0)$ is the same as that of $(S_n, S_{n-1}, \ldots, S_1, S_0 = 0)$, we get

Corollary 7.4. $W_n \stackrel{\mathscr{D}}{=} \max(W_0 + S_n, M_{n-1})$. *In particular if* $W_0 = 0$, *then* $W_n \stackrel{\mathscr{D}}{=} M_n$. It should be noted that this holds in the sense of one-dimensional distributions only and not processes. For example in the case $W_0 = 0$, the paths of $\{M_n\}$ are non-decreasing, those of $\{W_n\}$ not.

Suppose now $\mathbb{E}|X_n| < \infty$ and let $\mu = \mathbb{E}X_n$.

Corollary 7.5. *If* $\mu < 0$, *then* $M < \infty$ *a.s. and* $W_n \stackrel{\mathscr{D}}{\to} M$.

Proof. From $S_n/n \to \mu$ we have $S_n \to -\infty$. This implies in particular $M < \infty$.

Also $W_0 + S_n \to -\infty$ and $M_n \uparrow M$ a.s. and in distribution, and thus $\max(W_0 + S_n, M_{n-1}) = M_{n-1}$ eventually and $W_n \xrightarrow{\mathcal{D}} M$ follows. □

For the limiting behaviour for $\mu \geqslant 0$, see Problem 7.7.

Corollary 7.6. *If $\mu < 0$, then $M \stackrel{\mathcal{D}}{=} (M + X)^+$ where X is independent of M and governed by F. Furthermore $H(m) = \mathbb{P}(M \leqslant m)$ is the unique distribution function on $[0, \infty)$ which solves Lindley's integral equation*

$$H(m) = \int_{-\infty}^{m} H(m-x)\,\mathrm{d}F(x), \qquad m \geqslant 0 \tag{7.6}$$

Proof. The first statement can be proved by a limiting argument or from

$$(M+X)^+ = \max(0, M+X) = \max(0, X, X+X_0, X+X_0+X_1, \ldots)$$

$$\stackrel{\mathcal{D}}{=} \max(0, X_0, X_0+X_1, X_0+X_1+X_2, \ldots) = M.$$

Furthermore the r.h.s. of (7.6) is just $\mathbb{P}(M+X)^+ \leqslant m = \mathbb{P}(M + X \leqslant m)$ evaluated by conditioning upon $X = x$ and thus (7.6) is equivalent to $M \stackrel{\mathcal{D}}{=} (M+X)^+$, i.e. H being stationary for $\{W_n\}$. Thus if H_1, H_2 solve (7.6), we may consider the two stationary chains with initial distributions H_1 and H_2 respectively. From the fact that they both converge in distribution to M we conclude that $H_1 = H_2$. □

It is non-trivial, even in such simple models as Example 7.2 or the doubly exponential $M/M/1$ case in Example 7.1, to derive the distribution H of M and we return to the problem in Chapters VII and IX (of course, if one guessed H one can just check whether H solves Lindley's integral equation).

Problems

7.1. Consider $GI/M/m$ and let W_n be the queue length just before the nth arrival. Show that $\{W_n\}$ is ergodic for $\rho < 1$ [hint: Foster's criteria].

7.2. Let $\{A_n\}_0^\infty$, $\{B_n\}_0^\infty$ be independent sequences of independent identically distributed random variables and define $W_0 = B_0$, $W_{n+1} = (W_n - A_n)^+ + B_{n+1}$. Show that $\{W_n - B_n\}$ is a Lindley process corresponding to $X_n = B_n - A_n$. Show that if $\mathbb{E}B_n < \mathbb{E}A_n$, then in the limit W_n is distributed as $M + B$, where M has the distribution of $\sup S_n$, B that of B_n and M, B are independent.

7.3. Consider the *fixed cycle traffic light*, with the cycle divided into the *green period* where customers (say cars or pedestrians) can pass and the *red period* where they cannot. Let W_n^{G} be number of customers just after the start of the nth green period and B_n^{G} the number of customers arriving during the nth green period (similar conventions define $W_n^{\mathrm{R}}, B_n^{\mathrm{R}}$). Assuming that the maximal number of customers which can pass during a green period is some fixed number p, show that

$$W_{n+1}^{\mathrm{R}} = (W_n^{\mathrm{R}} + B_{n+1}^{\mathrm{G}} + B_n^{\mathrm{R}} - p)^+,$$

$$W_{n+1}^{\mathrm{G}} = (W_n^{\mathrm{G}} + B_n^{\mathrm{G}} - p)^+ + B_n^{\mathrm{R}}.$$

7.4. Consider the $M/G/1$ queue and let W_n be the queue length just after the nth departure, B_n the number of customers arriving during the nth service period. Show that $W_{n+1} = (W_n - 1)^+ + B_{n+1}$ and find the distribution of B_n in terms of a formula similar to (7.3). Show also in the $M/M/1$ case with $\rho < 1$ that the stationary distribution is $\pi_n = (1-\rho)\rho^n$.

7.5. Assume that F has negative mean and the density $\mathrm{d}F(x)/\mathrm{d}x = p\delta\,\mathrm{e}^{-\delta x}$ $(p = 1 - F(0))$ on $(0, \infty)$. Let $\eta \in (0, \delta)$ satisfy $\int_{-\infty}^{\infty} \mathrm{e}^{\eta x}\,\mathrm{d}F(x) = 1$. Show by direct calculation that $H(m) = 1 - (1 - \eta/\delta)\mathrm{e}^{-\eta m}$, $m \geqslant 0$, is the unique solution to Lindley's integral equation. Show hereby that the steady-state $GI/M/1$ actual waiting-time distribution is of this form.
7.6. Compute the matrix (7.4) for the $M/M/1$ case and show that $\pi_n = (1 - \rho)\rho^n$ is stationary for $\rho < 1$.
7.7. Consider a Lindley process with $\mu > 0$. Show that $W_n/n \to \mu$ a.s. Show also that the process is a null-recurrent Markov chain in the lattice case with $\mu = 0$ [hint: let $\tau = \inf\{n \geqslant 1 : S_n \leqslant 0\}$ and show that $\mathbb{E}\tau < \infty$ contradicts Wald's identity].

Notes

Among textbooks with systematic treatments of Lindley processes we mention in particular Feller (1971, VI.9) and Borovkov (1976).

8. RANDOM WALKS AND LINDLEY PROCESSES IN CONTINUOUS TIME

A random walk in continuous time is naturally defined as a process $\{S_t\}_{t \geqslant 0}$ with $S_0 = 0$ and having stationary independent increments. The structure of such processes (frequently called *Lévy processes*) is by now well known and closely related to the Lévy–Khinchine representation of infinitely divisible distributions. In fact, one knows that without loss of generality the process may be taken to have D-paths and to be of the form

$$S_t = \theta t + \sigma B_t + J_t \tag{8.1}$$

where $\{B_t\}$ is standard Brownian motion, $\sigma^2 \geqslant 0$, and J_t is a pure jump process, say with jump measure ν. This is more precisely to be interpreted as jumps of size x occurring at rate $\nu(\mathrm{d}x)$ in a certain sense. For example, if $\beta = \|\nu\| < \infty$, then jumps occur at rate β according to a Poisson process, and the sizes of jumps are independent identically distributed and governed by $G = \nu/\beta$. That is, $\{J_t\}$ is a compound Poisson process. The second simplest example is

$$\int_{-\infty}^{\infty} |y| \wedge 1\ \nu(\mathrm{d}y) < \infty, \tag{8.2}$$

which is known to correspond to $\{J_t\}$ having paths of locally bounded variation. We may then take

$$J_t = \sum_{i: t_i \leqslant t} y_i \tag{8.3}$$

where $\{(y_i, t_i)\}$ are the points in a bivariate Poisson process with intensity measure $\nu(\mathrm{d}y) \otimes \mathrm{d}t$, the assumption (8.2) ensuring (8.3) to converge absolutely. The theory may be generalized by requiring only ν-integrability of $y^2 \wedge 1$, but becomes then substantially more involved, and in fact (8.2) is more than sufficient for our purposes.

The definition of the corresponding Lindley process $\{W_t\}$ now requires that in some heuristical sense $W_{t+\mathrm{d}t} = (W_t + \mathrm{d}S_t)^+$. We shall not attempt to attach rigorous sense to this relation, but take a short cut inspired by Proposition 7.3 by simply defining

$$W_t = (W_0 + S_t) \vee \max_{0 \leqslant s \leqslant t} (S_t - S_s). \tag{8.4}$$

Proposition 8.1. *The Lindley process* $\{W_t\}_{t \geqslant 0}$ *defined by* (8.4) *is strong Markov.*

Proof. If for fixed t we let $\tilde{W}_0 = W_t$, $\tilde{S}_u = S_{t+u} - S_t$, we get

$$\begin{aligned}
&(\tilde{W}_0 + \tilde{S}_s) \vee \max_{0 \leqslant u \leqslant s} (\tilde{S}_s - \tilde{S}_u) \\
&\quad = (\tilde{W}_0 + S_{t+s} - S_t) \vee \max_{t \leqslant v \leqslant t+s} (S_{t+s} - S_v) \\
&\quad = (W_0 + S_{t+s}) \vee \max_{0 \leqslant v \leqslant t} (S_{t+s} - S_v) \vee \max_{t \leqslant v \leqslant t+s} (S_{t+s} - S_v)
\end{aligned}$$

which according to (8.4) is the same as W_{t+s}. That is, W_{t+s} is constructed pointwise from W_t and $\{\tilde{S}_u\}$ in the same way as W_s is constructed from W_0 and $\{S_u\}$, and the independence of $\{\tilde{S}_u\}$ and $\{W_v\}_{0 \leqslant v \leqslant t}$ then easily yields

$$\mathbb{E}[f(W_{t+s}) \mid W_v; 0 \leqslant v \leqslant t] = \mathbb{E}_{W_t} f(W_s)$$

and the Markov property. The strong Markov property can be easily seen from I.6.3. □

If the mean is well defined, we have $\mathbb{E}S_t = t\mu$ for some μ, and defining $M_T = \max_{0 \leqslant t \leqslant T} S_t$ we get exactly as in discrete time:

Proposition 8.2. $W_t \overset{\mathscr{D}}{=} (W_0 + S_t) \vee M_t$. *If* $\mu < 0$, *then* $M < \infty$ *and* $W_t \overset{\mathscr{D}}{\to} M$.

The problem that is left is to relate the development of the processes $\{W_t\}$ and $\{S_t\}$ in more intuitive terms, and thereby to identify $\{W_t\}$ in some main examples. We first note:

Proposition 8.3. *Define* $\omega = \inf\{t \geqslant 0 : W_0 + S_t \leqslant 0\}$. *Then also* $\omega = \inf\{t \geqslant 0 : W_t = 0\}$ *and* $W_t = W_0 + S_t$ *for* $t < \omega$.

Proof. For $s \leqslant t < \omega$ we have $W_0 + S_s > 0$, hence $W_0 + S_t > S_t - S_s$, so that by (8.4) $W_t = W_0 + S_t$. It is also clear that $S_\omega \leqslant S_s$ for all $s < \omega$, and (8.4) with $t = \omega$ then yields the second expression for ω. □

It thus remains to specify the development of $\{W_{\omega+t}\}_{t \geqslant \omega}$, i.e. in view of the strong Markov property that of $\{W_t\}$ subject to the starting condition $W_0 = 0$, which will be assumed for the following. This problem can in general be quite complicated, as is seen from the examples below (in particular 8.9 but to some extent also 8.6).

Example 8.4. Suppose that the positive increments of $\{S_t\}$ occur according to a compound Poisson process $\{S_t^{\uparrow}\}$ specified by β, B. That is, in (8.1) we have $\theta \leqslant 0$, $\sigma = 0$ and $\nu(\mathrm{d}x) = \beta B(\mathrm{d}x)$, $x > 0$, with $B(0) = 0$ ($\nu(\mathrm{d}x)$ may be arbitrary for $x < 0$). Then, letting τ be the time of the first upward jump of S, we have $\tau > 0$ and $\{S_t\}$ is non-increasing on $[0, \tau)$, i.e. $S_t \leqslant S_s$, $s \leqslant t < \tau$. Hence $W_t = 0$ for $t < \tau$, and it can also be easily seen from (8.4) that at time τ $\{W_t\}$ jumps from 0 to $S_\tau - S_{\tau-} = S_\tau^\uparrow$. That is, the process exits from 0 according to the intensity β and chooses the value in $(0, \infty)$ according to B. Here are two main cases:

(a) $\theta = -1$, $\nu(\mathrm{d}x) = 0$, $x < 0$. In conjunction with Proposition 8.3 it is seen that $\{W_t\}$ has the same upward jumps as $\{S_t^\uparrow\}$ and decreases at a linear rate in states $w > 0$. This description shows that $\{W_t\}$ *is simply the* $M/G/1$ *virtual waiting-time process*, cf. Section 1(d).

(b) $\theta = 0$, B is concentrated at 1, ν has an atom of size δ at -1 and is otherwise zero on $(-\infty, 0)$. That is, $S_t = S_t^\uparrow - S_t^\downarrow$, where $S_t^\uparrow, S_t^\downarrow$ are Poisson processes with intensities β and δ respect. Considering only initial values $W_0 = 0, 1, 2, \ldots$, we have a Markov process with states $0, 1, 2, \ldots$ with intensities $\lambda(i, i-1) = \delta$, $i = 1, 2, \ldots$, $\lambda(i, i+1) = \beta$, $i = 0, 1, \ldots$. That is, $\{W_t\}$ *is the* $M/M/1$ *queue length process*, cf. Section 3(a). □

In the next example as well as in later chapters we shall need:

Lemma 8.5. *If* $\{S_t\}$ *is standard Brownian motion, then the joint distribution of* S_t *and* M_t *is given by*

$$\mathbb{P}(S_t \leqslant x - y, M_t \geqslant x) = \mathbb{P}(S_t \geqslant x + y) \tag{8.5}$$

for $x, y \geqslant 0$. *Also* $M_t \stackrel{\mathscr{D}}{=} |S_t|$.

Proof. Define $\tau(x) = \inf\{t \geqslant 0 : S_t \geqslant x\}$. Since $M_t \geqslant x$ is equivalent to $\tau(x) \leqslant t$ and is automatic if $S_t \geqslant x + y$, we may rewrite (8.5) as

$$\mathbb{P}(S_t \leqslant x - y, \tau(x) \leqslant t) = \mathbb{P}(S_t \geqslant x + y, \tau(x) \leqslant t).$$

The truth of this relation follows by the *reflection principle* which states that Brownian motion, being symmetric ($S_t \stackrel{\mathscr{D}}{=} -S_t$), is equally likely to proceed from $S_{\tau(x)} = x$ to levels $\geqslant x + y$ or $\leqslant x - y$ within $s = t - \tau(x)$ units of time (here we have also used the strong Markov property). We then get

$$\begin{aligned}\mathbb{P}(M_t \geqslant x) = \mathbb{P}(S_t \leqslant x \leqslant M_t) + \mathbb{P}(S_t > x) &= \mathbb{P}(S_t \geqslant x) + \mathbb{P}(S_t > x)\\ &= 2\mathbb{P}(S_t \geqslant x) = \mathbb{P}(|S_t| \geqslant x),\end{aligned}$$

using (8.5) with $y = 0$ in the second step. □

Example 8.6. Here $\{S_t\}$ is standard Brownian motion, i.e. $\theta = 0$, $\nu = 0$, $\sigma^2 = 1$ in (8.1). Here $\{W_t\}$ *reduces to Brownian motion with reflection at zero*, which we shall assume as known to the reader to be representable as $\{|W_0 + S_t|\}$ and being a strong Markov process. To verify this claim, it is sufficient to see that the transition functions are the same. According to Proposition 8.3 and the strong Markov property, we may assume $W_0 = 0$, and the desired conclusion $W_t \stackrel{\mathscr{D}}{=} |S_t|$ then follows from $W_t \stackrel{\mathscr{D}}{=} M_t$ and Lemma 8.5. □

When discussing processes with unbounded jump measures, we need:

Lemma 8.7. *If $\{J_t\}$ is a pure jump process with jump measure ν satisfying* (8.2), *then $J_t/t \to 0$ a.s. as $t \downarrow 0$.*

Proof. Denote by $Z(\nu)$ the limit (if existing) of J_t/t as $t \downarrow 0$ so that we must show $Z(\nu) = 0$ a.s. This is obvious if ν is concentrated on $[\varepsilon, \infty)$ since then $J_t = 0$ for all sufficiently small t, and since also clearly $Z(\nu)$ is additive in ν, we may assume that ν is concentrated on $(0, \varepsilon)$ for some $\varepsilon > 0$. Then in particular

$$\mathbb{E}J_t = t\int_0^\infty y\nu(\mathrm{d}y) < \infty \tag{8.6}$$

and we shall need the fact that $\{J_t/t\}_{t>0}$ is a backwards martingale, i.e.

$$\mathbb{E}\left(\frac{J_s}{s}\middle|\mathscr{F}_t\right) = \frac{J_t}{t}, \qquad 0 < s < t, \tag{8.7}$$

where $\mathscr{F}_t = \sigma(J_s; s \geqslant t)$. To show this, assume first that s/t is rational, say $s/t = n/m$ with $n, m \in \mathbb{N}$. We may then just mimic the proof of the backwards martingale property of sample means in discrete time: letting $X_i = J_{it/m} - J_{(i-1)t/m}$, $S_k = X_1 + \cdots + X_k$, we have $S_n = J_s$, $S_m = J_t$. Furthermore by a symmetry argument, $\mathbb{E}(X_i|\mathscr{F}_t)$ does not depend on $i = 1, \ldots, m$, hence must be $\mathbb{E}(S_m|\mathscr{F}_t)/m = S_m/m$, and from this (8.7) is immediately apparent. For the case of general s, t, fix s and let $t(n) \downarrow t$ in such a way that $s/t(n)$ is rational. Then by a standard martingale result we have a.s.

$$\mathbb{E}\left(\frac{J_s}{s}\middle|\mathscr{F}_t\right) = \lim \mathbb{E}\left(\frac{J_s}{s}\middle|\mathscr{F}_{t(n)}\right) = \lim \frac{J_{t(n)}}{t(n)} = \frac{J_t}{t}$$

(the last identity by right continuity). Since $\{J_t/t\}$ has D-paths and is non-negative, it thus follows by the martingale convergence theorem in continuous time that $Z(\nu)$ exists and satisfies $\mathbb{E}Z(\nu) \leqslant \mathbb{E}J_1$.

Now if $\nu^{(n)}$ denotes the restriction of ν to $(0, 1/n)$ and the corresponding $J_t^{(n)}$ is constructed as in (8.3) from the same bivariate Poisson process by neglecting terms with $y_i > 1/n$, it is clear that $J_t^{(n)} = J_t$ for all sufficiently small t, since there are only a finite number of jumps $> 1/n$ in each bounded interval. Hence $Z(\nu) = Z(\nu^{(n)})$, and by (8.6),

$$\mathbb{E}Z(\nu) = \lim_{n\to\infty} \mathbb{E}Z(\nu^{(n)}) \leqslant \lim_n \mathbb{E}J_1^{(n)} = \lim_n \int_0^{1/n} y\nu(\mathrm{d}y) = 0,$$

so that $Z(\nu) = 0$. □

Example 8.8. Let $S_t = \theta t - J_t$, where $\theta > 0$ and the jump measure of $\{J_t\}$ is infinite, concentrated on $(0, \infty)$ and satisfies (8.2). Then by Lemma 8.7 we have $S_s > 0$ for sufficiently small s, say $0 < s \leqslant t_0$, and hence $S_t - S_s \leqslant S_t$ and $W_t = S_t$ for $t \leqslant t_0$ (recalling that $W_0 = 0$). That is, *$\{W_t\}$ leaves state 0 instantaneously and in the same way as $\{S_t\}$.* □

Example 8.9. Let $S_t = J_t - \theta t$, where $\theta > 0$, say $\theta = 1$, and the jump measure of $\{J_t\}$ is infinite, concentrated on $(0, \infty)$ and satisfies (8.2). We shall see that this case is much more complicated than Example 8.8 in the sense that $\tau_+ = \inf\{t > 0: W_t > 0\}$ *is still zero but that for any* $\varepsilon > 0$ *the Lebesgue measure of* $A_\varepsilon = \{t \in [0, \varepsilon]: W_t = 0\}$ *is non-zero.* To this end, we let $\tilde{J}_t^{(n)}$, $\tilde{W}_t^{(n)}$, etc. refer to the case where all jumps $\leqslant 1/n$ have been truncated (i.e. in (8.3) we neglect the terms with $y_i \leqslant 1/n$). Clearly, $\tilde{J}_t^{(n)} - \tilde{J}_s^{(n)} \uparrow J_t - J_s$ for $s < t$ and hence $\tilde{W}_t^{(n)} \uparrow W_t$. Now since $\{\tilde{J}_t^{(n)}\}$ is compound Poisson, each $\{\tilde{W}_t^{(n)}\}$ is simply of the $M/G/1$ virtual waiting-time form in Example 8.4, i.e. decreases at a unit rate in states > 0 and has the same upward jumps as $\{\tilde{J}_t^{(n)}\}$. From this it is obvious that

$$\tilde{W}_t^{(n)} = \tilde{J}_t^{(n)} - \int_0^t I(\tilde{W}_s^{(n)} > 0)\mathrm{d}s \qquad (\tilde{W}_0^{(n)} = 0)$$

and we may pass to the limit $n \to \infty$ to get

$$W_t = J_t - \int_0^t I(W_s > 0)\mathrm{d}s \qquad (W_0 = 0). \tag{8.8}$$

Now let $\varepsilon > 0$ satisfy $J_t \leqslant t/2$, $t \leqslant \varepsilon$, and assume $|A_\varepsilon| = 0$. Then (8.8) yields $W_\varepsilon = J_\varepsilon - \varepsilon < 0$ which is impossible. Also, $\tau_+ = 0$ a.s. follows, since if $\{J_t\}$ jumps say δ at time t, then (8.4) shows that $W_t \geqslant \delta$, and the jump times have 0 as the accumulation point. □

Problems

8.1. Show that the process of number of stages in $M/E_k/1$ is a continuous-time Lindley process.

Notes

Lévy processes are a standard topic. One of the standard survey treatments is Fristedt (1974). The literature on Lindley processes in continuous time is somewhat more scarce and scattered. Aspects of the topic are, however, closely related to dam and storage models, see Prabhu (1980) and Chapter XIII and references therein.

9. TIME-DEPENDENT PROPERTIES OF $M/M/1$

9a. The doubly infinite queue and its maximum

The key to our more refined study of the $M/M/1$ queue length process $\{X_t\}_{t \geqslant 0}$ is the Lindley process representation in Example 8.4. Combining with Proposition 8.2, we have:

Proposition 9.1. *The distribution of the $M/M/1$ queue length X_t at time t given*

$X_0 = i$ is that of $\max(i + S_t, M_t)$ *where $S_t = B_t - D_t$ is the difference between two Poisson processes with intensities β and δ, and $M_t = \sup_{0 \leqslant s \leqslant t} S_s$.*

The process $\{S_t\}_{t \geqslant 0}$ is frequently denoted as the *doubly infinite queue*. It models, for example, a queueing situation with taxis and passengers in front of a railway station, with B_t denoting the number of customers arriving before t and D_t the number of taxis. Thus if at time t $S_t > 0$, there is a queue of length S_t of passengers, whereas if $S_t < 0$ there is a queue of length $-S_t$ of taxis.

Letting $M = \sup_{0 \leqslant t < \infty} S_t$, the following simple observation will be useful in the following:

Proposition 9.2. *Let $\rho = \beta/\delta$. Then a.s.*: (i) $S_t \to -\infty$, $M < \infty$ *for* $\rho < 1$; (ii) $S_t \to +\infty$, $M = \infty$ *for* $\rho > 1$; (iii) $\overline{\lim}_{t\to\infty} S_t = +\infty$, $\underline{\lim}_{t\to\infty} S_t = -\infty$, $M = \infty$ *for* $\rho = 1$.

Proof. Let T_n be the value of S_t just after the nth jump, $T_0 = 0$. Then $\{T_n\}$ is a Bernoulli random walk with

$$p = \frac{\beta}{\beta + \delta} = \frac{\rho}{1 + \rho}, \qquad q = 1 - p = \frac{\delta}{\beta + \delta} = \frac{1}{1 + \rho}.$$

Hence if $\rho < 1$, $\mathbb{E}T_1 = p - q < 0$ and by the LLN $T_n/n \to p - q$, implying $T_n \to -\infty$, $S_t \to -\infty$ and hence $M < \infty$. The case $\rho > 1$ is treated similarly. For $\rho = 1$, we can appeal, say, to the LIL for T_n (a direct proof is also easy, cf. Problem 9.1).

□

Despite the simple relation to the Poisson distribution, the explicit form of the point probabilities of S_t is not elementary. Define the *modified Bessel function of integer order n* by

$$I_n(x) = \sum_{k=0}^{\infty} \frac{(x/2)^{2k+n}}{k!(k+n)!}, \qquad I_{-n}(x) = I_n(x), \qquad n \in \mathbb{N},$$

and let $\mu = \sqrt{\beta\delta}$, $\rho = \beta/\delta$. The argument of I_n will be $x = 2\mu t$ throughout, so unless otherwise stated I_n just denotes $I_n(2\mu t)$, and we shall let $\iota_n = \exp\{-(\beta + \delta)t\}\rho^{n/2} I_n$, $n \in \mathbb{Z}$, so that

$$\iota_{-n} = \rho^{-n}\iota_n, \qquad n \in \mathbb{Z}. \tag{9.1}$$

As a technical tool (a particular case of a method studied in Chapter XII), we shall use a process with β, δ both replaced by μ. This is denoted by $\mathbb{P}_0$ and has traffic intensity 1 and the same value of μ, and we have

$$\begin{aligned} \mathbb{P}(B_t = l, D_t = k) &= \exp\{-(\beta + \delta)t\} \frac{(\beta t)^l}{l!} \frac{(\delta t)^k}{k!} \\ &= \exp\{(2\mu - \beta - \delta)t\} \rho^{(l-k)/2} \mathbb{P}_0(B_t = l, D_t = k) \end{aligned} \tag{9.2}$$

Proposition 9.3.

$$\mathbb{P}(S_t = n) = \exp\{(2\mu - \beta - \delta)t\} \rho^{n/2} \mathbb{P}_0(S_t = n) = \iota_n, \qquad n \in \mathbb{Z}.$$

Proof. For $n \geqslant 0$ we get

$$\begin{aligned}\mathbb{P}(S_t = n) &= \sum_{k=0}^{\infty} \mathbb{P}(B_t = n+k, D_t = k) = \exp\{(2\mu - \beta - \delta)t\}\rho^{n/2}\mathbb{P}_0(S_t = n)\\ &= \exp\{(2\mu - \beta - \delta)t\}\rho^{n/2}\sum_{k=0}^{\infty} \mathrm{e}^{-\mu t}\frac{(\mu t)^{n+k}}{(n+k)!}\mathrm{e}^{-\mu t}\frac{(\mu t)^k}{k!}\\ &= \exp\{-(\beta+\delta)t\}\rho^{n/2}I_n = \iota_n\end{aligned}$$

The case $n \leqslant 0$ is treated similarly or by a symmetry argument (when passing from S_t to $-S_t$, the arrival and service intensities are interchanged and ρ changed to ρ^{-1}). □

As a corollary, which will be used in formula manipulations in the following, we note the identity

$$1 = \sum_{n=-\infty}^{\infty} \iota_n. \tag{9.3}$$

In order to apply Proposition 9.1, we have more generally to find the joint distribution of S_t and M_t:

Lemma 9.4. *For* $n + m \geqslant 0$, $m \geqslant 0$

$$\mathbb{P}(S_t = n, M_t \geqslant n+m) = \rho^{-m}\iota_{n+2m} \tag{9.4}$$

$$\mathbb{P}(S_t = n, M_t = n+m) = \rho^{-m}(\iota_{n+2m} - \rho^{-1}\iota_{n+2m+2}). \tag{9.5}$$

Proof. Let $F = \{S_t = n, M_t \geqslant n+m\}$, $G_k = \{B_t = n+k, D_t = k\}$. Then conditionally upon G_k it depends solely on the order of the $n+k$ increments of S_t and the k decrements whether or not F occurs, cf. Fig. 9.1. But by well-known properties of the Poisson process, this ordering is determined by two independent samples of sizes $n+k$, k from the uniform distribution on $[0, t]$. Hence $\mathbb{P}(F \mid G_k)$ is independent of the intensities and in particular, $\mathbb{P}(F \mid G_k) = \mathbb{P}_0(F \mid G_k)$ so that using

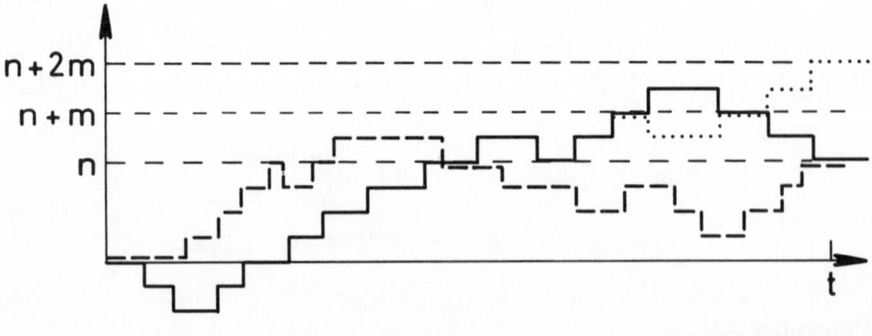

Fig. 9.1 The path (---) has the same B_t and D_t as (———), but a different M_t. The path (...) is obtained by reflection after $n+m$ has been reached

(9.2) we get

$$\mathbb{P}F = \sum_{k:k\geqslant 0, n+k\geqslant 0} \mathbb{P}(FG_k) = \sum \frac{\mathbb{P}G_k}{\mathbb{P}_0 G_k} \mathbb{P}_0(FG_k) = \exp\{(2\mu - \beta - \delta)t\}\rho^{n/2}\mathbb{P}_0 F.$$

Now since the $\mathbb{P}_0$-process is symmetric, it follows by the reflection principle just as for Brownian motion in Lemma 8.5 (cf. also Fig. 9.1) that

$$\mathbb{P}_0 F = \mathbb{P}_0(S_t = n + 2m) = \mathrm{e}^{-2\mu t} I_{n+2m}.$$

Hence

$$\mathbb{P}F = \exp\{(2\mu - \beta - \delta)t\}\rho^{n/2}\mathrm{e}^{-2\mu t} I_{n+2m} = \rho^{-m}\iota_{n+2m}$$

and (9.4) follows. Finally (9.5) is a consequence of (9.4). □

9b. The transition probabilities

Proposition 9.1 and Lemma 9.4 solve in principle the problem of evaluating

$$p_{ij}^t = \mathbb{P}(X_t = j \mid X_0 = i) = \mathbb{P}F, \qquad F = \{\max[i + S_t, M_t] = j\},$$

and it only remains to collect terms. Now F is the disjoint union of

$$F_1 = \{S_t = j - i \leqslant M_t \leqslant j\} \quad \text{and} \quad F_2 = \{S_t < j - i, M_t = j\}.$$

But $\mathbb{P}F_1, \mathbb{P}F_2$ are given by

$$\sum_{m=0}^{i} \mathbb{P}(S_t = j - i, M_t = j - i + m) = \sum_{m=0}^{i} \rho^{-m}\{\iota_{j-i+2m} - \rho^{-1}\iota_{j-i+2m+2}\}$$

$$= \iota_{j-i} - \rho^{-i-1}\iota_{j+i+2} = \iota_{j-i} - \rho^{j+1}\iota_{-i-j-2}$$

and

$$\sum_{n=-\infty}^{j-i-1} \rho^{n-j}\{\iota_{n+2(j-n)} - \rho^{-1}\iota_{n+2(j-n)+2}\} = \rho^j \sum_{n=-\infty}^{-j-i-1} \iota_n - \rho^{j+1} \sum_{n=-\infty}^{-j-i-1} \iota_{n-2}$$

$$= \rho^{-i-1}\iota_{i+j+1} + \rho^j \sum_{n=-\infty}^{-j-i-2} \iota_n - \rho^{j+1} \sum_{n=-\infty}^{-j-i-3} \iota_n,$$

where we have used (9.1) repeatedly. Summing, we have proved:

Theorem 9.5. *In the $M/M/1$ queue with $0 < \rho < \infty$, $p_{ij}^t = \mathbb{P}(X_t = j \mid X_0 = i)$ is given by*

$$\iota_{j-i} + \rho^{-i-1}\iota_{i+j+1} + (1 - \rho)\rho^j \sum_{n=-\infty}^{-j-i-2} \iota_n. \tag{9.6}$$

Analytical manipulations or alternative derivations provide a number of alternative expressions for p_{ij}^t, for example

$$(1 - \rho)\rho^j + \iota_{j-i} - \rho^{(j-i)/2} J, \tag{9.7}$$

where

$$J = \int_t^\infty \exp\{-(\beta+\delta)s\}\{I_{i+j}(2\mu s) - 2\rho^{1/2}I_{i+j+1}(2\mu s) + \beta I_{i+j+2}(2\mu s)\}\mathrm{d}s,$$

$$(1-\rho)\rho^j I(\rho<1) + 2\exp\{-(\beta+\delta)t\}\rho^{(j-i)/2}$$
$$\times \frac{1}{\pi}\int_0^\pi \frac{\exp\{2\mu t\cos\theta\}}{(1-2\rho^{1/2}\cos\theta+1)} g_i(\theta)g_j(\theta)\mathrm{d}\theta, \tag{9.8}$$

where $g_i(\theta) = \sin i\theta - \rho^{1/2}\sin(i+1)\theta$ (see Cohen, 1982, p. 178, for (9.7) and Takács, 1962, p. 23, for (9.8)). That trigonometric functions come in, may be understood from the standard analytical identity

$$I_n(t) = \frac{1}{\pi}\int_0^\pi \mathrm{e}^{t\cos\theta}\cos n\theta\,\mathrm{d}\theta.$$

The message of these formulae, as well as those for the busy-period distribution to be derived shortly, is perhaps not so much their particular form, but rather that they are extremely complicated. The $M/M/1$ queue being the very simplest queueing system, this probably suggests that time-dependent explicit solutions are in general not possible and indeed this is the case. Even the numerical evaluation on a computer requires some care and the most feasible approach may be to apply numerical integration of formulae like (9.8), thereby avoiding manipulation of infinite sums or integrals of Bessel functions of high order.

9c. The busy-period distribution

We first prove

Proposition 9.6. *The first passage time* $\tau = \inf\{t>0: S_t = 1\}$ *of the doubly infinite queue from* 0 *to* 1 *has density*

$$f(t) = \beta\exp\{-(\beta+\delta)t\}[I_0(2\mu t) - I_2(2\mu t)]$$
$$= \frac{\rho^{1/2}}{t}\exp\{-(\beta+\delta)t\}I_1(2\mu t). \tag{9.9}$$

Note that if $\rho<1$, then by Proposition 9.2 S_t drifts to $-\infty$ and hence (9.9) is defective, $\mathbb{P}(\tau<\infty) = \int f < 1$. More precisely, from $\mathbb{P}(\tau<\infty) = \mathbb{P}(M\geqslant 1) = \mathbb{P}_e(X_t\geqslant 1)$ it follows that $\int f = \rho$, $\rho<1$.

Proof. Using (9.5) we get

$$\mathbb{P}(\tau>t) = \mathbb{P}(M_t=0) = \sum_{n=-\infty}^{0}\mathbb{P}(S_t=n, M_t=0)$$
$$= \sum_{n=-\infty}^{0}\rho^n(\iota_{-n} - \rho^{-1}\iota_{2-n}) = C_0(t) - \rho C_2(t),$$

where

$$C_N(t) = \sum_{n=N}^{\infty} \exp\{-(\beta+\delta)t\}\rho^{-n/2} I_n(2\mu t).$$

Hence $f(t) = \rho C_2'(t) - C_0'(t)$ and to evaluate the derivatives, we need the formulae

$$I_0'(t) = I_1(t), \quad I_n'(t) = \tfrac{1}{2}[I_{n-1}(t) + I_{n+1}(t)] \qquad n = 1, 2, \ldots, \tag{9.10}$$

$$I_{n-1}(t) - I_{n+1}(t) = \frac{2n}{t} I_n(t) \qquad n = 1, 2, \ldots, \tag{9.11}$$

which may easily be seen from the power series definition of I_n. Using (9.10), we get for $N \geqslant 1$

$$\begin{aligned} C_N'(t) &= -(\beta+\delta)C_N(t) + \sum_{n=N}^{\infty} \exp\{-(\beta+\delta)t\}\rho^{-n/2}\mu(I_{n-1} + I_{n+1}) \\ &= -(\beta+\delta)C_N(t) + \delta C_{N-1}(t) + \beta C_{N+1}(t) \\ &= \exp\{-(\beta+\delta)t\}\{\delta\rho^{-(N-1)/2} I_{N-1} - \beta\rho^{-N/2} I_N\}. \end{aligned}$$

Hence

$$\begin{aligned} f(t) &= \rho C_2'(t) - C_1'(t) - \frac{d}{dt}\exp\{-(\beta+\delta)t\} I_0(2\mu t) \\ &= \exp\{-(\beta+\delta)t\}[\delta\rho^{1/2} I_1 - \beta I_2 - \delta I_0 + \beta\rho^{-1/2} I_1 \\ &\qquad + (\beta+\delta)I_0 - 2\mu I_1] \\ &= \exp\{-(\beta+\delta)t\}\beta(I_0 - I_2) \end{aligned}$$

and combining with (9.11), the proof is complete. □

We understand the *busy period* G of the queue to be the time from when a customer enters an empty system until the system is empty again. The busy period is followed by an interval of length H where the system is empty, the *idle period*, and $G + H$ constitute the *busy cycle*, see Fig. 9.2. (the notation G, H is used only at this place). In the $M/M/1$ case, it is clear that G and H are independent and H exponentially distributed with intensity β. Furthermore, we may identify G with the time of passage of the doubly infinite queue from 0 to -1. The distribution of this follows by a symmetry argument since we just have to interchange β and δ in Proposition 9.6, and thus:

Corollary 9.7. *The busy-period distribution* (defective if $\rho > 1$) *of the* $M/M/1$ *queue*

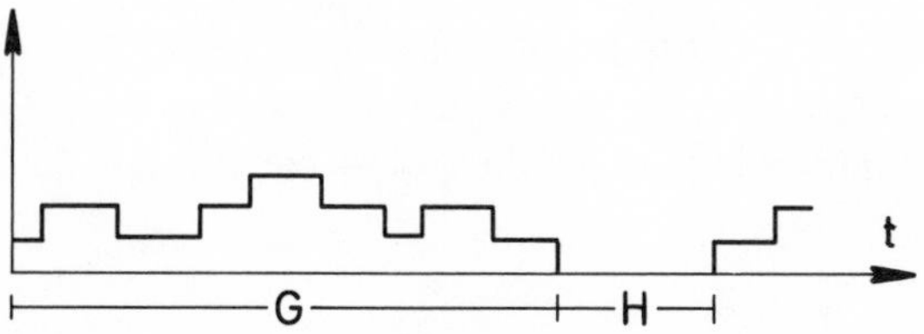

Fig. 9.2

is given by the density

$$g(t) = \delta \exp\{-(\beta+\delta)t\}[I_0(2\mu t) - I_2(2\mu t)]$$
$$= \frac{\rho^{-1/2}}{t} \exp\{-(\beta+\delta)t\} I_1(2\mu t). \tag{9.12}$$

As above, g is defective for $\rho > 1$. Moments will be derived shortly.

9d. Transform methods

In some cases, where quantities like the p_{ij}^t cannot be derived in closed analytical form, it may still be possible to find explicit expressions for transforms or double (bivariate) transforms. As an example, we quote for the present $M/M/1$ case the formula (e.g. Prabhu, 1965, Chapter 1.2a)

$$\varphi_i(\theta, s) = \int_0^\infty \sum_{j=0}^\infty e^{-\theta t} s^j p_{ij}^t \, dt = \frac{s^{i+1} - (1-s)\xi^{i+1}/(1-\xi)}{\beta(s-\xi)(\eta - s)} \tag{9.13}$$

where $\xi = \xi(\theta)$, $\eta = \eta(\theta)$ are the two roots of

$$\beta z^2 - (\beta + \delta + \theta) z + \delta = 0, \tag{9.14}$$

i.e.

$$\xi = \frac{\beta + \delta + \theta - R(\theta)}{2\beta}, \qquad \eta = \frac{\beta + \delta + \theta + R(\theta)}{2\beta},$$
$$R(\theta) = \sqrt{(\beta + \delta + \theta)^2 - 4\beta\delta}. \tag{9.15}$$

This can be proved, for example, by careful manipulation of the differential equations for the p_{ij}^t. Though more generally applicable, this method has serious drawbacks, however. In more complex cases, the expressions are even less transparent than (9.13) and their derivation may require much ingenuity. Also, even for the present case, the inversion of (9.13) is not easy whatever the purpose is formulae like (9.6)–(9.8) or numerical values.

We shall consider only one example where the transform methods work out in a quite elegant fashion, namely the distribution of the busy period G and its moments. We first note that

$$\mathbb{E}\, e^{\alpha S_t} = \mathbb{E} \exp\{\alpha(\beta_t - D_t)\} = \sum_{k,l=0}^\infty \exp\{\alpha(k-l)\}$$
$$\times \exp\{-(\beta+\delta)t\} \frac{(\beta t)^k}{k!} \frac{(\delta t)^l}{l!} = e^{t\psi(\alpha)},$$
$$\psi(\alpha) = \log \mathbb{E}\, e^{\alpha S_1} = \beta(e^\alpha - 1) + \delta(e^{-\alpha} - 1), \tag{9.16}$$

and we have:

Lemma 9.8. *For any* α, $\{Y_t\}_{t \geq 0} = \{\exp[\alpha S_t - t\psi(\alpha)]\}_{t \geq 0}$ *is a continuous time* *martingale and we have for* $\rho \leqslant 1$ *and* $\psi'(\alpha) \leqslant 0$ (i.e. $\alpha \leqslant -\log \rho/2$) *that*

$$1 = \mathbb{E}Y_0 = \mathbb{E}Y_G = \mathrm{e}^{-\alpha}\mathbb{E}\,\mathrm{e}^{-G\psi(\alpha)}. \qquad (9.17)$$

Proof. Since S_t has stationary independent increments, the martingale property follows from

$$\begin{aligned}&\mathbb{E}[\exp\{\alpha S_{t+s} - (t+s)\psi(\alpha)\} \,|\, (S_u)_{0\leqslant u\leqslant t}]\\ &= \exp\{\alpha S_t - t\psi(\alpha)\}\mathbb{E}[\exp\{\alpha(S_{t+s} - S_t) - s\psi(\alpha)\} \,|\, (S_u)_{0\leqslant u\leqslant t}]\\ &= \exp\{\alpha S_t - t\psi(\alpha)\}\mathbb{E}\exp\{\alpha S_s - s\psi(\alpha)\} = \exp\{\alpha S_t - t\psi(\alpha)\}.\end{aligned}$$

Now G, being the first passage time from 0 to -1 of $\{S_t\}$, is clearly a stopping time and the assumption $\rho \leqslant 1$ ensures $G < \infty$ a.s. Since clearly $Y_0 = 1$ and $S_G = -1$, the formula (9.17) thus follows by verifying the conditions of a suitable version of the optional stopping theorem. The details can be found in XII.4, where it also becomes apparent why the condition $\psi'(\alpha) \leqslant 0$ comes in. □

Proposition 9.9. *For* $\rho < 1$, *the Laplace transform of the* $M/M/1$ *busy period* G *is given by*

$$\mathbb{E}\,\mathrm{e}^{-\theta G} = \xi(\theta) = \frac{1}{2\beta}(\beta + \delta + \theta - \sqrt{(\beta + \delta + \theta)^2 - 4\beta\delta}), \qquad \theta \geqslant 0, \quad (9.18)$$

cf. (9.15). *In particular, the mean and variance are given by*

$$\mathbb{E}G = -\xi'(0) = \frac{1}{\delta(1-\rho)}, \qquad \mathbb{V}\mathrm{ar}\,G = \xi''(0) - \xi'(0)^2 = \frac{1+\rho}{\delta^2(1-\rho)^3}. \qquad (9.19)$$

Proof. Let $\tilde{\xi}(\theta) = \mathbb{E}\,\mathrm{e}^{-\theta G}$. It follows readily from (9.16) that $\psi(\alpha)$ decreases monotonically from ∞ to 0 on $(-\infty, 0]$. Hence for $\theta \geqslant 0$ we may find $\alpha \leqslant 0$ such that $\psi(\alpha) = \theta$ and (9.17) then yields $\tilde{\xi}(\theta) = \mathrm{e}^{\alpha}$. Now letting $\tilde{\xi} = \mathrm{e}^{\alpha}$ in (9.16) we get $\theta = \beta(\tilde{\xi} - 1) + \delta(\tilde{\xi}^{-1} - 1)$ which after some algebra shows that $\tilde{\xi}$ solves the quadratic (9.14). But in (9.15) we have $\eta(\theta) > 1$, $\theta > 0$, hence $\tilde{\xi} = \xi$. Finally (9.19) follows by some more algebra from $R(0) = \delta - \beta$ and

$$R'(\theta) = \frac{\beta + \delta + \theta}{R(\theta)}, \qquad \xi'(\theta) = \frac{1}{2\beta}\left(1 - \frac{\beta + \delta + \theta}{R(\theta)}\right),$$

$$\xi''(\theta) = \frac{1}{2\beta}\left[\frac{(\beta + \delta + \theta)^2 - R(\theta)^2}{R(\theta)^3}\right]. \qquad □$$

Distributional properties of the busy cycle follow easily as a corollary, since we just have to convolve the busy-period distribution with the distribution of the idle period which is exponential with intensity β. For example, the mean busy cycle is

$$\mathbb{E}G + \mathbb{E}H = \frac{1}{\delta(1-\rho)} + \frac{1}{\beta} = \frac{1}{\beta(1-\rho)}.$$

9e. The relaxation time

Suppose $\rho < 1$. We shall address ourselves to the question on the rate of approach of p^t_{ij}, as given by the formulae of Section 9(b), to its limiting value $(1-\rho)\rho^j$. To this end, we need the asymptotic properties of the Bessel functions:

Lemma 9.10. *As* $t \to \infty$

$$I_n(t) = \frac{e^t}{\sqrt{2\pi}}\left\{t^{-1/2} - t^{-3/2}\frac{4n^2-1}{8}\right\} + n^4 t^{-3/2} e^t o(1),$$

where the o(1) *terms here and in the proof are uniform in n.*

Proof. Letting $\beta = \delta = \mu = \frac{1}{2}$ for the moment, we shall appeal to the interpretation $I_n(t) = e^t \mathbb{P}(S_t = n)$, cf. Proposition 9.3, and apply the higher-order expansion in the local CLT for lattice distributions (see Bhattacharya and Rao, 1976, p. 231, or Gnedenko and Kolmogorov, 1954, p. 241; the result is stated there for discrete time random walks but is also valid in continuous time as may be seen by the same proof or by the method of discrete skeletons, cf. Appendix A9.2). To this end we need the cumulants κ_r of S_1, i.e. according to (9.16)

$$\kappa_r = \psi^{(r)}(0) = \frac{d^r}{d\alpha^r}\tfrac{1}{2}(e^\alpha + e^{-\alpha})|_{\alpha=0} = \begin{cases} 0 & r \text{ uneven} \\ 1 & r \text{ even} \end{cases}$$

Hence if $\varphi^{(r)}$ denotes the rth derivative of the standard normal density, evaluated at $y_n = n/\sqrt{t}$, we have

$$\begin{aligned}
\mathbb{P}_0(S_t = n) &= \frac{1}{\sqrt{t}}\left\{\varphi - \frac{1}{6\sqrt{t}}\frac{\kappa_3}{\kappa_2^{3/2}}\varphi^{(3)} + \frac{1}{24t}\left[\frac{\kappa_4}{\kappa_2^2}\varphi^{(4)} + \frac{\kappa_3^2}{3\kappa_2^3}\varphi^{(6)}\right]\right\} + t^{-3/2}o(1) \\
&= \frac{1}{\sqrt{t}}\left\{\varphi + \frac{\varphi^{(4)}}{24t}\right\} + t^{-3/2}o(1) = \frac{\varphi(y_n)}{\sqrt{t}}\left\{1 + \frac{y_n^4 - 6y_n^2 + 3}{24t}\right\} + t^{-3/2}o(1) \\
&= \frac{1}{\sqrt{2\pi t}}\left(1 - \frac{n^2}{2t} + n^4 t^{-2}o(1)\right)\left\{1 + \frac{1}{8t} + n^4 t^{-2}o(1)\right\} + t^{-3/2}o(1) \\
&= \frac{1}{\sqrt{2\pi t}}\left(1 - \frac{4n^2-1}{8t}\right) + n^4 t^{-3/2}o(1). \qquad \square
\end{aligned}$$

We can now evaluate the desired rate of convergence. Using (9.6), (9.3) and Lemma 9.10 we get

$$\begin{aligned}
(1-\rho)\rho^j - p^t_{ij} &= (1-\rho)\rho^j \sum_{n=-j-i-1}^{\infty} \iota_n - \iota_{j-i} - \rho^{-i-1}\iota_{i+j+1} \\
&= \exp\{-(\beta+\delta)t\}\left\{(1-\rho)\rho^j \sum_{n=-j-i-1}^{\infty} \rho^{n/2} I_n(2\mu t)\right. \\
&\qquad \left. - \rho^{(j-i)/2} I_{j-i}(2\mu t) - \rho^{(j-i+1)/2} I_{i+j+1}(2\mu t)\right\}
\end{aligned}$$

$$= \frac{\exp\{(2\mu-\beta-\delta)t\}}{\sqrt{2\pi}} \Big\{ (2\mu t)^{-1/2}[C_1(i,j) - \rho^{(j-i)/2} - \rho^{(j-i+1)/2}] + (2\mu t)^{-3/2} C_2(i,j) \Big\} + o(t^{-3/2}),$$

$$C_1(i,j) = (1-\rho)\rho^j \sum_{n=-j-i-1}^{\infty} \rho^{n/2} = (1-\rho)\rho^j \frac{\rho^{-(j+i+1)/2}}{1-\rho^{1/2}}$$

$$= \rho^{(j-i)/2} + \rho^{(j-i-1)/2},$$

$$C_2(i,j) = (1-\rho)\rho^j \sum_{n=-j-i-1}^{\infty} \rho^{n/2} \frac{4n^2-1}{8}$$

$$- \rho^{(j-i)/2} \frac{4(j-i)^2-1}{8} - \rho^{(j-i-1)/2} \frac{4(j+i+1)^2-1}{8}$$

(C_2 can be reduced, but we shall not carry out the tedious algebra). Hence the $t^{-1/2}$ term vanishes, and noting that $\beta + \delta - 2\mu = (\sqrt{\delta} - \sqrt{\beta})^2$, we have proved

Theorem 9.11. *If $\rho < 1$, then as $t \to \infty$*

$$p_{ij}^t = (1-\rho)\rho^j + \frac{\exp\{-(\sqrt{\beta}-\sqrt{\delta})^2 t\}}{4\sqrt{\pi(\beta\delta)^{3/2}}} t^{-3/2} C_2(i,j) + o\left(\frac{\exp\{-(\sqrt{\beta}-\sqrt{\delta})^2 t\}}{t^{3/2}}\right).$$

It is seen that the remainder term decreases essentially exponentially at rate $(\sqrt{\beta}-\sqrt{\delta})^2$ independently of i, j, and for this reason $(\sqrt{\beta}-\sqrt{\delta})^{-2}$ (or some multiple) is frequently denoted as the *relaxation time* of the system, measuring in some appropriate sense the time needed for the initial condition $X_0 = i$ to become unimportant and the system to relax in the steady state.

Problems

9.1. Show that if $\rho = 1$ in the proof of Proposition 9.2, then $\{T_n\}$ is a recurrent Markov chain (a) by explicit calculation of $\sum_0^\infty p_{00}^{(2n)}$; (b) by Foster's criteria and $\mathbb{E}_i|T_1| = |i|$, $i \neq 0$.
9.2. Find asymptotic expressions for the tails $\mathbb{P}(G > g)$ and $\mathbb{P}(G + H > f)$ of the busy period and busy cycle distribution respectively, in $M/M/1$.
9.3. Evaluate the Laplace transform (or generating function) of the number N of customers served in a busy period. Check the formula by $\mathbb{E}N = \delta \mathbb{E}G$ (and explain why this is true!).
9.4. Consider $M/M/1$ with $\rho \geqslant 1$ and let $S_t = \sum_1^{N_t} U_n - t$ with $\{N_t\}$ the arrival process, $U_1, U_2, \ldots$ the service times, and define $\tau(u) = \inf\{t \geqslant 0: S_t > u\}$, $B(u) = S_{\tau(u)} - u$. Evaluate the cumulant generating function $\kappa(\alpha) = \log \mathbb{E}\, e^{\alpha S_1}$. Explain heuristically that $B(u)$ is independent of $\tau(u)$ with $\mathbb{P}(B(u) > b) = e^{-\delta b}$ and that $\mathbb{E}\, e^{\alpha S_{\tau(u)} - \tau(u)\kappa(\alpha)} = 1$. Evaluate thereby the Laplace transform of $\tau(u)$. How is $\tau(u)$ related to the virtual waiting time?
9.5. Let $\tau = \inf\{t > 0: |S_t| = 2\}$ with $\{S_t\}$ the doubly infinite queue. Evaluate in the symmetric case $\beta = \delta$ the Laplace transform $\hat{G}$ of τ by a similar method as used for the busy period, and check with Proposition 6.3.

Notes

Treatments of time dependence for queues in general and for $M/M/1$ in particular can be found in (among many) Takács (1962), Cox and Smith (1961), Prabhu (1965) and Cohen (1982). Cohen's book is also a monumental treatise of the general area of transform methods in queueing theory. Standard references for Bessel functions are Olver (1965) and Watson (1958).

10. WAITING TIMES AND QUEUE DISCIPLINES IN $M/M/1$

10a. Virtual and actual waiting times in the FIFO case

So far queue length processes have received considerably more attention than the waiting times. The main reason has simply been that as discrete state space processes they are easier to handle by Markovian methods than the continuous waiting times, but we have by now actually collected enough results to be able to say something about waiting times. We shall concentrate on $M/M/1$ in equilibrium and look at the effects of changing the queue discipline. The notation is the same as in the preceding section or Section 3(a). We start with the usual (and simplest) FIFO case.

Theorem 10.1. *Consider the equilibrium* $M/M/1$ FIFO *queue. Then the virtual and the actual waiting time have a common distribution which is a mixture with weights* $1-\rho, \rho$ *of an atom at* 0 *and an exponential distribution with intensity* $\delta - \beta$,

$$\begin{aligned} \mathbb{P}_e(V_t \leqslant y) = \mathbb{P}_e(W_n \leqslant y) &= 1 - \rho + \rho(1 - \exp\{-(\delta - \beta)y\}) \\ &= 1 - \rho \exp\{-(\delta - \beta)y\}. \end{aligned} \tag{10.1}$$

Before embarking into the proof, we stress that when talking about 'the equilibrium $M/M/1$ queue', we must distinguish between the customer- and the time-dependent version. More precisely, a time-stationary version $\{V_t^*\}_{t\geqslant 0}$ of the virtual waiting time is not customer-stationary since, for example, the first customer to arrive has a waiting time with distribution different from (10.1). Indeed, his waiting time is $V_{\tau-}^*$ where τ is the first arrival instant, and obviously $V_{\tau-}^* = (V_0^* - \tau)^+$ is effectively stochastically smaller than the representative V_0^* for (10.1). Put in different terms, the particular customer is not 'sampled at random'. This is an example of phenomena related to the so-called waiting-time paradox, to which we return in IV.3.

Proof. The waiting time $V(t)$ of a customer arriving at time t is 0 if the system is empty which occurs with probability $1-\rho$. If $X_t = n > 0$ customers are present it is the residual service time Y_1 of the customer being served plus the total service times $Y_2, \ldots, Y_n$ of the remaining customers. But by the memoryless property of the exponential distribution, Y_1 has the same distribution as $Y_2, \ldots, Y_n$ and conditionally upon $X_t = n$, the Y_k are independent identically distributed and exponential with intensity δ. Hence

$$\mathbb{P}_e(V_t \leqslant y) = 1 - \rho + \sum_{n=1}^{\infty} (1-\rho)\rho^n \mathbb{P}(Y_1 + \cdots + Y_n \leqslant y)$$

which reduces to the r.h.s. of (10.1), cf. II.3.2.

We shall give three proofs (a), (b), (c), of the assertion concerning W_n (some routine calculations are omitted in (a), (b)):

(a) Apply the uniqueness of the solution to Lindley's integral equation (7.6) and show directly that if F is doubly exponential as in Example 7.1, then the r.h.s. of (10.1) is a solution (cf. also Problem 7.5).

(b) Letting $M(k) = T_0 + \cdots + T_{k-1}$ be the arrival time of customer k, we have $W_k = V_{M(k)-}$. Hence it suffices to show that in the limit $X_{M(k)-}$ follows the same distribution $\pi_n = (1-\rho)\rho^n$ as X_t since we may then condition on $X_{M(k)-}$ and proceed as for V_t above. But $\{X_{M(k)-}\}$ is the Markov chain studied in Example 7.2. Inserting $\mathrm{d}A(t) = \beta \,\mathrm{e}^{-\beta t}\,\mathrm{d}t$ yields $q_k = \rho/(1+\rho)^{k+1}$ and it is then immediately checked that π is stationary for the transition matrix (7.4) (cf. Problem 7.6).

(c) We use the Lindley process representation of random variables W_n, V_t having the equilibrium distributions

$$W_n \stackrel{\mathscr{D}}{=} \max\{0, U_0 - T_0, U_0 + U_1 - T_0 - T_1, \ldots\}, \qquad V_t \stackrel{\mathscr{D}}{=} \max_{0 \leqslant t < \infty} \{S_t^{\uparrow} - t\}, \tag{10.2}$$

cf. Examples 7.1, 8.4(a). With $M(k) = T_0 + \cdots + T_{k-1}$ as above, we may realize $S_t^{\uparrow}$ as $\sum_0^{\infty} U_k I(M(k+1) \leqslant t)$. Then $S_t^{\uparrow} - t$ increases only at times $M(1)$, $M(2), \ldots$, thus the maximum is attained either at time 0 or at some $M(k)$, and the desired conclusion follows from

$$\max_{0 \leqslant t < \infty} \{S_t^{\uparrow} - t\} = \max_{n=0,1,2,\ldots} \{S_{M(n)}^{\uparrow} - M(n)\}$$
$$= \max\{0, U_0 - T_0, U_0 + U_1 - T_0 - T_1, \ldots\}. \qquad \square$$

In connection with argument (b), we observe that in the literature it is often asserted that it can be inferred from the Poisson arrivals alone and without calculations that the equilibrium distribution of $X_{M(k)-}$ is as for X_t, the reason heuristically being that the state of the process seen by the arriving customer is chosen 'at random'. Such reasoning is obviously important for intuition, but also requires some tightening, and we return to the problem in a more general setting in V.3. Note also that (c) applies equally well to the $M/G/1$ case.

10b. The LIFO case

The basic observation is that passing from the FIFO to the LIFO queue discipline neither changes the distribution of the queue length at a fixed instant t

nor prior to an arrival. Hence exactly as above we may evaluate $\mathbb{P}_e(W_k \leqslant y)$ by conditioning upon the events $\{X_{M(k)-} = n\}$ having probabilities $(1-\rho)\rho^n$. Now clearly $W_k = 0$ if $n = 0$. Otherwise k must wait for the server to finish the customer presently in front of him and to clear customers arriving later than k. Thus k can start service at time

$$M(k) + W(k) = \inf\{t \geqslant M(k): X_t = n - 1\}.$$

But this shows that independently of $n \geqslant 1$ $W(k)$ is distributed as the time of first passage of the doubly infinite queue from 0 to -1, or equivalently as the busy period, cf. Section 9(c). Hence by Corollary 9.7:

Theorem 10.2. *Consider the equilibrium* $M/M/1$ LIFO *queue. Then*

$$\mathbb{P}_e(W_n \leqslant y) = 1 - \rho + \rho^{1/2} \int_0^y \frac{\exp\{-(\beta+\delta)t\}}{t} I_1(2\mu t)\mathrm{d}t. \tag{10.3}$$

10c. The SIRO case

We say that a customer is of n-type if he meets n other customers in the system at the time of his arrival (this occurs in equilibrium with probability $\pi_n = (1-\rho)\rho^n$, cf. proof (b) of Theorem 10.1).

Considering first the equilibrium SIRO case, let $H_n(y)$ denote the probability that the waiting time of a n-type customer is at least y. Then, noting that the queue length distribution is again as for the FIFO case, the unconditional waiting-time distribution satisfies

$$\mathbb{P}_e(W_n > y) = \sum_{n=0}^{\infty} \pi_n H_n(y) = (1-\rho) \sum_{n=0}^{\infty} \rho^n H_n(y) \tag{10.4}$$

(here obviously $H_0(y) = 0, y \geqslant 0$).

Consider now an n-type customer arriving at time zero and let u be the time of first exit from state n. If u is a departure time, then the customer is selected for service with probability $1/n$. That is, he continues to wait with probability $(n-1)/n$ and behaves then as an $(n-1)$-type customer who has already waited u time units. It clearly follows that

$$H_n(y+t) = \exp\{-(\beta+\delta)t\}H_n(y) + \int_0^t \left\{\beta H_{n+1}(y+t-u) + \delta H_{n-1}(y+t-u)\frac{n-1}{n}\right\}\exp\{-(\beta+\delta)u\}\mathrm{d}u. \tag{10.5}$$

Replacing t by $\mathrm{d}t$ yields heuristically

$$H_n(y) + H_n'(y)\mathrm{d}t = (1-(\beta+\delta)\mathrm{d}t)H_n(y) + \beta H_{n+1}(y)\mathrm{d}t + \delta H_{n-1}(y)\frac{n-1}{n}\mathrm{d}t,$$

$$H_n'(y) = \delta H_{n-1}(y)\frac{n-1}{n} - (\beta+\delta)H_n(y) + \beta H_{n+1}(y). \tag{10.6}$$

In fact, the rigorous proof of (10.6) (with $H_n'(0)$ meaning right derivative) from (10.5) follows just as when deriving say the backwards equation in II.3(d). It follows further by induction from (10.6) that H_n is C^∞ on $[0, \infty)$ with

$$|H_n^{(k)}(y)| \leqslant (\beta+\delta)^k \tag{10.7}$$

This is sufficient to ensure the validity of the series expansions

$$H_n(y) = \sum_{k=0}^{\infty} h_n^{(k)} \frac{y^k}{k!}, \qquad \mathbb{P}_e(W_n > y) = \sum_{k=0}^{\infty} h^{(k)} \frac{y^k}{k!} \tag{10.8}$$

where $h^{(k)} = \sum_0^\infty \pi_n h_n^{(k)}$ and

$$h_n^{(k)} = H_n^{(k)}(0) = \delta h_{n-1}^{(k-1)} \frac{n-1}{n} - (\beta+\delta) h_n^{(k-1)} + \beta h_{n+1}^{(k-1)}. \tag{10.9}$$

At least for numerical purposes, this presents a complete solution of the problem: the obvious relations $H_0(y) = 0, H_n(0) = 1$ $n \geqslant 1$ yields the boundary conditions

$$h_0^{(k)} = 0, \qquad h_n^{(0)} = 1 \tag{10.10}$$

and (10.9) can then be solved recursively (numerically (10.7) can even be used to assess how many terms that are needed in (10.8) to obtain a given accuracy). In summary:

Theorem 10.3. *The equilibrium $M/M/1$ SIRO waiting-time distribution is given by* (10.8) *with the $h_n^{(k)}$ determined by* (10.9) *and* (10.10).

Notes

A comparatively extensive treatment of the SIRO discipline (covering also $M/G/1$ and $GI/M/1$) and references are given in Cooper (1981). See also Gross and Harris (1974).

10d. The PS sojourn time

In the PS case, the concept of waiting time cannot be given its previous sense (or at least it is then always zero): a customer k starts service as soon as he enters the queue (though perhaps at a very low rate if the queue is long). Instead we shall be interested in his sojourn time W_k^*. This is always at least the service time U_k and so $W_k = W_k^* - U_k$ may be interpreted as the delay caused by the possible presence of other customers.

We let $K_n(y)$ denote the probability that the sojourn time of a n-type customer exceeds y. A service event within $\mathrm{d}t$ time units after the arrival of a n-type customer will terminate the sojourn of any particular customer with probability $1/(n+1)$. It easily follows that (with $K_{-1}(y) = 0$)

$$K_n(y+\mathrm{d}t) = (1-(\beta+\delta)\mathrm{d}t)K_n(y) + \beta K_{n+1}(y)\mathrm{d}t + \delta K_{n-1}(y)\frac{n}{n+1}\mathrm{d}t,$$

$$K_n'(y) = \delta K_{n-1}(y)\frac{n}{n+1} - (\beta+\delta)K_n(y) + \beta K_{n+1}(y). \tag{10.11}$$

This means that $\tilde{H}_0 = K_{-1}, \tilde{H}_1 = K_0, \tilde{H}_2 = K_1, \ldots$ satisfies exactly the same set of equations (10.6), (10.9), (10.10) as $H_0, H_1, H_2, \ldots$ in the SIRO case. The solution being unique, we get

Theorem 10.4. *The equilibrium* $M/M/1$ PS *sojourn-time distribution is given by*

$$\mathbb{P}_e(W_n^* > y) = \sum_{n=0}^{\infty} \pi_n K_n(y) = \sum_{n=0}^{\infty} \pi_n H_{n+1}(y) = (1-\rho) \sum_{k=0}^{\infty} \frac{y^k}{k!} \sum_{n=0}^{\infty} \rho^n h_{n+1}^{(k)}$$

It is also of interest to ask for the conditional distribution of the sojourn time W_n^* given the service time U_n. For example in a time-sharing computer, U_n represents the size of the job (ideal execution time) and W_n^* the actual execution time. The following result then intuitively states that the PS (and thereby presumably also RR with a small quantum) discipline is fair since the average sojourn time is proportionally dependent on U_n:

Theorem 10.5. $\mathbb{E}(W_n^* \mid U_n) = U_n/(1-\rho)$.

Proof. We let $m_n(u)$ denote the conditional expectation of the sojourn time of an n-type customer given that his service time is u and define $m(u) = \sum_0^\infty \pi_n m_n(u)$ so that we have to show $m(u) = u/(1-\rho)$. Now if no service or arrival events occur within t time units of the arrival of a n-type customer with service time u, he will have attained $t/(n+1)$ units of service and hence behave like a n-type customer with service time $u - t/(n+1)$. Also the sojourn time increases by t unless the particular customer is served. The intensity is initially $\delta/(n+1)$ for this and $n\delta/(n+1)$ for service of some other customer. Letting $t = \mathrm{d}t$, we get

$$\begin{aligned} m_n(u) &= \mathrm{d}t\left(1 - \delta\frac{1}{n+1}\mathrm{d}t\right) + m_n\left(u - \frac{\mathrm{d}t}{n+1}\right)\left\{1 - \beta\,\mathrm{d}t - \frac{n}{n+1}\delta\,\mathrm{d}t\right\} \\ &\quad + m_{n+1}(u)\beta\,\mathrm{d}t + m_{n-1}(u)\frac{n}{n+1}\delta\,\mathrm{d}t \\ &= \mathrm{d}t + m_n(u) - m_n'(u)\frac{\mathrm{d}t}{n+1} - m_n(u)\beta\,\mathrm{d}t - m_n(u)\frac{n}{n+1}\delta\,\mathrm{d}t \\ &\quad + m_{n+1}(u)\beta\,\mathrm{d}t + m_{n-1}(u)\frac{n}{n+1}\delta\,\mathrm{d}t, \end{aligned}$$

$$m_n'(u) = n + 1 - \beta(n+1)m_n(u) - n\delta m_n(u) + \beta(n+1)m_{n+1}(u) + \delta n m_{n-1}(u), \tag{10.12}$$

$$\begin{aligned} m'(u) &= \sum_{n=0}^{\infty} \pi_n m_n'(u) \\ &= \sum_{n=0}^{\infty} (n+1)\pi_n - \beta \sum_{n=0}^{\infty} (n+1)\pi_n m_n(u) - \delta \sum_{n=0}^{\infty} n\pi_n m_n(u) \\ &\quad + \beta \sum_{n=1}^{\infty} n\pi_{n-1} m_n(u) + \delta \sum_{n=0}^{\infty} (n+1)\pi_{n+1} m_n(u) \end{aligned} \tag{10.13}$$

which using $\pi_{n+1} = \rho\pi_n$, $\sum n\pi_n = \rho/(1-\rho)$ reduces to

$$\frac{\rho}{1-\rho} + 1 - \beta \sum_{n=0}^{\infty} n\pi_n m_n(u) - \beta m(u) - \delta \sum_{n=0}^{\infty} n\pi_n m_n(u)$$

$$+ \frac{\beta}{\rho} \sum_{n=0}^{\infty} n\pi_n m_n(u) + \rho\delta \sum_{n=0}^{\infty} n\pi_n m_n(u) + \rho\delta m(u)$$

$$= \frac{1}{1-\rho} + \sum_{n=0}^{\infty} n\pi_n m_n(u) \left\{ -\beta - \delta + \frac{\beta}{\rho} + \rho\delta \right\} + m(u)\{ -\beta + \rho\delta \}.$$

But since $\rho = \beta/\delta$, the two $\{\cdots\}$ are zero. Thus $m'(u) = 1/(1-\rho)$ and since clearly $m(0) = 0$, we have $m(u) = u/(1-\rho)$ as desired. □

It should be noted that the rigorous justification of this argument requires somewhat more care, as for Theorem 10.4. For example, to differentiate under the sum sign in (10.13) we need some bound on the $m_n'(u)$ or equivalently the $m_n(u)$ (cf. (10.12)), but we shall not go into the details.

Notes

The proof of Theorem 10.5 carries over to finding also the conditional Laplace transform of W_n^* given $U_n = u$, see Coffman *et al.* (1970). Of further studies of queues with PS we mention Schassberger (1984) and Ott (1984) for the $M/G/1$ case and Ramaswami (1984) and Cohen (1984) for the $GI/M/1$ case.

We have concentrated here on the derivation of the various waiting-time distributions and not on a comparison of their properties. One may note here that in view of Little's formula (Problem 1.5), the means are all the same. A comparison of variances can be found in Cohen (1982). From an application point of view, the study of particular queue disciplines (also including priority queueing which we have not treated here) is of main importance. A further text with much material is Heyman and Sobel (1982).

Part B: Basic Mathematical Tools

CHAPTER IV

Basic Renewal Theory

1. RENEWAL PROCESSES

Let $S_0 \leqslant S_1 < \cdots$ be the times of occurrences of some phenomenon and $Y_n = S_n - S_{n-1}$, $Y_0 = S_0$, see Fig. 1.1. Then $\{S_n\}_{n \in \mathbb{N}}$ is called a *renewal process* if Y_0, Y_1, $Y_2, \ldots$ are independent and $Y_1, Y_2, \ldots$ (but not necessarily Y_0) have the same distribution.

The S_n are called the *renewals* or sometimes the *epochs* of the renewal process. The common distribution F of $Y_1, Y_2, \ldots$ is the *interarrival* or *waiting-time distribution*. To avoid more than one renewal at a time we always assume that the Y_k have no mass at zero, $F(0) = 0$. The renewal process is *pure* or *zero-delayed* if $Y_0 = S_0 = 0$ a.s. Otherwise it is *delayed* and the *delay distribution* is the distribution of Y_0. One also sometimes considers *terminating* or *transient* renewal processes, where the interarrival distribution is *defective*, i.e. may have an atom at $+\infty$, $\|F\| = \lim_{t \to \infty} F(t) < 1$, and in which case $S_n = \infty$ eventually. If $\|F\| = 1$, the renewal process is *proper*.

A main case of a renewal process is, of course, the Poisson process, where the interarrival distribution is exponential. For example, the Poisson process provides an adequate description of the emission of particles from a radioactive source. We next list some further phenomena, which it has been suggested can be modelled by renewal processes.

Example 1.1. Consider an item, say an electric bulb, which fails at times S_0, $S_1, \ldots$ and is replaced at the time of failure by a new item of the same sort. Then F is the distribution of the lifetime of an item. The process is delayed if the item present at time $t = 0$ is not new, so that its lifetime need not have distribution F. □

Example 1.2. Consider a road on which vehicles are driving in one direction only and all with the same constant velocity. Two interpretations are possible: (i) the S_n are the instants when vehicles pass a certain point on the road, (ii) the time-scale $[0, \infty)$ is a map of the road and the S_n the positions of the vehicles at a certain instant. In both cases, the form of the interarrival distribution to be expected depends in an essential manner on whether there is little or much traffic on the road. In the first case (say on a rural road) F might be

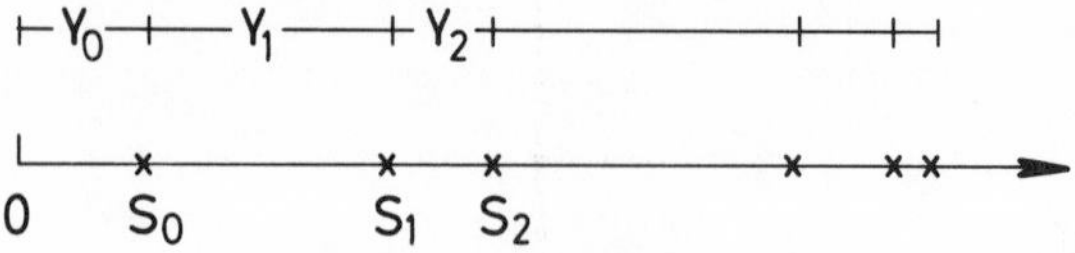

Fig. 1.1

taken to be exponential, while in the second case (say on a main street in a city) the vehicles would rather be equally spaced, i.e. F concentrated in one point. □

Example 1.3. Consider a continuous-time Markov process $\{X_t\}_{t \geqslant 0}$ or a discrete-time Markov chain $\{X_n\}_{n \in \mathbb{N}}$ with discrete or continuous state space, and let i be some fixed state. Then the instants S_n of visits to i form a renewal process (provided that in continuous time the strong Markov property holds and that $\mathbb{P}_i(\tau(i)=0)=0$; the last requirement fails say for Brownian motion). The process is pure if and only if $X_0 = i$ and transient if and only if state i is visited only finitely often (which is the definition of transience in the case of a discrete state space). Figure 1.2 illustrates the case of a two-state Markov jump process with $i=2$. If the exponential holding time of state k has mean μ_k, then F is the convolution of two exponential distributions with means μ_1 and μ_2 respectively. □

Though the main probabilistic object describing a renewal process is nothing but a sum of non-negative independent identically distributed random variables, the point of view of renewal theory is somewhat different from the one usually taken when studying such sums. In particular, rather than in the behaviour of S_n as a function of n we are interested in the number of steps needed to reach size t, i.e.

$$N_t = \#\{n=0, 1, 2, \ldots : S_n \leqslant t\} = \inf\{n : S_n > t\},$$

the number of renewals up to time t. Note that

$$N_t \leqslant n \Leftrightarrow S_n > t, \tag{1.1}$$

$$S_{N_t - 1} \leqslant t < S_{N_t}, \tag{1.2}$$

$$\{N_t = n\} = \{S_{n-1} \leqslant t < S_n\}. \tag{1.3}$$

These equations state, loosely speaking, that $t \to N_t$ is the inverse function of

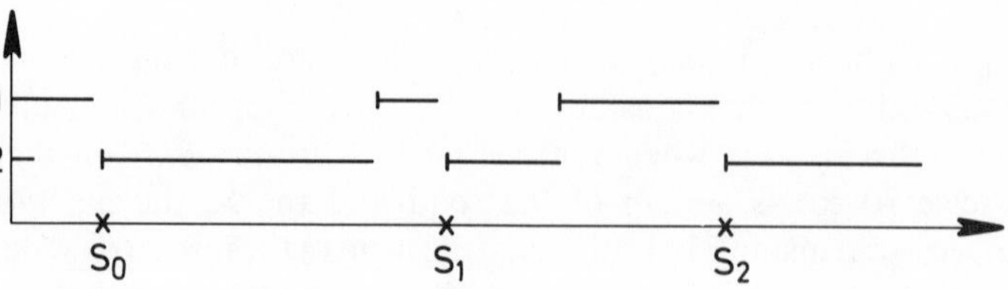

Fig. 1.2

$n \to S_n$, and suggest that classical results on $\{S_n\}_{n\in\mathbb{N}}$ could be converted to results on $\{N_t\}_{t\geqslant 0}$. For example:

Proposition 1.4. *Let* $\mu = \mathbb{E}Y_1 = \int_0^\infty x\,\mathrm{d}F(x)$ *be the mean of the interarrival distribution. Then* (irrespective of the distribution of Y_0 or whether $\mu < \infty$ or $\mu = \infty$)

$$\frac{N_t}{t} \to \mu^{-1} \text{ a.s.} \quad \text{as } t \to \infty, \tag{1.4}$$

$$\mathbb{E}N_t/t \to \mu^{-1} \quad \text{as } t \to \infty. \tag{1.5}$$

Proof. Since $S_n/n \to \mu$ a.s. and $N_t \to \infty$, it follows by dividing (1.2) by N_t that $t/N_t \to \mu$, i.e. $N_t/t \to \mu^{-1}$ a.s. From this also the $\underline{\lim} \geqslant -$ part of (1.5) follows by Fatou's lemma. To get $\overline{\lim} \leqslant$, consider a renewal process $\{\tilde{S}_n\}_{n\in\mathbb{N}}$ where the interarrival times are $\tilde{Y}_0 = Y_0$, $\tilde{Y}_n = Y_n \wedge a, n = 1, 2, \ldots,$ and let $\tilde{N}_t, \tilde{\mu}$, etc. be defined in the obvious way. Now by (1.3), $\tilde{N}_t$ is a stopping time with respect to the $\tilde{Y}_k$ for any fixed value of Y_0. Hence we may apply Wald's identity conditionally upon Y_0 and get

$$\mathbb{E}(\tilde{Y}_1 + \cdots + \tilde{Y}_{\tilde{N}_t}) = \mathbb{E}\mathbb{E}(\tilde{Y}_1 + \cdots + \tilde{Y}_{\tilde{N}_t} | Y_0) = \mathbb{E}\mathbb{E}(\tilde{N}_t | Y_0)\tilde{\mu} = \mathbb{E}\tilde{N}_t\tilde{\mu}.$$

Clearly, $\tilde{N}_t \geqslant N_t$ and by (1.2), $\tilde{S}_{\tilde{N}_t} = \tilde{S}_{\tilde{N}_t - 1} + \tilde{Y}_{\tilde{N}_t} \leqslant t + a$. Thus

$$\begin{aligned}\overline{\lim_{t\to\infty}} \frac{\mathbb{E}N_t}{t} &\leqslant \overline{\lim_{t\to\infty}} \frac{\mathbb{E}\tilde{N}_t}{t} = \overline{\lim_{t\to\infty}}\, \tilde{\mu}^{-1}\mathbb{E}(\tilde{Y}_1 + \cdots + \tilde{Y}_{\tilde{N}_t})/t \\ &\leqslant \overline{\lim_{t\to\infty}}\, \tilde{\mu}^{-1}\mathbb{E}\tilde{S}_{\tilde{N}_t}/t \leqslant \tilde{\mu}^{-1}.\end{aligned}$$

As $a \uparrow \infty$, then $\tilde{\mu} \uparrow \mu$ and (1.5) follows. □

A CLT analogue is given in VI.4.

One of the main points of renewal theory turns out to be obtaining refinements of (1.5). For this reason, (1.5) is sometimes called the *elementary renewal theorem.*

In the same way as in I.2, we shall now define the *backwards recurrence-time process* $\{A_t\}_{t\geqslant 0}$ and the *forwards recurrence process* $\{B_t\}_{t\geqslant 0}$ associated with the renewal process. In the language of Example 1.1, A_t is the *age* of the current item and B_t its *residual* or *excess lifetime.* That is, A_t is the time elapsed since the last renewal and B_t the waiting time until the next renewal epoch $> t$,

$$A_t = t - S_{N_t - 1}, \qquad B_t = S_{N_t} - t$$

(note that $B_0 = Y_0$ on $\{Y_0 > 0\}$ and $B_0 = Y_1$ on $\{Y_0 = 0\}$). The paths have the form illustrated in Fig. 1.3 and are right-continuous by definition.

For a given renewal process, it is only possible to attach sense to A_t if $t \geqslant Y_0$. However, for any $a \geqslant 0$ with $1 - F(a) > 0$ we can define a renewal process by 'starting with a renewal at $-a$', i.e. letting Y_0 have the conditional distribution of Y_1 given $Y_1 > a$,

$$\mathbb{P}(Y_0 > y) = \mathbb{P}(Y_1 > a + y | Y_1 > a) = \frac{1 - F(a + y)}{1 - F(a)}. \tag{1.6}$$

Letting $A_t = a + t$, $t < Y_0$, we then get a version of $\{A_t\}$ which is defined for all $t \geq 0$ and has $A_0 = a$. In fact, we shall show:

Proposition 1.5. *The processes $\{A_t\}_{t\geq 0}$ and $\{B_t\}_{t\geq 0}$ are time-homogeneous strong Markov processes.*

Proof. The Markov property is intuitively obvious from the construction in the same way as in discrete time in I.2: B_t decreases linearly (and deterministically) until 0 is hit, then jumps to Y_1, decreases linearly to 0, jumps to Y_2 and so on, and this motion clearly has the asserted properties. The motion of A_t is also linear (but the jump times are not deterministic). Given $\{A_s\}_{0\leq s\leq t}$, the evolution of the process after time t depends on the past only through the distribution G of the waiting time until the next jump. But the tails $1 - G(y)$ are given by (1.6) with $a = A_t$, which implies the Markov property.

For the strong Markov property for $\{B_t\}$, let f be continuous and bounded and let $g(b) = \mathbb{E}_b f(B_s)$. An inspection of the paths immediately shows that $g(b-t) = \mathbb{E}_b f(B_{s+t})$, $0 < t < b$. As $t \downarrow 0$, we have $f(B_{s+t}) \to f(B_s)$ and hence by dominated convergence $g(b-t) \to g(b)$, i.e. g is left-continuous. For $u \downarrow t$ we have $B_u \uparrow B_t$ so that $g(B_u) \to g(B_t)$. Thus $\{g(B_t)\}$ has right-continuous paths, and the strong Markov property for $\{B_t\}$ follows by I.6.3. For $\{A_t\}$, we have for $0 < t < s$

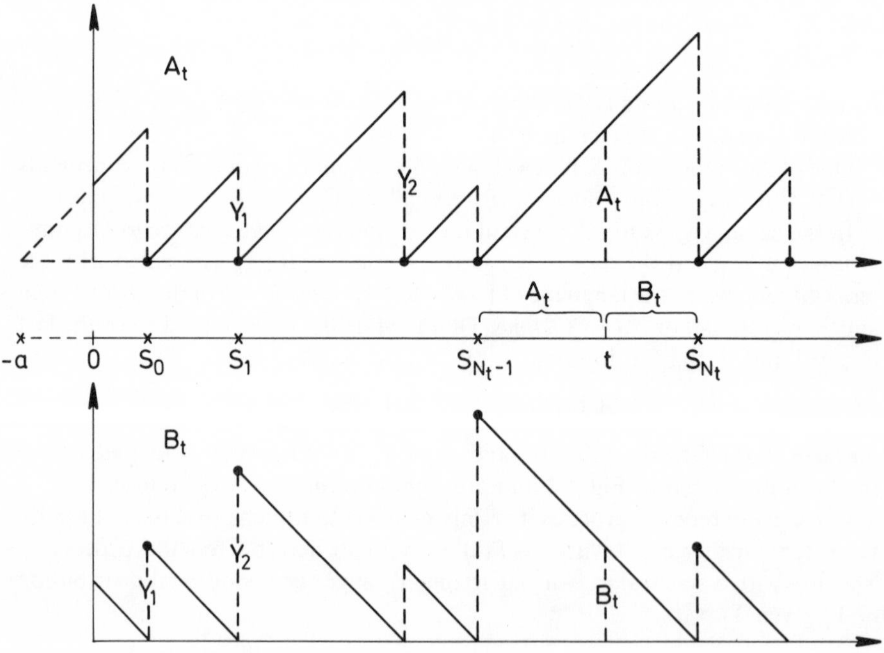

Fig. 1.3

that

$$\mathbb{E}_a f(A_{s+t}) = \frac{1}{1-F(a)} \left\{ (1 - F(a+t)) \mathbb{E}_{a+t} f(A_s) + \int_0^t \mathbb{E}_0 f(A_{s+t-y}) F(a + \mathrm{d}y) \right\}$$
$$= \mathbb{E}_{a+t} f(A_s) + \mathrm{o}(1).$$

Therefore $h(a) = \mathbb{E}_a f(A_s)$ is right-continuous, which in view of $A_u \downarrow A_t$, $u \downarrow t$, implies the paths of $\{h(A_t)\}$ to be so. □

We note that a number of Markov processes associated with queues and related models (see e.g. Problems 1.3, VIII.3.2, XIII.3.1) have paths of a similar shape as $\{A_t\}$, $\{B_t\}$ and that the strong Markov property in such cases follows by small variants of the proof of Proposition 1.5.

It was remarked in Example 1.3 that the recurrence times of some points i in a Markov process $\{X_t\}_{t \geqslant 0}$ form a renewal process. Proposition 1.5 shows that *any* renewal process is of this type (with $X_t = A_t$ and $i = 0$).

Problems

1.1 (*the type I counter*). The incoming particles constitute a Poisson process, but the registrations do not, since for technical reasons the counter cannot register the second of two particles emitted at almost the same time. Suppose that each *registered* particle locks the counter for a time with distribution G, that particles arriving in a locked period have no effect and that locking times of different particles are independent mutually and of the Poisson process. Show that the registrations constitute a renewal process and find the interarrival distribution.

1.2 (*the pedestrian delay problem*). At time 0, a pedestrian arrives at a road and wants to cross. Crossing is possible when the gap between two cars is at least ξ. Find the distribution of the waiting time until crossing is performed.

1.3. Show that $\{(A_t, B_t)\} t \geqslant 0$ has the strong Markov property [hint: consider $\mathbb{E}_{a+t,b-t} f(A_s) g(B_s)$].

Notes

A main standard reference for renewal theory is Chapter XI of Feller (1971), but the topic is treated in a number of other places such as Çinlar (1975, Ch. 9) and Karlin and Taylor (1975, Ch. 5) where, however, the proof of the renewal theorem is omitted, Cox (1962) and Chapter 5 of Jagers (1975).

In view of the basic importance of renewal theory it is not surprising that several generalizations have been extensively studied. We mention renewal theory for Markov chains (X.2 and references therein), for random walks (surveyed in Gut, 1986), non-linear renewal theory (a term used in a different sense say by Chover and Ney, 1968, and Woodroofe, 1982), multivariate renewal theory, a topic which is also rather diverse and where we mention for example Bickel and Yahav (1965), Stam (1969/71) and Carlsson and Wainger (1984), and finally renewal theory with infinite mean, see Garsia and Lamperti (1962) and Erickson (1970).

2. RENEWAL EQUATIONS AND THE RENEWAL MEASURE

The *renewal equation* is the convolution equation $Z = z + F * Z$ (for the convolution notation, see the Notes at the end of Section IV.2), i.e.

$$Z(t) = z(t) + \int_0^t Z(t-u) \, \mathrm{d}F(u), \qquad t \geqslant 0. \tag{2.1}$$

Here one thinks of Z as an unknown function on $[0, \infty)$, z as a known function on $[0, \infty)$ and F as a known non-negative (Radon) measure on $(0, \infty)$. In current terminology, it is often assumed that F is a probability, i.e. $\| F \| = 1$, in which case (2.1) is *proper*. If $\| F \| < 1$, the renewal equation (2.1) is *defective*, but we shall also consider the *excessive* case $1 < \| F \| \leqslant \infty$. We always assume that $F(0) = 0$. We shall first give some examples.

Example 2.1. Consider a pure renewal process with interarrival distribution F and the recurrence times A_t, B_t defined as in Section 1. Let $\xi \geqslant 0$ be fixed and define

$$Z_A(t) = \mathbb{P}(A_t \leqslant \xi), \qquad Z_B(t) = \mathbb{P}(B_t \leqslant \xi).$$

Then Z_A, Z_B satisfy the renewal equations

$$Z_A = z_A + F * Z_A, \qquad z_A(t) = \mathbb{P}(A_t \leqslant \xi, Y_1 > t) = I(t \leqslant \xi)(1 - F(t)) \tag{2.2}$$

$$Z_B = z_B + F * Z_B, \qquad z_B(t) = \mathbb{P}(B_t \leqslant \xi, Y_1 > t) = F(t + \xi) - F(t). \tag{2.3}$$

The proof of this is carried out by the *renewal argument*, i.e. (i) conditioning on the value u of Y_1, which yields

$$Z_A(t) = \mathbb{P}(A_t \leqslant \xi, Y_1 > t) + \int_0^t \mathbb{P}(A_t \leqslant \xi \mid Y_1 = u) \, \mathrm{d}F(u), \tag{2.4}$$

and (ii) remarking that the process starts from scratch at time Y_1, which yields

$$\mathbb{P}(A_t \leqslant \xi \mid Y_1 = u) = \mathbb{P}(A_{t-u} \leqslant \xi) = Z_A(t - u)$$

for $u \leqslant t$. Thus, since $A_t = t$ on $\{Y_1 > t\}$, (2.2) and (2.4) are the same equation. Equation (2.3) is derived in a similar manner using $B_t = Y_1 - t$ on $\{Y_1 > t\}$. □

Example 2.2. (*Lotka's integral equation*). This comes from classical deterministic or semi-deterministic population theory associated with the names of Sharpe and Lotka. Consider the female part of a population, where women aged a give birth (to a single daughter) at rate $\lambda(a)$ and survive to age $a + t$ in a proportion of ${}_tp_a$ (in traditional demographic notation). We are interested in $Z(t)$, the overall birth-rate at time t, which can be split into the rates $z(t)$, $Z_0(t)$ of births from women born respectively before and after time zero. To determine $z(t)$, we must know the age structure of the population at time zero, which will be represented by its density $f_0(a)$ (thus $\int_0^\infty f_0(a) \, \mathrm{d}a$ is the initial population size, not necessarily $= 1$). Then women aged a at time zero have density $f_0(a) {}_tp_a$ at time t and are aged $a + t$, and hence

$$z(t) = \int_0^\infty f_0(a) {}_tp_a \lambda(a + t) \, \mathrm{d}a.$$

Similarly, women born at time $t - s$ provide a contribution $Z(t - s) {}_sp_0 \lambda(s) \, \mathrm{d}s$ to $Z_0(t)$ so that

$$Z(t) = z(t) + \int_0^t Z(t - s) {}_sp_0 \lambda(s) \, \mathrm{d}s \tag{2.5}$$

and we have a renewal equation with $\mathrm{d}F(s) = {}_s p_0 \lambda(s)\,\mathrm{d}s$. Note that $\|F\| = \int_0^\infty {}_s p_0 \lambda(s)\,\mathrm{d}s$ is the average number of daughters born to a woman, the so-called *net reproduction rate*. This could have values both < 1, $= 1$ and > 1, but the typical case is that of a growing population with $\|F\| > 1$, where (2.5) is thus excessive. Note also that other quantities of interest, like the density

$$\begin{cases} f_0(a-t)_t p_{a-t} & t \leqslant a \\ Z(t-a)_a p_0 & t > a \end{cases}$$

of women aged a at time t and the total population size

$$N(t) = \int_0^t Z(t-a)_a p_0\,\mathrm{d}a + \int_0^\infty f_0(a)_t p_a\,\mathrm{d}a \tag{2.6}$$

are readily expressed in terms of Z. □

Example 2.3. (the *ruin problem* of insurance mathematics). Assume that the claims incurred by an insurance company arrive according to a Poisson process $\{N_t\}_{t \geqslant 0}$ with intensity λ, that the sizes of the claims are independent identically distributed non-negative random variables $X_1, X_2, \ldots$, and that the inflow of premium up to time t is ct. Thus the risk reserve at time t is

$$U_t = u + ct - \sum_{n=1}^{N_t} X_n,$$

with $u = U_0$ the initial value (cf. Fig. 2.1). We are interested in the probabilities

$$Z(u) = \mathbb{P}(U_t \geqslant 0 \quad \forall t \,|\, U_0 = u), \qquad 1 - Z(u) = \mathbb{P}\left(\inf_{0 \leqslant t < \infty} U_t < 0 \,|\, U_0 = u \right)$$

of ultimate survival and ultimate ruin of the company, say for the purpose of assessing whether c has been chosen sufficiently large compared to u.

To study Z we note that the process renews itself at the time s of the first claim, holding the new initial fortune $u + cs - X_1$ (terminates if $X_1 > u + cs$). It follows that

$$Z(u) = \int_0^\infty \lambda\,\mathrm{e}^{-\lambda s} \int_0^{u+cs} Z(u + cs - x)\,\mathrm{d}G(x)\,\mathrm{d}s$$

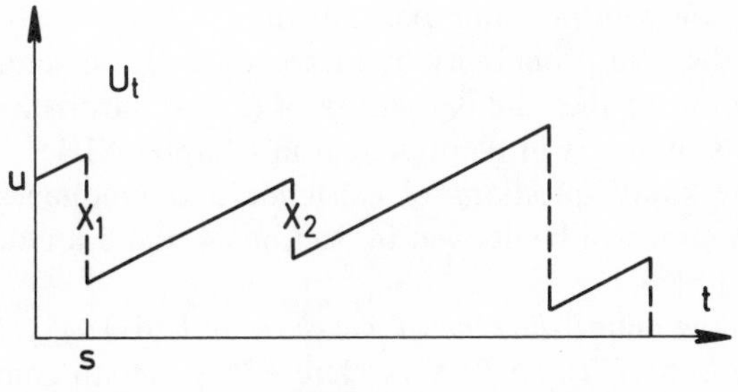

Fig. 2.1

with $G(x) = \mathbb{P}(X_1 \leqslant x)$. Letting $t = u + cs$, we get

$$Z(u)\mathrm{e}^{-\lambda u/c} = \frac{\lambda}{c}\int_u^\infty \mathrm{e}^{-\lambda t/c}\int_0^t Z(t-x)\,\mathrm{d}G(x)\,\mathrm{d}t.$$

This representation shows that Z is differentiable, and differentiating with respect to u yields

$$\mathrm{e}^{-\lambda u/c}\left(Z'(u) - \frac{\lambda}{c}Z(u)\right) = -\frac{\lambda}{c}\mathrm{e}^{-\lambda u/c}\int_0^u Z(u-x)\,\mathrm{d}G(x),$$

$$Z'(u) = \frac{\lambda}{c}Z(u) - \frac{\lambda}{c}\int_0^u Z(u-x)\,\mathrm{d}G(x)$$

Integrating in $\mathrm{d}u$ from 0 to t and letting

$$h(y) = \int_0^{t-y} Z(u)\,\mathrm{d}u, \quad 0 \leqslant y \leqslant t, \qquad h(y) = 0,\, y > t,$$

yields

$$\begin{aligned} Z(t) - Z(0) - \frac{\lambda}{c}h(0) &= -\frac{\lambda}{c}\int_0^t\int_0^u Z(u-x)\,\mathrm{d}G(x)\,\mathrm{d}u \\ &= -\frac{\lambda}{c}\int_0^t h(x)\,\mathrm{d}G(x) = -\frac{\lambda}{c}\int_0^\infty h(x)\,\mathrm{d}G(x) \\ &= -\frac{\lambda}{c}h(0) - \frac{\lambda}{c}\int_0^\infty h'(x)(1-G(x))\,\mathrm{d}x, \end{aligned}$$

i.e.

$$Z(t) = Z(0) + \frac{\lambda}{c}\int_0^t Z(t-x)(1-G(x))\,\mathrm{d}x \tag{2.7}$$

This is of the form (2.1), with $\mathrm{d}F(x) = \lambda/c(1-G(x))\,\mathrm{d}x$. We note that $\|F\| = \lambda v/c$, where

$$v = \int_0^\infty (1-G(x))\,\mathrm{d}x = \int_0^\infty y\,\mathrm{d}G(y)$$

is the mean claim size, and that λv is the mean size of the claims received per unit time, c the inflow of premium per unit time.

The study of the ruin probability by means of (2.7) is a standard approach, but one might observe that the derivation of (2.7) is surprisingly lengthy and alternative points of view will be presented in Chapter XIII. □

We shall now study questions of existence and uniqueness of solutions. Asymptotic estimates will be derived in Sections 4 and 5 for the case $\|F\| = 1$ and in VI.5 for $\|F\| \neq 1$.

Given F, we now define *the renewal measure* by $U(\mathrm{d}x) = \sum_0^\infty F^{*n}(\mathrm{d}x)$ and *the renewal function U* by $U(t) = \sum_0^\infty F^{*n}(t)$. Here F^{*n} is the nth convolution power of F, i.e. the probability distribution degenerate at zero for $n = 0$, while for

$n = 0, 1, 2, \dots$

$$F^{*(n+1)}(t) = \int_0^t F^{*n}(t-u)\,\mathrm{d}F(u) = \int_0^t F(t-u)\,\mathrm{d}F^{*n}(u).$$

In particular, $\| F^{*n} \| = \| F \|^n$.

Theorem 2.4. *(i) The renewal function $U(t)$ is finite for all $t < \infty$; (ii) if the function z in the renewal equation (2.1) is bounded on finite intervals, then $Z = U * z$ (i.e. $Z(t) = \int_0^t z(t-x)\,\mathrm{d}U(x)$) is well defined, solution to (2.1) and the unique solution to (2.1) which is bounded on finite interval (i.e. $\sup_{0 \leqslant t \leqslant T} |Z(t)| < \infty$ for all $T < \infty$); (iii) if $\| F \| = 1$ then $U(t)$ is the expected number $\mathbb{E}N_t$ of renewals up to time t in a pure renewal process with interarrival distribution F. More generally, in any renewal process with interarrival distribution F, the expected number of renewals in $(t, t+a]$ is*

$$\mathbb{E}(N_{t+a} - N_t) = \int_0^a U(a-\xi)\,\mathrm{d}G_t(\xi) = G_t * U(a) = U * G_t(a) \tag{2.8}$$

where $G_t(\xi) = \mathbb{P}(B_t \leqslant \xi)$, and the expression (2.8) cannot exceed $U(a)$.

Proof. (i) Since $U(t)$ does not involve the restriction of F to (t, ∞), we may put $\mathrm{d}F(x) = 0$, $x > t$ if necessary to ensure that the Laplace transform $\hat{F}(\beta) = \int_0^\infty \mathrm{e}^{-x\beta}\,\mathrm{d}F(x)$ is well defined for $\beta > 0$. Then $\widehat{F^{*n}} = \hat{F}^n$ and hence

$$F^{*n}(t) \leqslant \mathrm{e}^{\beta t} \int_0^t \mathrm{e}^{-\beta s}\,\mathrm{d}F^{*n}(s) \leqslant \mathrm{e}^{\beta t} \hat{F}(\beta)^n.$$

By monotone convergence we can choose β such that $\hat{F}(\beta) < 1$ and $U(t) < \infty$ then follows by summing over n (alternatively, if $\| F \| = 1$ then (i) follows from (iii) and Section 1). For (ii), it is now obvious that $Z = U * z$ is well defined and bounded on finite intervals. Defining $U_N = \sum_0^N F^{*n}$, $Z_N = U_N * z$ we have $Z_{N+1} = z + F * Z_N$ and that $Z = \lim Z_N$ is a solution follows as $N \to \infty$. Given two solutions of the type considered, their difference V satisfies $V = F * V = \cdots = F^{*n} * V$ so that

$$|V(t)| = \left| \int_0^t V(t-x)\,\mathrm{d}F^{*n}(x) \right| \leqslant \sup_{0 \leqslant y \leqslant t} |V(y)| \cdot F^{*n}(t)$$

and $V(t) = 0$ follows as $n \to \infty$. For (iii), it follows from (1.1) that in a pure renewal process with interarrival distribution F

$$\mathbb{E}N_t = \sum_{n=0}^{\infty} \mathbb{P}(N_t > n) = \sum_{n=0}^{\infty} \mathbb{P}(S_n \leqslant t)$$

$$= \sum_{n=0}^{\infty} \mathbb{P}(Y_1 + \cdots + Y_n \leqslant t) = \sum_{n=0}^{\infty} F^{*n}(t) = U(t),$$

and the more general (2.8) then follows by conditioning on $\{B_t = \xi\}$ and noting that a pure renewal process starts at time $t + \xi$. Finally, an upper bound for

(2.8) is obtained by replacing G_t with the distribution degenerate at zero and this yields $U(a)$. □

Example 2.5. In many examples, the form $Z = U * z$ of the solution to $Z = z + F * Z$ can be seen directly and sometimes this is even the most natural approach. Consider as an example a *shot-noise process*

$$W_t = \sum_{n=0}^{N_t - 1} f(t - S_n, X_n)$$

where $X_0, X_1, \ldots$ are independent identically distributed and independent of the renewal process. Then $Z(t) = \mathbb{E}W_t$ satisfies

$$Z(t) = \sum_{n=0}^{\infty} \mathbb{E}[f(t - S_n, X_n); S_n \leqslant t]$$
$$= \sum_{n=0}^{\infty} \int_0^t z(t-u)\,\mathrm{d}F^{*n}(u) = U * z(t),$$

where $z(t) = \mathbb{E}f(t, X_1)$.

The shot-noise process is used to describe certain electrical tubes, where primary impulses of sizes $X_0, X_1, \ldots$ are emitted at the epochs of a renewal process. An impulse of size X then creates secondary effects which are of size $f(t, X)$ after time t. Similar phenomena occur in road traffic noise, where the renewal process describes the passing of the cars, X_n is the noise level of the nth car and $f(t, X)$ is the actual noise at a distance of t. □

It is clear by the same argument as in the proof of Theorem 2.4 that if the renewal process is terminating, $\|F\| < 1$, then $U(t)$ can still be interpreted as the expected number of renewals in $[0, t]$. In particular, the expected number of renewals within finite time is

$$\|U\| = \lim_{t\to\infty} U(t) = \sum_{n=0}^{\infty} \|F^{*n}\| = \sum_{n=0}^{\infty} \|F\|^n = \frac{1}{1 - \|F\|},$$

cf. also I.2.4.

However, the renewal measure has a different important interpretation in the terminating case. Define the *lifetime* or *maximum* by $M = \sup\{S_n : S_n < \infty\}$. Then

Proposition 2.6. *If* $\|F\| < 1$, *then the distribution of* M *is given by* $\mathbb{P}(M \leqslant x) = (1 - \|F\|)U(x)$.

Proof. We give two arguments, (a) and (b). In (a), put $\sigma = \inf\{n \geqslant 0 : Y_{n+1} = \infty\}$. Then

$$M = Y_1 + \cdots + Y_\sigma, \qquad \mathbb{P}(\sigma = n) = (1 - \|F\|)\|F\|^n$$

and conditionally upon σ, we have $\mathbb{P}(Y_k \leqslant y | \sigma) = G(y)$ for $k \leqslant \sigma$, where $G(y) =$

$F(y)/\|F\|$. Hence

$$\mathbb{P}(M \leqslant y) = \sum_{n=0}^{\infty} \mathbb{P}(\sigma = n)\mathbb{P}(Y_1 + \cdots + Y_n \leqslant y \mid \sigma = n)$$

$$= \sum_{n=0}^{\infty} (1 - \|F\|)\|F\|^n G^{*n}(y) = (1 - \|F\|) \sum_{n=0}^{\infty} F^{*n}(y) = (1 - \|F\|)U(y).$$

In (b), let $Z(x) = \mathbb{P}(M \leqslant x)$, $x \geqslant 0$. Now if $Y_1 = \infty$, then $M = 0$ and hence $M \leqslant x$, whereas if $0 < Y_1 = y < \infty$, then for $\{M \leqslant x\}$ to occur we must have $y \leqslant x$ and that the lifetime of the renewal process starting at y_1 is at most $x - y$. Hence

$$Z(x) = 1 - \|F\| + \int_0^x Z(x - y)\,\mathrm{d}F(y)$$

which implies

$$Z(x) = U * (1 - \|F\|)(x) = (1 - \|F\|)U(x). \qquad \square$$

Problems

2.1. Find a renewal equation for the joint distribution of the recurrence times (A_t, B_t).

2.2. (*the type II counter*). As in Problem 1.1, we assume that the particles arrive at the counter according to a Poisson process with intensity λ, but use a different model for the locking mechanism, namely that locking times of different particles are independent identically distributed with common distribution G and that each particle arriving at the counter cancels the after-effect (if any) of its predecessors. Show that the probability $Z(t)$ of the duration of a locked period to exceed t satisfies the renewal equation

$$Z(t) = (1 - G(t))\mathrm{e}^{-\lambda t} + \int_0^t Z(t - x)(1 - G(x))\lambda \mathrm{e}^{-\lambda x}\,\mathrm{d}x.$$

2.3. Show that $Z(t) = Z(0)\cdot U(t) = 0$ in the ruin problem with $\lambda v/c \geqslant 1$.

2.4. Show that if $\hat{F}$ is well defined (e.g. if $\|F\| < \infty$), then $\hat{U} = (1 - \hat{F})^{-1}$ and that in (2.1) $\hat{Z} = \hat{z}/(1 - \hat{F})$.

2.5. Show that if the measure H has density h, then $U * H$ has density $U * h$. Explain in the case $H = F$ how this implies the existence of the *renewal density* $\mathrm{d}U(x)/\mathrm{d}x$.

Notes

In the notation, we have used the convention that the convolution $F * Z$ of a function Z and a measure F is a function. When identifying the measure $\mathrm{d}F(x)$ with the function $F(t) = \int_0^t \mathrm{d}F(x)$, this is consistent with the usual convolution of measures. Indeed, if Z itself corresponds to $\mathrm{d}Z(x)$, then $F * Z(t) = \int_0^t \mathrm{d}(F * Z)(x)$. The proof of this is elementary as well as that of formulae like $F * G = G * F$, $F * (G * Z) = (F * G) * Z$ used without further notice in the text (here G is another measure).

More material on the demographic model in Example 2.2 can be found in Pollard (1973). For further (more probabilistic) applications of renewal theory to population models, see Keiding and Hoem (1976), Asmussen and Hering (1983, X.2–3) and Jagers and Nerman (1984).

3. STATIONARY RENEWAL PROCESSES

The definition of a renewal process (or a general point process on $[0, \infty)$) being stationary has the obvious meaning: if for any t we shift the origin to t, then the distribution of the epochs should be left unchanged. Clearly, this will hold if the distribution of the forward recurrence time B_t does not depend on t, i.e. if the Markov process $\{B_t\}$ is stationary. Conversely, this is also necessary since otherwise the waiting time until the first renewal epoch following t has a distribution depending on t.

We assume throughout $\|F\| = 1$ and we let μ denote the mean of F, F_0 the measure with density $\mathrm{d}F_0(x)/\mathrm{d}x = (1 - F(x))/\mu$. Then F_0 is a distribution on $(0, \infty)$ if $\mu < \infty$, whereas F_0 vanishes on $(0, \infty)$ if $\mu = \infty$ (it may then sometimes be convenient intuitively to think of F_0 as the distribution degenerate at ∞). Recall from Appendix A.4 that the *intensity measure* is the measure counting the expected number of renewals.

Theorem 3.1 *If $\mu < \infty$, then F_0 has the following properties: (i) A renewal process with delay distribution F_0 is stationary; (ii) F_0 is stationary for the Markov process $\{A_t\}_{t \geqslant 0}$; (iii) F_0 is stationary for the Markov process $\{B_t\}_{t \geqslant 0}$; (iv) The intensity measure of the renewal process with delay distribution F_0 is stationary, i.e. Lebesgue measure times a constant* [necessarily μ^{-1} by the elementary renewal theorem]. *Conversely, let F_0^* be a given distribution, let* (i)* *be* (i) *with F_0 replaced by F_0^* and so on. Then if either of* (i)*-(iv)* *holds, we have* (v) $\mu < \infty$, $F_0^* = F_0$.

We separate two main steps of the proof into the following lemmas:

Lemma 3.2. (i) *If the distribution of A_t is independent of t, then so is the distribution of B_t*; (ii) *if $\mu < \infty$ and A_t has distribution F_0, then so has B_t.*

Proof. Part (i) follows since

$$\mathbb{P}(B_t > b \mid A_t = a) = \mathbb{P}(Y_k > a + b \mid Y_k > a) = \frac{1 - F(a+b)}{1 - F(a)} \tag{3.1}$$

does not depend on t. In part (ii), it follows from (3.1) that

$$\mathbb{P}(B_t > b) = \int_0^\infty \frac{1 - F(a+b)}{1 - F(a)} F_0(\mathrm{d}a) = \frac{1}{\mu}\int_0^\infty (1 - F(a+b))\,\mathrm{d}a$$

$$= \frac{1}{\mu}\int_b^\infty (1 - F(a))\,\mathrm{d}a = 1 - F_0(b). \qquad \square$$

Lemma 3.3. *A measure H on $(0, \infty)$ satisfies $U * H(\mathrm{d}x) = \mathrm{d}x$ (Lebesgue measure) if and only if H has density $1 - F(x)$.*

Proof. If H has density $h(x) = 1 - F(x)$, then (Problem 2.5) it is easily seen that $U * H$ has density

$$U * h = U - U * F = F^{*0} = 1,$$

i.e. $U * H(\mathrm{d}x) = \mathrm{d}x$. If also $U * H_1(\mathrm{d}x) = \mathrm{d}x$, then $U * H(a) = a = U * H_1(a)$. Thus the solutions of $Z = H + F * Z$, $Z = H_1 + F * Z$ are the same, and this immediately implies $H = H_1$. $\square$

Proof of Theorem 3.1. It follows from Lemma 3.2 and the introductory remarks that the easy part of Theorem 3.1 is (ii)* $\Rightarrow$ (iii)* $\Leftrightarrow$ (i)* $\Rightarrow$ (iv)*. It is thus sufficient to show that (iv)* $\Rightarrow$ (v) and that (ii) holds. For (iv)* $\Rightarrow$ (v), we note that by Theorem 2.4 (iii), a renewal process with delay distribution F_0^* has intensity measure $U * F_0^*$. If (iv)* holds, we thus have $U * F_0^*(\mathrm{d}x) = \mathrm{d}x/\mu$. Obviously this excludes $\mu = \infty$, and Lemma 3.3 then yields $\mu F_0^*(\mathrm{d}x) = (1 - F(x))\,\mathrm{d}x$, i.e. $F_0^* = F_0$. For (ii) consider a renewal process with A_0 distributed according to F_0 and thus (Lemma 3.2) with delay distribution F_0. According to Lemma 3.3, the intensity measure is $U * F_0(\mathrm{d}a) = \mathrm{d}a/\mu$, and conditioning upon the time a of last renewal before t then yields

$$\mathbb{P}(A_t \leqslant \xi) = \int_0^{(\xi - t)+} \frac{1 - F(a + t)}{1 - F(a)} F_0(\mathrm{d}a) + \int_{(t-\xi)+}^{t} (1 - F(t - a)) \frac{1}{\mu} \mathrm{d}a$$

(with the first term corresponding to $Y_0 > t$). Inserting $F_0(\mathrm{d}a) = (1 - F(a))\,\mathrm{d}a/\mu$ and substituting $y = a + t$ and $y = t - a$ respectively, this reduces to

$$\int_t^{t \vee \xi} (1 - F(y)) \frac{1}{\mu} \mathrm{d}y + \int_0^{t \wedge \xi} (1 - F(y)) \frac{1}{\mu} \mathrm{d}y = F_0(\xi). \qquad \square$$

Corollary 3.4. *A (delayed) renewal process is stationary if and only if $\mu < \infty$ and the distribution of the initial delay B_0 is F_0. In that case $\{B_t\}$ is stationary.*

This is an obvious consequence of (i), (iii) and the converse part of Theorem 3.1. One might think from (ii) that the same result with B_0 and $\{B_t\}$ replaced by A_0 and $\{A_t\}$ would hold. Indeed, it is clearly sufficient for stationarity that A_0 has distribution F_0. An obvious counter-example to necessity is provided by the Poisson process, where $F_0 = F$, but by the memoryless property of the exponential distribution B_0 is always distributed according to $F = F_0$ if one constructs the process by assigning the distribution of A_0.

Now define $C_t = A_t + B_t$ as (in the terminology of Example 1.1) the *current lifetime* of the item at time t.

Corrollary 3.5. *In a stationary renewal process with initial age distribution F_0, $\{C_t\}$ is also stationary. The common distribution of the C_t is then the one with density x/μ with respect to* $\mathrm{d}F(x)$.

Proof. Stationarity is clear, and using integration upon parts we get

$$\mathbb{P}(C_0 > t) = \int_0^{\infty} \mathbb{P}(Y_1 > t \mid Y_1 > a)\,\mathrm{d}F_0(a)$$

$$= \int_0^t (1 - F(t))\,\mathrm{d}a/\mu + \int_t^{\infty} (1 - F(a))\,\mathrm{d}a/\mu = \int_t^{\infty} x/\mu\,\mathrm{d}F(x). \qquad \square$$

This result is known as the *inspection* or *waiting-time paradox*, stating that the item at time t is not typical in the sense that its distribution is not F. The reason is loosely that the item is likely to be one of those with long lifetimes. The paradox is important, not only for its own sake but also as a warning for intuition in many similar situations.

Problems

3.1. Show the above claim that $F_0 = F$ in the Poisson process.
3.2. Evaluate F_0 if F is degenerate at 1.
3.3. Evaluate F_0 for the cases $F = E_k$, $F = H_k$ and $F \in \mathscr{P}\mathscr{H}_{\mathrm{ME}}$, cf. III.6.
3.4. Find the density of $F_0 * F^{*n}$.
3.5. By considering times $t, t + \mathrm{d}t$, give an alternative heuristical (and somewhat easier!) proof of the stationarity of F_0 for $\{A_t\}$, $\{B_t\}$.
3.6. Show that the current life distribution in Corollary 3.2 is stochastically larger than F, $\mathbb{P}(C_0 > t) \geqslant 1 - F(t)$.

Notes

For a discussion of some problems related to the discussion following Corollary 3.4, see Berndtsson and Jagers (1979).

4. THE RENEWAL THEOREM IN ITS EQUIVALENT VERSIONS

The renewal theorem is one of the very fundamental results of probability theory, perhaps not so much because of its intrinsic interest but rather because of the applicability to, and strong implications for, a number of other areas. It has several versions, one analytical giving asymptotic estimates for the solutions of (proper) renewal equations, and various probabilistic ones which all in some way state that as $t \to \infty$, then a (pure or delayed) renewal process behaves like a stationary one if $\mu < \infty$ and has a behaviour like null recurrence if $\mu = \infty$. In the present section, we state the various versions and prove their equivalence. The proof of the renewal theorem is then given in Section 5.

From now on it becomes necessary to distinguish between F being lattice or non-lattice. In the lattice case, one may rescale time so as to make F aperiodic on $\mathbb{N}$ and a number of aspects of renewal theory for that case has already been studied in I.2 (including the Problems). We shall therefore almost entirely concentrate on the non-lattice case and only state a few selected results for the lattice case.

Before being able to state all versions of the renewal theorem, we need a definition. Suppose for a while that z in the renewal equation $Z = z + F * Z$ is non-negative, and define for some $h > 0$

$$\bar{z}_h(x) = \sup_{y \in I_h^n} z(y), \qquad \underline{z}_h(x) = \inf_{y \in I_h^n} z(y), \qquad x \in I_h^n = (nh, (n+1)h],$$

cf. Fig. 4.1. Then we call z *directly Riemann integrable* (d.R.i.) if $\int \bar{z}_h = \int_0^\infty \bar{z}_h(x)\,\mathrm{d}x$ is finite for some (and then all) h, and $\int \bar{z}_h - \int \underline{z}_h \to 0$ as $h \to 0$. For functions with compact support this concept is the same as Riemann integrability, whereas in the general case it will be seen to be somewhat stronger than Lebesgue integrability. If z can also attain negative values, we say that z is d.R.i. if both z^+ and z^- are so.

Proposition 4.1. *Suppose $z \geqslant 0$. Then if z is d.R.i., z is Lebesgue integrable and*

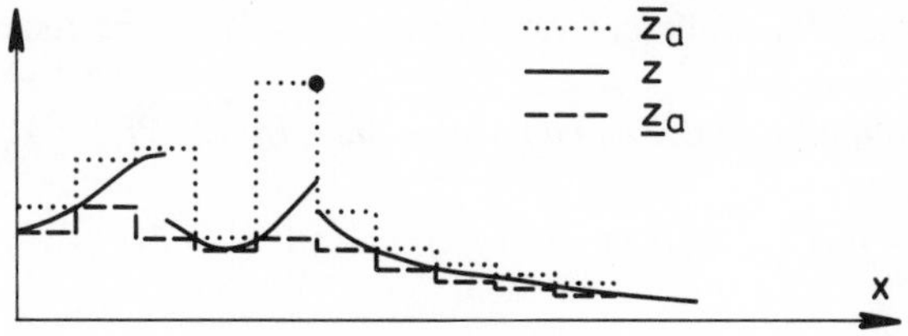

Fig. 4.1

$\int \bar{z}_h$, $\int \underline{z}_h$ have the common limit $\int z$ as $h \downarrow 0$. A necessary condition for z being d.R.i. is

(i) *z is bounded and continuous almost everywhere with respect to Lebesgue measure. Sufficient conditions are:*
(ii) *$\int \bar{z}_h < \infty$ for all h and (i) holds;*
(iii) *z has bounded support and (i) holds;*
(iv) *$z \leqslant z^*$ with z^* d.R.i. and (i) holds for z;*
(v) *z is non-increasing and Lebesgue integrable.*

Proof. Boundedness is necessary for $\int \bar{z}_h < \infty$. Suppose z is bounded but not almost everywhere continuous. Then if we let $\bar{z}(x) = \overline{\lim}_{y \to x} z(y)$, $\underline{z}(x) = \underline{\lim}_{y \to x} z(y)$ we have for some $\varepsilon > 0$ that the Lebesgue measure of $\{x: \bar{z}(x) > \underline{z}(x) + \varepsilon\}$ is non-zero, say $\delta > 0$. But except possibly for $x = nh$ we have

$$\bar{z}_h(x) \geqslant \bar{z}(x) \geqslant \underline{z}(x) \geqslant \underline{z}_h(x),$$

hence

$$\int \bar{z}_h - \int \underline{z}_h \geqslant \int \bar{z} - \int \underline{z} \geqslant \varepsilon\delta$$

for any h, and the necessity of (i) follows. In particular, if z is d.R.i. then by (i) $\bar{z}_h \downarrow z$ almost everywhere and $\lim_{h \downarrow 0} \bar{z}_h$ is Lebesgue integrable by monotone convergence. Hence z is so too, and $\int z = \lim \int \bar{z}_h$. Similarly, $\int \underline{z}_h \to \int z$. The same argument gives the sufficiency of (ii), and obviously (iii)$\Rightarrow$(ii). If (iv) holds, then $\int \bar{z}_h \leqslant \bar{z}_h^* < \infty$ and (ii) holds. Finally by a standard argument (v)$\Rightarrow$(ii). □

Example 4.2. We shall show that

$$z_A(t) = I(t \leqslant \xi)(1 - F(t)), \qquad z_B(t) = F(t + \xi) - F(t)$$

in Example 2.1 are d.R.i. Here F, having only a finite number of jumps, is a.e.c., and thus the assertion for z_A follows from (iii). If $\mu < \infty$, we may apply (iv) to $z = z_B$, with $z^* = 1 - F(t)$ being d.R.i. according to (v). If $\mu = \infty$, let $m \in \mathbb{N}$ satisfy $\xi \leqslant (m-1)h$, use (ii) and

$$\int \bar{z}_h \leqslant h \sum_{h=0}^{\infty} \{F((n+1)h + \xi) - F(nh)\} \leqslant h \sum_{n=0}^{\infty} \{F((n+m)h) - F(nh)\}$$

$$= h \lim_{N \to \infty} \sum_{n=0}^{N} \cdots \leqslant h \lim_{N \to \infty} \sum_{k=N+1}^{N+m} F(kh) \leqslant hm.$$ □

We can now state four different versions ((4.1)–(4.4)) of the renewal theorem:

Theorem 4.3. *Suppose that F is non-lattice and proper ($\|F\| = 1$), and let*

$$\mu = \int_0^\infty x\,\mathrm{d}F(x), \qquad F_0(t) = \mu^{-1}\int_0^t (1 - F(y))\,\mathrm{d}y$$

(i.e. $F_0 = 0$ in case $\mu = \infty$). *Then*

(4.1) (BLACKWELL'S RENEWAL THEOREM)

Let $U = \sum_{n=0}^{\infty} F^{*n}$ be the renewal function. Then for all a,

$$\lim_{t\to\infty} U(t+a) - U(t) = \frac{a}{\mu}.$$

More generally, in any (possibly delayed) *renewal process with interarrival distribution F the expected number $V_t(a)$ of renewals in $(t, t+a]$ tends to a/μ as $t \to \infty$.*
(4.2) *Let $\{A_t\}_{t\geqslant 0}$ be the backward recurrence-time process in a* (possibly delayed) *renewal process with interarrival distribution F. Then* (irrespective of the distribution of A_0) *for all ξ, $\lim_{t\to\infty} \mathbb{P}(A_t \leqslant \xi) = F_0(\xi)$.*
(4.3) *Let $\{B_t\}_{t\geqslant 0}$ be the forward recurrence-time process in a* (possibly delayed) *renewal process with interarrival distribution F. Then* (irrespective of the distribution of B_0) *for all ξ, $\lim_{t\to\infty} \mathbb{P}(B_t \leqslant \xi) = F_0(\xi)$.*
(4.4) (KEY RENEWAL THEOREM)
Suppose that the function z in the renewal equation (2.1) *is d.R.i. Then*

$$\lim_{t\to\infty} Z(t) = \lim_{t\to\infty} U * z(t) = \frac{1}{\mu}\int_0^\infty z(x)\,\mathrm{d}x.$$

We note that in the case $\mu < \infty$, (4.1)–(4.3) state that the renewal process becomes asymptotic stationary as $t \to \infty$, cf. Theorem 3.1. In particular, (4.2) and (4.3) are equivalent to $A_t \xrightarrow{\mathscr{D}} F_0$ and $B_t \xrightarrow{\mathscr{D}} F_0$. If $\mu = \infty$, (4.2) and (4.3) states that the mass in the distributions of A_t and B_t drifts off to $+\infty$.

Proof of (4.4)$\Rightarrow$(4.2). Consider first the case of a pure renewal process, where according to Example 2.1 we have to show

$$Z_A(t) = U * z_A(t) \to F_0(\xi), \qquad z_A(t) = I(t \leqslant \xi)(1 - F(t)).$$

But z_A being d.R.i. was shown in Example 4.2, hence by (4.4) the limit of $Z_A(t)$ exists and is $\mu^{-1}\int z_A = F_0(\xi)$. In case of a general delay distribution F_0^*, replace F by F_0^* in (2.4), let $t \to \infty$ and note that the first term tends to zero and $F_0^* * Z_A(t)$ to $\lim Z_A(t) = F_0(\xi)$ by dominated convergence. □

Proof of (4.2)$\Leftrightarrow$(4.3). This may be easily seen from the identity

$$\{B_t \leqslant \xi\} = \{\text{renewal in } (t, t+\xi]\} = \{A_{t+\xi} < \xi\} \tag{4.5}$$

□

Proof of (4.3)$\Rightarrow$(4.1). Let $h(\xi)=U(a-\xi)I(\xi\leqslant a)$. Then h is bounded and a.e.c. w.r.t. $\mathrm{d}\xi$, hence a.e.c. w.r.t. $\mathrm{d}F_0(\xi)$, and since $G_t \to F_0$ in the sense of weak convergence of bounded measures (here $G_t(\xi)=\mathbb{P}(B_t\leqslant\xi)$), we get from (2.8)

$$V_t(a)=U*G_t(a)=\int_0^\infty h(\xi)\,\mathrm{d}G_t(\xi)\to\int_0^\infty h(\xi)\,\mathrm{d}F_0(\xi)=U*F_0(a)=\frac{a}{\mu}. \qquad \square$$

Proof of (4.1)$\Rightarrow$(4.4). Let $nh<x\leqslant(n+1)h$ and define $I_k=I_k(x)=(x-(k+1)h, x-kh]$. Then

$$Z(x)=\int_0^x z(x-y)\,\mathrm{d}U(y)=\int_0^{x-nh} z(x-y)\,\mathrm{d}U(y)+\sum_{k=0}^{n-1}\int_{I_k} z(x-y)\,\mathrm{d}U(y).$$

Since $z(t)\to 0$ as $t\to\infty$, the first term tends to 0. The second is at most

$$\sum_{k=0}^{n-1}\bar{z}_h(kh)(U(x-kh)-U(x-(k+1)h))$$

$$\leqslant\sum_{k=0}^{M}\bar{z}_h(kh)(U(x-kh)-U(x-(k+1)h))+U(h)\sum_{k=M+1}^{n-1}|\bar{z}_h(kh)|,$$

since by Theorem 2.4(iii) $U(t+h)-U(t)\leqslant U(h)$ for all t. Letting first $n, x\to\infty$ with M fixed, next $M\to\infty$ and finally $h\to 0$ yields

$$\overline{\lim_{x\to\infty}}\, Z(x)\leqslant\frac{h}{\mu}\sum_{k=0}^{M}\bar{z}_h(kh)+U(h)\sum_{k=M+1}^{\infty}|\bar{z}_h(kh)|,$$

$$\overline{\lim_{x\to\infty}}\, Z(x)\leqslant\frac{1}{\mu}\int\bar{z}_h+0, \qquad \overline{\lim_{x\to\infty}}\, Z(x)\leqslant\frac{1}{\mu}\int_0^\infty z(t)\,\mathrm{d}t.$$

$\underline{\lim}\geqslant$ is proved similarly. $\square$

Suitable versions of the renewal theorem also exist in the lattice case. Mathematically, this case is somewhat easier and has to a large extent already been treated in I.2. For example:

Proposition 4.4. *Suppose F is lattice with span δ. Then if $\varphi(y)=\sum_{n=0}^\infty z(y+n\delta)$ converges absolutely for some $y\in[0,\delta)$, it holds that $U*z(y+n\delta)\to(\delta/\mu)\varphi(y)$ as $n\to\infty$.*

Proof. The renewal measure U is supported by $\{0,\delta,2\delta,\ldots\}$, with mass say $u_{k\delta}$ at $k\delta$. By I.2.2, $u_{k\delta}\to\delta/\mu$ as $k\to\infty$, and hence by dominated convergence,

$$Z(y+n\delta)=U*z(y+n\delta)=\sum_{k=0}^{n} z(y+n\delta-k\delta)u_{k\delta}$$

$$=\sum_{k=0}^{n} z(y+k\delta)u_{(n-k)\delta}\to\frac{\delta}{\mu}\varphi(y). \qquad \square$$

The connection between Blackwell's renewal theorem and the key renewal theorem may in more abstract terms be rephrased as follows. Consider for each

t the measure v_t on $[0,\infty)$ obtained by time reversion of the renewal measure restricted to $[0,t]$, i.e.

$$\int_0^\infty f(x)\,\mathrm{d}v_t(x) = \int_0^t f(t-y)\,\mathrm{d}U(y).$$

Then Blackwell's theorem asserts that $v_t([0,a)) \to a/\mu$ for each a which by general results from measure theory is equivalent to $v_t(\mathrm{d}a) \to \mathrm{d}a/\mu$ vaguely, i.e. $\int f(a)v_t(\mathrm{d}a) \to \int f(a)\,\mathrm{d}a/\mu$ whenever f has compact support and is continuous (or more generally bounded and a.e. continuous). Any such f is d.R.i. and hence we may view the key renewal theorem as an extension of Blackwell's theorem to involve also convergence of certain f with unbounded support, a case of major importance for the applications.

Problems

4.1. Show that the stationary distribution of the current life in Corollary 3.5 is also a limiting distribution irrespective of the delay distribution.
4.2. Show that if the renewal density u exists (Problem 2.5), then $u(x) \to 1/\mu$.
4.3. Show that the z in Problem 2.2 (the type II counter) is d.R.i. and express $\lim Z(t)$ in terms of the Laplace transform $\hat{G}(\beta)$.
4.4. Find examples of functions which are Lebesgue integrable but not d.R.i.
4.5. Give a simplified proof of Theorem 3.1 by invoking the renewal theorem.

5. PROOF OF THE RENEWAL THEOREM

The formulation of the renewal theorem in terms of the recurrence-time processes shows that for the lattice case the situation is essentially settled by the analysis of Chapter I. The non-lattice case is considerably more involved, and no really short and elementary approach is known. In the present text two proofs of the renewal theorem will be presented, in this section a standard (in part analytical) one developed largely by Feller (1971), and in VI.2 a more recent probabilistic one based upon the concept of a coupling. Some ingredients are common, in particular the way in which the non-lattice property comes in:

Lemma 5.1. *Suppose that F is non-lattice on $(0,\infty)$ and define $U = \sum_0^\infty F^{*n}$, $S = \mathrm{supp}(U)$. Then S is asymptotically dense at infinity in the sense that $d(x,S) \to 0$, $x \to \infty$.*

Proof. We may assume $1 \in \mathrm{supp}(F)$ and shall show that $\overline{\lim}\, d(x,S) \leqslant q^{-1}$ for any $q \in \mathbb{N}$. Define $I_p = [(p-1)/q, p/q]$. If an irrational $\theta \in \mathrm{supp}(F)$ exists, then (see Problem 5.1) we can find integers k_p, m_p with $k_p\theta - m_p \in I_p$. Let $n_0 = m_1 \vee \cdots \vee m_q$ and $x \in n + I_p$ with $n \geqslant n_0$. Then

$$y = k_p\theta - m_p + n \in \mathrm{supp}(F^{*k_p + (n-m_p)}) \subseteq S, \qquad |x-y| \leqslant \frac{1}{q},$$

and thus indeed $d(x,S) \leqslant q^{-1}$ for $x \geqslant n_0$. If on the other hand all $\theta \in \mathrm{supp}(F)$ are rational, then θ can be chosen on the form q_1/q_2 with q_1, q_2 relative prime and $q_2 \geqslant q$. Let $r = 0, 1, \dots, q_2$. Then r/q_2 is of the form $k\theta - m$ for some $k, m \in \mathbb{N}$, and the same argument as before applies. $\qquad\square$

Proposition 5.2. (CHOQUET-DENY) *If F is a non-lattice distribution on $(0, \infty)$ and φ a bounded continuous function satisfying $\varphi = F * \varphi$, then φ is necessarily constant.*

Proof. Since

$$\mathbb{E}[\varphi(x - S_{n+1}) \mid Y_1, \dots, Y_n] = \int_0^\infty \varphi(x - S_n - y) \, dF(y)$$

$$= F * \varphi(x - S_n) = \varphi(x - S_n),$$

the sequence $\{\varphi(x - S_n)\}$ is a bounded martingale and hence converges a.s. and in L^2. By the Hewitt–Savage 0–1 law the limit is almost surely constant which by L^2-theory for martingales implies that $\varphi(x) = \varphi(x - S_1) = \dots = \varphi(x - S_n) = \dots$ a.s. Thus $\varphi(x - u) = \varphi(x)$ for F^{*n} — a.a. u, which by the continuity of φ shows that $\varphi(x - u) = \varphi(x)$ for all $u \in \operatorname{supp} F^{*n}$. Now let a, b be given. Then by Lemma 5.1 we can choose sequences $\{a_n\}$, $\{b_n\}$ with $n - a_n \to a$, $n - b_n \to b$ and $a_n, b_n \in \bigcup_0^\infty \operatorname{supp} F^{*m}$. Then

$$\varphi(a) = \lim \varphi(n - a_n) = \lim \varphi(n) = \lim \varphi(n - b_n) = \varphi(b) \qquad\square$$

Let now $\lambda^{(t)}(dy) = U(t - dy)$, $\lambda(dy) = dy/\mu$, $t \geqslant 0$, so that the renewal theorem is equivalent to $\lambda^{(t)} \to \lambda$ vaguely (cf. the concluding remarks in Section 4). To show this, it is sufficient to show that each sequence $\{s_n\}$ has a subsequence $\{t_n\}$ with $\lambda^{(t_n)} \to \lambda$. By Theorem 2.4 (iii), $\sup_n \lambda^{(s_n)}(K) < \infty$ for each compact set K which by standard facts from measure theory implies that $\{\lambda^{(s_n)}\}$ is vaguely compact. Thus $\lambda^{(t_n)} \to \nu$ for some subsequence $\{t_n\}$ and some ν, and we have to prove $\nu = \lambda$.

Lemma 5.3. $\lambda^{(t_n + x)} \to \nu$ *for each $x \in \mathbb{R}$.*

Proof. With $\nu^{(x)}(dy) = \nu(dy - x)$, it is clear that $\lambda^{(t_n + x)} \to \nu^{(x)}$ for each $x \in \mathbb{R}$. To get $\nu^{(x)} = \nu$, it is sufficient to show that continuous functions z with compact support have the same integral, i.e. that

$$\varphi(x) = \int z(y)\nu^{(x)}(dy) = \lim_{n \to \infty} \int z(y)\lambda^{(t_n + x)}(dy)$$

$$= \lim_{n \to \infty} U * z(t_n + x) \qquad (5.1)$$

is independent of x. But $Z = U * z$ being the solution of a renewal equation, we may write (5.1) as

$$\lim_{n \to \infty} \left\{ z(t_n + x) + \int_0^{t_n + x} Z(t_n + x - y) \, dF(y) \right\} = 0 + \int_0^\infty \varphi(x - y) \, dF(y)$$

(using dominated convergence). Hence by the Choquet–Deny theorem, we have only to show that φ is continuous which will follow if Z is uniformly continuous.

But let z be supported by $[0, T]$ and define $\kappa(\varepsilon) = \sup\{|z(x) - z(y)| : |x - y| \leqslant \varepsilon\}$. Then for $|x_1 - x_2| \leqslant \varepsilon$,

$$|Z(x_1) - Z(x_2)| = \left| \int_0^{x_1} z(x_1 - y)U(\mathrm{d}y) - \int_0^{x_2} z(x_2 - y)U(\mathrm{d}y) \right|$$
$$\leqslant \kappa(\varepsilon) \cdot U(T + \varepsilon) \to 0, \qquad \varepsilon \downarrow 0. \qquad \square$$

Proof of the renewal theorem. The conclusion of Lemma 5.3 means that v is translation invariant, i.e. $v(\mathrm{d}x) = \gamma \, \mathrm{d}x$ for some γ, and we have to show $\gamma = 1/\mu$. From

$$U(t_n - a) - U(t_n - a - h) = \lambda^{(t_n)}[a, a + h) \to \gamma h$$

for all $h > 0$, it then follows by the same argument as in the proof of (4.1)$\Rightarrow$(4.4) that $U * z(t_n) \to \gamma \int z$ whenever z is d.R.i. If $\mu < \infty$, let $z(x) = 1 - F(x)$. Then $U * z \equiv 1$, cf. Lemma 3.3, and $\int z = \mu$ so that $1 = \gamma\mu$. If $\mu = \infty$, let $z(x) = (1 - F(x))I(x \leqslant x_0)$. Then similarly we get $\gamma \int z \leqslant 1$ which, letting $x_0 \to \infty$, yields $\gamma\mu = \gamma \cdot \infty \leqslant 1$ and $\gamma = 0 = 1/\mu$. $\square$

Problems

5.1. Let θ be irrational and $\theta_n = n\theta \bmod 1$. Show that $\{\theta_n\}$ is dense in $[0, 1]$ [hint: choose β as limit of a subsequence, choose $n_1 > n_2$ with $|\theta_{n_i} - \beta| < \varepsilon/2$ and show that $d(x, \{\theta_{k(n_1 - n_2)}\}_{k \in \mathbb{N}}) < \varepsilon$ for all $x \in [0, 1]$].

Notes

The present proof of the renewal theorem follows texts like Feller (1971) and Jagers (1975). For alternative proofs, we mention the coupling proof in VI.2, a Markov chain proof due to McDonald (1975) and finally the Fourier analytical proofs which can be found, for example, in Woodroofe (1982).

CHAPTER V

Regenerative Processes

1. BASIC LIMIT THEORY

The classical definition of a stochastic process $\{X_t\}$ to be *regenerative* means in intuitive terms that the process can be split into independent identically distributed *cycles* (sometimes also denoted as *tours*). A basic example is the $GI/G/1$ queue and its busy cycles, that is, the time intervals separated by the instants S_n with a customer entering an empty system, cf. Fig. 1.1. At each such instant the queue *regenerates*, i.e. starts completely from scratch independently of the past. Different cycles are independent and are all governed by the same probability law.

This structure is found in the majority of examples and can safely be used as a guide for intuition in the present chapter (except for Section 2(e)), but we shall use a slightly wider definition. Assume that $\{X_t\}$ has state space E and discrete or continuous-time parameter, $t \in T = \mathbb{N}$ or $t \in T = [0, \infty)$. We then call $\{X_t\}_{t \in T}$ (pure or delayed) *regenerative* if there exists a (pure or delayed) renewal process $\{S_n\} = \{Y_0 + \cdots + Y_n\}$ with the following property: for each $n \geq 0$, the post-S_n-process

$$\{Y_{n+1}, Y_{n+2}, \ldots, \{X_{S_n + t}\}_{t \in T}\}$$

is independent of $S_0, S_1, \ldots, S_n$ (or equivalently of Y_0, $Y_1, \ldots, Y_n$) and its distribution does not depend on n. We call $\{S_n\}$ the *imbedded renewal process* and refer to the S_n as *regeneration points.*

Concerning the definition, we note the following points:

(a) Cycles are still well defined, all governed by the same probability law but some dependence between cycles may occur (for main examples, see Sections 2(e) and VI.3).

(b) The imbedded renewal process is by no means unique. For example, we may as well use $\{S_{2n}\}$ or, in the $M/G/1$ case of the queueing example, take the instants of initiation of idle periods as regeneration points.

(c) The S_n need not be given as function of $\{X_t\}$. In particular, there are some examples (e.g. VI.3) where we need to enlarge the probability space and introduce randomization before the regenerative properties of $\{X_t\}$ can be recognized. It may, however, in some cases be convenient to have the S_n given

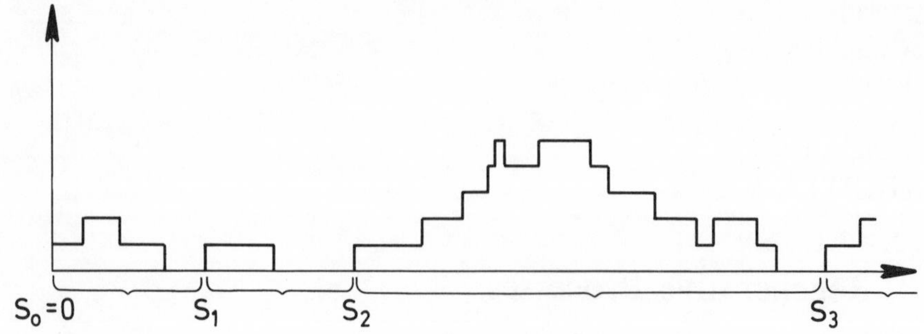

Fig. 1.1

as stopping times for $\{X_t\}$. One can then just enlarge the state space to $E \times [0, \infty)$ and consider $\{\tilde{X}_t\} = \{X_t, B_t\}$, with $\{B_t\}$ the forward recurrence-time process of $\{S_n\}$ (clearly, $\{\tilde{X}_t\}$ is regenerative with the same imbedded renewal process).

To a given delayed regenerative process, there clearly corresponds a pure or zero-delayed on with a uniquely given probability law (e.g. $\{X_{S_0+t}\}_{t\in T}$). We let $\mathbb{P}_0, \mathbb{E}_0$ correspond to the zero-delayed case and then write $Y = Y_1$ for the length of the first cycle, $\mu = \mathbb{E}_0 Y$.

A trivial but noteworthy property is that the regenerative property is preserved under mappings (nothing like that is true, say, for a Markov process!):

Proposition 1.1 *If $\{X_t\}_{t\in T}$ is regenerative and $\varphi: E \to E_0$ any measurable mapping, then $\{\varphi(X_t)\}_{t\in T}$ is regenerative with the same imbedded renewal process.*

The power of the concept of regenerative processes lies in the existence of a limiting distribution under conditions which are very mild and usually easy to verify. For example, in continuous time it is only required that $\mu < \infty$, that the cycle length distribution is non-lattice and that the sample paths satisfy some conditions which are automatic in any concrete example:

Theorem 1.2. *Assume that a* (possibly delayed) *continuous-time regenerative process $\{X_t\}_{t\geqslant 0}$ has metric state space E, right-continuous paths and non-lattice cycle length distribution with finite mean μ. Then the limiting distribution, say $\mathbb{P}_e$, of X_t exists, is given by*

$$\mathbb{E}_e f(X_t) = \frac{1}{\mu}\mathbb{E}_0 \int_0^Y f(X_s)\,\mathrm{d}s \tag{1.1}$$

and for all bounded functions $f: E \to \mathbb{R}$ which are continuous a.e. with respect to the distribution (1.1), *it holds that $\mathbb{E}f(X_t) \to \mathbb{E}_e f(X_s)$ as $t \to \infty$.*

Proof. It is immediately checked that

$$A \to \frac{1}{\mu}\mathbb{E}_0 \int_0^Y I(X_s \in A)\,\mathrm{d}s$$

defines a probability measure on the Borel σ-algebra on E and hence by standard facts on weak convergence it is sufficient to prove that $\mathbb{E}f(X_t)\to\mathbb{E}_e f(X_t)$ whenever f is continuous and $0\leqslant f\leqslant 1$. Letting $Z(t)=\mathbb{E}_0 f(X_t)$, $z(t)=\mathbb{E}_0[f(X_t); t<Y]$, $F_0^*(x)=\mathbb{P}(Y_0\leqslant x)$ it follows by the usual renewal argument and conditioning upon Y_0 that

$$\mathbb{E}f(X_t)=\mathbb{E}[f(X_t); t<Y_0]+\int_0^t Z(t-x)\,\mathrm{d}F_0^*(x), \tag{1.2}$$

$$Z(t)=z(t)+\int_0^t Z(t-x)\,\mathrm{d}F(x). \tag{1.3}$$

Hence letting $t\to\infty$ in (1.2) it follows that it is sufficient to show

$$Z(t)\to\mathbb{E}_e f(X_t)=\frac{1}{\mu}\int_0^\infty \mathbb{E}_0[f(X_t); t<Y]\,\mathrm{d}t=\frac{1}{\mu}\int_0^\infty z(t)\,\mathrm{d}t,$$

i.e. by the key renewal theorem to show that z is directly Riemann integrable (d.R.i.). But z is right-continuous, hence continuous a.e. by Proposition A3.1 of the Appendix. Also $z(t)\leqslant z^*(t)=\mathbb{P}(Y>t)=1-F(t)$ where z^* is d.R.i. by IV.4.1 (v). Part (iv) of IV.4.1 completes the proof. □

The basic renewal argument in the above proof may be given in various closely related ways. For example, the following result is frequently useful:

Proposition 1.3. *Let $\{X_t\}_{t\in T}$ be regenerative and $\{A_t\}_{t\geqslant Y_0}$ the backwards recurrence-time process of the imbedded renewal process. Further let $f:E\to\mathbb{R}$ be bounded and measurable, and define $g(t)=\mathbb{E}_0[f(X_t)\mid Y>t]$. Then*

$$\mathbb{E}f(X_t)=\mathbb{E}[g(A_t); Y_0\leqslant t]+\mathbb{E}[f(X_t); Y_0>t]. \tag{1.4}$$

In particular, in the zero-delayed case $\mathbb{E}_0 f(X_t)=\mathbb{E}_0 g(A_t)$.

Proof. Conditioning upon Y_0 shows that it is sufficient to consider the zero-delayed case. Define $Z(t)$, $z(t)$ as above and let

$$Z_1(t)=\mathbb{E}_0 g(A_t),\quad z_1(t)=\mathbb{E}_0[g(A_t); Y>t].$$

Then $Z=U*z$, $Z_1=U*z_1$ and the desired conclusion $Z(t)=Z_1(t)$ follows since

$$\mathbb{E}_0[g(A_t); Y>t]=g(t)\mathbb{P}(Y>t)=\mathbb{E}_0[f(X_t); Y>t]$$

implies that $z_1=z$. □

Proposition 1.3 yields an alternative proof of the limit result of Theorem 1.2, see Problem 1.3, and we also note the following strengthening (for total variation convergence, see Appendix A6):

Corollary 1.4. *If $\{A_t\}_{t\geqslant Y_0}$ converges to F_0 in total variation, then also a total variation limit of $\{X_t\}_{t\in T}$ exists and is given by* (1.1) *for $T=[0,\infty)$, whereas for*

$T = \mathbb{N}$

$$\mathbb{E}_e f(X_n) = \frac{1}{\mu}\mathbb{E}_0 \sum_{k=0}^{Y-1} f(X_k) = \frac{1}{\mu}\mathbb{E}_0 \sum_{k=1}^{Y} f(X_k). \tag{1.5}$$

Proof. We must show that $\mathbb{E}f(X_t)$ converges to the asserted limit uniformly in the bounded measurable f with $\|f\| \leqslant 1$. But since a uniform bound on the last term in (1.4) is $\mathbb{P}(Y_0 > t)$, the uniformity is immediate from the total variation convergence of A_t and (1.4). Also for $T = [0, \infty)$ (the case $T = \mathbb{N}$ is entirely similar), the limit is indeed given by

$$\mathbb{E}_e f(X_t) = \int_0^\infty g(t) F_0(\mathrm{d}t) = \frac{1}{\mu}\int_0^\infty \mathbb{E}_0[f(X_t) \mid Y > t]\mathbb{P}_0(Y > t)\,\mathrm{d}t$$

$$= \frac{1}{\mu}\mathbb{E}_0 \int_0^\infty f(X_t) I(Y > t)\,\mathrm{d}t. \qquad \square$$

Corollary 1.5. *Let $\{X_n\}_{n\in\mathbb{N}}$ be regenerative in discrete time with $\mu = \mathbb{E}_0 Y < \infty$ and let d be the period of the cycle length distribution. Then:*

(i) *In the aperiodic case $d = 1$, a total variation limit exists and is given by* (1.5);
(ii) *If $d > 1$, then as $n \to \infty$*

$$\frac{1}{d}\sum_{j=0}^{d-1} \mathbb{E}f(X_{nd+j}) \to \frac{1}{\mu}\mathbb{E}_0 \sum_{k=0}^{Y-1} f(X_k). \tag{1.6}$$

Proof. The process $\{A_n\}$ is Markovian and if $\mu < \infty$, $d = 1$, it follows from I.2–4 or Chapter IV that $A_n \to F_0$ weakly, hence also (since the state space is discrete) in total variation. Thus (i) follows from Corollary 1.4, whereas (ii) is a similar application of Proposition 1.3 and I.(4.2) (or (3.3) below). $\square$

We return to total variation convergence for $T = [0, \infty)$ in VI.1–2.

Problems

1.1. Let $\{A_t\}$ be the backwards recurrence-time process of a renewal process with interarrival distribution F with $\mu < \infty$ and let $X_t = I(A_t \in \mathbb{Q})$. Show that $\{X_t\}$ is regenerative, but that X_t need not converge in distribution if say F is concentrated on $\mathbb{Q}$.

1.2. Show by an example that $\mu < \infty$ is not necessary for convergence in distribution of a regenerative process [hint: $X_t \equiv 0$].

1.3. In Proposition 1.3, show that g is a.e. continuous provided f is continuous and the paths right-continuous. Give hereby an alternative proof of the limit result of Theorem 1.2.

1.4. Let $\{X_t\}_{t\in T}$ be regenerative and satisfying the conditions for existence of a limit distribution π, and let $f: E \to [0, \infty)$ be π-integrable. Show that $\mathbb{E}f(X_t) \to \int f(x)\pi(\mathrm{d}x)$ holds always when $T = \mathbb{N}$ but not when $T = [0, \infty)$ [hint: backwards recurrence times in renewal processes, f chosen such that $z(x) = f(x)(1 - F(x))$ is Lebesgue integrable with $\overline{\lim}\, z(x) = \infty$].

Notes

Some main references in the theory of regenerative processes are Feller (1966, Ch. XIII; 1971, XI.8), Smith (1955) and Miller (1972). Stidham (1972) has also frequently been cited, and Cohen (1976) surveys some main applications to queueing theory. There is occasionally some confusion about the precise definition of a stochastic process to be regenerative. Versions like the present one, allowing some dependence between cycles, seem to have been noticed independently by a number of people. Sometimes the term 'synchronous' instead of the present 'zero-delayed' is used.

Of concepts related to regenerative processes we mention regenerative phenomena (Kingman, 1972), renewing events (Borovkov, 1984) and Palm theory for point processes, a readable introduction to which can be found for example in Berbee (1979).

2. FIRST EXAMPLES AND APPLICATIONS

Examples and applications of regenerative processes to queues and related models will abound in Part C, so here we shall only consider a few topics of a somewhat different flavour.

2a. Renewal processes

Consider a renewal process with non-lattice interarrival distribution F. If $\mu < \infty$, the stationary limiting distributions of the recurrence times A_t, B_t and of the current life $C_t = A_t + B_t$ have been found in IV.3.1, IV.3.5. However, their particular form may be easier to understand from the basic formula (1.1) as follows. For $0 \leqslant t < Y$ we have $A_t = t$, $B_t = Y - t$, $C_t = Y$. In particular,

$$\begin{aligned}\mathbb{P}_e(A_t \leqslant \xi) &= \frac{1}{\mu}\mathbb{E}_0 \int_0^Y I(A_t \leqslant \xi)\,\mathrm{d}t = \frac{1}{\mu}\mathbb{E}_0 \int_0^Y I(t \leqslant \xi)\,\mathrm{d}t \\ &= \frac{1}{\mu}\mathbb{E}_0 \int_0^Y I(Y - s \leqslant \xi)\,\mathrm{d}s = \mathbb{P}_e(B_t \leqslant \xi)\end{aligned}$$

and the common value is

$$\frac{1}{\mu}\mathbb{E}_0 \int_0^\infty I(t \leqslant \xi, t < Y)\,\mathrm{d}t = \frac{1}{\mu}\int_0^\xi \mathbb{P}_0(t < Y)\,\mathrm{d}t = \frac{1}{\mu}\int_0^\xi (1 - F(t)\,\mathrm{d}t = F_0(\xi))$$

(from results of VI.3 it will also follow that the density $1 - F(x)$ is stationary for $\{A_t\}$, $\{B_t\}$ even if $\mu = \infty$). Finally

$$\begin{aligned}\mathbb{P}_e(C_t \leqslant \xi) &= \frac{1}{\mu}\mathbb{E}_0 \int_0^Y I(C_t \leqslant \xi)\,\mathrm{d}t = \frac{1}{\mu}\mathbb{E}_0 \int_0^Y I(Y \leqslant \xi)\,\mathrm{d}t \\ &= \frac{1}{\mu}\mathbb{E}_0[Y; Y \leqslant \xi] = \frac{1}{\mu}\int_0^\xi x\,\mathrm{d}F(x). \qquad \square\end{aligned}$$

Problems

2.1. Explain that $D_t = A_t/C_t$ is the relative position of the current time in the renewal interval. Show that in the limit D_t is uniformly distributed on $(0, 1)$ and is independent of C_t.

2b. Alternating renewal process

A point process on $[0, \infty)$ with first epoch at Y_0 and interarrival times $Y_1, Y_2, \ldots$ is called an *alternating renewal process* if $Y_0, Y_1, \ldots$ are independent and

$$\mathbb{P}(Y_{2k-1} \leqslant y) = F_1(y), \qquad \mathbb{P}(Y_{2k} \leqslant y) = F_2(y), \qquad k = 1, 2, \ldots,$$

for suitable distributions F_1, F_2 on $(0, \infty)$. Such processes arise, for example, in reliability theory where Y_{2k-1} could be the lifetime of the kth item and Y_{2k} the time needed to replace it. Here one might ask, for example, for the probability $p(t)$ that the system is operating at time t, for the distribution of the remaining lifetime of the current item and so on. These quantities are easily obtained by observing that the system regenerates at every second renewal. For example, for $p(t)$ we can define $X_t \in \{0, 1\}$ by

$$X_t = 0 \quad \text{on} \bigcup_{k=0}^{\infty} [Y_0 + \cdots + Y_{2k-1}, Y_0 + \cdots + Y_{2k}),$$

$$X_t = 1 \quad \text{on} \bigcup_{k=0}^{\infty} [Y_0 + \cdots + Y_{2k-2}, Y_0 + \cdots + Y_{2k-1}),$$

Then $p(t) = \mathbb{P}(X_t = 1)$ and the $Y_0 + \cdots + Y_{2k-2}$ are regeneration points for $\{X_t\}$. The cycle length distribution is the one $F = F_1 * F_2$ of $Y_1 + Y_2$, and if F is non-lattice with $\mu = \mathbb{E}Y_1 + \mathbb{E}Y_2 < \infty$ and $\mathbb{E}_0$ refers to the case $Y_0 = 0$, we get

$$\lim_{t \to \infty} p(t) = \frac{1}{\mu} \mathbb{E}_0 \int_0^{Y_1 + Y_2} I(X_t = 1)\, \mathrm{d}t = \frac{1}{\mu} \mathbb{E}Y_1 = \frac{\mathbb{E}Y_1}{\mathbb{E}Y_1 + \mathbb{E}Y_2}.$$

Further characteristics of the system are easily computed in just the same manner, see Problem 2.2. □

Problems

2.2. Consider an alternating renewal process as in Section 2(b). Show that, conditionally upon the system to be operating, the past life, the residual life and the total life of the current item all have the same limit distributions as in a renewal process with interarrival distribution F_1.

2c. A pacemaker neuron model

Consider a 'pacemaker' neuron firing signals whenever the membrane potential reaches a threshold value w_T. After a firing, the potential is reset to a constant value $w_0 < w_T$ and then increases linearly until either w_T is reached and a new firing occurs, or the cell itself receives an inhibiting signal in which case the

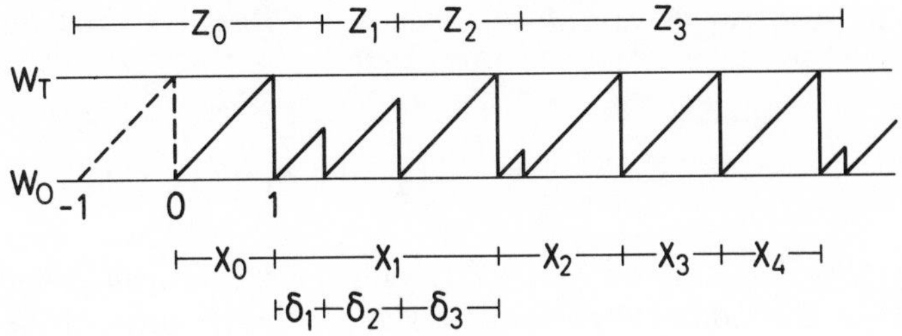

Fig. 2.1

potential is also reset to w_0. It is assumed that the input signals arrive according to a renewal process with interarrival times $Z_0, Z_1, \ldots$ with common distribution G, and we are interested in the lengths $X_0, X_1, \ldots$ of the intervals between successive firings of the pacemaker.

An illustration of the process is given in Fig. 2.1. The scale there and in the following is chosen such that $w_T - w_0 = 1$ and the linear rate of increase is 1. It is seen that the initial conditions have been chosen such that the last input before $t = 0$ has been received at $t = -1$. That is, the delay distribution of the renewal process is that of $Z - 1$ conditionally upon $Z > 1$. To avoid ambiguities when defining the process in cases such as input and signals at the same time, we assume that G has a density g. Furthermore, $0 < p = \int_0^1 g(x)\,\mathrm{d}x < 1$ will ensure that all $X_n < \infty$.

It is seen that the X_n can be divided into two types, long ones ($X_n > 1$) and short ones ($X_n = 1$). Each Z with $Z > 1$ induces a long X which can be split into three independent phases $\delta_1, \delta_2, \delta_3$. The first δ_1 is the time $Z - [Z]$ to the next inhibiting signal. The next δ_2 is a sum of Z_n with $Z_k < 1$, and finally $\delta_3 = 1$. After a long X_n, the next X_{n+1} is thus determined by the same initial conditions as in Fig. 2.1 and we have a zero-delayed regenerative process with independent identically distributed cycles and imbedded renewal process

$$S_0 = Y_0 = 0, \qquad S_{k+1} = \inf\{n > S_k : X_{n-1} > 1\}.$$

In Fig. 2.1, X_1 is long corresponding to Z_0, and X_2 is long corresponding to Z_2.

We shall verify the conditions of Corollary 1.5 for the existence of a limiting distribution of X_n, and as an example of the applicability of (1.5) show how one can evaluate $\mathbb{E}_e X_n$ which is of importance because $1/\mathbb{E}_e X_n$ is the average firing rate (as is made clear in the next section).

Define $M = [Z]$ and assume $\int_1^2 g(x)\,\mathrm{d}x > 0$. Then

$$f_1 = \mathbb{P}(Y = 1) = \mathbb{P}(X_0 > 1) = \mathbb{P}(Z_0 < 2 | Z_0 > 1) > 0$$

and aperiodicity is immediate. Furthermore the first cycle consists of $[Z_0 - 1] = [Z_0] - 1$ short values and a long one. Thus $Y = [Z_0]$, and with $q = 1 - p$ we get

$$f_n = \mathbb{P}(Y = n) = \mathbb{P}([Z] = n | Z > 1|) = \frac{\mathbb{P}(M = n)}{q}.$$

Hence if $\mathbb{E}Z < \infty$, we have $\mathbb{E}M < \infty$, hence $\mu = \sum n f_n < \infty$ and the desired conclusion. Then

$$\mathbb{E}_e X_n = \frac{1}{\mu}\mathbb{E}_0 \sum_{n=0}^{Y-1} X_n = \frac{1}{\mu}\mathbb{E}_0(Y - 1 + \delta_1 + \delta_2 + \delta_3).$$

Now δ_2 is distributed as $U_1 + \cdots + U_N$, with the U_n independent identically distributed as Z conditionally upon $Z < 1$ and N independent of the U_n with $\mathbb{P}(N = n) = qp^n$. Hence by Wald's identity, $\mathbb{E}\delta_2 = \mathbb{E}[Z|Z < 1]p/q$, and since δ_1 is distributed as $Z_0 - [Z_0]$, we get

$$\begin{aligned}\mathbb{E}_e X_n &= \frac{1}{\mu}\mathbb{E}_0(Y + \delta_1 + \delta_2) = 1 + \frac{q}{\mathbb{E}M}\mathbb{E}_0(\delta_1 + \delta_2)\\ &= 1 + \frac{q}{\mathbb{E}M}\left\{\mathbb{E}[Z - [Z]|Z > 1] + \mathbb{E}[Z|Z < 1]\frac{p}{q}\right\}\\ &= 1 + \frac{q}{\mathbb{E}M}\left\{\frac{1}{q}\int_1^\infty xg(x)\,\mathrm{d}x - \frac{\mathbb{E}M}{q} + \frac{1}{q}\int_0^1 xg(x)\,\mathrm{d}x\right\}\\ &= \frac{\mathbb{E}Z}{\mathbb{E}M}.\end{aligned}$$

□

Notes

The example is from Henningsen and Liestøl (1983). A survey of the general area of neuron models is given by Fienberg (1974).

2d. Regenerative simulation

As explained in III.1, many practical situations call for numerical values of a parameter of the form $\theta = \mathbb{E}_e f(X_t)$. For example, $\{X_t\}$ could be a queue length process so that $f(x) = x$ would correspond to θ being the mean queue length in the steady state, $f(x) = I(x \geqslant N)$ to θ being the probability of queue length at least N in steady state and so on (similar remarks apply to waiting-time processes in discrete time). Now Theorem 1.2 states that θ is indeed well defined, but to use formula (1.1) to express θ in terms of the interarrival and service-time distributions may be difficult or impossible. Hence an alternative method could be required, and here we shall look at simulation.

Now the standard simulation (Monte Carlo) technique for estimating θ would be to design a simulation experiment giving as outcome a response variable R having the $\mathbb{P}_e$-distribution of $f(X_t)$. One then would perform N replicates giving independent identically distributed variables $R_1, \ldots, R_N$ distributed as R, estimate θ by the empirical mean $\hat{\theta} = \bar{R}_N = (R_1 + \cdots + R_N)/N$ and give the uncertainty on $\hat{\theta}$, say, in the form of asymptotic $1 - \alpha$ confidence intervals $[\hat{\theta} - sf_\alpha/\sqrt{N}, \hat{\theta} + sf_\alpha/\sqrt{N}]$, where f_α is the $(1 - \alpha/2)$-fractile of the standard normal distribution and s^2 is the empirical variance,

$$s^2 = \frac{1}{N-1}\sum_{n=1}^{N}(R_n - \bar{R}_N)^2.$$

This method is not feasible here since we may well simulate the queue starting from any given set of initial conditions but not in the unknown steady state. A partial solution would be to simulate the queue in $[0, T]$ starting from, say, an empty queue and choose T so large that hopefully $R = f(X_T)$ would have a distribution close to the required steady-state distribution. However, with T large each replication of the experiment becomes time-consuming and one is faced with the uncertainty inherent in the choice of T.

Instead we focus on the basic formula (1.1) and estimate the unknown $\mu = \mathbb{E}_0 Y$, $\nu = \mathbb{E}_0 \int_0^Y f(X_t)\,\mathrm{d}t$ by simulation of a regenerative cycle. That is, the simulation experiment consists in running one cycle and observing a two-dimensional response variable $R = (R(1), R(2))$ given by

$$R(1) = Y, \qquad R(2) = \int_0^Y f(X_t)\,\mathrm{d}t.$$

We then create independent identically distributed replicates $R_1, \ldots, R_N$ and estimate $\mu, \nu, \theta = \nu/\mu$ by

$$\hat{\mu} = \frac{1}{N}\sum_{n=1}^{N} R_n(1), \qquad \hat{\nu} = \frac{1}{N}\sum_{n=1}^{N} R_n(2), \qquad \hat{\theta} = \frac{\hat{\nu}}{\hat{\mu}}.$$

By the LLN, $\hat{\mu}$ and $\hat{\nu}$ are strongly consistent for μ, ν ($\hat{\mu} \to \mu$, $\hat{\nu} \to \nu$ a.s. as $N \to \infty$) and hence $\hat{\theta}$ is so for θ. Confidence bands can also be obtained. To this end, let

$$\Sigma = \mathbb{V}\mathrm{ar}_0 R = \begin{pmatrix} \mathbb{V}\mathrm{ar}_0 R(1) & \mathbb{C}\mathrm{ov}_0(R(1), R(2)) \\ \mathbb{C}\mathrm{ov}_0(R(1), R(2)) & \mathbb{V}\mathrm{ar}_0 R(2) \end{pmatrix}.$$

Then $(\hat{\mu}, \hat{\nu}) = \bar{R}_N$ is two-dimensional asymptotically normally distributed with mean (μ, ν) and covariance matrix Σ/N. Letting $\varphi(x, y) = y/x$, it thus follows by a standard transformation result that $\hat{\theta} = \varphi(\hat{\mu}, \hat{\nu})$ is asymptotically normally distributed with mean $\varphi(\mu, \nu) = \theta$ and variance σ^2/N, where

$$\sigma^2 = (D\varphi)'\Sigma D\varphi$$

with

$$D\varphi = \begin{pmatrix} \partial\varphi/\partial x \\ \partial\varphi/\partial y \end{pmatrix} = \begin{pmatrix} -y/x^2 \\ 1/x \end{pmatrix}$$

evaluated at $(x, y) = (\mu, \nu)$. Now the empirical covariance matrix S with elements

$$S_{ij} = \frac{1}{N-1}\sum_{n=1}^{N}(R_n(i) - \bar{R}_N(i))(R_n(j) - \bar{R}_N(j)) \qquad i, j = 1, 2$$

is strongly consistent for Σ and $(-\hat{\nu}/\hat{\mu}^2, 1/\hat{\mu})$ for $D\varphi$. Hence

$$s^2 = \frac{\hat{\nu}^2}{\hat{\mu}^4}S_{11} + \frac{1}{\hat{\mu}^2}S_{22} - \frac{2\hat{\nu}}{\hat{\mu}^3}S_{12}$$

is strongly consistent for σ^2 and $[\hat{\theta} - sf_\alpha/\sqrt{N}, \hat{\theta} + sf_\alpha/\sqrt{N}]$ is an asymptotic confidence interval for θ at level $1-\alpha$. □

Notes

Regenerative simulation has become a standard tool within the last 10 years. Some basic references are Crane and Lemoine (1977) and Iglehart and Shedler (1979). Among many texts, for example Cooper (1981) also has a chapter on simulation of queues.

Problems

2.3. The $M/M/1$ queue length process $\{X_t\}_{t\geqslant 0}$ with $\beta = 70$, $\delta = 100$, $\rho = 0.7$ was simulated in 11 busy cycles and the following values of the cycle length $R(1) = Y$ and $R(2) = \int_0^Y X_t \, dt$ recorded:

	$R(1)$	$R(2)$
	0.1494	0.5023
	0.0320	0.0104
	0.0124	0.0036
	0.0114	0.0019
	0.0212	0.0046
	0.0271	0.0169
	0.0142	0.0103
	0.0145	0.0003
	0.0243	0.0094
	0.0122	0.0001
	0.1175	0.2332
Sum	0.4363	0.7930

$$\left(\sum_{n=1}^{11} R_n(i)R_n(j)\right) = \begin{pmatrix} 0.0398 & 0.1038 \\ 0.1038 & 0.3073 \end{pmatrix}$$

Check whether the deviation of the corresponding estimate for $\mathbb{E}_e X_t$ from the true value $\rho/(1-\rho)$ is within the statistical uncertainty.

2e. Functionals of regenerative processes

In a variety of contexts, one is interested in more general functionals of the paths of a regenerative process $\{X_t\}_{t\in T}$ than just the value of a single X_t. For example, for $T = \mathbb{N}$, it would be of interest to say something not only about X_n but also about the dependence between consecutive values (X_n, X_{n+1}). Other examples could be $\max_{0\leqslant s\leqslant S} X_{t+s}$, $\int_0^\infty e^{-\beta s} X_{t+s} \, ds$ and so on. For such cases, the classical independent cycle property does not carry over to the functionals. For instance, for the (X_n, X_{n+1}) example (X_{Y-1}, X_Y) and (X_Y, X_{Y+1}) belong to distinct cycles but may clearly be dependent. However, *the slightly weaker definition that we have given of a regenerative process also includes such cases* since we have required the post-S_n-process to be independent only of $S_0, S_1, \ldots, S_n$, not of the whole pre-S_n-process. A convenient formalism for expressing this is what could be called the lifting of the regenerative process to function space. Let $\{X_t\}_{t\in T}$ be regenerative (not necessarily with independent identically distributed cycles) with imbedded renewal process $\{S_n\}$. If $T = [0, \infty)$, we assume in addition that the state space E is Polish and that $\{X_t\}_{t\geqslant 0}$ has paths in $\tilde{E} = D([0,\infty), E)$. For $T = \mathbb{N}$ we let $\tilde{E} = E^{\mathbb{N}}$. It is then an immediate consequence of the regenerativity of $\{X_t\}$ that the $\tilde{E}$-

valued process $\{\theta_t X\}_{t\in T}$ defined by $\theta_t X = \{X_{t+s}\}_{s\in T}$ is again regenerative with the same imbedded renewal process $\{S_n\}$. For $T=[0,\infty)$ it is also easily checked that the paths of $\{\theta_t X\}$ are right-continuous (they are not in $D([0,\infty),\tilde{E})$, however, since $\lim_{s\uparrow t}\theta_s x$ will fail to exist in this space if $\{x_s\}$ is a D-function with a jump at t). Hence if $\{S_n\}$ satisfies the condition for existence of a limit of $\{X_t\}$, then also $\{\theta_t X\}$ has a limit $X^{(s)}$ (in total variation for $T=\mathbb{N}$ and weakly for $T=[0,\infty)$) given by

$$\mathbb{E}f(X^{(s)}) = \frac{1}{\mathbb{E}_0 C}\mathbb{E}_0\int_0^C f(\theta_t X)\,\mathrm{d}t \quad \left(\sum_0^{C-1} \text{ if } T=\mathbb{N}\right).$$

As a random element of $\tilde{E}$, the process $X^{(s)} = \{X_t^{(s)}\}_{t\in T}$ represents a strictly stationary version of the given regenerative process. Indeed, the stationarity follows from

$$\begin{aligned}\theta_u X^{(s)} &\overset{\mathscr{D}}{=} \theta_u(\lim \theta_t X) \overset{\mathscr{D}}{=} \lim \theta_u\theta_t X \overset{\mathscr{D}}{=} \lim \theta_{u+t}X \\ &\overset{\mathscr{D}}{=} \lim \theta_t X \overset{\mathscr{D}}{=} X^{(s)}. \end{aligned} \tag{2.1}$$

Note the peculiarity of the process $\{\theta_t X\}_{t\in T}$ that it is deterministic given its initial value $\theta_0 X = X$: for any $t\in T$, $\theta_t X$ is a function of $\theta_0 X$. Note also that for $T=[0,\infty)$ we obtain convergence of $\theta_t X$ in function space without as usual having to invoke tightness. □

Problems

2.4. Show in the neuron model of Section 2(c) that the limits of $\mathbb{P}(X_n = X_{n+1} = 1)$, $\mathbb{P}(X_n = 1, X_{n+1} > 1)$, $\mathbb{P}(X_n > 1, X_{n+1} = 1)$ and $\mathbb{P}(X_{n+1} > 1)$ all exist, and compute these limits [Check: e.g. the first one is $\mathbb{E}(M-2)^+/\mathbb{E}M$].

2.5. Give a rigorous proof that indeed the limit and the shift in (2.1) may be interchanged [hint: in continuous time, check that $X^{(s)}$ has no fixed discontinuities].

Notes

For a somewhat different approach to path-functionals of regenerative processes, see Miller (1974).

3. TIME-AVERAGE PROPERTIES

We shall state and prove the results only in continuous time, the modifications to discrete time being obvious.

We are interested in what is sometimes called *cumulative processes*, i.e. functionals of the form $\int_0^t f(X_s)\,\mathrm{d}s$ for regenerative processes with independent identically distributed (i.i.d.) cycles, and the basic idea is to write

$$\int_0^t f(X_s)\,\mathrm{d}s = U_0(t) + U_1 + \cdots + U_{N_t-1} + \Delta_t, \tag{3.1}$$

where

$$U_0(t) = \int_0^{t\wedge S_0} f(X_s)\,\mathrm{d}s, \qquad U_n = \int_{S_{n-1}}^{S_n} f(X_s)\,\mathrm{d}s,$$

$$\Delta_t = \int_{S_{N_t-1}}^{t} f(X_s)\,\mathrm{d}s.$$

Here $U_0(t)=U_0(S_0)$ eventually and becomes negligible in the limit $t\to\infty$, $U_1+\cdots+U_{N_t-1}$ is a random sum of i.i.d. summands distributed as $U=\int_0^Y f(X_s)\,\mathrm{d}s$ and can be handled by standard tools, and the only problem turns out to be to bound Δ_t. To this end, define

$$V=\max_{0\leqslant t<Y}\left|\int_0^t f(X_s)\,\mathrm{d}s\right|, \qquad V_n=\max_{S_{n-1}\leqslant t<S_n}\left|\int_{S_{n-1}}^t f(X_s)\,\mathrm{d}s\right|.$$

Then the V_n are i.i.d. with $V_n\overset{\mathscr{D}}{=}V$ and

$$\max_{S_{n-1}\leqslant t<S_n}|\Delta_t|=V_n. \tag{3.2}$$

Clearly, $\int_0^t f(X_s)\,\mathrm{d}s$ is well defined for all t if and only if $V<\infty$ $\mathbb{P}_0$ a.s. and $U_0(t)$ is well defined for all t, and this will be assumed throughout.

We start by a LLN which contains as a special case results stated in I.4 and II.4.

Theorem 3.1. *Suppose that the regenerative process $\{X_t\}_{t\geqslant 0}$ has i.i.d. cycles and finite mean cycle length $\mu<\infty$. Then*

$$\frac{1}{t}\int_0^t f(X_s)\,\mathrm{d}s\to\frac{1}{\mu}\,\mathbb{E}_0U \tag{3.3}$$

if and only if $\mathbb{E}_0V<\infty$. If $f\geqslant 0$, then (3.3) *holds without further conditions.*

Proof. By the usual LLN,

$$\frac{U_1+\cdots+U_{N_t-1}}{t}\approx\frac{N_t-1}{t}\,\mathbb{E}_0U\approx\frac{1}{\mu}\,\mathbb{E}_0U \quad \text{a.s.}$$

Also obviously $U_0(t)/t\to 0$ a.s. and hence (3.3) holds if and only if $\Delta_t/t\to 0$. But when $S_{n-1}\leqslant t<S_n$, $t\approx n\mu$ and hence by (3.2), this is equivalent to $V_n/n\to 0$ a.s. which in turn is well known to hold if and only if $\mathbb{E}_0V<\infty$. Suppose next $f\geqslant 0$. Then if $\mathbb{E}_0U<\infty$, $0\leqslant V\leqslant U$ implies $\mathbb{E}_0V<\infty$ and (3.3). The case $\mathbb{E}_0U=\infty$ follows easily by monotonicity. □

It may be noted that if f is bounded, then $V\leqslant\|f\|\,Y$ and hence $\mathbb{E}_0V<\infty$ is automatic from $\mu<\infty$. Also if the conditions for the existence of $\mathbb{P}_e$ are satisfied, we may simply write the limit $\mathbb{E}_0U/\mu$ as $\mathbb{E}_ef(X_s)$.

We next prove a CLT analogue:

Theorem 3.2. *Suppose that the regenerative process $\{X_t\}_{t\geqslant 0}$ has i.i.d. cycles and $\mu<\infty$. Then if in addition $\mathbb{V}\mathrm{ar}_0\,Y<\infty$, $\mathbb{V}\mathrm{ar}_0\,U<\infty$, it holds that the limiting distribution of*

$$\frac{\displaystyle\int_0^t f(X_s)\,\mathrm{d}s-\frac{t}{\mu}\,\mathbb{E}_0U}{\sqrt{t}}$$

exists and is normal with mean zero and variance σ^2/μ, where

$$\sigma^2 = \mathbb{V}\mathrm{ar}_0\left(U - \frac{\mathbb{E}_0 U}{\mu} Y\right) = \mathbb{V}\mathrm{ar}_0 U + \left(\frac{\mathbb{E}_0 U}{\mu}\right)^2 \mathbb{V}\mathrm{ar}_0 Y - \frac{2\mathbb{E}_0 U}{\mu} \mathbb{C}\mathrm{ov}_0(U, Y).$$

Proof. We again use (3.1) where clearly $U_0(t)/\sqrt{t} \xrightarrow{\mathbb{P}} 0$. Also $\Delta_t/\sqrt{t} \xrightarrow{\mathbb{P}} 0$ is clear as above from $V_n/\sqrt{n} \xrightarrow{\mathbb{P}} 0$ (moments of V are not needed for this). Thus it remains to prove that

$$\frac{1}{\sqrt{t}}\left(U_1 + \cdots + U_{N_t - 1} - \frac{t}{\mu}\mathbb{E}_0 U\right) \tag{3.4}$$

has the desired limit distribution. But letting

$$T(n) = U_1 + \cdots + U_n - \frac{\mathbb{E}_0 U}{\mu}\{Y_1 + \cdots + Y_n\},$$

we may write (3.4) as

$$\frac{T(N_t - 1)}{\sqrt{t}} + \frac{1}{\sqrt{t}} \frac{\mathbb{E}_0 U}{\mu}\{Y_1 + \cdots + Y_{N_t - 1} - t\}. \tag{3.5}$$

Now clearly $T(n)/\sqrt{n}$ is asymptotically normal with mean zero and variance σ^2. Since $(N_t - 1)/t \xrightarrow{\mathbb{P}} \mu^{-1}$ it thus follows by Anscombe's theorem (Chung, 1974, p. 216) that the first term of (3.5) has the desired limit distribution, and it remains only to check that the second term vanishes in the limit. But

$$t - Y_1 - \cdots - Y_{N_t - 1} = t - S_{N_t - 1} + Y_0 = A_t + Y_0.$$

Here the backwards recurrence time A_t converges in distribution in the non-lattice case, whereas $A_{n\delta}$ converges in distribution if we have a lattice distribution with span δ. Since $A_t \leqslant A_{n\delta} + \delta$ for $n\delta \leqslant t < (n+1)\delta$, we thus always have $A_t/\sqrt{t} \xrightarrow{\mathbb{P}} 0$ and since clearly $Y_0/\sqrt{t} \xrightarrow{\mathbb{P}} 0$, the proof is complete. □

In some applications, it is more convenient to identify the form of the limiting distribution by means of the LLN 3.1 than to use the formula (1.1). In particular, this may be the case when a discrete-time process $\{Z_n\}$ is imbedded in a continuous-time one $\{X_t\}_{t \geqslant 0}$, and one wishes to relate the limiting distributions. We shall exemplify this by establishing the important PASTA property (*P*oisson *A*rrivals *S*ee *T*ime *A*verages). Here $\{X_t\}$ is *any* E-valued stochastic process (not necessarily regenerative) with D-paths, and $Z_n = X_{\tau(n)-}$, where the $\tau(n)$ are the epochs of a Poisson process $\{N_t\}_{t \geqslant 0}$ which is not anticipated by $\{X_t\}$ in the sense that $\{N_{t+s} - N_t\}_{s \geqslant 0}$ is independent of $\mathscr{F}_t = \sigma(X_s, N_s : s \leqslant t)$ for any t. Write α for the intensity of $\{N_t\}$.

Theorem 3.3. *Let f be any bounded measurable function on E. Then $\lim_{N\to\infty} \sum_1^N f(Z_n)/N$ exists if and only if $\lim_{T\to\infty} \int_0^T f(X_t)\,\mathrm{d}t/T$ does so, and in that case the limits are equal.*

Lemma 3.4. *Define*

$$Y'_t = \sum_{n=1}^{N_t} f(X_{\tau(n)-}), \qquad Y''_t = \alpha \int_0^t f(X_s)\,\mathrm{d}s, \qquad Y_t = Y'_t - Y''_t.$$

Then $\{Y_n\}_{n\in\mathbb{N}}$ *is a martingale with respect to* $\{\mathscr{F}_n\}$. *Furthermore,* $\mathbb{E}(Y_{n+1} - Y_n)^2 \leqslant \alpha \| f \|^2$.

Proof. It suffices to consider a continuous f with $0 \leqslant f \leqslant 1$. The martingale property (even in continuous time) is a standard fact in the modern theory of counting processes, but we shall give a self-contained proof. Define $M_n = Y_{n+1} - Y_n$

$$M^{(k,l)} = f(X_{(l-1)/k})\{I(N_{l/k} - N_{(l-1)/k} > 0) - 1 + \mathrm{e}^{-\alpha/k}\},$$

$$M_n^{(k)} = \sum_{l=nk+1}^{nk+k} M^{(k,l)}.$$

Then $M_n^{(k)} \to M_n$ as $k \to \infty$ and since $|M_n^{(k)}|$ is bounded by a constant times $N_{n+1} - N_n$, it is sufficient to show $\mathbb{E}(M_n^{(k)} | \mathscr{F}_n) = 0$. But this is clear since the $M^{(k,l)}$ are themselves martingale increments with respect to $\{\mathscr{F}_{l/k}\}_{l \geqslant nk}$ and we also get

$$\mathbb{E}M_n^2 = \lim_{k\to\infty} \mathbb{E}M_n^{(k)2} = \lim_{k\to\infty} \sum_{l=nk+1}^{(n+1)k} \mathbb{E}M^{(k,l)2}$$

$$\leqslant \lim_{k\to\infty} k\,\mathbb{V}\mathrm{ar}(I(N_{1/k} > 0)) = \lim_{k\to\infty} k\,\mathrm{e}^{-\alpha/k}(1 - \mathrm{e}^{-\alpha/k}) = \alpha. \qquad \square$$

Proof of Theorem 3.3. Let $U_n = \sum_1^n (Y_k - Y_{k-1})/k$. It follows from Lemma 3.4 that $\sup_n \mathbb{E}U_n^2 \leqslant \alpha \sum_1^\infty k^{-2} < \infty$. Hence by the convergence theorem for L^2-bounded martingales U_n has an a.s. limit and Kronecker's lemma then yields $Y_n/n \to 0$. Therefore the limits of Y'_n/n and Y''_n/n exist at the same time and are then equal, which (using $(N_{n+1} - N_n)/n \to 0$) implies that the limits of Y'_t/t and Y''_t/t exist at the same time and are then equal. But

$$Y'_t/t = \frac{N_t}{t}\frac{1}{N_t}\sum_{n=1}^{N_t} f(Z_n) = \alpha \frac{1}{N_t}\sum_{n=1}^{N_t} f(Z_n) + \mathrm{o}(1)$$

and thus the limits of Y'_t/t and $\sum_1^N f(Z_n)/N$ exist at the same time and differ only by the factor α. This completes the proof. $\square$

Example 3.5. Consider the $M/G/1$ queue with traffic intensity $\rho < 1$ and arrival epochs $\{\tau(k)\}$. It is then easy to see (and will be proved in VIII and IX) that $\{W_n\}, \{V_t\}, \{Q_t\}$ as defined in III.1 are all regenerative and satisfy the conditions of Section 1 for the existence of weak limits W, V, Q. Hence time-averages also exist and are given by the limit distributions (e.g. $\sum_1^N f(W_n)/N \to \mathbb{E}f(W)$). Since $W_n = V_{\tau(n)-}$, we conclude from the PASTA property that $\mathbb{E}f(W) = \mathbb{E}f(V)$ for all bounded measurable f. That is, we have established the fundamental property

$W \stackrel{\mathscr{D}}{=} V$ of the $M/G/1$ queue. A similar argument applies to the sequence $Z_n = Q_{\tau(n)-}$ of queue length just before instants of arrival and shows that $Z \stackrel{\mathscr{D}}{=} Q$.

□

Example 3.6 Consider instead the $GI/M/1$ queue with $\rho < 1$. We can then represent the service events by a Poisson process $\{N_t\}$, where epochs of $\{N_t\}$ are redundant when the queue is empty. With $Z_n = Q_{\tau(n)-}$, we find, as previously, that limits exist and satisfy $Z \stackrel{\mathscr{D}}{=} Q$. For the queueing interpretation, the sequence Z'_n of queue lengths just before departures is, however, of greater intrinsic interest. Obviously, $\{Z'_n\}$ is obtained from $\{Z_n\}$ by cancelling the n with $Q_{\tau(n)-} = 0$. These occur with asymptotic frequency $\mathbb{P}(Q = 0)$ and hence $f(0) = 0$ implies

$$\mathbb{E}f(Z') = \lim_{N\to\infty} \frac{1}{N} \sum_{n=1}^{N} f(Z'_n) = \frac{1}{\mathbb{P}(Q>0)} \lim_{N\to\infty} \frac{1}{N} \sum_{n=1}^{N} f(Z_n)$$

$$= \frac{\mathbb{E}f(Z)}{\mathbb{P}(Q>0)} = \frac{\mathbb{E}f(Q)}{\mathbb{P}(Q>0)} = \mathbb{E}[f(Q)|Q>0].$$

That is, the limit Z' has the same distribution as Q conditioned upon $\{Q > 0\}$.

□

Notes

Theorem 3.1 and 3.2 are from Smith (1955), though he appears to have some superfluous conditions for the CLT. Another much-cited paper in the area is Brown and Ross (1972). The PASTA property has been studied by a number of authors, the present material being taken from Wolff (1982). For a general discussion of Poisson processes and martingales, see e.g. Bremaud (1981).

CHAPTER VI

Further Topics in Renewal Theory and Regenerative Processes

1. SPREAD-OUT DISTRIBUTIONS

By a *component* of a distribution F on $\mathbb{R}$ we understand a non-negative G with the property $0 \neq G \leqslant F$. We say that F is *spread out* if, for some n, F^{*n} has a component G which is absolutely continuous, i.e. has a density g with respect to Lebesgue measure.

In applied contexts, situations where F is non-lattice and spread out (even with $n = 1$ in the above definition), are virtually the same. Strengthening the non-lattice assumption of renewal theory and regenerative processes to F being spread out does not therefore appear terribly restrictive, and the theory then gains some simplifications and strengthenings, rather in the spirit of the discrete time case. The basic tool is *Stone's decomposition* of the renewal measure:

Theorem 1.1. *If the interarrival distribution F of a renewal process is spread out, then we can write $U = U_1 + U_2$, where U_1, U_2 are non-negative measures on $[0, \infty), U_2$ is bounded ($\| U_2 \| < \infty$) and U_1 has a bounded continuous density $u_1(x) = \mathrm{d}U_1(x)/\mathrm{d}x$ satisfying $u_1(x) \to 1/\mu$ as $x \to \infty$.*

The proof is based on smoothness properties of the convolution. Most of these are easy to check and are used without further reference. However, we shall prove:

Lemma 1.2. *Let $g(x)$ be bounded and integrable, $g \in L_\infty \cap L_1$. Then $g^{*2}(x) = \int g(x-y)g(y)\mathrm{d}y$ is continuous.*

Proof. Choose continuous bounded functions $g_n \in L_1$ with $\| g - g_n \|_1 = \int |g - g_n| \to 0$. Then $g_n * g(x) = \int g_n(x-y)g(y)\mathrm{d}y$ is continuous by dominated convergence. Furthermore, $\| g^{*2} - g_n * g \|_\infty \leqslant \| g \|_\infty \| g - g_n \|_1 \to 0$, where $\| \cdot \|_\infty$ is the supremum norm. Thus g^{*2} is continuous as the uniform limit of continuous functions. □

Proof of Theorem 1.1. Consider first the case where F has a component G with a continuous density g with compact support and $g(0) = 0$, and let $H = F - G$,

$U_2 = \sum_0^\infty H^{*n}$. Then

$$F^{*n} = G * \sum_{k=0}^{n-1} F^{*(n-k-1)} * H^{*k} + H^{*n},$$

$$U = G * \sum_{k=0}^{\infty} H^{*k} * \sum_{n=k+1}^{\infty} F^{*(n-k-1)} + U_2 = G * U_2 * U + U_2.$$

Since $\|H\| = 1 - \|G\| < 1$, we have $\|U_2\| < \infty$, and we must show that $U_1 = G * U_2 * U$ has the desired properties. Now $G * U$ has density $U * g(x) = \int_0^x g(x-y)\,U(\mathrm{d}y)$ which is bounded and continuous by dominated convergence ($g(0) = 0$ is needed for this since otherwise a discontinuity at x arises when U has an atom at x). Also $U * g(0) = 0$, and hence by the same argument $U_1 = U_2 * (G * U)$ has the bounded continuous density $u_1 = U_2 * (U * g)$. Finally g, having compact support, is directly Riemann integrable, hence $U * g(x)$ has a limit, say β, as $x \to \infty$ and therefore $u_1(x) \to \gamma = \beta \|U_2\|$. But from $U = U_1 + U_2$ and $\|U_2\| < \infty$ we then get $U(x, x+a] \to a\gamma$, hence $\gamma = 1/\mu$ by Blackwell's theorem.

In the general case, let m be such that F^{*m} has a component with a bounded density. Hence by Lemma 3.2, F^{*2m} has a component with a continuous density, hence also a component with a density g which is continuous with compact support and has $g(0) = 0$. Define $U^{(k)} = F^{*k} \sum_0^\infty F^{*2nm}$. Then from above, $U^{(0)} = U_1^{(0)} + U_2^{(0)}$ with $\|U_2^{(0)}\| < \infty$ and $U_1^{(0)}$ having a bounded continuous density $u_1^{(0)}$ with $u_1^{(0)}(x) \to 1/\mathbb{E}S_{2m} = 1/2m\mu$. This is readily seen to imply a similar decomposition of $U^{(k)} = F^{*k} * U^{(0)}$ and since $U = \sum_0^{2m-1} U^{(k)}$, $U_1 = \sum_0^{2m-1} U_1^{(k)}$ and $U_2 = \sum_0^{2m-1} U_2^{(k)}$ have the desired properties. □

We proceed to give some main consequences of Stone's decomposition. The first is a version of the key renewal theorem $U * z(x) \to \int z/\mu$, where the strengthened assumption on F permits a weakening of the conditions on z, in particular to avoid reference to direct Riemann integrability.

Corollary 1.3. *Let z be bounded and Lebesgue integrable with $z(x) \to 0$, $x \to \infty$. Then $U * z(x) \to (1/\mu)\int_0^\infty z(y)\mathrm{d}y$ provided F is spread out.*

Proof. By dominated convergence,

$$Z(x) = U * z(x) = U_1 * z(x) + U_2 * z(x) = \int_0^x z(y) u_1(x-y)\mathrm{d}y + \int_0^x z(x-y)\mathrm{d}U_2(y)$$

$$\to \int_0^\infty z(y)\frac{1}{\mu}\mathrm{d}y + \int_0^\infty 0 \cdot \mathrm{d}U_2(y).$$ □

Corollary 1.4. *Consider a regenerative process $\{X_t\}_{t \geqslant 0}$ with the cycle length distribution F being spread out with finite mean μ. Suppose as the only path regularity condition that $X_t(\omega)$ is measurable jointly in (t, ω). Then, no matter the initial conditions, the limiting distribution $\mathbb{P}_e$ of X_t exists in the sense of total variation convergence and is given by*

$$\mathbb{E}_e f(X_t) = \frac{1}{\mu}\int_0^\infty \mathbb{E}_0[f(X_s); Y > s]\mathrm{d}s.$$

Proof. It is easily seen that it is sufficient to consider the zero-delayed case. Define

$$Z(t) = \mathbb{P}_0(X_t \in A), \qquad z(t) = \mathbb{P}_0(X_t \in A, Y > t).$$

Then $z(t)$ is Lebesgue measurable, and, because of $z(t) \leqslant \mathbb{P}(Y > t)$, also integrable with limit 0 at ∞. As in V.1, $Z = z + F * Z = U * z$. Here

$$U_2 * z(x) \leqslant \int_0^x \mathbb{P}(Y > x - y) \mathrm{d}U_2(y),$$

$$|U_1 * z(x) - \mathbb{P}_e(X_x \in A)|$$

$$= \left| \int_0^x z(y) u_1(x-y) \mathrm{d}y - \frac{1}{\mu} \int_0^\infty z(y) \mathrm{d}y \right|$$

$$\leqslant \frac{1}{\mu} \int_x^\infty \mathbb{P}_0(Y > t) \mathrm{d}t + \int_0^x \mathbb{P}_0(Y > t) \left| u_1(x-y) - \frac{1}{\mu} \right| \mathrm{d}y$$

and both these bounds are uniform in A and tend to zero as $x \to \infty$ (using dominated convergence). This proves t.v. convergence. □

A somewhat easier proof can be obtained using coupling, see the next section.

In many cases, it is also necessary for total variation convergence that F is spread out. For example:

Corollary 1.5. *Let $\{B_t\}$ be the forwards recurrence-time process of a renewal process with interarrival distribution F with finite mean μ, and define G_t, F_0 by $G_t(x) = \mathbb{P}(B_t \leqslant x)$, $F_0(\mathrm{d}x) = (1/\mu)(1 - F(x))\,\mathrm{d}x$. Then $\| G_t - F_0 \| \to 0$ for any distribution of the initial delay if and only if F is spread out.*

Proof. The sufficiency follows from Corollary 1.4. Suppose F is not spread out so that for each n, F^{*n} is concentrated on a Lebesgue null set N_n, and consider the zero-delayed case. On $S_n \leqslant t < S_{n+1}$, we have $B_t = S_{n+1} - t \in N_{n+1} - t$. Hence G_t is concentrated on the null set $\{y - t : y \in N_0 \cup N_1 \cup \cdots\}$ and the absolute continuity of F_0 yields $\| G_t - F_0 \| = 1$. □

Finally we mention that instead of spread-out distributions one frequently works with distributions which are *strongly non-lattice*, i.e. satisfy *Cramér's condition* (C) $\overline{\lim}_{|s| \to \infty} |\hat{F}(s)| < 1$, where $\hat{F}$ is the characteristic function of F. We have:

Proposition 1.6. *F is spread out $\Rightarrow F$ is strongly non-lattice $\Rightarrow F$ is non-lattice.*

Proof. If F is itself absolutely continuous, then $\hat{F}(s) \to 0$, $|s| \to \infty$, according to the Riemann–Lebesgue lemma. Thus if $F^{*n} \geqslant \varepsilon G$ with G absolutely continuous, we have

$$\overline{\lim_{|s| \to \infty}} |\hat{F}(s)| = \overline{\lim_{|s| \to \infty}} |\hat{F}^{*n}(s)|^{1/n} = \overline{\lim_{|s| \to \infty}} |\hat{F}^{*n}(s) - \varepsilon \hat{G}(s)|^{1/n} \leqslant (1 - \varepsilon)^{1/n}$$

and (C) holds. Finally, if F is lattice, say concentrated on $\{0, \pm\delta, \pm 2\delta, \dots\}$, then $\hat{F}(s) = 1$ for $s = 2n\pi/\delta$ and (C) cannot hold. □

In fact some results in renewal theory and regenerative processes require distributions which are only strongly non-lattice rather than spread out. However, the disadvantage of (C) is that the probabilistic significance is not clear and thus one has to rely on analytical methods.

Notes

The theory was initiated by Stone (1966). For further aspects, a main more recent reference is Arjas *et al.* (1978). An example where F is singular but F^{*2} not is given in Feller (1971, p. 146). It can be shown in continuation of Proposition 1.6 that discrete distributions cannot satisfy (C). See for example Bhattacharya and Rao (1976, p. 207).

2. THE COUPLING METHOD

By a *coupling* of two stochastic processes $\{X_t\}_{t\in T}$, $\{X'_t\}_{t\in T}$ having the same state space E and the same time parameter set $T = \mathbb{N}$ or $T = [0, \infty)$, we understand a realization of $\{X_t\}, \{X'_t\}$ on a common probability space in conjunction with a random time $T < \infty$ with the property

$$X_t = X'_t \quad \text{on} \quad \{T \leqslant t\}. \tag{2.1}$$

The concept is useful in particular when one wants to show that for large t the distributions of X_t and X'_t are close or to obtain explicit estimates of the deviations. To this end, the basic tool is the *coupling inequality*

$$\| \mathbb{P}(X_t \in \cdot) - \mathbb{P}(X'_t \in \cdot) \| \leqslant \mathbb{P}(T > t) \tag{2.2}$$

estimating the total variation distance. This follows simply by taking any measurable $A \subseteq E$ and writing

$$\begin{aligned} \mathbb{P}(X_t \in A) - \mathbb{P}(X'_t \in A) = \mathbb{P}(X_t \in A, T \leqslant t) - \mathbb{P}(X'_t \in A, T \leqslant t) \\ + \mathbb{P}(X_t \in A, T > t) - \mathbb{P}(X'_t \in A, T > t). \end{aligned} \tag{2.3}$$

Here the two first terms in (2.3) are equal because of (2.1) and an obvious upper bound on each of the two last terms is $\mathbb{P}(T > t)$.

Here we shall only be concerned with the case where $\{X'_t\}$ is strictly stationary, i.e. $\pi(A) = \mathbb{P}(X'_t \in A)$ is independent of t, so that (2.3) reduces to

$$\| \mathbb{P}(X_t \in \cdot) - \pi(\cdot) \| \leqslant \mathbb{P}(T > t). \tag{2.4}$$

Two main applications of the coupling method arise in this way: first, if $\{X_t\}$ can be coupled to $\{X'_t\}$ at all, we have $\mathbb{P}(T > t) \to 0$, and it is thus immediate that the distribution of X_t converges to π in total variation. Next, we get explicit rates of the speed of convergence. For example, if we can prove that $\mathbb{E}\mathrm{e}^{\varepsilon T} < \infty$ for some $\varepsilon > 0$, then $\mathbb{P}(T > t) = \mathrm{O}(\mathrm{e}^{-\varepsilon t})$ and (2.4) shows that the convergence is exponentially fast. Similarly, $\mathbb{E}T^\alpha < \infty$ yields the bound $\mathrm{O}(t^{-\alpha})$ in (2.4) and so on.

As our first example, we shall give a different proof of the ergodic theorem for Markov chains. Thus let an irreducible positive recurrent aperiodic transition matrix P on a countable state space E be given, and let π be its stationary distribution. We want to show $\mathbb{P}_i(X_n = j) \to \pi_j$ for all j, which will follow from (2.4) if we can couple a version of $\{X_n\}$ starting from i to a stationary version $\{X'_n\}$. This is done by first realizing $\{X'_n\}$ and next letting $\{X_n\}$ develop independently of $\{X'_n\}$ and according to P until $T = \inf\{n: X_n = X'_n\}$. For $n > T$, we then let $X_n = X'_n$. That $\{X_n\}$ becomes a Markov chain governed by P is an easy consequence of the strong Markov property. To show $T < \infty$, we may identify T by the hitting time of the diagonal by a bivariate Markov chain with independent coordinates governed by P, i.e. transition matrix Q with elements

$$q_{j_1 j_2, k_1 k_2} = p_{j_1 k_1} p_{j_2 k_2}$$

Since $p_{ij}^n > 0$ for all $n \geqslant n_0(i,j)$, it is clear that Q is irreducible and aperiodic. Also $\pi \otimes \pi$ is obviously a stationary distribution, the existence of which implies positive recurrence and the finiteness of hitting times.

The Markov chain $\{X_n\}$ is called *geometrically ergodic* if p_{ij}^n tends to π_j at a geometric rate η^n (with $\eta < 1$ independent of i,j). As remarked above, this property can be established by showing $\mathbb{E}\eta^{-T} < \infty$. For example:

Proposition 2.1. *Any ergodic Markov chain with a finite state space is geometrically ergodic.*

Proof. Since the state space for the bivariate chain is also finite, there exists $\delta > 0$ and m such that for all $i,j \in E$

$$\mathbb{P}_{ij}(T \leqslant m) \geqslant \sum_{k \in E} q_{ij,kk}^m \geqslant \delta.$$

From this a geometric trial argument easily yields $\mathbb{E}\eta^{-T} < \infty$ whenever $\eta^m > 1 - \delta$. □

We next turn to renewal theory. It is tempting to try to prove the renewal theorem for the case $\mu < \infty$ by coupling a zero-delayed and a stationary renewal process, or equivalently, say, their backwards recurrence-time chains $\{A_t\}$, $\{A'_t\}$. A little reflection shows that a literal translation of the Markov chain procedure does not apply. This is because if, say, the interarrival distribution F is absolutely continuous, then independent realizations cannot lead to simultaneous renewals and thereby a finite coupling epoch. We return to this problem later and construct first for each $\delta > 0$ a 'coupling' epoch T_δ with the weaker property that the renewals following T_δ in the two processes are not equal but only separated at most δ. Nevertheless, this turns out to be sufficient for proving the renewal theorem.

Thus let two independent renewal processes be given, one which is stationary and has epochs $S'_0, S'_1, \ldots$, and one which is zero-delayed and has epochs $S''_0 = 0, S''_1, \ldots$, and define for some $\delta > 0$

$$\sigma = \inf\{n: S'_k \in [S''_n, S''_n + \delta] \quad \text{for some } k\} \quad \text{(cf. Fig. 2.1).}$$

Lemma 2.2. *If F is non-lattice with $\mu < \infty$, then $\sigma < \infty$ a.s. for all $\delta > 0$.*

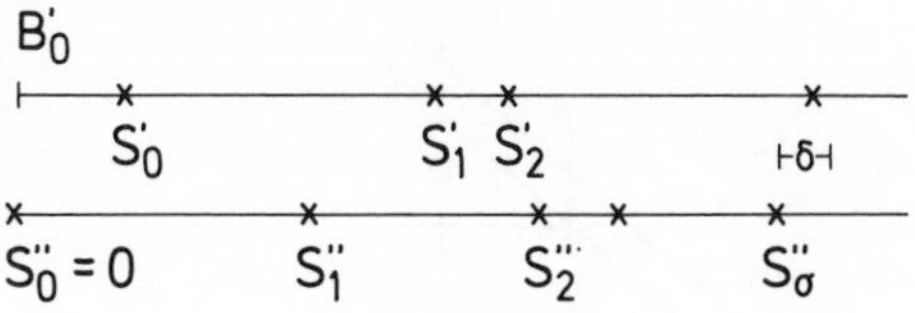

Fig. 2.1

Proof. Let for $m=0,1,2,\ldots$

$$A_m=\{S'_k\in[S''_n,S''_n+\delta] \text{ for some } k \text{ and some } n\geqslant m\},$$

$A_\infty=\{S'_k\in[S''_n,S''_n+\delta]$ for some $k=k(n)$ and infinitely many $n\}$.
Then $\{\sigma<\infty\}=A_0\supseteqq A_1\supseteqq\cdots\downarrow A_\infty$. By stationarity and independence, $\mathbb{P}A_0=\mathbb{P}A_1=\cdots$ so that $\mathbb{P}A_0=\mathbb{P}A_\infty$. Hence, since $\mathbb{P}(A_0|B'_0)\geqslant\mathbb{P}(A_\infty|B'_0)$,

$$\mathbb{P}(A_0|B'_0)=\mathbb{P}(A_\infty|B'_0)\quad\text{a.s.}\tag{2.5}$$

According to IV.5.1, we can choose T such that $U(t+\delta)-U(t)>0$ for all $t\geqslant T$. Then if n satisfies $\mathbb{P}(S''_n-b\geqslant T)>0$, we have

$$\mathbb{P}(Y'_1+\cdots+Y'_k\in(S''_n-b,S''_n-b+\delta]\quad\text{for some}\quad k|B'_0=b)>0$$

and hence $\mathbb{P}(A_0|B'_0=b)>0$, implying by (2.5) that $\mathbb{P}(A_\infty|B'_0)>0$ a.s. But A_∞ is invariant under finite permutations of the independent identically distributed elements of the sequence $\{(Y'_n,Y''_n)\}$, hence $\mathbb{P}(A_\infty|B'_0)\in\{0,1\}$ a.s. by the Hewitt–Savage 0–1 law. Therefore $\mathbb{P}(A_\infty|B'_0)=1$ a.s. so that $\mathbb{P}(\sigma<\infty)=\mathbb{P}A_\infty=1$. □

We can now give an alternative version of the

Proof of the renewal theorem, non-lattice case, $\mu<\infty$. It suffices to consider the zero-delayed case. Define

$$\tau=\inf\{k:S'_k\in[S''_\sigma,S''_\sigma+\delta]\},$$

$$Y_n=Y''_n,\qquad n\leqslant\sigma,\qquad Y_{\sigma+m}=Y'_{\tau+m},\qquad m=1,2,\ldots,\qquad S_n=Y_1+\cdots+Y_n$$

(see Fig. 2.2 for an illustration). Then the Y_n are independent identically distributed with common distribution F so that $S_0=0,S_1,S_2,\ldots$ are the epochs of a pure renewal process with interarrival distribution F. Let $\{A_t\}_{t\geqslant0}$ and $\{A'_t\}_{t\geqslant0}$ be the associated backwards recurrence-time processes. Then if $\delta<\varepsilon$,

$$A'_{t+\delta}>\varepsilon+\delta\Rightarrow A_t>\varepsilon\Rightarrow A'_t>\varepsilon-\delta\quad\text{on } \{t>S_\sigma\},\tag{2.6}$$

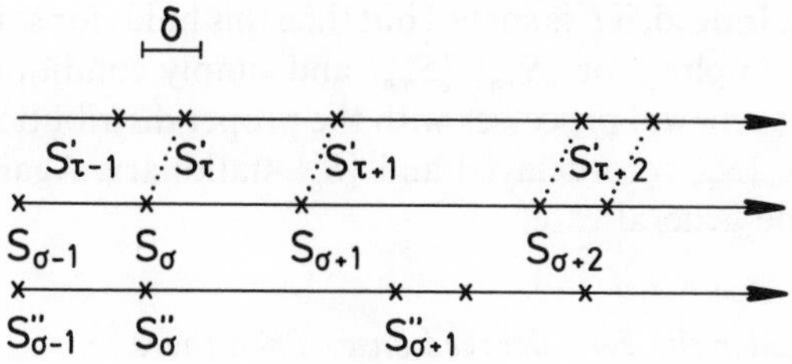

Fig. 2.2

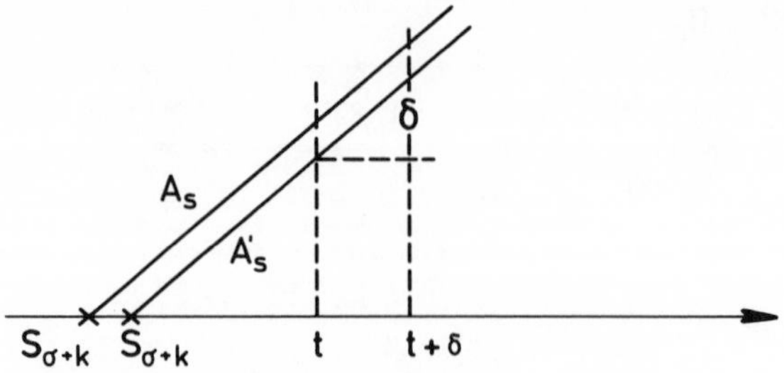

Fig. 2.3

cf. Fig. 2.3, and thus

$$\begin{aligned}\overline{\lim_{t\to\infty}}\ \mathbb{P}(A_t>\varepsilon)&\leqslant\overline{\lim_{t\to\infty}}\ \mathbb{P}(t\leqslant S_\sigma)+\mathbb{P}(A'_t>\varepsilon-\delta,t>S_\sigma)\\&=\overline{\lim_{t\to\infty}}\ \mathbb{P}(A'_t>\varepsilon-\delta)=1-F_0(\varepsilon-\delta)\end{aligned}$$

since $\{t\leqslant S_\sigma\}\downarrow\varnothing$ a.s. as $t\to\infty$. Letting $\delta\downarrow 0$ yields $\overline{\lim}\ \mathbb{P}(A_t>\varepsilon)\leqslant 1-F_0(\varepsilon)$ and $\underline{\lim}\geqslant$ is proved similarly, using the l.h. implication in (2.6) instead of the r.h. This proves the renewal theorem in the form IV.(4.2). □

We shall say that a renewal process with interarrival distribution F *admits coupling* if for any initial delays Y_0, Y'_0 it is possible to construct corresponding renewal processes $\{S_n\},\{S'_n\}$ such that $S_n=S'_n$ eventually. With $\sigma=\inf\{n:S_n=S'_n\}$, the coupling epoch is then $T=S_\sigma=S'_\sigma$ and this T serves also as a coupling epoch for the backwards and forwards recurrence-time processes (the one for the forward process may even be taken slightly smaller).

Theorem 2.3. *A non-lattice renewal process with $\mu<\infty$ admits coupling if and only if F is spread out.*

The necessity follows from Corollary 1.5: the coupling inequality applied to $\{B_t\}$, $\{B'_t\}$ with $\{B_t\}$ arbitrary and $\{B'_t\}$ stationary shows that we have total variation convergence to F_0, hence F must be spread out. Sufficiency can also easily be obtained from Corollary 1.5 (see to this end Proposition 3.13 of the next section). A direct construction of the coupling epoch is, however, instructive and will also be needed below for the rate results.

For the proof, it is no restriction to assume that $F(\mathrm{d}x)\geqslant\eta\,\mathrm{d}x$ on $(a,a+2b)$ for suitable $a,b,\eta>0$. Indeed, if F is spread out then this holds for some F^{*m}. We may then construct a coupling for $\{S_{mn}\},\{S'_{mn}\}$ and simply condition in the missing points S_k, S'_k to get renewal processes with the proper distributions. In the proof, we shall also take $\{S_n\}$ zero-delayed and $\{S'_n\}$ stationary. Again, easy modifications apply to the general case.

Lemma 2.4. *Consider the zero-delayed case. Then there exists $A<\infty$ such that distributions of the B_t with $t\geqslant A$ have a common uniform component on $(0,b)$. That* *is, for some* $\delta \in (0,1)$ *and all* $t \geqslant A$,

$$\mathbb{P}(u < B_t \leqslant v) \geqslant \delta \frac{v-u}{b}, \qquad 0 < u < v < b \tag{2.7}$$

Proof. As in IV.2.1,

$$z(t) = \mathbb{P}(u < B_t \leqslant v, Y > t) = F(t+v) - F(t+u),$$

$$\mathbb{P}(u < B_t \leqslant v) = U * z(t) \geqslant \int_0^t z(y) u_1(t-y) \mathrm{d}y$$

$$\geqslant \int_a^{a+b} z(y) u_1(t-y) \mathrm{d}y \geqslant \eta(v-u) \int_a^{a+b} u_1(t-y) \mathrm{d}y,$$

where u_1 is as in Stone's decomposition. From $u_1(t) \to 1/\mu > 0$ it follows that the last integral is $\geqslant b/2\mu$ for t sufficiently large, proving the lemma. □

We can now easily complete the

Proof of Theorem 2.3. The coupled zero-delayed and stationary renewal processes, $\{S_n\}$ and $\{S'_n\}$ respectively, are constructed in steps, say for $k = 0, 1, 2, \ldots$. After step k, the renewal processes have been realized in a certain random interval $[0, t_k]$, as also have the overshots B_{t_k}, B'_{t_k}. To get started, one lets $t_0 = 0$, $B_{t_0} = 0$ and chooses B'_{t_0} according to F_0. In step $k+1$, one then lets

$$L_k = \max\{B_{t_k}, B'_{t_k}\}, \qquad t_{t+1} = t_k + L_k + A$$

with A from Lemma 2.4 and

$$s_k = L_k + A - B_{t_k}, \qquad s'_k = L_k + A - B'_{t_k},$$

cf. Fig. 2.4. Then $s_k, s'_k \geqslant A$, and by Lemma 2.4 we can choose U_k, V_k, R_k, R'_k such that $\mathbb{P}(U_k = 1) = 1 - \mathbb{P}(U_k = 0) = \delta$, that V_k is uniform on $(0, b)$ and that

$$B_{t_{k+1}} = U_k V_k + (1 - U_k) R_k, \qquad B'_{t_{k+1}} = U_k V_k + (1 - U_k) R'_k$$

have the overshot distribution corresponding to s_k and s'_k respectively (U_k, V_k, R_k, R'_k are taken independent of all preceding U_l, V_l, R_l, R'_l). The renewals for $\{S_n\}$ in $[t_{k+1} - s_k, t_{k+1}]$ are then just taken according to the conditional distribution of the renewal process given that its overshot at time s_k has the value

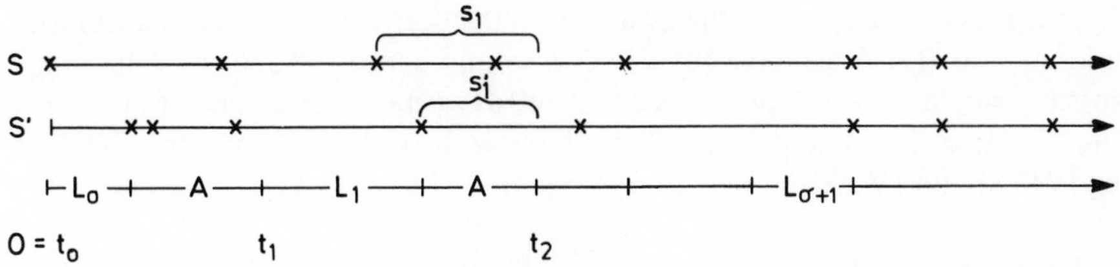

Fig. 2.4

of the constructed $B_{t_{k+1}}$, and similarly for $\{S'_n\}$. The procedure is stopped at step $\sigma = \inf\{n \geq 0: U_n = 1\}$. Then the two processes have a renewal in common, namely at epoch $T = t_\sigma + L_{\sigma+1}$, and we just let the renewals after T be the same. It is clear that this procedure produces renewal processes $\{S_n\}$, $\{S'_n\}$ with the desired properties and that $\mathbb{P}(\sigma = n) = \delta(1-\delta)^n$, $T < \infty$ a.s. □

Turning to the rate results, we first note a further property of the coupling epoch T:

Lemma 2.5. *If* $\int_0^\infty e^{\eta x}\, dF(x) < \infty$ *for some* $\eta > 0$, *then also* $\mathbb{E}\, e^{\varepsilon T} < \infty$ *for some* (possibly smaller) $\varepsilon > 0$.

Proof. Define $Z(t) = \mathbb{E}\, e^{\eta B_t}$. Then $Z(\infty) = \int_0^\infty e^{\eta x}\, dF_0(x) < \infty$, and representing Z as $U * z$ in the usual way, it is easily seen that $Z(t) \to Z(\infty)$. In particular, Z is bounded, say $Z(t) \leq c_1$. Now

$$\begin{aligned} &\mathbb{E}[\exp\{\eta(A + L_{k+1})\} \mid B_{t_k}, B'_{t_k}] \\ &\leq \mathbb{E}[\exp\{\eta(A + B_{t_{k+1}})\} + \exp\{\eta(A + B'_{t_{k+1}})\} \mid B_{t_k}, B'_{t_k}] \\ &= \mathbb{E}[\exp\{\eta(A + B_{s_k})\} + \exp\{\eta(A + B'_{s'_k})\}] \leq 2c_1 e^{\eta A} = c(\eta) \end{aligned}$$

(say). Similarly $c_2 = \mathbb{E} e^{\eta(A + L_0)} < \infty$ and letting $T_n = \sum_0^n (A + L_k)$, it follows easily by induction that $\mathbb{E} e^{\eta T_n} \leq c_2 c(\eta)^n$. Now for some (large) p and some q (close to 1) satisfying $1/p + 1/q = 1$ it holds that $c(\eta)^{1/p}(1-\delta)^{1/q} < 1$. With $\varepsilon = \eta/p$ Hölder's inequality then yields

$$\begin{aligned} \mathbb{E} e^{\varepsilon T} &\leq \mathbb{E} e^{\varepsilon T_{\sigma+1}} = \sum_{n=0}^{\infty} \mathbb{E}[e^{\varepsilon T_{n+1}}; \sigma = n] \\ &\leq \sum_{n=0}^{\infty} (\mathbb{E} e^{\eta T_{n+1}})^{1/p} \mathbb{P}(\sigma = n)^{1/q} \leq c_2^{1/p} \delta^{1/q} \sum_{n=0}^{\infty} c(\eta)^{(n+1)/p}(1-\delta)^{n/q} < \infty. \end{aligned}$$ □

Theorem 2.6. *Suppose* $\int_0^\infty e^{\eta x}\, dF(x) < \infty$ *for some* $\eta > 0$. *Then for some* $\varepsilon > 0$

(i) $\|\mathbb{P}(B_t \in \cdot) - F_0(\cdot)\| = O(e^{-\varepsilon t})$, cf. *Corollary 1.5*;

(ii) *in Stone's decomposition,* $U_2[x, \infty) = O(e^{-\varepsilon x})$ *and* $u_1(x) = 1/\mu + O(e^{-\varepsilon x})$;

(iii) if z *is measurable with* $z(x) = O(e^{-\delta x})$ *for some* $\delta > \varepsilon$, *then*

$$U * z(t) = \frac{1}{\mu}\int_0^\infty z(x)\, dx + O(e^{-\varepsilon t}).$$

Proof. Here (i) is immediate by apparent from (2.4) and Lemma 2.5. The proof of (ii) proceeds by reinspecting the estimates in the derivation of Stone's decomposition. From $H < F$ we have $\int e^{\eta x} dH(x) < \infty$ and hence $\int e^{\varepsilon x} dH(x) < 1$ for some possibly smaller $\varepsilon > 0$. This implies that $\int e^{\varepsilon x} dU_2(x) < \infty$ and hence $U_2(x, \infty) = O(e^{-\varepsilon x})$. Also since $g(x) = 0$, $x > T$, we have with G_t the distribution of B_t that (cf. IV.(2.8) and IV.3.3)

$$|U * g(x) - \beta| = \left| \int_{x-T}^{x} g(x-y) U(dy) - \frac{1}{\mu}\int_0^T g(y) dy \right|$$

$$= \left| \int_0^T g(T-y) G_{x-T} * U(\mathrm{d}y) - \frac{1}{\mu} \int_0^T g(T-y) \mathrm{d}y \right|$$

$$= \left| \int_0^T g(T-y) \{G_{x-T} - F_0\} * U(\mathrm{d}y) \right| \leqslant \| g \|_\infty \| G_{x-T} - F_0 \| U(T) = \mathrm{O}(\mathrm{e}^{-\varepsilon x}),$$

$$u_1(x) = \int_0^x U * g(x-y) \mathrm{d}U_2(y) = \int_0^x [\beta + \mathrm{O}(\mathrm{e}^{-\varepsilon(x-y)})] \mathrm{d}U_2(y)$$

$$= \beta \| U_2 \| - \beta U_2(x, \infty) + \mathrm{e}^{-\varepsilon x} \int_0^x \mathrm{O}(\mathrm{e}^{\varepsilon y}) \mathrm{d}U_2(y) = \frac{1}{\mu} + \mathrm{O}(\mathrm{e}^{-\varepsilon x}).$$

Finally, in (iii)

$$U * z(x) = \int_0^x z(x-y) U_2(\mathrm{d}y) + \int_0^x z(y) u_1(x-y) \mathrm{d}y$$

$$= \mathrm{e}^{-\varepsilon x} \int_0^x \mathrm{O}(\mathrm{e}^{\varepsilon y}) U_2(\mathrm{d}y) + \int_0^x z(y) \left\{ \frac{1}{\mu} + \mathrm{O}(\mathrm{e}^{-\varepsilon(x-y)}) \right\} \mathrm{d}y$$

$$= \mathrm{O}(\mathrm{e}^{-\varepsilon x}) + \int_0^\infty z(y) \mathrm{d}y/\mu - \int_x^\infty z(y) \mathrm{d}y/\mu + \mathrm{e}^{-\varepsilon x} \int_0^x \mathrm{O}(\mathrm{e}^{-(\delta-\varepsilon)y}) \mathrm{d}y$$

$$= \mathrm{O}(\mathrm{e}^{-\varepsilon x}) + \int_0^\infty z(y) \mathrm{d}y/\mu.$$

Corollary 2.7. *If, in addition to the conditions of Corollary 1.4, a regenerative process has* $\int_0^\infty \mathrm{e}^{\eta x} \mathrm{d}F(x) < \infty$ *for some* $\eta > 0$, *then the total variation convergence is exponentially fast. That is,* $\| \mathbb{P}(X_t \in \cdot) - \mathbb{P}_e(X_t \in \cdot) \| = \mathrm{O}(\mathrm{e}^{-\varepsilon t})$ *for some* $\varepsilon > 0$.

Proof. If $z(t) = \mathbb{P}(X_t \in A, t < Y)$, then $z(t) \leqslant \mathbb{P}(Y > t) = \mathrm{O}(\mathrm{e}^{-\eta t})$ uniformly in A, and a check of the above proof of (iii) shows that we have uniformly in A in the relation

$$\mathbb{P}(X_t \in A) = U * z(t) = \frac{1}{\mu} \int_0^\infty z(y) \mathrm{d}y + \mathrm{O}(\mathrm{e}^{-\varepsilon t}) = \mathbb{P}_e(X_t \in A) + \mathrm{O}(\mathrm{e}^{-\varepsilon t}). \qquad \square$$

Problems

2.1. Let $\{X_n\}, \{X'_n\}$ be independent Markov chains with initial distributions λ, λ' and the same transition matrix P. Assume that P is irreducible aperiodic null recurrent with stationary measure ν. Show that $\{Z_n\} = \{(X_n, X'_n)\}$ is irreducible aperiodic. Define now $T = \inf\{n: X_n = X'_n = i\}$ for some fixed state i. Show that $\{Z_n\}$ is recurrent if and only if $\mathbb{P}(T < \infty) = 1$ irrespective of i, λ, λ'. In the transient case, let $\lambda = \lambda'$ and show hereby $\mathbb{P}_\lambda(X_n = i) \leqslant \mathbb{P}(Z_n = (i, i))^{1/2} \to 0$. In the recurrent case, let $B \subseteq E$ be finite with $i \in B$, and define $\lambda'_j = \nu_j / \nu(B)$ for $j \in B$, $\lambda'_j = 0$ for $j \notin B$. Show that

$$\overline{\lim}\, \mathbb{P}_\lambda(X_n = i) = \overline{\lim}\, \mathbb{P}_{\lambda'}(X'_n = i) \leqslant \overline{\lim} \frac{(\nu P^n)_i}{\nu(B)} = \nu_i / \nu(B)$$

and hereby that $\mathbb{P}_\lambda(X_n = i) \to 0$.

2.2. Let $\{X_n^{(1)}\}, \{X_n^{(2)}\}, \ldots$ be independent Bernoulli random walks on $\mathbb{Z}$ with $p = 1/2$, and define $Y_n^{(N)} = (X_n^{(1)}, \ldots, X_n^{(N)})$. Use the estimate $\mathbb{P}(X_{2n}^{(1)} = 0) \approx cn^{-1/2}$ to show that $\{Y_n^{(N)}\}$ is transient for $N = 3, 4, \ldots$ and recurrent for $N = 1, 2$, and show hereby that both the recurrent and the transient case in Problem 2.1 may arise.

Notes

The idea of a coupling goes back to Doeblin more than 40 years ago, but todays interest in the subject was largely initiated by Pitman (1974). The coupling proof of the renewal theorem was given independently by Lindvall (1977) and Ney (unpublished), whereas Theorem 2.3 is from Lindvall (1982). Further applications of coupling to related problems are in Griffeath (1978), Berbee (1979) and Thorisson (1983), and further rate results in renewal theory and references in Grübel (1983) and Carlsson (1983).

Though powerful and elegant, the coupling method also has some limitations. In particular, the rate results come out as upper bounds and not precise approximations like those derived for $M/M/1$ in III.9. Also the null recurrent case in Markov chains and renewal theory with infinite mean are not trivial to handle (Problem 2.1 was communicated by Hermann Thorisson, December 1985).

3. MARKOV PROCESSES: REGENERATION AND HARRIS RECURRENCE

From the point of view of regenerative processes, the Markov case is a rather special one. Without doubt, a major force of the concept of a regenerative process is precisely that neither the Markov property nor other restrictions need to be put on the evolution in between regeneration points. Conversely, from the point of view of Markov processes on a general state space E, regeneration appears at first sight as a severe restriction. There is no apparent choice of regeneration points since, for example, the renewal processes of entrances to a fixed state $x \in E$, so important in the discrete case, will only be non-terminating in quite special cases.

Nevertheless, the connection between Markov processes and regenerative processes has in recent years proved out to be of basic importance, and in fact ergodic theory for Markov processes in a simple and satisfying form is hardly known outside the set-up to be developed below.

We consider as in I.6 a Markov process $\{X_t\}_{t \in T}$ on $(E, \mathscr{E})$ with $T = \mathbb{N}$ or $T = [0, \infty)$ and let $\mathscr{F}_t = \sigma(X_s; s \leqslant t)$. If $T = [0, \infty)$, it is assumed that E is Polish, $\mathscr{E}$ the Borel σ-algebra, that $\{X_t\}$ has D-paths and the strong Markov property holds. Write $P^t(x, A) = \mathbb{P}_x(X_t \in A)$.

Letting $\tau(R) = \inf\{t \geqslant 1; X_t \in R\}$, we call $R \in \mathscr{E}$ *recurrent* if $\mathbb{P}_x(\tau(R) < \infty) = 1$ for all $x \in E$ (if $T = [0, \infty)$, we need in addition to assume that $\tau(R)$ is measurable). By the strong Markov property, this is equivalent to $\{t: X_t \in R\}$ being unbounded with probability 1, irrespective of initial conditions. We call R a *regeneration set* if R is recurrent and for some $r > 0$ the $P^r(x, \cdot)$, $x \in R$, contain a common component, i.e. for some $\varepsilon \in (0, 1)$ and some probability measure λ on E,

$$P^r(x, B) \geqslant \varepsilon \lambda(B), \qquad x \in R \tag{3.1}$$

for all $B \in \mathscr{E}$. For example this holds for a one-point set $R = \{x\}$ if and only if x is a recurrent state since then we may just take an arbitrary $r > 0$, $\varepsilon = \frac{1}{2}$ and $\lambda(B) = P^r(x, B)$. The following example is typical of applications and shows that regeneration sets exist in far more general situations:

Example 3.1. Assume that the transition functions contain components with smooth densities, i.e. for some μ and r we have

$$P^r(x, B) \geqslant \int_B f^r(x, y)\mu(\mathrm{d}y),$$

$$E_0 = \left\{x \in E: \int f^r(x, y)\mu(\mathrm{d}y) > 0\right\} \neq \varnothing, \tag{3.2}$$

where $f^r(x, y)$ is jointly continuous in (x, y) in a suitable topology on E. Then a regeneration set exists, provided that for some $x_0 \in E_0$ every neighbourhood of x_0 is recurrent. Indeed, choose $y_0 \in \operatorname{supp} \mu$ with $\delta = f^r(x_0, y_0) > 0$ and let R, S be neighbourhoods of x_0, y_0 with $f^r(x, y) \geqslant \delta/2, x \in R, y \in S$. Then if $\lambda(B) = \mu(BS)/\mu(S)$, we have for $x \in R$ that

$$P^r(x, B) \geqslant \int_{BS} f^r(x, y)\mu(\mathrm{d}y) \geqslant \frac{\delta}{2}\mu(BS) = \frac{\delta\mu(S)}{2}\lambda(B). \qquad \square$$

We call a Markov chain with a regeneration set *Harris recurrent* or just a *Harris chain* (the traditional equivalent definition looks somewhat different, see the end of the section; in continuous time the terminology is less well established). We shall justify the term 'regeneration set' by showing that it is possible to construct $\{X_t\}$ simultaneously with a renewal process $S_0, S_1, \ldots$ with respect to which the Markov process becomes regenerative. The idea is to randomize by, with probability ε, letting a regeneration occur r time units after a visit to R and then restart according to λ. Choose an initial value $X_0 = x$ and just take the usual version of the process up to the time $\tau(R)$ where R is hit. Then realize $X_{\tau(R)+r}$ by, with probability ε, letting the distribution be λ and a renewal epoch occur at $\tau(R) + r$, with probability $1 - \varepsilon$, letting the distribution be

$$[P^r(X_{\tau(R)}, \cdot) - \varepsilon\lambda(\cdot)]/(1 - \varepsilon).$$

After that, realize the whole segment $\{X_{\tau(R)+s}\}_{0<s<r}$ by choosing it according to the conditional distribution of $\{X_s\}_{0<s<r}$ given that the boundary values X_0, X_r are the constructed $X_{\tau(R)}, X_{\tau(R)+r}$. Now repeat the procedure with the new initial value $X_{\tau(R)+r} = x$ and so on. That we get a Markov process with the given transition probabilities is intuitively obvious and easily verified. Also the distribution of $X_t = X_{S_n}$ at a renewal epoch $t = S_n$ is λ for all n and independent of $S_1, \ldots, S_n$. Hence the post-S_n-process evolves in the same way for all n and is independent of $S_1, \ldots, S_n$. Thus we indeed have a regenerative process in the general sense of V.1 where we do not require independent cycles. In fact, $X_{\tau(R)+s}$ need not be independent of $X_{\tau(R)+r}$ if $0 < s < r$ and hence the last $r - 1$ values in a cycle need not be independent of the next cycle. At least, the construction ensures that cycles are one-dependent (cycles $1, \ldots, n-1$ are independent of cycles $n+1, n+2, \ldots$).

We denote by $\mathbb{P}_\lambda$ the zero-delayed case where X_0 is chosen according to λ, by Y the length of the first cycle of the P_λ-process.

The regeneration points obviously behave rather like stopping times, but are not so in the strict sense since in addition to $\mathscr{F}_\infty = \sigma(X_t; t\in T)$ they also depend on the 0–1 variables determining the randomizations. However, they fall into the framework of so-called *randomized stopping times.* We shall not go into a discussion of this subject but mention only that one of the possible definitions of τ being a randomized stopping time in the Markov case is

$$\mathbb{E}_x[g(X_n, X_{n+1},\ldots); \tau > n] = \mathbb{E}_x[\mathbb{E}_{X_n} g(X_0, X_1,\ldots); \tau > n], \tag{3.3}$$

$$\mathbb{E}_x[g(X_s; s\geqslant t); \tau > t] = \mathbb{E}_x[\mathbb{E}_{X_t} g(X_s, s\geqslant 0); \tau > t], \tag{3.4}$$

for $T = \mathbb{N}$ and $T = [0,\infty)$ respectively. This relation will be needed below so we shall give a proof. Let $T = \mathbb{N}$ (the continuous case is entirely similar) and τ be any of $S_0, S_1,\ldots$. By the Markov property,

$$\mathbb{E}_x g(X_n, X_{n+1},\ldots) = \mathbb{E}_x \mathbb{E}_{X_n} g(X_0, X_1,\ldots)$$

and therefore it is sufficient to prove (3.3) with $\tau > n$ replaced by $\tau \leqslant n$. But conditionally upon τ, it holds on $\{\tau \leqslant n\}$ that the Markov process is restarted according to λ at time τ. Thus on $\{\tau\leqslant n\}$

$$\mathbb{E}_x[g(X_n, X_{n+1},\ldots)|\tau, X_n] = \mathbb{E}_{X_n} g(X_0, X_1,\ldots)$$

and (3.3) follows easily.

A measure ν on $(E,\mathscr{E})$ is called *stationary* if $\nu\geqslant 0$, $\nu\neq 0$, ν is σ-finite and $\nu P^s = \nu$ for all $s\in T$.

Theorem 3.2. *For a Markov process with a regeneration set, a stationary measure* ν *can be defined by*

$$\int f\,\mathrm{d}\nu = \begin{cases} \mathbb{E}_\lambda \sum_{n=0}^{Y-1} f(X_n) & T=\mathbb{N} \\ \mathbb{E}_\lambda \int_0^Y f(X_t)\,\mathrm{d}t & T=[0,\infty). \end{cases}$$

Proof. It is clear that $\nu\geqslant 0$ and $\nu\neq 0$. Also, a geometric trial argument easily shows that if $E_{n,m} = \{x\in E: \mathbb{P}_x(\tau(R)\leqslant n)\geqslant 1/m\}$ then $\nu(E_{n,m})<\infty$. Since $\bigcup_{n,m} E_{n,m} = E$, ν is σ-finite. To show $\nu P^s = \nu$, let $T=[0,\infty)$ (the discrete time case differs only in notation). Let f be fixed and define $g(x) = \mathbb{E}_x f(X_s)$. Then

$$\int f(y)\nu P^s(\mathrm{d}y) = \int \mathbb{E}_x f(X_s)\nu(\mathrm{d}x) = \int g(x)\nu(\mathrm{d}x)$$

$$= \mathbb{E}_\lambda \int_0^\infty g(X_t) I(Y>t)\,\mathrm{d}t = \int_0^\infty \mathbb{E}_\lambda[\mathbb{E}_{X_t} f(X_s); Y>t]\,\mathrm{d}t.$$

But according to (3.4), this is the same as

$$\int_0^\infty \mathbb{E}_\lambda[f(X_{s+t}); Y>t]\,\mathrm{d}t$$

$$= \mathbb{E}_\lambda \int_s^{Y+s} f(X_u)\,\mathrm{d}u = \mathbb{E}_\lambda\left\{\int_s^Y f(X_u)\,\mathrm{d}u + \int_Y^{Y+s} f(X_u)\,\mathrm{d}u\right\}$$

$$= \mathbb{E}_\lambda\left\{\int_s^Y f(X_u)\mathrm{d}u + \int_0^s f(X_u)\mathrm{d}u\right\} = \mathbb{E}_\lambda \int_0^Y f(X_u)\mathrm{d}u = \int f(y)\nu(\mathrm{d}y),$$

using the regeneration at Y. Since this holds for all f, the proof is complete □

Corollary 3.3. *Let ν be as in Theorem 3.2 and $T = \mathbb{N}$. Then $A \in \mathscr{E}$ is recurrent if and only if $\nu(A) > 0$, or equivalently if and only if $\mathbb{P}_\lambda(X_n \in A) > 0$ for some n.*

Proof. Starting from $X_0 = x$, we eventually end up with a regeneration, and thus A is recurrent if and only if $\mathbb{P}_\lambda(X_n \in A$ infinitely often$) = 1$. Since the cycles are independent identically distributed, this is in turn equivalent to $\mathbb{P}_\lambda(X_n \in A$ for some $n < Y) > 0$ which again holds if and only if the expected number $\nu(A)$ of visits to A before Y is > 0. The last characterization now follows easily by a renewal argument. □

To investigate whether the stationary measure is unique, we first look at the case $T = \mathbb{N}$. Let F be recurrent, $\{X_n^F\}$ the Markov chain restricted to F, cf. I.3, and let ν^F denote the restriction of ν to F (i.e. $\nu^F(A) = \nu(A)$, $A \in \mathscr{E}$, $A \subseteq F$). Then:

Proposition 3.4. *If ν is stationary for $\{X_n\}$ and $0 < \nu(F) < \infty$, then ν^F is stationary for $\{X_n^F\}$.*

Proof. We may assume that $\nu^F(F) = \nu(F) = 1$. Letting $\mathbb{P}_\nu$ denote the measure defined for finite segments by

$$\mathbb{P}_\nu(X_0 \in A_0, \ldots, X_n \in A_n) = \int \mathbb{P}_x(X_0 \in A_0, \ldots, X_n \in A_n)\nu(\mathrm{d}x),$$

it is easily seen that $\mathbb{P}_\nu$ can be handled by the same formal rules as if ν was a proper probability (e.g. we have by stationarity that $\mathbb{P}_\nu(X_k \in A) = \nu(A)$). Let $A \subseteq F$ and define $c_n(A) = \mathbb{P}_\nu(X_0 \notin F, \ldots, X_{n-1} \notin F, X_n \in A)$. Then

$$\begin{aligned} c_n(A) &= \mathbb{P}_\nu(X_1 \notin F, \ldots, X_n \notin F, X_{n+1} \in A) \\ &= c_{n+1}(A) + \mathbb{P}_\nu(X_0 \in F, X_1 \notin F, \ldots, X_n \notin F, X_{n+1} \in A) \\ \nu(A) &= \mathbb{P}_\nu(X_1 \in A) = \mathbb{P}_\nu(X_0 \in F, X_1 \in A) + c_1(A) \\ &\vdots \\ &= \sum_{k=1}^{n} \mathbb{P}_\nu(X_0 \in F, X_1 \notin F, \ldots, X_{k-1} \notin F, X_k \in A) + c_n(A). \end{aligned}$$

Letting $n \to \infty$ yields

$$c_n(A) \to \nu(A) - \mathbb{P}_\nu(X_0 \in F, X_{\tau(F)} \in A) = \nu(A) - \mathbb{P}_{\nu^F}(X_1^F \in A).$$

But for $A = F$ the r.h.s. is just $1 - 1 = 0$. Thus $c_n(A) \leqslant c_n(F) \to 0$, and $\nu(A) = \mathbb{P}_{\nu^F}(F_1^F \in A)$ and stationarity follows. □

Theorem 3.5. *A* (discrete-time) *Harris chain has a stationary measure which is unique up to a multiplicative constant.*

Proof. Suppose first that $\mathbb{E}_\lambda Y = \nu(E) < \infty$ and let $\tilde{\nu}$ be a different stationary measure with $\tilde{\nu}(E) < \infty$. Then $\pi = \nu/\nu(E)$, $\tilde{\pi} = \tilde{\nu}/\tilde{\nu}(E)$ are stationary distributions and by V.1.5(ii) we have that

$$\frac{1}{d}\sum_{j=1}^{d} \mathbb{P}_{\tilde{\pi}}(X_{nd+j} \in A) \to \frac{1}{\mathbb{E}_\lambda Y}\mathbb{E}_\lambda \sum_{n=0}^{Y-1} I(X_n \in A) = \pi(A),$$

where d is the period of Y. But by stationarity of $\tilde{\pi}$, the l.h.s. is just $\tilde{\pi}(A)$ for all n. Hence $\pi = \tilde{\pi}$ and $\nu, \tilde{\nu}$ are proportional.

In the general case where not necessarily $\nu(E) < \infty$, $\tilde{\nu}(E) < \infty$, proportionality follows if we can show that $\nu(A)/\nu(F) = \tilde{\nu}(A)/\tilde{\nu}(F)$ whenever $A \subseteq F$ and $\nu(F)$, $\tilde{\nu}(F) \in (0, \infty)$. Here F is recurrent according to Corollary 3.3 since $\nu(F) > 0$ and we can consider the chain $\{X_n^F\}$ for which $\nu^F, \tilde{\nu}^F$ are both stationary by Proposition 3.4. Thus if we can prove that $\{X_n^F\}$ has a regeneration set, we have from above that $\nu^F = c\tilde{\nu}^F$ and the desired conclusion follows. To this end, choose $m, k, \delta > 0$ such that

$$R^F = \left\{ x \in F : \mathbb{P}_x\left(X_m \in R, S_0 = m + r, \sum_{n=0}^{m+r} I(X_n \in F) = k \right) \geqslant \delta \right\}$$

has positive ν-measure. Then R^F is recurrent for $\{X_n\}$, hence for $\{X_n^F\}$ and for $x \in R^F$

$$\mathbb{P}_x(X_k^F \in A) \geqslant \delta \mathbb{P}_\lambda(X_{\tau(R)} \in A). \qquad \square$$

The uniqueness problem for $T = [0, \infty)$ is deferred for a time.

For $T = \mathbb{N}$, we call the chain *aperiodic* if the $\mathbb{P}_\lambda$-distribution of Y is aperiodic (it follows from Proposition 3.10 below that this property does not depend on the choice of R, λ, ε). For $T = [0, \infty)$, terminology such as 'non-lattice cycles' or 'spread-out cycles' refer to the $\mathbb{P}_\lambda$-distribution of Y in a similar manner. We call $\{X_t\}$ *positive recurrent* if $\mathbb{E}_\lambda Y = \|\nu\| < \infty$ and *null recurrent* if $\|\nu\| = \infty$ (an aperiodic positively recurrent Harris chain is simply called *Harris ergodic*). With $\pi = \nu/\|\nu\|$, the basic limit theorems for regenerative processes then give:

Theorem 3.6. *For a Harris ergodic chain, the $\mathbb{P}_x$-distribution of X_n converges to π in total variation. In particular, $P^n(x, A) \to \pi(A)$ for all $A \in \mathscr{E}$. For a continuous-time positive recurrent Markov process with non-lattice cycles, the $\mathbb{P}_x$-distribution of X_t converges weakly to π.*

Theorem 3.6 is the main ergodic theorem for Harris chains, and we proceed to miscellaneous complements and extensions. First, since the LLN holds for identically distributed one-dependent variables $C_1, C_2, \ldots$ (consider $\{C_{2n}\}$ and $\{C_{2n+1}\}$ separately!), the same proof as in V.3 yields:

Proposition 3.7. *In the positive recurrent case, the time-averages $\sum_0^N f(X_n)/N$ $(T = \mathbb{N})$, $\int_0^t f(X_s)\mathrm{d}s/t$ $(T = [0, \infty))$ converge to $\int f\,\mathrm{d}\pi$ for any bounded measurable f.*

Proposition 3.8. *Suppose that a continuous-time Markov process $\{X_t\}_{t \geqslant 0}$ has* *either spread-out cycles or that* (3.1) *holds for all r in an open interval. Then:*

(i) *Every discrete skeleton* $\{X_{n\delta}\}_{n\in\mathbb{N}}$ *is a Harris chain;*
(ii) *The stationary measure* ν *is unique up to a constant;*
(iii) *In the positive recurrent case* $\|\nu\| < \infty$, *the* $\mathbb{P}_x$*-distribution of* X_t *converges to* $\pi = \nu/\|\nu\|$ *in total variation;*
(iv) *In the null recurrent case,* $\mathbb{P}_x(X_t \in F) \to 0$ *for any set* $F \in \mathscr{E}$ *with* $\nu(F) < \infty$.

Proof. In the second case, we can impose an additional randomization by letting the regenerations occur at times after visits to R which are not fixed at r but uniformly distributed on say (a, b). Then it is immediately clear that the cycle length distribution is absolutely continuous, and we may proceed exactly as in the first case. First (iii) follows by Corollary 1.4. For (i), we first show that R is recurrent for $\{X_{n\delta}\}$. Letting B_t be the forwards recurrence time of the imbedded renewal process, it follows first easily from $\{\delta\}$ being recurrent for $\{B_t\}$ that $[0, \delta]$ is recurrent for $\{B_{n\delta}\}$. Also when cycles are spread out, it is easy to see by a renewal argument that $g(t) = \mathbb{P}_\lambda(X_t \in R) \geqslant \varepsilon > 0$ for t in an interval of length $> 2\delta$, therefore for $t \in [(m-1)\delta, m\delta]$ with m suitably chosen. Hence, if $\mathscr{G}_t = \sigma(X_s, B_s : s \leqslant t)$,

$$\sum_{n=0}^{\infty} \mathbb{P}(X_{(n+m)\delta} \in R \mid \mathscr{G}_{n\delta}) \geqslant \sum_{n=0}^{\infty} g(m\delta - B_{n\delta}) I(B_n \leqslant \delta)$$

$$\geqslant \varepsilon \sum_{n=0}^{\infty} I(B_n \leqslant \delta) = \infty$$

and R being recurrent for $\{X_{n\delta}\}$ follows by the conditional Borel–Cantelli lemma. That R is a regeneration set for $\{X_{m\delta}\}$ is then easily proved: if $m_0\delta > r$, then for $x \in R$ we get from (3.1) that

$$\mathbb{P}^{m_0\delta}(x, A) \geqslant \varepsilon \mathbb{P}_\lambda(X_{m_0\delta - r} \in A).$$

This proves (i), and (ii) is a consequence of (i) and Theorem 3.5. For (iv), check that $z(t) = \mathbb{P}_\lambda(X_t \in F, Y > t)$ satisfies the assumptions of Corollary 1.3. Hence $\mathbb{P}_\lambda(X_t \in F) = U * z(t)$ converges to $\int z/\mu = \nu(F)/\mu = 0$. That $\mathbb{P}_x(X_t \in F)$ also does so follows by the obvious conditioning argument. □

Clearly, the proof of (iv) applies to the case $T = \mathbb{N}$ case as well, and thus:

Corollary 3.9. *In the discrete time null recurrent case,* $\mathbb{P}_x(X_n \in F) \to 0$ *whenever* $\nu(F) < \infty$.

Turning to the periodicity problem, we have the following result concerning the existence of cyclic classes.

Proposition 3.10. *For* $T = \mathbb{N}$, *there exists a* $d = 1, 2, \ldots$ *and a partitioning* $E = E_1 \cup \cdots \cup E_d$ *such that* $P(x, E_{i+1}) = 1$ *for all* $x \in E_i \setminus N$, *where N is a* ν*-null set* (here we identify E_{d+1} with E_1 and so on). *Furthermore, such a partitioning is unique in the sense that if a different one is given in terms of* $\tilde{d}, \tilde{E}_1, \ldots, \tilde{E}_{\tilde{d}}, \tilde{N}$, *then d is a multiple of* $\tilde{d}$, $d = c\tilde{d}$, *and after a cyclic rotation of* E_j *one can achieve* $\tilde{E}_j = \bigcup_{r=1}^{c} E_{j+r\tilde{d}}$ *up to* ν-

null sets. Finally d can be characterized as the $\mathbb{P}_\lambda$-period of Y for some [and therefore all] *choice of R, λ, ε in* (3.1).

Proof. We start from one representation (3.1) and define d to be the $\mathbb{P}_\lambda$-period of Y,

$$F_i = \{x \in E : P^{nd-r-i}(x, R) > 0 \quad \text{for some} \quad n = 0, 1, 2, \ldots\}.$$

Since R is recurrent, $E = F_0 \cup \cdots \cup F_{d-1}$. The F_i need not be disjoint but, however, $\nu(F_i F_j) = 0$ for $i \neq j$. In fact, otherwise there is a m with $\mathbb{P}_\lambda(X_m \in F_i F_j) > 0$, implying that for some n_1, n_2 both $m + n_1 d - i$ and $m + n_2 d - j$ are in the support of Y which is impossible.

A similar argument shows that $\mathbb{P}_\lambda(X_{nd+i} \in F_i) = 1$ for all n, i. Noting that if $\nu(A) = 0$, then $\mathbb{P}_\lambda(X_n \in A) = 0$ for all n, it follows that if we define $E_0 = F_0$, $E_i = F_i - E_0 - \cdots - E_{i-1}$ then E is the disjoint union of the E_i and $\mathbb{P}_\lambda(X_{nd+i} \in E_i) = 1$ for all n, i. To show that $E_{i,j} = \{x \in E_i : P(x, E_j) > 0\}$ is a N-null set for $j \neq i + 1$, note similarly that otherwise $\mathbb{P}_\lambda(X_m \in E_{i,j}) > 0$ for some m. Here m must be of the form $nd + i$ and then $\mathbb{P}_\lambda(X_{nd+i+1} \in E_j) > 0$ which is only possible if $j = i + 1$.

Now let $\tilde{E}_0, \ldots, \tilde{E}_{\tilde{d}-1}$ be a different set of cyclic classes, fix j and choose i with $\nu(E_i \tilde{E}_j) > 0$. Let ψ be a probability measure which is equivalent to the restriction of ν to $E_i \tilde{E}_j$. Then it is easy to see that if $A \subseteq E_i$, $\nu(A) > 0$, then $\mathbb{P}_\psi(X_{nd} \in A) > 0$ for all sufficiently large n. Letting first $A = E_i \tilde{E}_j$, it follows that for some n both nd and $(n+1)d$ are multiples of $\tilde{d}$. Hence $d = c\tilde{d}$. Next with $A = E_i \backslash \tilde{E}_j$, $\mathbb{P}_\psi(X_{nd} \in A) > 0$ would imply that nd is not a multiple of $\tilde{d}$ which is impossible. Hence $\nu(E_i \backslash \tilde{E}_j) = 0$. That is, if $\nu(E_i \tilde{E}_j) > 0$, then $E_i \subseteq \tilde{E}_j$ up to a ν-null set. Choose the numbering such that $E_0 \subseteq \tilde{E}_0$. Then $E_i \subseteq \tilde{E}_i$ $i = 0, \ldots, c-1$, $E_c \subseteq \tilde{E}_0$, $E_{c+1} \subseteq \tilde{E}_1, \ldots, E_{2c} \subseteq \tilde{E}_0, \ldots$ and $\tilde{E}_j = \bigcup_0^{c-1} E_{j+rc}$ follows. □

Now let φ be a non-trivial σ-finite measure on $(E, \mathscr{E})$. We call $\{X_n\}$ *φ-recurrent* if any $F \in \mathscr{E}$ with $\varphi(F) > 0$ is recurrent, and *φ-irreducible* if to any $x \in E$ and $F \in \mathscr{E}$ with $\varphi(F) > 0$ we can find n with $\mathbb{P}_x(X_n \in F) > 0$ (obviously, φ-recurrence implies φ-irreducibility). Then:

Theorem 3.11 (OREY'S C-SET THEOREM) *Let $\{X_n\}$ be φ-irreducible and $\varphi(F) > 0$. Then we can find $C \subseteq F$ with $\varphi(C) > 0$, $\varepsilon > 0$ and r such that $\mathbb{P}_x(X_r \in A) \geqslant \varepsilon\varphi(AC)$ for all $x \in C$.*

The proof is a highly technical application of differentiation results for set functions and will not be given here (see the Notes for text book references). However, the results permits us to characterize Harris chains (as defined here by existence of regeneration sets) in the following more traditional way:

Corollary 3.12. *A Markov chain is a Harris chain if and only if for some φ it is φ-recurrent. In that case, any set F with $\nu(F) > 0$ contains a regeneration set.*

Proof. If a regeneration set exists, the construction of the imbedded renewal process immediately shows that $\{X_n\}$ is λ-recurrent. Suppose, conversely, that

$\{X_n\}$ is φ-recurrent, in particular φ-irreducible, and let $\varphi(F)>0$. Choosing C as in the C-set theorem, we see that C is a regeneration set. Thus the stationary measure ν exists, the chain is ν-recurrent (Corollary 3.3) and we may repeat the argument to see that any F with $\nu(F)>0$ contains a regeneration set. □

We remark that in practical cases the existence of regeneration sets seems far more easy to check than φ-recurrence. For example for $E=\mathbb{R}$, the obvious choice of φ is frequently Lebesgue measure (possibly restricted to some interval) and it may be fairly easy to check that every interval is recurrent. But one needs to show recurrence of every Borel set of positive Lebesgue measure, and since such a set A can have a very complicated structure (e.g. A need not have interior points), this is a considerable task.

With the obvious Markov chain analogue of a definition from Section 3, we finally have

Proposition 3.13. *A Markov chain $\{X_n\}$ with a stationary distribution π admits coupling if and only if it is Harris ergodic. In that case, for any set F with $\pi(F)>0$ and any two initial distributions μ, μ' it is possible to construct coupled versions $\{X_n\}$, $\{X'_n\}$ with the property $X_T=X'_T \in F$, where T is the coupling epoch.*

Proof. Suppose first that $\{X_n\}$ is Harris ergodic, let μ, μ', F be given and choose a regeneration set $R\subseteq F$ as in Corollary 3.12. We construct $\{X_n\}$, $\{X'_n\}$ by first realizing coupled versions of the imbedded renewal process (this is possible according to the discussion of coupling of ergodic Markov chains on a discrete state space given at the beginning of Section 2). Let T be the epoch of the first common renewal. Then we may choose $X_T=X'_T$ distributed according to λ (so that in particular $X_T=X'_T\in R\subseteq F$) and independent of the renewal process up to T, and condition in the remaining X_n $(n\neq T)$ in such a way that $X_n=X'_n$, $n>T$.

Suppose, conversely, that $\{X_n\}$ admits coupling so that we have total variation convergence to π. Let $A\subseteq E$, $\pi(A)=2\varepsilon>0$ and define $\tau=\inf\{n\geqslant 1: X_n\in A\}$. For any ψ, we may find $m(\psi)$ such that $\mathbb{P}_\psi(X_{m(\psi)}\in A)>\varepsilon$ and hence $\mathbb{P}_\psi(\tau\leqslant m(\psi))>\varepsilon$. For a fixed $x\in E$, we now successively define integers $n(1)<n(2)<\cdots$ by $n(1)=m(\delta_x)$ and $n(k+1)=n(k)+m(\psi_k)$, where ψ_k is the conditional $\mathbb{P}_x$-distribution of $X_{n(k)}$ given $\{\tau>n(k)\}$. Then $\mathbb{P}_x(\tau>n(k))\leqslant(1-\varepsilon)^k$ so that $\mathbb{P}_x(\tau<\infty)=1$ and π-recurrence follows. Null recurrence is excluded by the existence of a stationary distribution, and periodicity by total variation convergence as is easily seen from Proposition 3.10. □

Problems

3.1. The *Ornstein–Uhlenbeck process* $\{X_t\}_{t\geqslant 0}$ with parameter $\xi>0$ may be described as a Markov process with state space $\mathbb{R}$ and the $\mathbb{P}_x$-distribution of X_t being normal with mean $e^{-\xi t}x$ and variance $(1-e^{-2\xi t})/2\xi$. Show that (a) the normal distribution with mean zero and variance $1/2\xi$ is stationary; (b) any discrete skeleton $\{X_{n\delta}\}_{n\in\mathbb{N}}$ is Harris recurrent [hint: Foster's criteria]; (c) X_t converges in total variation to the stationary distribution.

3.2. Show that if (3.1) holds for $r+1$ (with the same R, λ!) as well as r, then the chain is aperiodic.

Notes

The theory was initiated largely by Harris in the 1950s (though Doeblin had some early results) and further main work done by Orey and others in the 1960s. Expositions of the theory in that setting are in Orey (1971) and Revuz (1975). The role of regenerative processes and minorization conditions like (3.1) was realized by Nummelin (1978, 1984) and Athreya and Ney (1978a). The problems arising when $r > 1$ are sometimes treated a little carelessly! Our terminology 'regeneration set' is not standard—Nummelin (1984) uses 'small set' and one should also beware of confusion with sets which are regenerative in the sense of Maisonneuve (1974).

4. SECOND-ORDER PROPERTIES

We are concerned with certain refinements of renewal theory, which require the existence of the second moment $\mathbb{E}Y^2$ of the interarrival distribution F, or equivalently that $\sigma^2 = \mathbb{V}\text{ar}\, Y < \infty$. For simplicity, only the non-lattice case is considered.

The first (and simplest) problem to be studied is to look for expansions of the renewal function $U(t)$ more detailed than the one $U(t) \cong t/\mu$ provided by the elementary renewal theorem. This can be obtained by noting that

$$S_{N_t} = t + B_t \tag{4.1}$$

and taking expectations: by Wald's identity,

$$\mathbb{E}S_{N_t} = \mu\mathbb{E}N_t = \mu U(t), \qquad U(t) = \frac{t}{\mu} + \frac{\mathbb{E}B_t}{\mu}. \tag{4.2}$$

Furthermore, from $B_t \xrightarrow{\mathscr{D}} F_0$ one expects that

$$\mathbb{E}B_t \to \int_0^\infty x\, \mathrm{d}F_0(x) = \frac{1}{\mu}\int_0^\infty x(1 - F(x))\mathrm{d}x = \frac{\mathbb{E}Y^2}{2\mu} = \frac{\sigma^2 + \mu^2}{2\mu}. \tag{4.3}$$

To see that this is indeed the case, evaluate e.g. $Z(t) = \mathbb{E}B_t$ by the renewal argument and check that

$$z(t) = \mathbb{E}[B_t; Y_1 > t] = \mathbb{E}[Y_1 - t; Y_1 > t]$$

is directly Riemann integrable with integral $\mathbb{E}Y^2/2$. Combining (4.1)–(4.3), we have proved

Proposition 4.1.

$$U(t) = \frac{t}{\mu} + \frac{\mathbb{E}Y^2}{2\mu^2} + \mathrm{o}(1), \qquad t \to \infty.$$

One might expect that (assuming higher order moments) the o(1) term could be further expanded as $c_1/t + c_2/t^2 + \cdots$. However, under suitable regularity conditions the rate of decay is in fact exponentially fast, cf. Problem 4.3.

Ignoring B_t in (4.1) yields the lower bound $U(t) \geqslant t/\mu$. We shall next find an upper bound somewhat related to the asymptotic expression in Proposition 4.1:

Proposition 4.2.

$$U(t) \leqslant \frac{t}{\mu} + \frac{\mathbb{E}Y^2}{\mu^2}.$$

Proof. According to (4.2), we must show that $\mathbb{E}B_t \leqslant \mathbb{E}Y^2/\mu$ for all t. It follows from IV.(2.8) that $U(t+s) - U(s) \leqslant U(t)$. From this it follows by (4.1), (4.2) that

$$\mathbb{E}B_t + \mathbb{E}B_s = \mu(U(t) + U(s)) - t - s \geqslant \mu U(t+s) - t - s = \mathbb{E}B_{t+s},$$

$$\begin{aligned}\mathbb{E}B_t &\leqslant \inf\{\mathbb{E}B_s + \mathbb{E}B_{t-s} : 0 \leqslant s \leqslant t/2\} \\ &\leqslant \frac{2}{t}\int_0^{t/2} [\mathbb{E}B_s + \mathbb{E}B_{t-s}]\mathrm{d}s = \frac{2}{t}\int_0^t \mathbb{E}B_s\,\mathrm{d}s.\end{aligned} \tag{4.4}$$

Now an inspection of the paths shows that

$$\int_0^t B_s\,\mathrm{d}s = \frac{1}{2}\sum_{n=1}^{N_t} Y_n^2 - \tfrac{1}{2}B_t^2.$$

Thus

$$\begin{aligned}\int_0^t \mathbb{E}B_s\,\mathrm{d}s &= \tfrac{1}{2}\mathbb{E}N_t\mathbb{E}Y^2 - \tfrac{1}{2}\,\mathbb{E}B_t^2 = \tfrac{1}{2}U(t)\mathbb{E}Y^2 - \tfrac{1}{2}\mathbb{E}B_t^2 \\ &= \tfrac{1}{2}(t + \mathbb{E}B_t)EY^2/\mu - \tfrac{1}{2}\mathbb{E}B_t^2.\end{aligned} \tag{4.5}$$

Letting $\alpha = \mathbb{E}B_t$ we have $\mathbb{E}B_t^2 \geqslant \alpha^2$ and combining (4.4), (4.5) yields

$$t\alpha \leqslant (t + \alpha)\mathbb{E}Y^2/\mu - \alpha^2,$$

i.e.

$$\alpha^2 + \alpha(t - \mathbb{E}Y^2/\mu) - t\mathbb{E}Y^2/\mu \leqslant 0. \tag{4.6}$$

But the l.h.s. of (4.6) is a quadratic with roots $-t$ and $\mathbb{E}Y^2/\mu$. Thus $-t \leqslant \alpha \leqslant \mathbb{E}Y^2/\mu$. □

Our next objective is to establish the CLT for the number of renewals and the corresponding expansion of the variance:

Proposition 4.3. (a) *As $t \to \infty$, N_t is asymptotically normal with mean t/μ and variance σ^2/μ^3;*

(b)
$$\mathbb{V}\mathrm{ar}\, N_t \cong \frac{t\sigma^2}{\mu^3} + \mathrm{o}(t).$$

Proof. The results are also valid for general delay distributions, but will for simplicity only be proved for the zero-delayed case. Here (a) can easily be shown by applying Anscombe's theorem to

$$U_t = \frac{S_{N_t} - N_t\mu}{t^{1/2}} = \frac{B_t + t - N_t\mu}{t^{1/2}} \approx \frac{t - N_t\mu}{t^{1/2}} \tag{4.7}$$

in just the same way as in V.3.2. An elementary direct argument is as follows: let y

be fixed and let $n = n(t)$ depend on t in such a way that

$$\frac{t}{\mu} + \left(\frac{t\sigma^2}{\mu^3}\right)^{1/2} y \in (n-1, n].$$

Then

$$t(1+o(1)) = n\mu + O(1), \qquad t = n\mu + o(n),$$

$t^{1/2} = (n\mu)^{1/2} + o(n^{1/2})$ (by Taylor expansion) so that

$$t = n\mu + O(1) - \left(\frac{t\sigma^2}{\mu}\right)^{1/2} y = n\mu - \sigma y n^{1/2} + o(n^{1/2}).$$

Therefore the CLT for S_n yields

$$\mathbb{P}\left(\frac{N_t - t/\mu}{(t\sigma^2/\mu^3)^{1/2}} \leqslant y\right) = \mathbb{P}(N_t \leqslant n) = \mathbb{P}(S_n > t)$$

$$= \mathbb{P}\left(\frac{S_n - n\mu}{\sigma n^{1/2}} > -y + o(1)\right) \to 1 - \Phi(-y) = \Phi(y),$$

proving (a). The proof of (b) can be carried out in a number of ways, none of which are entirely brief. Recalling (4.7), we let

$$V_t = \mu \frac{\mathbb{E}N_t - N_t}{t^{1/2}}, \qquad W_t = U_t - V_t = \frac{B_t - \mathbb{E}B_t}{t^{1/2}}$$

and have to prove $\mathbb{E}V_t^2 \to \sigma^2/\mu$. We first note that by a renewal argument,

$$\mathbb{E}B_t^2 = z(t) + \int_0^t \mathbb{E}B_{t-s}^2 \,\mathrm{d}F(s) = \int_0^t z(t-s)\mathrm{d}U(s), \tag{4.8}$$

where $z(t) = \mathbb{E}[B_t^2; t < Y_1]$. Since $B_t \leqslant Y_1$ on $\{t < Y_1\}$, we have $z(t) \to 0$ so that (4.8) and the elementary renewal theorem yield $\mathbb{E}B_t^2 = o(t)$ and therefore $\mathbb{E}W_t^2 \to 0$. Now by a standard stopping time identity, $\mathbb{E}U_t^2 = \sigma^2 \mathbb{E}N_t/t \to \sigma^2/\mu$. Hence by the Cauchy–Schwarz inequality, $\mathbb{E}U_t W_t \to 0$ and

$$\mathbb{E}V_t^2 = \mathbb{E}U_t^2 + \mathbb{E}W_t^2 - 2\mathbb{E}U_t W_t \to \frac{\sigma^2}{\mu} + 0 - 0. \qquad \square$$

Problems

4.1. Check the details in the proof of (4.3) outlined in the text.

4.2. Give an alternative proof of Proposition 4.1 by showing that $Z(t) = U(t) - t/\mu$ satisfies a renewal equation with $z(t) = 1 - F_0(t)$.

4.3. Show that $U(t) = t/\mu + \mathbb{E}Y^2/2\mu^2 + O(e^{-\varepsilon t})$ for some $\varepsilon > 0$ provided that F is spread out and $\int_0^\infty e^{\delta x} \mathrm{d}F(x) < \infty$ for some $\delta > 0$.

4.4. Show that $\mathbb{E}N_t^2 = 2U * U(t) - U(t)$, that $U * U(t) = t^2/2\mu^2 + t\mathbb{E}Y^2/\mu^2 + o(t)$ and give hereby a different derivation of the expression for $\mathbb{V}\mathrm{ar}\, N_t$.

Notes

Proposition 4.2 is from Lorden (1970). Proposition 4.3 is standard, but a variety of approaches are available. The present proof of part (b) is essentially that of Siegmund (1969), whereas for alternatives we mention in particular Chow *et al.* (1979).

5. EXCESSIVE AND DEFECTIVE RENEWAL EQUATIONS.

Recall that the renewal equation $Z = z + F * Z$ is called *excessive* if $\| F \| > 1$ and *defective* if $\| F \| < 1$. We still have $Z = U * z$ (provided Z, z are bounded on finite intervals), but Blackwell's renewal theorem does not apply to determine the asymptotic behaviour of U and thereby Z. However, by a trick we may reduce to the case $\| F \| = 1$:

Theorem 5.1. *Assume that $Z = z + F * z$ with Z, z bounded on finite intervals and that for some real β*

$$\hat{F}(\beta) = \int_0^\infty e^{\beta x}\, dF(x) = 1. \tag{5.1}$$

Define

$$\tilde{Z}(x) = e^{\beta x} Z(x), \qquad \tilde{z}(x) = e^{\beta x} z(x), \qquad d\tilde{F}(x) = e^{\beta x}\, dF(x).$$

Then $\tilde{Z} = \tilde{z} + \tilde{F} * \tilde{Z} = \tilde{U} * z$ and if $\tilde{z}$ is directly Riemann integrable (d.R.i.)

$$\lim_{x \to \infty} e^{\beta x} Z(x) = \lim_{x \to \infty} \tilde{Z}(x) = \frac{1}{\tilde{\mu}} \int_0^\infty \tilde{z}(t) dt = \frac{\displaystyle\int_0^\infty e^{\beta t} z(t) dt}{\displaystyle\int_0^\infty t\, e^{\beta t}\, dF(t)} \tag{5.2}$$

Proof. Clearly $\tilde{Z}, \tilde{z}$ are bounded on finite intervals and $\| \tilde{F} \| = 1$ by (5.1). Also

$$\begin{aligned} \tilde{Z}(x) &= e^{\beta x} \left\{ z(x) + \int_0^x z(x-y) dF(y) \right\} \\ &= \tilde{z}(x) + \int_0^x e^{\beta(x-y)} z(x-y) e^{\beta y}\, dF(y) = \tilde{z}(x) + \tilde{F} * \tilde{Z}(x) \end{aligned}$$

and the remaining statements are immediately apparent from Chapter IV. □

For a closer study of the assumption (5.1) and its implication (5.2), we need to treat the excessive and the defective case separately.

Proposition 5.2. *Consider the excessive case $1 < \| F \| \leqslant \infty$. Then a solution β to* (4.1) *is necessarily strictly negative, $\beta < 0$. A sufficient condition for the existence of β is $1 < \hat{F}(\delta) < \infty$ for some δ and then always $\tilde{\mu} < \infty$. Thus holds in particular if $\| F \| < \infty$.*

Proof. Since $\hat{F}(\beta) \geqslant \| F \|$ for $\beta \geqslant 0$, it is clear that (5.1) implies $\beta < 0$. If δ exists, then by monotone convergence $\hat{F}(\beta)$ is a continuous function of $\beta \in (-\infty, \delta]$ with limits 0 as $\beta \to -\infty$ and $\hat{F}(\delta) > 1$ as $\beta \uparrow \delta$. Hence the value 1 is obtained and the F-integrability of $e^{\delta x}$ implies that of $x\, e^{\beta x}$, i.e. $\tilde{\mu} < \infty$. Finally if $\| F \| < \infty$, we can just take $\delta = 0$. □

Example 5.3. Consider Lotka's integral equation (IV.2.2) for the density $Z(t)$ of births in a population at time t and assume that, as will typically be the case, that the net reproduction rate $\|F\|$ is >1. The assumption $\|F\|<\infty$ is innocent from the demographic point of view and hence we may conclude that β exists, is <0 and that

$$\tilde{\mu}=\int_0^\infty s\,\mathrm{e}^{\beta s}\,\mathrm{d}F(s)=\int_0^\infty s\,\mathrm{e}^{\beta s}{}_s p_0\lambda(s)\mathrm{d}s<\infty.$$

Also the assumption of $\tilde{z}$ being d.R.i. is innocent. In fact, inspection of the expression for z shows this to hold if only the birth intensity $\lambda(u)$ is bounded and continuous and the survival rate ${}_tp_a$ is continuous (then $z(t)$ is bounded and continuous, hence $\tilde{z}(t)=\mathrm{e}^{\beta t}z(t)$ d.R.i. because of IV.4.1(iv), (v)). Thus under these assumptions, $Z(t)$ grows asymptotically exponentially, $Z(t)\cong C\mathrm{e}^{-\beta t}$, where

$$C=\frac{1}{\tilde{\mu}}\int_0^\infty \tilde{z}(t)\mathrm{d}t=\frac{1}{\tilde{\mu}}\int_0^\infty\int_0^\infty \mathrm{e}^{\beta t}f_0(a){}_tp_a\lambda(a+t)\mathrm{d}t\,\mathrm{d}a$$

(the rate $-\beta$ is usually called the *Malthusian parameter* of the population). From this the limiting behaviour of other quantities is easily obtained. For example, for the total population size $N(t)$ we easily obtain from IV.(2.6) that

$$\mathrm{e}^{\beta t}N(t)=\int_0^t \mathrm{e}^{\beta(t-a)}Z(t-a)\mathrm{e}^{\beta a}{}_ap_0\,\mathrm{d}a+\mathrm{e}^{\beta t}\int_0^\infty f_0(a){}_tp_a\,\mathrm{d}a$$

$$\to C\int_0^\infty \mathrm{e}^{\beta a}{}_ap_0\,\mathrm{d}a+0.$$ □

In the defective case, a simple conclusion can be obtained without reference to condition (5.1):

Proposition 5.4. *If in the defective case z is bounded and $z(\infty)=\lim_{t\to\infty}z(t)$ exists, then $Z(t)\to z(\infty)/(1-\|F\|)=Z(\infty)$ (say).*

Proof. Using dominated convergence and IV.(2.9) we get

$$Z(t)=U*z(t)=\int_0^t z(t-y)U(\mathrm{d}y)\to\int_0^\infty z(\infty)U(\mathrm{d}y)=\frac{z(\infty)}{1-\|F\|}.$$ □

If $z(\infty)=0$, this result is rather imprecise, and also in some cases with $z(\infty)\neq 0$ it is of substantial interest to estimate the rate of convergence of $z(t)$ to $z(\infty)$. To this end (5.1) comes in. However, as (5.1) already shows, the conditions for the existence of β are rather stronger than in the excessive case and require the existence of exponential moments. We have the following analogue of Proposition 5.2:

Proposition 5.5. *Consider the defective case $\|F\|<1$. Then a solution β to (5.1) is necessarily strictly positive, $\beta>0$. A sufficient condition for the existence of β is $1<\hat{F}(\delta)<\infty$ for some δ, and then always $\hat{\mu}<\infty$. This holds in particular if $\hat{F}(\delta)<\infty$ for all $\delta\in\mathbb{R}$.*

Proof. Exactly as for Proposition 5.2, except that for the last step one notes that if $\hat{F}(\delta) < \infty$ for all δ, then $\hat{F}(\delta) \to \infty$ as $\delta \to \infty$. □

Now define $Z_1(t) = Z(t) - Z(\infty)$,

$$z_1(t) = z(t) - z(\infty) + z(\infty)\frac{F(t) - \|F\|}{1 - \|F\|}.$$

Proposition 5.6. *Suppose that in the defective case* $\|F\| < 1$ *a solution* β *to* (5.1) *satisfying* $\tilde{\mu} < \infty$ *exists. Then if* $z(\infty) = \lim_{t\to\infty} z(t)$ *exists and* $e^{\beta t}(z(t) - z(\infty))$ *is d.R.i*

$$Z(t) - Z(\infty) \cong e^{-\beta t}\frac{1}{\tilde{\mu}}\left\{\int_0^\infty e^{\beta t}[z(t) - z(\infty)]dt - \frac{z(\infty)}{\beta}\right\}, \qquad t \to \infty. \tag{5.3}$$

Proof. Since $U * F = U - 1$, we get

$$U * z_1 = U * z - z(\infty)U + z(\infty)\frac{U - 1 - \|F\|U}{1 - \|F\|}$$

$$= U * z - \frac{z(\infty)}{1 - \|F\|} = Z - Z(\infty) = Z_1.$$

Now since $\beta > 0$,

$$e^{\beta t}(\|F\| - F(t)) \leqslant \int_t^\infty e^{\beta s}\, dF(s) = 1 - \tilde{F}(t).$$

The r.h.s. in non-increasing with integral $\tilde{\mu}$. Thus the l.h.s. is d.R.i., hence $e^{\beta t}z_1(t)$ is so and (5.2) yields

$$e^{\beta t}Z_1(t) \to \frac{1}{\tilde{\mu}}\int_0^\infty e^{\beta t}z_1(t)dt.$$

Thus (5.3) follows from

$$\int_0^\infty e^{\beta t}[\|F\| - F(t)]dt = \int_0^\infty e^{\beta t}\int_t^\infty dF(s) = \int_0^\infty \frac{1}{\beta}(e^{\beta s} - 1)dF(s) = \frac{1}{\beta}(1 - \|F\|).$$

□

Example 5.7. Consider the ruin problem (IV.2.3) of insurance mathematics and assume that, as will typically be the case, the premium exceeds the expected claims. Recalling that the probability $Z(u)$ of ultimate survival with initial reserve u satisfies $Z = z + F * Z$ with $z(u) = Z(0)$, $dF(x) = \lambda/c(1 - G(x))dx$, where λ is the arrival intensity, c the premium per unit time and G the claim size distribution, this amounts to $\lambda v/c < 1$ where $v = \int_0^\infty x\, dG(x)$.

It remains to evaluate $Z(0)$. First note that

$$ct - \sum_{n=1}^{N_t} X_n \cong t\left(c - \frac{N_t}{t}v\right) \cong t(c - \lambda v)$$

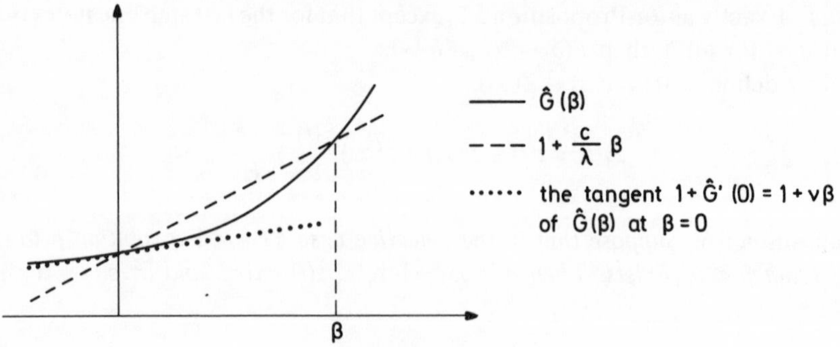

Fig. 5.1

so that $ct - \sum_1^{N_t} X_n$ tends to infinity, hence attains a minimum $m > -\infty$ and thus $Z(u) = \mathbb{P}(u + m \geqslant 0) \to 1$ as $u \to \infty$. Using Proposition 5.4 we therefore get $1 = Z(0)/(1 - \lambda v/c)$, i.e. $Z(0) = 1 - \lambda v/c$. Since for $\beta > 0$

$$\hat{F}(\beta) = \frac{\lambda}{c}\int_0^\infty e^{\beta x}(1 - G(x))dx = \frac{\lambda}{c\beta}\int_0^\infty (e^{\beta x} - 1)dG(x) = \frac{\lambda}{c\beta}(\hat{G}(\beta) - 1),$$

the assumption $\hat{F}(\beta) = 1, \tilde{\mu} < \infty$ amounts to

$$\hat{G}(\beta) = 1 + \frac{c}{\lambda}\beta \tag{5.4}$$

for some $\beta > 0$ satisfying $\hat{G}'(\beta) < \infty$, cf. Fig. 5.1. Since $z(t) = z(\infty)$, we have thus from Proposition 5.6 derived the well-known *Cramér–Lundberg approximation* for the probability $1 - Z(u) = Z(\infty) - Z(u)$ of ultimate ruin,

$$1 - Z(u) \cong e^{-\beta u}\frac{1 - \lambda v/c}{\tilde{\mu}\beta}, \qquad u \to \infty \tag{5.5}$$

The equation (5.4) is known as the *Lundberg equation.* □

Notes

The material is standard, see Feller (1971, Ch. XI) or Jagers (1975, Ch. 5). Further related material will be developed in Chapters XII and XIII.2.

CHAPTER VII

Random Walks

1. BASIC DEFINITIONS

We consider throughout this chapter a random walk $S_0 = 0$, $S_n = X_1 + \cdots + X_n$ where the X_n are independent identically distributed with common distribution F. The case where F has support contained in a half-line $(-\infty, 0]$ or $[0, \infty)$ is to a large extent covered by renewal theory, and hence we assume that supp(F) contains points of both positive and negative sign (in particular, F is non-degenerate). In statements concerning the mean $\mathbb{E}X$, it is understood that this is well defined, i.e. that $\mathbb{E}X^+$ and $\mathbb{E}X^-$ are not both infinite (thus we may have $\mathbb{E}X = +\infty$ or $\mathbb{E}X = -\infty$).

The relevance of random walks for queueing theory should already be clear from the discussion of Lindley processes in III.7. A main point was found there to be the study of the distribution of the maximum, but a number of further functionals are important, both as technical tools and because of other queueing relations and interpretations (e.g. the so-called *ladder epochs* and *ladder heights*, a terminology arising from path decompositions to be discussed in Section 2, will be found in Chapter VIII to be closely related to busy periods and busy cycles). For the sake of easy reference, we shall give a list here of the functionals to play an important role in the following. For a graphical illustration, see Fig. 1.1.

M_n the *partial maximum* $\max_{0 \leqslant k \leqslant n} S_k$ of the first n partial sums.

M the (total) *maximum* $\sup_{0 \leqslant k < \infty} S_k$ (which may be infinite). Clearly, $M_n \uparrow M$ as $n \to \infty$.

τ_+ the first (strict) *ascending ladder epoch* $\tau_+ = \tau_+^{\mathrm{s}} = \inf\{n \geqslant 1 : S_n > 0\}$ or the *entrance time* to $(0, \infty)$. The distribution of τ_+ may be defective, i.e. $\mathbb{P}(\tau_+ = \infty) = \mathbb{P}(S_n \leqslant 0 \text{ for all } n \geqslant 1) > 0$.

S_{τ_+} the first (strict) *ascending ladder height* (defined on $\{\tau_+ < \infty\}$ only).

G_+ the (strict) *ascending ladder height distribution* $G_+(x) = \mathbb{P}(S_{\tau_+} \leqslant x, \tau_+ < \infty)$. Here G_+ is concentrated on $(0, \infty)$ and may be defective, i.e. $\|G_+\| = \mathbb{P}(\tau_+ < \infty) < 1$.

τ_- the first (weak) *descending ladder epoch* $\tau_- = \tau_-^{\mathrm{w}} = \inf\{n \geqslant 1 : S_n \leqslant 0\}$ or the *entrance time* to $(-\infty, 0]$.

S_{τ_-} the first (weak) *descending ladder height* (defined on $\{\tau_- < \infty\}$ only).

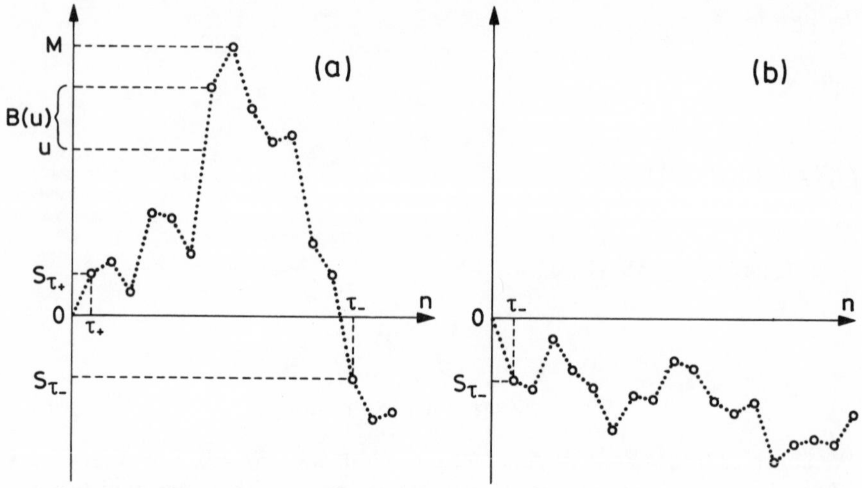

Fig. 1.1 Two paths of the random walk. Path (b) has $M=0$ and $\tau_+=\infty$

G_- the (weak) descending *ladder height distribution* $G_-(x)=\mathbb{P}(S_{\tau_-}\leqslant x, \tau_-<\infty)$. Here G_- is concentrated on $(-\infty,0]$ and may be defective, i.e. $\|G_-\|=\mathbb{P}(\tau_-<\infty)<1$.

$\tau(u)$ the time $\inf\{n\geqslant 1: S_n>u\}$ of *first passage* to level $u\geqslant 0$ or the *entrance time* to (u,∞). The distribution of $\tau(u)$ may be defective. Clearly, $\tau(0)=\tau_+$.

$B(u)$ the *overshot* $S_{\tau(u)}-u$ (defined on $\{\tau(u)<\infty\}$ only). Clearly, $B(0)$ is the ascending ladder variable S_{τ_+}.

$B(\infty)$ a random variable having the limiting distribution (if existent) of $B(u)$ as $u\to\infty$.

It is seen that there is a slight asymmetry between positive and negative values in that we have taken the ascending ladder epochs and ladder heights strict and the descending ones weak, corresponding to different treatments of values n with $S_n=0$. Weak ascending and strict descending ladder epochs can be defined in the obvious way by

$$\tau_+^{\mathrm{w}}=\inf\{n\geqslant 1: S_n\geqslant 0\},\qquad \tau_-^{\mathrm{s}}=\inf\{n\geqslant 1: S_n<0\}.$$

The corresponding ladder heights are $S_{\tau_+^{\mathrm{w}}}$, $S_{\tau_-^{\mathrm{s}}}$, with distributions say G_+^{w}, G_-^{s}. These quantities may be needed in arguments involving sign changes, say if we want to study the minimum $\inf_{0\leqslant k<\infty}S_k$ rather than the maximum M. Fortunately, a separate treatment can almost always be avoided by reference to the following result:

Proposition 1.1. *Define* $\zeta=\mathbb{P}(S_{\tau_-}=0)=\mathbb{P}(\tau_-\neq\tau_-^{\mathrm{s}})$. *Then* $\zeta<1$ *and* $\zeta=\mathbb{P}(S_{\tau_+^{\mathrm{w}}}=0)=\mathbb{P}(\tau_+\neq\tau_+^{\mathrm{w}})$, *and if* δ_0 *is the distribution degenerate at zero, then*

$$G_+^{\mathrm{w}}=\zeta\delta_0+(1-\zeta)G_+,\qquad G_-=\zeta\delta_0+(1-\zeta)G_-^{\mathrm{s}}. \tag{1.1}$$

The proof is based upon a trivial but important observation:

Lemma 1.2. *For any n,*

$$\{S_k\}_{k=0}^n = \{X_1 + \cdots + X_k\}_{k=0}^n \stackrel{\mathscr{D}}{=} \{X_{n-k+1} + \cdots + X_n\}_{k=0}^n = \{S_n - S_{n-k}\}_{k=0}^n$$

Proof of Proposition 1.1. For any n, we get by Lemma 1.2 that

$$\begin{aligned}\mathbb{P}(S_{\tau_+^w} = 0, \tau_+^w = n) &= \mathbb{P}(S_0 = 0, S_k < 0\, k = 1,\ldots,n-1, S_n = 0)\\ &= \mathbb{P}(S_n - S_n = 0, S_n - S_{n-k} < 0\, k = 1,\ldots,n-1, S_n - S_0 = 0)\\ &= \mathbb{P}(S_k > 0\, k = 1,\ldots,n-1, S_n = 0) = \mathbb{P}(S_{\tau_-^w} = 0, \tau_-^w = n)\end{aligned}$$

and $\zeta = \mathbb{P}(S_{\tau_+^w} = 0)$ follows by summation over n. The relation (1.1) is obvious and finally $1 - \zeta \geqslant \mathbb{P}\,(X_1 < 0) > 0$. □

We shall also need:

Lemma 1.3. *G_+ is lattice with span d if and only if F is so, and in particular non-lattice if and only if F is so. The same statements hold with G_+ replaced by any of G_+^w, G_-, G_-^s.*

Proof. It has to be shown that G_+ is concentrated on $\{0, \pm d, \pm 2d, \ldots\}$ if and only if F is so. The 'if' part is obvious and for the converse we may assume $d = 1$. That is, we have to show that if G_+ is concentrated on $\mathbb{N}$, then F is concentrated on $\mathbb{Z}$. Obviously, $\operatorname{supp}(F) \cap (0, \infty) \subseteq \operatorname{supp}(G_+) \subseteq \mathbb{N}$ so we have to show that $X_1 I(X_1 < 0) \in \mathbb{Z}$. But this follows since a path with $X_1 < 0$, $X_k > 0$, $k = 2,\ldots,\tau_+$, has positive probability and satisfies $n = X_{\tau_+} = X_1 + m$ for some $n, m \in \mathbb{N}$. The last assertion of the lemma follows by symmetry arguments and (1.1). □

Problems

1.1. Show that $\tau_+ \stackrel{\mathscr{D}}{=} T_0 + T_1 + \cdots + T_{\sigma-1}$, where T_0, $T_1 \ldots$ are independent, T_0 is distributed as τ_+^w given $S_{\tau_+^w} > 0$, $T_1, T_2, \ldots$ are distributed as τ_+^w given $S_{\tau_+^w} = 0$ and σ is independent of the T_k with $\mathbb{P}(\sigma = n) = \zeta^{n-1}(1-\zeta)$ $n = 1, 2, \ldots$

Notes

Random walks are of course one of the classical areas of probability theory and give rise to a broad spectrum of problems, of which the present treatment covers only a rather narrow range.

2. LADDER PROCESSES AND CLASSIFICATION

By iterating the definitions of τ_+, τ_- we can define whole sequences $\{\tau_+(n)\}$, $\{\tau_-(n)\}$ of ladder epochs by $\tau_+(1) = \tau_+$, $\tau_-(1) = \tau_-$,

$$\tau_+(n+1) = \inf\{k > \tau_+(n): S_k > S_{\tau_+(n)}\},$$

$$\tau_-(n+1) = \inf\{k > \tau_-(n): S_k \leqslant S_{\tau_-(n)}\}.$$

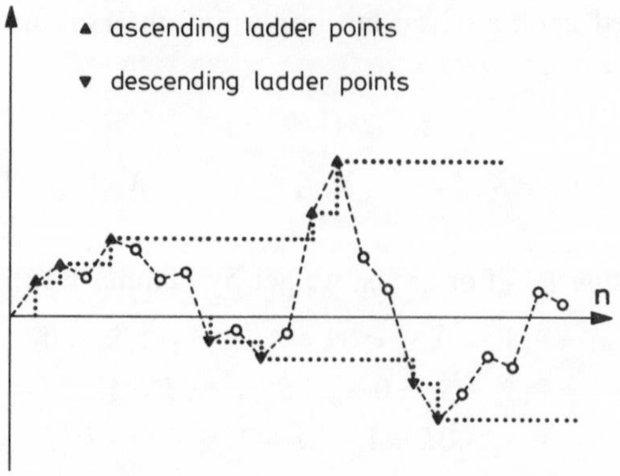

Fig. 2.1 The ladder point processes

The points in the plane of the form $(\tau_+(n), S_{\tau_+(n)})$ are called the *ascending ladder points.* Similarly, the $(\tau_-(n), S_{\tau_-(n)})$ are the *descending ladder points*, $\{S_{\tau_+(n)}\}$ the *ascending ladder height process* and so on.

The importance of these concepts is due to the fact that the sections of the random walk between (say) ascending ladder points are just independent identically distributed replicates. For example, the ascending ladder epoch process $\{\tau_+(n)\}$ is a discrete time renewal process with governing probabilities $f_n = \mathbb{P}(\tau_+ = n)$, thus terminating if and only if $\|G_+\| < 1$. Similarly, the ascending ladder height process is a renewal process governed by G_+, hence proper if and only if $\|G_+\| = 1$. Further, its forward recurrence-time process is readily seen to coincide with the overshot process $\{B(u)\}_{u\geqslant 0}$ of the random walk. Hence the renewal theorem IV.(4.3) and Lemma 1.3 yield the following result (the lattice version is obvious and omitted):

Theorem 2.1. *The overshot $B(u)$ is proper if and only if the ascending ladder height process is non-terminating, i.e. $\|G_+\| = 1$. In that case, it holds as $u \to \infty$ that $B(u) \xrightarrow{\mathscr{D}} \infty$ if $\mathbb{E}S_{\tau_+} = \infty$, whereas if $\mathbb{E}S_{\tau_+} < \infty$ and F is non-lattice, then $B(u) \xrightarrow{\mathscr{D}} B(\infty)$ with $B(\infty)$ having density $(1 - G_+(x))/\mathbb{E}S_{\tau_+}$.*

Also, it is clear that the maximum M equals the lifetime $\sup\{S_{\tau_+(n)} : \tau_+(n) < \infty\}$ of the ascending ladder height process. Hence if we let $U_+ = \sum_0^\infty G_+^{*n}$ denote the corresponding renewal measure, IV.2.6 yields the elementary but important:

Theorem 2.2. *The maximum M is finite if and only if $\|G_+\| < 1$. In that case, the distribution of M is the normalized ascending ladder height renewal measure $U_+/\|U_+\| = (1 - \|G_+\|)U_+$.*

The renewal measure U_+ is of basic importance in the following, and we proceed to give yet a third interpretation. Define

$$U_+^{(n)}(A) = \mathbb{P}(S_n > S_k, k = 0, \ldots, n-1, S_n \in A)$$

so that $U_+^{(n)}$ denotes the (defective) distribution of ladder heights corresponding to ladder epoch n and we have $U_+ = \sum_0^\infty U_+^{(n)}$.

Theorem 2.3.
(a) $U_+^{(n)}(A) = \mathbb{P}(S_k > 0, k = 1, \ldots, n, S_n \in A)$;
(b) $U_+(A) = \mathbb{E} \sum_{n=0}^{\tau_- - 1} I(S_n \in A)$;
(c) $\mathbb{E}\tau_- = \| U_+ \| = (1 - \| G_+ \|)^{-1}$.

Proof. Here (a) is an immediate consequence of Lemma 1.2, (b) is obtained by summing over n and using $\{S_k > 0, k = 1, \ldots, n\} = \{\tau_- > n\}$, and finally (c) follows by letting $A = [0, \infty)$ (or $\mathbb{R}$) in (b). □

We proceed to classify the random walk into several types. The first result is as follows:

Theorem 2.4. *For any random walk with F not degenerate at zero, one of the following possibilities occur*:

(i) Oscillating case: *G_+ and G_- are both proper, $\| G_+ \| = \| G_- \| = 1$ and $\overline{\lim}\, S_n = +\infty$, $\underline{\lim}\, S_n = -\infty$ a.s. Furthermore $\mathbb{E}\tau_+ = \mathbb{E}\tau_- = \infty$,*
(ii) Drift to $+\infty$: *G_+ is proper and G_- defective, and $S_n \to +\infty$ a.s. Furthermore $\mathbb{E}\tau_+ = (1 - \| G_- \|)^{-1} < \infty$.*
(iii) Drift to $-\infty$: *G_+ is defective and G_- proper, and $S_n \to -\infty$ a.s. Furthermore $\mathbb{E}\tau_- = (1 - \| G_+ \|)^{-1} < \infty$.*

A sufficient condition for (i) *is* $\mathbb{E}X = 0$, *for* (ii) $\mathbb{E}X > 0$ *and for* (iii) $\mathbb{E}X < 0$.

Necessary and sufficient (though rather intractable) conditions also covering the case $\mathbb{E}X^+ = \mathbb{E}X^- = \infty$ are given in Corollary 4.4. Note that if say $-\infty < \mathbb{E}X < 0$, then Wald's identify applies and yields

$$1 - \| G_+ \| = 1/\mathbb{E}\tau_- = \mathbb{E}X/\mathbb{E}S_{\tau_-} \tag{2.1}$$

Proof. Since $\overline{\lim}\, S_n$ is exchangeable, we have by the Hewitt–Savage 0–1 law that $\overline{\lim}\, S_n = a$ a.s. for some $a \in [-\infty, \infty]$. If $|a| < \infty$, then $\overline{\lim}\,(S_n - X_1) = a$ would imply $a + X_1 \overset{\mathscr{D}}{=} a$ which is impossible. Similarly, it is seen that $\mathbb{P}(|\underline{\lim}\, S_n| = \infty) = 1$ and hence indeed only the possibilities (i′) $\overline{\lim}\, S_n = \infty$, $\underline{\lim}\, S_n = -\infty$, (ii′) $S_n \to \infty$, (iii′) $S_n \to -\infty$ occur.

Since $S_{\tau_+} > 0$ and obviously $\overline{\lim}\, S_n = \overline{\lim}\, S_{\tau_+(n)}$, we see that $\overline{\lim}\, S_n = \infty$ if and only if the ascending ladder height renewal process is non-terminating, i.e. if and only if $\| G_+ \| = 1$. Similarly, $\underline{\lim}\, S_n = -\infty$ if and only if $\| G_-^{(s)} \| = 1$, i.e. if and only if $\| G_- \| = 1$ (Proposition 1.1). Noting that the expressions for $\mathbb{E}\tau_-$ is just Proposition 2.3(c) and that the one for $\mathbb{E}\tau_+$ is proved similarly, it is then seen that indeed (i)⇔(i′), (ii)⇔(ii′), (iii)⇔(iii′).

By the LLN, it is clear that $\mathbb{E}X > 0 \Rightarrow$(ii) and $\mathbb{E}X < 0 \Rightarrow$(iii). To see that (i) holds if $\mathbb{E}X = 0$, suppose that instead we are, for example, in case (iii). Then $\mathbb{E}\tau_- < \infty$ and Wald's identity yields $\mathbb{E}S_{\tau_-} = 0$ which is impossible since $\mathbb{P}(S_{\tau_-} < 0) > 0$. □

Now define $U = \sum_0^\infty F^{*n}$ so that for any Borel set $A \subseteq \mathbb{R}$

$$U(A) = \sum_{n=0}^{\infty} \mathbb{P}(S_n \in A) = \mathbb{E} \sum_{n=0}^{\infty} I(S_n \in A)$$

is the expected number of visits of the random walk to A (which may of course be infinite).

Lemma 2.5. *If F is non-lattice, then* $\operatorname{supp}(U) = \mathbb{R}$. *If F is aperiodic on $\mathbb{Z}$, then* $\operatorname{supp}(U) = \mathbb{Z}$.

Proof. Suppose first F is non-lattice. Let $x \in \mathbb{R}$ be fixed and choose to a given $\varepsilon > 0$ T such that $d(y, \operatorname{supp}(U_+)) < \varepsilon$ for all $y \geq T$ (this is possible in view of Lemma 1.3 and IV.5.1). Choose next $z < x - T$ with $z \in \operatorname{supp}(F^{*k})$ for some k and $u \in \operatorname{supp}(U_+)$ with $|u - (x - z)| < \varepsilon$. Then $z + u$ is clearly in $\operatorname{supp}(U)$ so that $d(x, \operatorname{supp}(U)) < \varepsilon$ which letting $\varepsilon \downarrow 0$ proves that $x \in \operatorname{supp}(U)$. The lattice case is similar. □

Now if F is lattice, say aperiodic on $\mathbb{Z}$, we may view $\{S_n\}$ as a Markov chain on $\mathbb{Z}$ and irreducibility follows from Lemma 2.5. Hence by I.1.2 we have recurrence if $\sum_0^\infty \mathbb{P}(S_n = 0) = \infty$ (which is easily seen to be equivalent to $U(A) = \infty$ for all finite non-empty subsets of $\mathbb{Z}$) and transience otherwise. Also in the non-lattice case a similar dichotomy holds:

Theorem 2.6. *For any non-lattice random walk, one of the following two possibilities occur:*

(i) TRANSIENT CASE: *for any bounded interval J, we have $U(J) < \infty$ and* $\mathbb{P}(S_n \in J$ infinitely often (i.o.)$) = 0$. *That is, $|S_n| \to \infty$ a.s.*
(ii) RECURRENT CASE: *for any non-degenerate interval J, we have $U(J) = \infty$ and* $\mathbb{P}(S_n \in J \text{ i.o.}) = 1$.

Proof. We shall show that: (i′) $U((-\varepsilon, \varepsilon)) < \infty$ for some $\varepsilon > 0$; (ii′) $U((-\varepsilon, \varepsilon)) = \infty$ for all $\varepsilon > 0$ imply the conclusions of (i), and (ii). Define

$$I = (x - \varepsilon/2, x + \varepsilon/2), \tau = \inf\{n : S_n \in I\}.$$

In case (i′), we have on $\{\tau < \infty\}$ that $S_\tau + (-\varepsilon, \varepsilon)$ covers I. Thus the strong Markov property (I.6.2) and the space homogeneity of the random walk implies that $U(I) \leq U((-\varepsilon, \varepsilon)) < \infty$. But then for any bounded interval J we have $U(J) < \infty$, since J can be covered by a finite number of intervals of lengths ε. Also, by the Borel–Cantelli lemma, only a finite number of the events $A_n = \{S_n \in J\}$ can occur since $\sum \mathbb{P}(A_n) = U(J) < \infty$. Thus (i) holds.

In case (ii′), it suffices to show that $U(I) = \infty$ and $\mathbb{P}(S_n \in I \text{ i.o.}) = 1$ (since x, ε are arbitrary). By Lemma 2.5 we have $\mathbb{P}(\tau < \infty) > 0$. Define

$$q(\delta) = \mathbb{P}(|S_n| < \delta \quad \text{for some } n > 1),$$
$$p_k(\delta) = \mathbb{P}(|S_n| < \delta \quad \text{at least } k \text{ times}).$$

Applying the strong Markov property at the time of the kth visit to $(-\delta,\delta)$ shows that $p_{k+1}(\delta)\leqslant p_k(\delta)q(2\delta)$, which in conjunction with $\sum p_k(\delta)=U((-\delta,\delta))=\infty$ shows that $q(2\delta)=1$. But $q(\delta)=1$ for all $\delta>0$ and the strong Markov property applied to τ easily yields $\mathbb{P}(S_n\in I$ for at least two $n|\tau<\infty)=1$. Repeating the argument, we get $\mathbb{P}(S_n\in I$ infinitely often (i.o.) $|\tau<\infty)=1$, i.e. $\mathbb{P}(S_n\in I \text{ i.o.})=\mathbb{P}(\tau<\infty)>0$. Since $\{S_n\in I \text{ i.o.}\}$ is an exchangeable event, the Hewitt–Savage 0–1 law implies $\mathbb{P}(S_n\in I \text{ i.o.})=1$ and therefore also $U(I)=\infty$. The lattice case is entirely similar. □

Corollary 2.7. *The random walk is transient if* $\mathbb{E}X\neq 0$ *and recurrent if* $\mathbb{E}X=0$ *or, more generally, if the weak LLN* $|S_n|/n\xrightarrow{\mathbb{P}}0$ *holds.*

Proof. That $\mathbb{E}X\neq 0$ implies transience is obvious since then S_n eventually has the same sign as $\mathbb{E}X$. Recalling the interpretation of U, we have

$$\sum_{k=-m}^{m-1} U([k,k+1])\geqslant U([-m,m]),$$

$$U([k,k+1])\leqslant \mathbb{P}(S_n\in[k,k+1] \quad \text{for some } n)U([-1,1])\leqslant U([-1,1]).$$

Hence letting $S_n^*=|S_n|/n$, we have for each K and m that

$$U([-1,1])\geqslant\frac{1}{2m}U([-m,m])\geqslant\frac{1}{2m}\sum_{n=1}^{Km}\mathbb{P}(|S_n|<m)$$
$$\geqslant\frac{1}{2m}\sum_{n=1}^{Km}\mathbb{P}(S_n^*<K^{-1})$$

But $S_n^*\xrightarrow{\mathbb{P}}0$ implies $\sum_1^N\mathbb{P}(S_n^*<\varepsilon)/N\to 1$ for all $\varepsilon>0$. Hence letting $m\to\infty$ yields $U([-1,1])\geqslant K/2$. Thus indeed $U([-1,1])=\infty$ and recurrence follows. □

Problems

2.1. Explain (in the case $\mathbb{E}X<0$) how the expression $(1-\|G_+\|)U_+$ for the distribution of M is connected to Theorem 2.3(b), (c) and the basic formula V.(1.1) for the limits of regenerative processes.

2.2. Let $Y_0^{(1)}$, $Y_0^{(2)}$ be initial delays for a discrete renewal process with infinite mean, $\mu=\sum_1^\infty nf_n=\infty$. Let $c=f_1+\cdots+f_N$ with N chosen such that at least two terms are non-zero, and define

$$g_n=f_n/c,\qquad n\leqslant N,\qquad h_n=f_n/(1-c),\qquad n>N.$$

Consider independent sequences $\{U_n^{(1)}\}$, $\{U_n^{(2)}\}$, $\{V_n\}$, $\{B_n\}$ of independent identically distributed random variables such that the $U_n^{(i)}$ are governed by $\{g_n\}$, the V_n by $\{h_n\}$ and $\mathbb{P}(B_n=1)=1-\mathbb{P}(B_n=0)=c$, and define

$$Y_n^{(i)}=B_nU_n^{(i)}+(1-B_n)V_n,\qquad S_n^{(i)}=\sum_0^n Y_k^{(i)}.$$

Show that $\{S_n^{(1)}\},\{S_n^{(2)}\}$ are renewal processes governed by $\{f_n\}$, and that $\sigma=\inf\{n: S_n^{(1)}=S_n^{(2)}\}$ is almost surely finite. [this construction is known as the *Ornstein coupling*].

2.3. Show that if $\mathbb{E}X>0$, then $\mathbb{E}\tau(u)\cong u/\mathbb{E}X$ as $u\to\infty$ [hint: the elementary renewal theorem applied to the ascending ladder height process].

Notes

The systematic use of ladder processes is largely due to Feller (1971).

3. WIENER–HOPF FACTORIZATION

The expressions given in Theorems 2.1 and 2.2 for the distributions of $B(\infty)$ and M respectively, indicate that it is of major interest to compute G_+ and, for symmetry reasons, G_-. This problem certainly does not appear to be easy, and in fact the known algorithms (to be presented in the next section) for expressing G_+, G_- in terms of F seem too complicated to be of much use. However, in some situations it is easy (or at least possible) to compute *one* of G_+, G_- (this is discussed in detail in Chapter IX, but for a simple example take F concentrated on $\mathbb{Z}$ with $\mathbb{E}X \geqslant 0$, $\mathbb{P}(X \geqslant 2) = 0$; then obviously G_+ is degenerate at 1). The set of formulae presented in the following main result then allows calculation of the other ladder height distribution and thereby the distributions of M, $B(\infty)$, etc.:

Theorem 3.1.

(a) $U_+ * F(A) = \begin{cases} U_+(A), & A \subseteqq (0,\infty) \\ G_-(A), & A \subseteqq (-\infty, 0] \end{cases}$'

(b) $F = G_+ + G_- - G_+ * G_-$.

In terms of characteristic functions we may rewrite (b) as $\hat{F} = \hat{G}_+ + \hat{G}_- - \hat{G}_+\hat{G}_-$ or equivalently

$$1 - \hat{F} = (1 - \hat{G}_+)(1 - \hat{G}_-) = (1 - \zeta)(1 - \hat{G}^s_+)(1 - \hat{G}^s_-). \tag{3.1}$$

This formula (and some generalizations like Problem 4.2) is known in the literature as the *Wiener–Hopf factorization identity*, and we shall refer to (b) in the same way. We see that, say, knowing G_+, we can solve (3.1) for $\hat{G}_-$. Alternatively, G_- can be computed by (a), and this is frequently more appealing.

Proof. Define $F_n = \{S_k > 0 \ k = 1, \ldots, n\}$,

$$G^{(n)}_-(A) = \mathbb{P}(S_{\tau_-} \in A, \tau_- = n) = \mathbb{P}(S_n \in A; F_{n-1}), \qquad A \subseteqq (-\infty, 0].$$

Since $U^{(n-1)}_+$ describes the distribution of S_{n-1} on F_{n-1}, cf. Theorem 2.3(a), we have thus

$$G^{(n)}_-(A) = U^{(n-1)}_+ * F(A), \qquad n = 1, 2, \ldots, \qquad A \subseteqq (-\infty, 0], \tag{3.2}$$

$$U^{(n)}_+(A) = U^{(n-1)}_+ * F(A), \qquad n = 1, 2, \ldots, \qquad A \subseteqq (0, \infty), \tag{3.3}$$

(3.3) being an obvious result of the definition of $U^{(n)}_+$. When $A \subseteqq (0, \infty)$, we have $G^{(n)}_-(A) = 0$ and similarly $U^{(n)}_+(A) = 0$ when $A \subseteqq (-\infty, 0]$. Thus the identity $G^{(n)}_- + U^{(n)}_+ = U^{(n-1)}_+ * F$ holds on the whole line, and summing over n and noting that $U^{(0)}_+ = \delta_0 = G^{*0}_+$, $U_+ = U_+ * G_+ + \delta_0$, we get

$$U_+ * F = G_- + U_+ - U^{(0)}_+ = G_- + U_+ * G_+. \tag{3.4}$$

But $U_+ * G_+$ being concentrated on $(0, \infty)$, we have $U_+ * G_+(A) = 0$ when

$A \subseteq (-\infty, 0]$. Similarly, $G_-(A) = 0$ and $U_+ * G_+(A) = U_+(A)$ when $A \subseteq (0, \infty)$, and thus (a) follows from (3.4). For (b), convolution of (3.4) with G_+ yields

$$U_+ * F - F = G_+ * G_- + U_+ * G_+ - G_+ \tag{3.5}$$

and inserting (3.4), (b) follows. □

If $\zeta > 0$, then obviously some asymmetry is inherent in Theorem 3.1 and variants of the formulae may be required. For example:

Corollary 3.2. (a) $G_+(A) = U_- * F(A)$, $A \subseteq (0, \infty)$; (b) $F = G_+^{\mathrm{w}} + G_-^{\mathrm{s}} - G_+^{\mathrm{w}} * G_-^{\mathrm{s}}$.

Proof. A sign reversion yields immediately the truth of (b) as well as $G_+^{\mathrm{w}}(A) = U_-^{\mathrm{s}} * F(A)$ for $A \subseteq [0, \infty)$ and in particular for $A \subseteq (0, \infty)$. Now clearly, the strict and the weak ascending ladder height process have renewals in the same points, those for the strict case being of multiplicity 1 and those for the weak case being of multiplicity distributed as say N where $\mathbb{P}(N = n) = \zeta^{n-1}(1-\zeta)$, $n = 1, 2, \ldots$ and hence $\mathbb{E}N = (1-\zeta)^{-1}$. Thus $U_-^{\mathrm{s}} = (1-\zeta)U_-$, and by (1.1) we get for $A \subseteq (0, \infty)$ that

$$G_+(A) = (1-\zeta)^{-1} G_+^{\mathrm{w}}(A) = (1-\zeta)^{-1} U_-^{\mathrm{s}} * F(A) = U_- * F(A). \quad \square$$

There is also a uniqueness statement corresponding to Theorem 3.1:

Proposition 3.3. *Let* $\tilde{G}_+$, $\tilde{G}_-$ *be* (possibly defective) *distribution on* $(0, \infty)$ *and* $(-\infty, 0]$ *respectively, and define* $\tilde{U}_+ = \sum_0^\infty \tilde{G}_+^{*n}$. *Then the following statements are equivalent*:

(a′) $\tilde{U}_+ * F = \tilde{G}_- + \tilde{U}_+ * \tilde{G}_+ = \tilde{G}_- + \tilde{U}_+ - \delta_0$;
(a″) $\tilde{U}_+ * F(A) = \tilde{G}_-(A)$ *for* $A \subseteq (-\infty, 0]$ *and* $\tilde{U}_+ * F(A) = \tilde{U}_+(A)$ *for* $A \subseteq (0, \infty)$;
(b′) $F = \tilde{G}_+ + \tilde{G}_- - \tilde{G}_+ * \tilde{G}_-$;
(c) $\tilde{G}_+ = G_+, \tilde{G}_- = G_-$.

Proof. Here (a″) is merely a reformulation of (a′), and (a′)⇒(b′) follows just as in the proof of Theorem 3.1. Conversely, if (b′) holds, then convolution with $\tilde{U}_+$ yields

$$\tilde{U}_+ * F = (\tilde{U}_+ - \delta_0) + \tilde{U}_+ * \tilde{G}_- - (\tilde{U}_+ * \tilde{G}_- - \tilde{G}_-) = \tilde{U}_+ - \delta_0 + \tilde{G}_-,$$

i.e. (a′) holds. Thus (a′), (a″), (b′) are equivalent, and it remains to show that (b′)⇒(c).

Suppose first $\| G_- \| = 1$. We claim that

$$\tilde{U}_+(A) \geqslant \sum_{n=0}^{N} U_+^{(n)}(A), \qquad A \subseteq [0, \infty). \tag{3.6}$$

Indeed, this is clear when $A = \{0\}$ or $N = 0$, and if (3.6) holds for N, then for $A \subseteq (0, \infty)$ (3.3) and (a″) yield

$$\sum_{n=0}^{N+1} U_+^{(n)}(A) = \left(\delta_0 + \sum_{n=0}^{N} U_+^{(n)} * F \right)(A) = \sum_{n=0}^{N} U_+^{(n)} * F(A) \leqslant \tilde{U}_+(A).$$

Thus letting $N \to \infty$ we get $\tilde{U}_+(\mathrm{d}x) \geqslant U_+(\mathrm{d}x)$ and, using (a″) and Theorem 3.1(a), $\tilde{G}_-(\mathrm{d}x) \geqslant G_-(\mathrm{d}x)$. Since $\|\tilde{G}_-\| \leqslant 1 = \|G_-\|$, this implies $\tilde{G}_- = G_-$, and $\tilde{G}_+ = G_+$ then easily follows, say by (3.1) and the corresponding result of (b′).

If $\|G_-\| < 1$, we have $\|G^w_+\| = \|G_+\| = 1$. Define $\tilde{\zeta} = \tilde{G}_-(\{0\})$,

$$\tilde{G}^w_+ = \tilde{\zeta}\delta_0 + (1-\tilde{\zeta})\tilde{G}_+, \quad \tilde{G}^s_- = \frac{1}{1-\tilde{\zeta}}(\tilde{G}_- - \tilde{\zeta}\delta_0).$$

One then readily checks that

$$\tilde{G}^w_+ + \tilde{G}^s_- - \tilde{G}^w_+ * \tilde{G}^s_- = \tilde{G}_+ + \tilde{G}_- - \tilde{G}_+ * \tilde{G}_- = F,$$

and thus a sign reversion argument completes the proof. □

Problems

3.1. Evaluate $U_- * F(A)$ for $A \subseteq (-\infty, 0]$.

3.2. Find the distribution of $\sup\{S_1, S_2, \ldots\}$ [hint: the forms on $(-\infty, 0]$ and $(0, \infty)$ are very different].

Notes

The exposition largely follows Feller (1971). The traditional approach is analytical (based on transforms) and short introductions can be found in Cohen (1975) and (for the continuous-time case) Prabhu (1974).

4. TRANSFORM IDENTITIES

The theory to be developed represents a digression from the rest of the book in the sense that (with one exception in XII.2) the results will not be used any further. Nevertheless, it is a classical corner-stone of probability theory and an instructive example of both the merits and deficits of transform methods. To illustrate the scope and flavour of the theory, we state two of the main results:

Theorem 4.1. *For $|r| < 1$ and $t \in \mathbb{R}$*

$$1 - \mathbb{E}r^{\tau_+} \exp\{itS_{\tau_+}\} = \exp\left\{-\sum_{n=1}^{\infty} \frac{r^n}{n} \mathbb{E}[\mathrm{e}^{itS_n}; S_n > 0]\right\} \tag{4.1}$$

(in expressions like the l.h.s. of (4.1), the integration is understood to be carried out on $\{\tau_+ < \infty\}$ only)

Theorem 4.2. (Spitzer's identity) *For $|r| < 1$ and $t \in \mathbb{R}$*

$$\sum_{n=0}^{\infty} r^n \mathbb{E}\mathrm{e}^{itM_n} = \exp\left\{\sum_{n=1}^{\infty} \frac{r^n}{n} \mathbb{E}\mathrm{e}^{itS_n^+}\right\} \tag{4.2}$$

and, provided $M < \infty$ a.s.,

$$\mathbb{E}\mathrm{e}^{itM} = \exp\left\{\sum_{n=1}^{\infty} \frac{1}{n}(\mathbb{E}\mathrm{e}^{itS_n^+} - 1)\right\}. \tag{4.3}$$

It is seen that in a certain sense a complete solution of the random walk problems has been provided: by unique theorems for transforms, the distributions of (τ_+, S_{τ_+}), M_n, M are in principle determined by (4.1)–(4.3), and the solutions are explicit in the sense that knowing F, we can in principle also evaluate the distribution F^{*n} of S_n, thereby expressions like $\mathbb{E}e^{itS_n^+}$ and finally by summation the desired transforms. However, the weaknesses of the approach should also be apparent. First, the inversion of transforms presents a major problem, and presumably even worse, the expressions in (4.1)–(4.3) are exceedingly difficult to evaluate in terms of F. As example, one needs only to think of the $M/M/1$ waiting time W, where $W \overset{\mathscr{D}}{=} M$ with F doubly exponential (cf. III.7), and it was found in III.10 that W is exponential. The simplicity of this result should be compared with the effort required to compute first the F^{*n} and next (4.3), and it is strongly indicated that for even only slightly more general $GI/G/1$ queues the computational difficulties are formidable.

Proof of Theorem 4.1. We let r be fixed throughout and define

$$\beta_n(t) = \mathbb{E}[e^{itS_n}; \tau_+ = n], \qquad \gamma_n(t) = \mathbb{E}[e^{itS_n}; \tau_+ > n]$$

$$\beta(t) = \sum_{n=1}^{\infty} r^n \beta_n(t) = \mathbb{E}r^{\tau_+} \exp\{itS_{\tau_+}\}, \qquad \gamma(t) = \sum_{n=1}^{\infty} r^n \gamma_n(t).$$

With $\hat{F}$ the characteristic function of F, we then have

$$\beta_n(t) + \gamma_n(t) = \mathbb{E}[e^{itS_n}; \tau_+ \geqslant n] = \hat{F}(t)\gamma_{n-1}(t),$$

and since $\gamma_0 = 1$, it follows by summation that $\beta + \gamma = r\hat{F}(1 + \gamma)$. Equivalently, $1 - r\hat{F} = (1 - \beta)/(1 + \gamma)$, and taking logarithms and expanding we get

$$\sum_{n=1}^{\infty} \frac{r^n}{n} \hat{F}^n = \sum_{n=1}^{\infty} \frac{\beta^n}{n} - \sum_{n=1}^{\infty} (-1)^n \frac{\gamma^n}{n} \tag{4.4}$$

if r is so small, say $|r| < r_0$, that $|\beta(t)| < 1$ and $|\gamma(t)| < 1$ for all t. Obviously, β and γ are the characteristic function of bounded measures $\tilde{\varphi}, \tilde{\psi}$ supported by $(0, \infty)$ and $(-\infty, 0]$. Thus also $\varphi = \sum_1^\infty \tilde{\varphi}^{*n}/n$ is supported by $(0, \infty)$ and $\psi = \sum_1^\infty (-1)^n \tilde{\psi}^{*n}/n$ by $(-\infty, 0]$, and we may rewrite (4.4) as $\hat{H} = \hat{\varphi} - \hat{\psi}$ where $H = \sum_1^\infty r^n F^{*n}/n$. By the uniqueness theorem, it therefore follows that H and φ agree on $(0, \infty)$. Taking transform yields

$$\sum_{n=1}^{\infty} \frac{r^n}{n} \mathbb{E}[e^{itS_n}; S_n > 0] = \hat{\varphi}(t) = \sum_{n=1}^{\infty} \frac{\beta^n}{n} = -\log(1 - \beta)$$

which is the same as (4.1). The truth of (4.1) for $r_0 \leqslant |r| < 1$ follows by an analytical continuation argument. □

Inspection of the proof immediately shows that just the same argument shows:

Corollary 4.3. *The formula* (4.1) *remains valid if the pair of qualifiers* $(\tau_+, S_n > 0)$ *is replaced by any of* $(\tau_+^w, S_n \geqslant 0)$, $(\tau_-, S_n \leqslant 0)$, $(\tau_-^s, S_n < 0)$.

Corollary 4.4. $1 - \mathbb{E}r^{\tau_+} = \exp\{-\sum_1^\infty (r^n/n)\mathbb{P}(S_n > 0)\}$, $|r| < 1$. *Furthermore*

$$\frac{1}{\mathbb{E}\tau_-} = 1 - \| G_+ \| = \exp\left\{ -\sum_{n=1}^{\infty} \frac{1}{n}\mathbb{P}(S_n > 0) \right\} \tag{4.5}$$

and the assertions (i) $S_n \to -\infty$, (ii) $M < \infty$, (iii) $\| G_+ \| < 1$ *and* (iv) $\sum_1^\infty \mathbb{P}(S_n > 0)/n < \infty$ *are equivalent.*

Proof. The first statement follows just by letting $t = 0$ in (4.1). The first identity in (4.5) has been shown in Theorem 2.4, and the second follows from $\| G_+ \| = \lim_{r\uparrow 1} \mathbb{E}r^{\tau_+}$. Finally the last statement follows from Theorem 2.4 and (4.5). □

Proof of (4.2). Define

$$A_{n,k} = \{S_k = M_n; S_l < M_n l = 0, \ldots, k-1\},$$
$$\psi_n(t) = \mathbb{E}[\mathrm{e}^{itS_n}; A_{n,0}] = \mathbb{E}[\mathrm{e}^{itS_n}; M_n = 0].$$

Then

$$\mathbb{E}\mathrm{e}^{itM_n}\exp\{iu(S_n - M_n)\} = \sum_{k=0}^{n} \mathbb{E}[\mathrm{e}^{itS_k}\exp\{iu(S_n - S_k)\}; A_{n,k}]$$
$$= \sum_{k=0}^{n} \psi_{n-k}(u)\mathbb{E}[\mathrm{e}^{itS_k}; A_{k,k}].$$

Letting $u = 0$ we obtain

$$\sum_{n=0}^{\infty} r^n \mathbb{E}\mathrm{e}^{itM_n} = \sum_{n=0}^{\infty} r^n \psi_n(0)\cdot \sum_{k=0}^{\infty} r^k \mathbb{E}[\mathrm{e}^{itS_k}; A_{k,k}] = A_1 \cdot A_2(t)$$

(say). Here

$$A_1 = \sum_{n=0}^{\infty} r^n \mathbb{P}(M_n = 0) = \sum_{n=0}^{\infty} r^n \mathbb{P}(\tau_+ > n) = \frac{1}{1-r}(1 - \mathbb{E}r^{\tau_+})$$
$$= \exp\left\{ \sum_{n=0}^{\infty} \frac{r^n}{n}(1 - \mathbb{P}(S_n > 0)) \right\} \text{ (using Corollary 4.4)},$$

$$A_2(t) = \sum_{n=0}^{\infty} \mathbb{E}r^{\tau_+(n)}\exp\{itS_{\tau_+(n)}\}$$
$$= \sum_{n=0}^{\infty} (\mathbb{E}r^{\tau_+}\exp\{itS_{\tau_+}\})^n = (1 - \mathbb{E}r^{\tau_+}\exp\{itS_{\tau_+}\})^{-1}$$
$$= \exp\left\{ \sum_{n=1}^{\infty} \frac{r^n}{n}\ \mathbb{E}[\mathrm{e}^{itS_n}; S_n > 0] \right\}$$

and the proof is completed by observing that

$$\mathbb{P}(S_n \leqslant 0) + \mathbb{E}[\mathrm{e}^{itS_n}; S_n > 0] = \mathbb{E}\mathrm{e}^{itS_n^+}.$$ □

Proof of (4.3). If $M < \infty$, then (iv) of Corollary 4.4 permits us to let $r \uparrow 1$ and use dominated convergence in (4.1) to get

$$1-\hat{G}_+(t)=1-\mathbb{E}\exp\{itS_{\tau_+}\}=\exp\left\{-\sum_{n=1}^{\infty}\frac{1}{n}\mathbb{E}[e^{itS_n};S_n>0]\right\} \tag{4.6}$$

and since the characteristic function of M is $(1-\hat{G}_+(0))/(1-\hat{G}_+(t))$, (4.3) follows easily □

In the further development of the theory, one discovers that a certain care is frequently need to give rigorous proofs of expected results. For example, one might ask whether (4.6) also holds when $\|G_+\|=1$, whether, say, the Laplace transform is obtained by replacing it by $\beta<0$ and whether the expressions for moments which come out by formal differentiation are correct. We shall not go into these points, but give a direct proof of a result of the last type:

Proposition 4.5.

$$\mathbb{E}M_n=\sum_{k=1}^{n}\frac{1}{k}\mathbb{E}S_k^+,\qquad \mathbb{E}M=\sum_{k=1}^{\infty}\frac{1}{k}\mathbb{E}S_k^+.$$

Proof. Letting $F_n=\{S_n>0\}$, $G_n=\{M_n>0,S_n\leqslant 0\}$ we have with $K=\max\{S_k-X_1:k=1,\ldots,n\}$ that

$$\begin{aligned}\mathbb{E}M_n&=\mathbb{E}[M_n;M_n>0]=\mathbb{E}[M_n;F_n]+\mathbb{E}[M_n;G_n]\\&=\mathbb{E}[X_1;F_n]+\mathbb{E}[K;F_n]+\mathbb{E}[M_n;G_n].\end{aligned}$$

By symmetry arguments, the two first terms are $\mathbb{E}[S_n;F_n]/n=\mathbb{E}S_n^+/n$ and $\mathbb{E}[M_{n-1};F_n]$ respectively, whereas the last is

$$\mathbb{E}[M_{n-1};G_n]=\mathbb{E}[M_{n-1};M_{n-1}>0,S_n\leqslant 0]=\mathbb{E}[M_{n-1};S_n\leqslant 0].$$

Hence $\mathbb{E}M_n=\mathbb{E}S_n^+/n+\mathbb{E}M_{n-1}$ and the desired expression for $\mathbb{E}M_n$ follows by iteration. For $\mathbb{E}M$, let $n\uparrow\infty$ and use monotone convergence. □

Note that even the conditions for $\mathbb{E}M$ to be finite are not at all apparent from Proposition 4.5. We return to the problem in VIII.2.

Problems

4.1. Show that (4.6) also holds if $\|G_-\|<1$ [hint: first find G_- and use Wiener–Hopf factorization].

4.2. Let $\hat{g}_+(r,t)$ denote the expression (4.1) and $\hat{g}_-(r,t)$ the same thing with τ_+ replaced by τ_-. Show that $1-r\hat{F}(t)=\hat{g}_+(r,t)\hat{g}_-(r,t)$.

Notes

The results of this section were found by Baxter and Spitzer around 1960. Good references are Chung (1974), Woodroofe (1982), Siegmund (1985) and Feller (1971, Chs. XII and XVIII).

Part C: Special Models and Methods

CHAPTER VIII

Steady-State Properties of $GI/G/1$

1. NOTATON. THE ACTUAL WAITING TIME

We consider the (FIFO) $GI/G/1$ queue in the notation of III.1(b). That is, the customers are numbered $n = 0, 1, 2, \ldots$, U_n is the service time of n, T_n the time between the arrivals of n and $n+1$ and $A(x) = \mathbb{P}(T_n \leqslant x)$ is the interarrival distribution, $B(x) = \mathbb{P}(U_n \leqslant x)$ the service-time distribution (we assume $A(0) = \mathbb{P}(T_k = 0) = 0$, $B(0) = \mathbb{P}(U_k = 0) = 0$). We let $\mu_A = \mathbb{E}T_n$ denote the interarrival mean and $\mu_B = \mathbb{E}U_n$ the mean service time (μ_A, μ_B are assumed finite throughout). Then $\rho = \mu_B/\mu_A$ is the traffic intensity. Unless otherwise stated, *it is assumed that customer* 0 *has just arrived at time* $t = 0$ *to an empty queue.*

The two basic tools in the analysis of the system are: random walks which yield information on the waiting-time distribution; and regenerative processes which permit conclusions to be made on the existence of limits of other functionals such as queue lengths, as well as providing expressions for the limits (which will turn out to be given essentially by the waiting-time distribution).

Some of the basic facts on the waiting times have already been touched upon, but will now be put together. Define $X_n = U_n - T_n$, $\mu = \mathbb{E}X_n = \mu_B - \mu_A$, $S_0 = 0$, $S_n = X_0 + \cdots + X_{n-1}$, $M_n = \max_{0 \leqslant k \leqslant n} S_n$. Then $\mu \gtreqless 0$ precisely when $\rho \gtreqless 1$ and III.7 yields:

Proposition 1.1. *The actual waiting-time process* $\{W_n\}$ *is a Lindley process generated by* $\{S_n\}$, $W_{n+1} = (W_n + X_n)^+$. *In particular,*

$$W_n = \max(S_n, S_n - S_1, \ldots, S_n - S_{n-1}, 0), \tag{1.1}$$

$$W_n \overset{\mathscr{D}}{=} M_n \tag{1.2}$$

and if $\rho < 1$, *then a limiting steady state exists and is given by* $\mathbb{P}_e(W_n \leqslant x) = \mathbb{P}(M \leqslant x)$.

[The formulae (1.1) and (1.2) require slight variants for $W_0 \neq 0$, cf. III.7. However, the limit result still holds true.]

Our interest in the following is centred around the so-called *stable case* $\rho < 1$ and we shall only briefly as a digression indicate the typical behaviour for $\rho \geqslant 1$.

Proposition 1.2. (i) *If* $\rho = 1$, $\sigma^2 = \mathbb{V}\mathrm{ar}\ X_n < \infty$ *then the limiting distribution of* $W_n/\sqrt{n}$ *exists and is that of the absolute value of a normal variate with mean zero and variance* σ^2; (ii) *if* $\rho > 1$, *then a.s.* $W_n/n \to \mu = \mu_A(\rho - 1)$.

Proof. In case (i), it is well known that $M_n/\sqrt{n}$ has the asserted limit properties (the easiest proof is presumably by Donsker's theorem, Billingsley, 1968, Chapter 2; for a direct proof, see Chung, 1974, pp. 217–222). In case (ii), we have $S_n/n \to \mu > 0$ a.s. Hence by (1.1), $W_n > 0$ eventually. Hence if η is the last n with $W_n = 0$, we have $W_n = S_n - S_\eta$ $n \geq \eta$ from which we get $W_n/n \approx S_n/n \approx \mu$. □

Now define

$$\sigma(0) = 0, \qquad \sigma = \inf\{n \geq 1 : W_n = 0\}, \qquad \sigma(1) = \sigma,$$
$$\sigma(k+1) = \inf\{n > \sigma(k) : W_n = 0\}.$$

Since $W_0 = 0$, we may interpret σ as the number of customers served in the first busy period and $\sigma(k)$ as the number of the customer initiating the kth busy cycle.

Proposition 1.3. *The* $\sigma(k)$ *are regeneration points for the actual waiting-time process. We have* $\mathbb{P}(\sigma < \infty) = 1$ *if and only if* $\rho \leq 1$. *Hence for* $\rho \leq 1$ $\{W_n\}$ *is aperiodic regenerative with imbedded renewal sequence* $\{\sigma(n)\}$. *Furthermore,* $\sigma = \sigma(1)$ *coincides with the weak descending ladder epoch,* $\sigma = \tau_- = \inf\{n \geq 1 : S_n \leq 0\}$. *We have*

$$W_n = S_n = U_0 + \cdots + U_{n-1} - T_0 - \cdots - T_{n-1} \qquad n = 0, \ldots, \sigma - 1, \quad (1.3)$$
$$-S_\sigma = -S_{\tau_-} = I \quad (1.4)$$

where I *is the idle period corresponding to the first busy cycle, and furthermore* $\mathbb{E}\sigma < \infty$ *if and only if* $\rho < 1$.

Proof. By the Lindley process property, we have $W_n = S_n$, $n = 0, \ldots, \tau_- - 1$, $W_{\tau_-} = 0$, and this makes it clear that $\sigma = \tau_-$. Also I is the amount by which the last interarrival time exceeds the residual work at the time of the last arrival in the cycle,

$$I = T_{\sigma-1} - (W_{\sigma-1} + U_{\sigma-1}) = -S_\sigma = -S_{\tau_-}.$$

It is clear that the $\sigma(k)$ are regeneration points, and by general random walk results we have finally $\sigma = \tau_- < \infty$ a.s. if and only if $\mu = \mathbb{E}X_n \leq 0$, i.e. $\rho \leq 1$, and $\mathbb{E}\tau_- = \mathbb{E}S_{\tau_-}/\mathbb{E}X_n < \infty$ if and only if $\mathbb{E}X_n < 0$, i.e. $\rho < 1$. Finally aperiodicity follows from $\mathbb{P}(\sigma = 1) = \mathbb{P}(U \leq T) > 0$. □

For the sake of easy reference, some of the main variates occurring in the rest of the chapter will now be introduced.

Definition 1.4. *Suppose* $\rho < 1$. *Then throughout this and the next chapter:*

(i) *W will denote a random variable having the steady-state distribution, say H, of W_n, $\mathbb{P}(W \leqslant x) = \mathbb{P}_e(W_n \leqslant x) = H(x)$; similarly,*

(ii) *V, Q have the equilibrium distribution of the virtual waiting time V_t and the queue length Q_t* (which will be shown to exist if the interarrival distribution A is non-lattice);

(iii) *$Q_n^{\mathrm{A}}, Q_n^{\mathrm{D}}$ denote the queue length just prior to the* nth *arrival and just after the* nth *departure, and $Q^{\mathrm{A}}, Q^{\mathrm{D}}$ the corresponding steady-state quantities;*

(iv) *U, T, X, $\{T^{(k)}\}_0^\infty$ have the distribution of U_n, T_n, $X_n = U_n - T_n$, $\{T_0 + \cdots + T_{k-1}\}_0^\infty$ respectively and are mutually independent and independent of W, V, Q, Q^{A}, etc.; similar conventions apply for*

(v) *U^*, T^* having densities* $\mathrm{d}B_0(x)/\mathrm{d}x = (1 - B(x))/\mu_{\mathrm{B}}$ *and* $\mathrm{d}A_0(x)/\mathrm{d}x = (1 - A(x))/\mu_{\mathrm{A}}$.

The distributions B_0, A_0 are familiar from renewal theory, IV.3. Also, from the independence of W_n and U_n, it is seen that we may identify $W + U$ by the sojourn time in the steady state.

A main problem for the study of the actual waiting time is obviously to study the distribution of W. Various expressions are available. From Proposition 1.1 and VII.2.2 we have

$$H(x) = \mathbb{P}(W \leqslant x) = (1 - \|G_+\|)U_+(x) = (1 - \|G_+\|)\sum_{n=0}^{\infty} G_+^{*n}(x), \tag{1.5}$$

whereas Proposition 1.3 and V.(1.5) yield

$$H(x) = \mathbb{P}(W \leqslant x) = \frac{1}{\mathbb{E}\sigma}\mathbb{E}\sum_{n=0}^{\sigma-1} I(W_n \leqslant x). \tag{1.6}$$

These formulae are, however, not intrinsically different in view of VII.2.3(b). A somewhat different characterization of H is as the unique solution to Lindley's integral equation III.(7.6) with

$$F(x) = \mathbb{P}(X_n \leqslant x) = \mathbb{P}(U_n - T_n \leqslant x) = \int_0^\infty B(x+y)\,\mathrm{d}A(y).$$

Also, the characteristic function has been found in VII.4 but is obviously quite complicated.

Proposition 1.5. *$W \stackrel{\mathscr{D}}{=} (W + X)^+$, whereas the conditional distribution of $(W + X)^-$ given $\{W + X \leqslant 0\}$ coincides with the common distribution of $-S_{\tau_-}$ and I. In particular, for $f:[0,\infty) \to [0,\infty)$*

$$\mathbb{E}f((W+X)^-) = \frac{\mathbb{E}f(-S_{\tau_-})}{\mathbb{E}\tau_-} = -\mathbb{E}X\frac{\mathbb{E}f(I)}{\mathbb{E}I}, \tag{1.7}$$

$$\mathbb{E}(W+X)^- = -\mathbb{E}X. \tag{1.8}$$

Proof. The first statement was noted previously in III.7.6 and yields in particular

$$\mathbb{P}(W + X \leqslant 0) = \mathbb{P}((W+X)^+ = 0) = \mathbb{P}(W = 0) = 1 - \|G_+\| = 1/\mathbb{E}\tau_-,$$

cf. VII.2.3(c). Also, by VII.3.1(a),

$$\mathbb{E}f((W+X)^-) = \int_{-\infty}^{0} f(-x)H * F(\mathrm{d}x) = (1-\|G_+\|)\int_{-\infty}^{0} f(-x)U_+ * F(\mathrm{d}x)$$

$$= (1-\|G_+\|)\int_{-\infty}^{0} f(-x)\,\mathrm{d}G_-(x) = \frac{1}{\mathbb{E}\tau_-}\mathbb{E}f(-S_{\tau_-})$$

$$= \mathbb{P}(W+X \leqslant 0)\mathbb{E}f(-S_{\tau_-}).$$

Recalling $\mathbb{E}S_{\tau_-} = \mathbb{E}\tau_-\mathbb{E}X$ and (1.4), the proof is complete. □

Problems

1.1. Give a direct proof of (1.8) by using $W+X=(W+X)^+-(W+X)^-$.

2. THE MOMENTS OF THE ACTUAL WAITING TIME

The problem is to study conditions for the existence of $\mathbb{E}W^p$, $p \geqslant 1$, and, as far as possible, to derive an explicit expression. In view of $W \overset{\mathscr{D}}{=} M$, this is really a random walk problem (as in the case for many other aspects of the behaviour of the waiting time, cf. e.g. Sections 5 and 6) and can therefore be formulated in that setting alone. The queueing interpretation may, however, require some slight reformulations: for example in the following existence result, $\mathbb{E}(X^+)^{p+1} = \mathbb{E}([U-T]^+)^{p+1}$ is readily seen to be equivalent to $\mathbb{E}U^{p+1} < \infty$, whereas $\mathbb{E}X^- < \infty$ is automatic in view of $\mathbb{E}T < \infty$.

Theorem 2.1. *Consider a random walk with $\mu = \mathbb{E}X < 0$. Then for $p \geqslant 1$, $\mathbb{E}M^p < \infty$ provided that $\mathbb{E}(X^+)^{p+1} < \infty$. Conversely, if $\mathbb{E}M^p < \infty$ and $\mathbb{E}X^- < \infty$, then $\mathbb{E}(X^+)^{p+1} < \infty$.*

Proof. We first note that the pth moment v_n of a sum $Y_1 + \cdots + Y_n$ of non-negative independent identically distributed summands with $\mathbb{E}Y_1^p < \infty$ is of the order $\mathrm{O}(n^p)$. Indeed, Jensen's inequality gives the lower bound $v_n \geqslant (n\mathbb{E}Y)^p$, whereas the upper bound comes from

$$v_n^{1/p} = [\mathbb{E}(Y_1 + \cdots + Y_n)^p]^{1/p} = \|Y_1 + \cdots + Y_n\|_p \leqslant n\|Y\|_p.$$

Hence if $\alpha = \mathbb{E}(S_{\tau_+}^p : \tau_+ < \infty) < \infty$,

$$\mathbb{E}M^p = (1-\|G_+\|)\sum_{n=0}^{\infty}\int_0^{\infty} x^p\,\mathrm{d}G_+^{*n}(x) = (1-\|G_+\|)\sum_{n=0}^{\infty}\|G_+\|^n\mathrm{O}(n^p)$$

will be finite in view of $\|G_+\| < 1$, whereas if $\alpha = \infty$ then the term corresponding to $n=1$ in the sum is infinite and hence $\mathbb{E}M^p = \infty$. Now let $\beta = \mathbb{E}[S_{\tau_+^{\mathrm{w}}}^p ; \tau_+^{\mathrm{w}} < \infty]$, $U(y) = U_-^{\mathrm{s}}[-y, 0]$. Then by VII.3.1(a)

$$\frac{1}{p}\beta = \int_0^{\infty} x^{p-1}\mathbb{P}(x < S_{\tau_+^{\mathrm{w}}} < \infty)\,\mathrm{d}x = \int_0^{\infty} x^{p-1}U_-^{\mathrm{s}} * F(x,\infty)\,\mathrm{d}x$$

$$= \int_0^\infty x^{p-1}\,\mathrm{d}x \int_x^\infty U^{\mathrm{s}}_{-}(x-y,0]\,\mathrm{d}F(y)$$

$$= \int_0^\infty \mathrm{d}F(y) \int_0^y x^{p-1} U(y-x)\,\mathrm{d}x. \tag{2.1}$$

By the elementary renewal theorem we have for suitable c_1, c_2 that $U(z) \leqslant c_1 + c_2 z$, and since for large y

$$\int_0^y x^{p-1}(c_1 + c_2(y-x))\,\mathrm{d}x = \frac{1}{p} y^p c_1 + \frac{1}{p} y^{p+1} c_2 - \frac{1}{p+1} y^{p+1} c_2 \cong \frac{c_2}{p(p+1)} y^{p+1} \tag{2.2}$$

it follows that $\mathbb{E}(X^+)^{p+1} < \infty$ implies $\beta < \infty$, hence $\alpha < \infty$ and $\mathbb{E}M^p < \infty$. Conversely, if $\mathbb{E}X^- < \infty$, then $\mathbb{E}S_{\tau_-} = \mathbb{E}\tau_- \mathbb{E}X > -\infty$ and hence $U(z) \geqslant d_1 + d_2 z$ with $d_2 > 0$. If $\mathbb{E}M^p < \infty$, then $\alpha < \infty$, $\beta < \infty$ and combining (2.1) and (2.2) yields $\mathbb{E}(X^+)^{p+1} < \infty$. □

Not even the moments of M (if they exist) can be found very explicitly. For example, VII.4.5 and (1.5) yield the expressions

$$\mathbb{E}M = \sum_{n=1}^\infty \frac{1}{n} \mathbb{E}S_n^+ = \frac{\mathbb{E}[S_{\tau_+}; \tau_+ < \infty]}{1 - \|G_+\|}. \tag{2.3}$$

A further important relation is the following:

Theorem 2.2. *If for some* $p = 1, 2, \ldots$ $\mathbb{E}|X|^{p+1} < \infty$, *then*

$$\sum_{q=0}^p \binom{p+1}{q} \mathbb{E}M^q \mathbb{E}X^{p+1-q} = \mathbb{E}[-(M+X)^-]^{p+1} = \frac{\mathbb{E}S_{\tau_-}^{p+1}}{\mathbb{E}\tau_-}. \tag{2.4}$$

[note that in the queueing setting, we may rewrite the r.h.s. of (2.4) as $(-1)^p \mathbb{E}X\, \mathbb{E}I^{p+1}/\mathbb{E}I$, cf. (1.7)].

Proof. The last identity in (2.4) follows from (1.7). To show the remaining part of the theorem, first suppose $\mathbb{E}(X^+)^{p+2} < \infty$. Then

$$\mathbb{E}M^{p+1} < \infty, \qquad \mathbb{E}[(M+X)^-]^{p+1} \leqslant \mathbb{E}|X|^{p+1} < \infty,$$

and since $(M+X)^+(M+X)^- = 0$ we get

$$\begin{aligned}(M+X)^{p+1} &= [(M+X)^+ - (M+X)^-]^{p+1} \\ &= [(M+X)^+]^{p+1} + [-(M+X)^-]^{p+1},\end{aligned}$$

$$\begin{aligned}\mathbb{E}(M+X)^{p+1} &= \sum_{q=0}^{p+1} \binom{p+1}{q} \mathbb{E}M^q \mathbb{E}X^{p+1-q} \\ &= \mathbb{E}[(M+X)^+]^{p+1} + \mathbb{E}[-(M+X)^-]^{p+1} \\ &= \mathbb{E}M^{p+1} + \mathbb{E}[-(M+X)^-]^{p+1}\end{aligned}$$

and (2.4) follows. In the general case, replace X_n by $X_n^{(k)} = X_n \wedge k$ and let $M^{(k)}$

be defined correspondingly. Then $\mathbb{E}[X^{(k)+}]^{p+2} < \infty$, hence

$$\sum_{q=0}^{p} \binom{p+1}{q} \mathbb{E}M^{(k)q}\mathbb{E}X^{(k)p+1-q} = \mathbb{E}[-(M^{(k)} + X^{(k)})^-]^{p+1}.$$

But clearly, $M^{(k)} \leqslant M$ and $M^{(k)} \uparrow M$ as $k \to \infty$. Hence letting $k \to \infty$ the desired conclusion follows by monotone convergence. □

Rewriting in queueing notation, we get in particular for the mean waiting time ($p = 1$) that

$$2\mathbb{E}(-X)\mathbb{E}W = \mathbb{E}X^2 - \mathbb{E}[(W+X)^-]^2 = \mathbb{V}\text{ar}\,X - \mathbb{V}\text{ar}\,(W+X)^-$$
$$= \mathbb{E}X^2 - \frac{\mathbb{E}S_{\tau_-}^2}{\mathbb{E}\tau_-} = \mathbb{E}X^2 - \frac{\mathbb{E}(-X)\mathbb{E}I^2}{\mathbb{E}I} \tag{2.5}$$

(here the second equality follows from (1.8)). Considerable effort has been put into converting these expressions into bounds or approximations which are more explicit in the sense that only the distribution of X (or U, T) is invoked, and preferably only even the first few moments. We return to the approximations in Section 6 and XII.6, and here present only some of the roughest bounds,

$$\mathbb{E}U^2 - \mathbb{E}U\mathbb{E}T \leqslant 2\mathbb{E}(-X)\mathbb{E}W \leqslant \mathbb{V}\text{ar}\,X = \mathbb{V}\text{ar}\,U + \mathbb{V}\text{ar}\,T \tag{2.6}$$

[the lower bound may be negative and hence trivial. The upper bound is in fact sharp in an asymptotic sense, cf. Section 6.] Here the upper bound is obvious from $\mathbb{V}\text{ar}\,(W+X)^- \geqslant 0$. For the lower bound, rewrite (2.5) as

$$\mathbb{E}U^2 - 2\mathbb{E}U\mathbb{E}T + \mathbb{E}T^2 - \mathbb{E}[(W+X)^-]^2 = \mathbb{E}U^2 - 2\mathbb{E}U\mathbb{E}T + \mathbb{E}CD,$$

where $C = T + (W+X)^-$, $D = T - (W+X)^-$. Here

$$D = T + W + X - (W+X)^+ = W + U - (W+X)^+$$

so that

$$\mathbb{E}CD = \mathbb{E}T(W - (W+X)^+) + \mathbb{E}T\mathbb{E}U + \mathbb{E}(W+X)^-(W+U). \tag{2.7}$$

The last term in (2.7) is obviously non-negative and thus it is sufficient to show that the first one is so too. But $f(T) = T$ and $g(T) = W - (W+U-T)^+$ are both non-decreasing in T for fixed W, U. Hence by a well-known inequality (Problem 2.2)

$$\mathbb{E}[f(T)g(T)|W,U] \geqslant \mathbb{E}[f(T)|W,U]\mathbb{E}[g(T)|W,U]$$
$$= \mathbb{E}T\mathbb{E}[(W-(W+X)^+))|W,U],$$
$$\mathbb{E}T(W-(W+X)^+) \geqslant \mathbb{E}T\mathbb{E}(W-(W+X)^+) = \mathbb{E}T\cdot 0 = 0.$$ □

Problems

2.1. Consider a random walk with $\|G_+\| = \|G_-\| = 1$. Show that $\mathbb{E}S_{\tau_+} < \infty$, $\mathbb{E}S_{\tau_-} > -\infty$ if and only if $\mathbb{E}X^2 < \infty$, $\mathbb{E}X = 0$, and that then $\mathbb{E}X^2 = -\mathbb{E}S_{\tau_+}\mathbb{E}S_{\tau_-}$ [hint: necessity and the stated identity follow by Wiener–Hopf factorization of the characteristic functions].
2.2 (Chebycheff's covariance inequality). Let X be a random variable and f, g non-decreasing functions. Show that $\mathbb{E}f(X)g(X) \geqslant \mathbb{E}f(X)\mathbb{E}g(X)$. [hint: first reduce to the case

$f \geq -a$ and next to $f \geq 0$, let $x_0 = g^{-1}(\mathbb{E}g(X))$ and look at the regions $X \geq x_0$, $X \leq x_0$ separately].

Notes

Theorem 2.1 goes back to Kiefer and Wolfowitz (1956) and there are various more or less intricate proofs around. The present one is new but observed independently by Wolff (1984). Theorem 2.2 is from Lemoine (1976); there are also some related formulae in Marshall (1968). As one of many applications of formula (2.5), we mention the observation by Minh and Sorli (1983) that when estimating $\mathbb{E}W$ by simulation, the only unknown quantity is $\mathbb{E}I^2$, and that simulating $\mathbb{E}I^2$ rather than $\mathbb{E}W$ increases precision. Finally, bounds for $\mathbb{E}W$ and related quantities are surveyed in Stoyan (1983).

3. THE VIRTUAL WAITING TIME

In continuous time, there is a regenerative structure similar to the one in Proposition 1.3: The instants with a customer entering an empty queue are regeneration points. Letting C be the first such instant and recalling that we start with customer 0 having just arrived, it is seen that C is just the length of the first busy cycle. Furthermore, $\mathbb{P}(C < \infty) = 1$ is equivalent to $\mathbb{P}(\sigma < \infty) = 1$, i.e. $\rho \leq 1$ (cf. Proposition 1.3). In fact, there is a close relation between σ, C and the first busy period G: since precisely the customers $0, 1, \ldots, \sigma - 1$ are served in the first busy period, we have $G = U_0 + \cdots + U_{\sigma-1}$ and the first busy cycle ends at the time $C = T_0 + \cdots + T_{\sigma-1}$ of arrival of customer σ. One checks immediately that $\{\sigma \leq n\}$ is independent of $T_n, T_{n+1}, \ldots, U_n, U_{n+1}, \ldots$, and hence Wald's identity yields the first part of

Proposition 3.1. *Suppose $\rho \leq 1$. Then the mean busy cycle is $\mathbb{E}C = \mu_A \mathbb{E}\sigma$, the mean busy period is $\mathbb{E}G = \mu_B \mathbb{E}\sigma$ and the mean idle period is $\mathbb{E}I = \mathbb{E}C - \mathbb{E}G = -\mu\mathbb{E}\sigma$. Furthermore the busy cycle distribution is non-lattice if and only if the interarrival distribution A is so, and spread out if and only if A is so.*

The second part is often stated to be obvious, but some care is needed (cf. Problem 3.3), and we shall prove the following general result:

Proposition 3.2. *Let $T_0, T_1, \ldots$ be independent identically distributed, governed by a distribution A on $(0, \infty)$ with finite mean, and let $\sigma \geq 1$ be a random time such that $\mathbb{E}\sigma < \infty$ and $T_n, T_{n+1}, \ldots$ are independent of $\{\sigma \leq n\}$ for any fixed n. Then the distribution K of $C = T_0 + \cdots + T_{\sigma-1}$ is non-lattice if and only if A is so, and spread out if and only if A is so.*

Proof. By Wald's identity, we have $\mathbb{E}C < \infty$. Also by an obvious iteration procedure it may be assumed that random times $\sigma(1) = \sigma < \sigma(2) < \cdots$ have been constructed such that $\{T_0 + \cdots + T_{\sigma(k)-1}\}$ is a renewal process governed by K. Then, in the obvious notation the renewal measures satisfy $U_A \geq U_K$. Suppose K was lattice (say aperiodic on $\mathbb{N}$) but A not. Then by Blackwell's renewal theorem,

$$\frac{h}{\mathbb{E}T} = \lim_{n\to\infty} \{U_A(n) - U_A(n-h)\} \geq \lim_{n\to\infty} \{U_K(n) - U_K(n-h)\} = \frac{1}{\mathbb{E}C}$$

for all $h < 1$ which is impossible. Similarly, if A was spread out but K not, U_K would be concentrated on a Lebesgue null set N. Using Stone's decomposition, we see that the U_A-measure of N is finite whereas the U_K-measure is infinite, contradicting $U_A \geqslant U_K$.

If, conversely, A is not spread out, then U_A is concentrated on a Lebesgue null set N. Hence U_K is concentrated on N, and K cannot be spread out. That K is lattice if A is so is even more trivial. □

The remaining part of Proposition 3.1 now follows for $\rho < 1$. For $\rho = 1$, replace B by an equivalent (in the sense of null sets) and stochastically smaller distribution $\tilde{B}$. Then the busy cycle distributions are equivalent, and since $\tilde{\rho} < 1$, Proposition 3.2 applies to $\tilde{C}$.

Corollary 3.3. *Suppose $\rho < 1$ and that A is non-lattice. Then a limiting equilibrium distribution of the virtual waiting time V_t exists and is given by*

$$\mathbb{E}f(V) = \frac{1}{\mathbb{E}C}\,\mathbb{E}\int_0^C f(V_s)\,\mathrm{d}s. \tag{3.1}$$

If A is spread out, then $V_t \to V$ in total variation.

Proof. For $\rho < 1$, we have $\mathbb{E}\sigma < \infty$. Hence Proposition 3.1 ensures that the basic limit results for regenerative processes in V.1 and VI.1 are applicable. □

The distribution in (3.1) will turn out to be closely related to the equilibrium actual waiting-time distribution, as will be proved below by careful sample path inspection of the r.h.s. of (3.1). To illustrate the method we start by two simple examples. In the following we let $\mathscr{F}_n = \sigma(U_k, T_k{:}0 \leqslant k \leqslant n)$. Then S_n, W_n are $\mathscr{F}_{n-1}$-measurable and $\{\sigma \leqslant n\} \in \mathscr{F}_{n-1}$.

First the time spent by $\{V_t\}_{t \geqslant 0}$ in state 0 in the time interval $[0, C)$ is just the idle period. Thus from Proposition 3.1 we get immediately

$$\mathbb{P}(V = 0) = \frac{1}{\mathbb{E}C}\,\mathbb{E}\int_0^C I(V_s = 0)\mathrm{d}s = \frac{\mathbb{E}I}{\mathbb{E}C} = \frac{(\mu_A - \mu_B)\mathbb{E}\sigma}{\mu_A \mathbb{E}\sigma} = 1 - \rho \tag{3.2}$$

(note that this is always explicit in contrast to $\mathbb{P}(W = 0) = 1/\mathbb{E}\sigma$). Next:

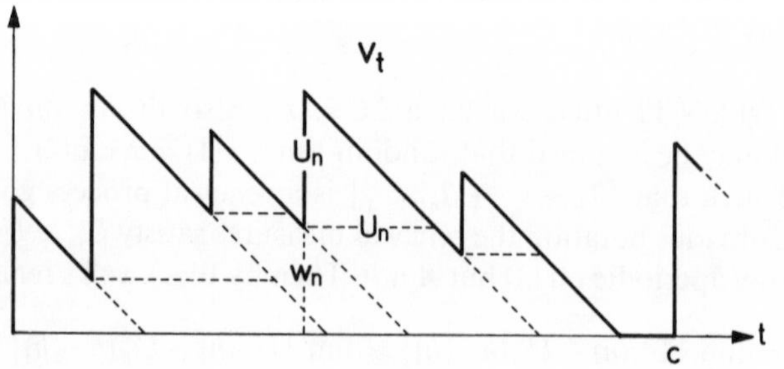

Fig. 3.1

Proposition 3.4. $\mathbb{E}V = \rho\left\{\dfrac{\mathbb{E}U^2}{2\mu_B} + \mathbb{E}W\right\}$.

Proof. Consider the partitioning of the subgraph of $\{V_t\}_{0\leq t<C}$ into the triangles and parallelograms in Fig. 3.1. The area of the nth triangle is $\frac{1}{2}U_n^2$ and of the nth parallelogram W_nU_n, hence

$$\begin{aligned}
\mathbb{E}V &= \frac{1}{\mathbb{E}C}\mathbb{E}\int_0^C V_s \mathrm{d}s = \frac{1}{\mathbb{E}C}\mathbb{E}\sum_{n=0}^{\sigma-1}[\tfrac{1}{2}U_n^2 + W_nU_n] \\
&= \frac{1}{\mathbb{E}C}\mathbb{E}\sum_{n=0}^{\infty}\mathbb{E}[\tfrac{1}{2}U_n^2 + W_nU_n; \sigma > n|\mathscr{F}_{n-1}] \\
&= \frac{1}{\mathbb{E}C}\sum_{n=0}^{\infty}\mathbb{E}[\tfrac{1}{2}\mathbb{E}U^2 + W_n\mathbb{E}U; \sigma > n] \\
&= \frac{1}{\mu_A\mathbb{E}\sigma}\left\{\tfrac{1}{2}\mathbb{E}U^2\mathbb{E}\sigma + \mu_B\mathbb{E}\sum_{n=0}^{\sigma-1}W_n\right\} \\
&= \rho\left\{\frac{\mathbb{E}U^2}{2\mu_B} + \frac{1}{\mathbb{E}\sigma}\mathbb{E}\sum_{n=0}^{\sigma-1}W_n\right\} = \rho\left\{\frac{\mathbb{E}U^2}{2\mu_B} + \mathbb{E}W\right\}. \qquad \square
\end{aligned}$$

We next give as the main result the relation between the distributions of W and V (for the meaning of U^*, T^*, see Definition 1.4).

Theorem 3.5. *The conditional distribution of V given $\{V > 0\}$ is the same as the distribution $H * B_0$ of $W + U^*$. Equivalently,*

$$\mathbb{P}(V \leq x) = 1 - \rho + \rho\mathbb{P}(W + U^* \leq x) = 1 - \rho + \rho H * B_0(x), \tag{3.3}$$

cf. (3.2). *An alternative characterization is $V \overset{\mathscr{D}}{=} (W + U - T^*)^+$.*

Proof. We show (3.3) in the equivalent form

$$\mathbb{P}_e(V_t > x) = \frac{1}{\mathbb{E}C}\mathbb{E}\int_0^C I(V_s > x)\mathrm{d}s = \rho\mathbb{P}(W + U^* > x). \tag{3.4}$$

In the manner indicated on Fig. 3.2, we can split the time $\{V_t\}_{0\leq t<C}$ spends above x into contributions I_n from the customers $n = 0, 1, \ldots, \sigma - 1$. To evaluate I_n, note that the virtual waiting time just after the arrival of n is $W_n + U_n$. Hence $I_n = 0$ if $W_n + U_n \leq x$. Otherwise the time from the arrival time to the right endpoint of I_n is $\delta_n + I_n = (W_n + U_n - x)^+$, and since $\delta_n = (W_n - x)^+$, we get

$$\begin{aligned}
\mathbb{P}(V > x) &= \frac{1}{\mathbb{E}C}\mathbb{E}\int_0^C I(V_s > x)\mathrm{d}s = \frac{1}{\mathbb{E}C}\mathbb{E}\sum_{n=0}^{\sigma-1}I_n \\
&= \frac{1}{\mathbb{E}C}\mathbb{E}\sum_{n=0}^{\sigma-1}\{(W_n + U_n - x)^+ - (W_n - x)^+\}.
\end{aligned}$$

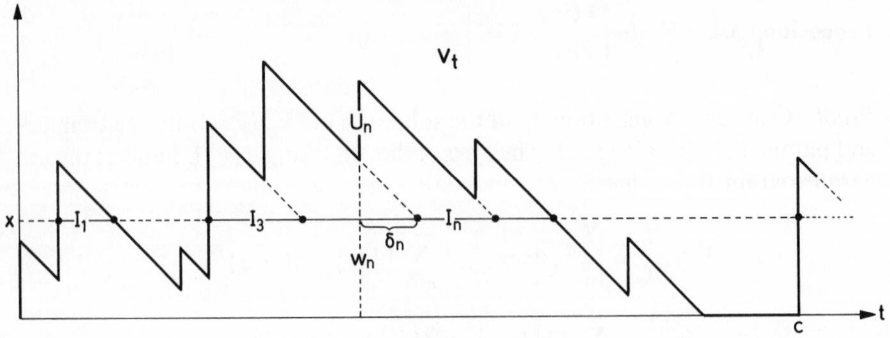

Fig. 3.2

Since U_n and U are both independent of $\mathscr{F}_{n-1}$, this becomes

$$\frac{1}{\mathbb{E}C}\sum_{n=0}^{\infty}\mathbb{E}[(W_n+U-x)^+-(W_n-x)^+;\sigma>n]. \tag{3.5}$$

Now using integration upon parts we have

$$\mathbb{E}[(U+a)^+-a^+]=\int_{a^-}^{\infty}(1-B(u))\mathrm{d}u=\mu_B\mathbb{P}(U^*>a^-).$$

Hence (3.5) is

$$\frac{1}{\mathbb{E}C}\sum_{n=0}^{\infty}\mu_B\mathbb{P}[U^*>(x-W_n)^+;\sigma>n]$$

$$=\frac{\rho}{\mathbb{E}\sigma}\mathbb{E}\sum_{n=0}^{\sigma-1}I(U^*>(x-W_n)^+)=\rho\mathbb{P}(U^*>(x-W)^+).$$

Since $U^*\geqslant 0$ a.s., this is the same as $\sigma\mathbb{P}(U^*>x-W)$ and (3.4) follows. The last assertion of the theorem is equivalent to $\mathbb{P}(V>x)=\mathbb{P}(W+U-T^*>x)$, $x>0$. Now between arrivals n and $n+1$ we have $V_t>x$ in an interval of length $T_n\wedge(W_n+U_n-x)^+$. Hence

$$\mathbb{P}(V>x)=\frac{1}{\mathbb{E}C}\mathbb{E}\sum_{n=0}^{\sigma-1}T_n\wedge(W_n+U_n-x)^+. \tag{3.6}$$

Using integration upon parts, we have

$$\mathbb{E}T\wedge a=\int_0^a(1-A(t))\,\mathrm{d}t=\mu_A\mathbb{P}(T^*<a).$$

Hence (3.6) is

$$\frac{1}{\mathbb{E}C}\sum_{n=0}^{\infty}\mathbb{E}[T\wedge(W_n+U-x)^+;\sigma>n]$$

$$= \frac{1}{\mathbb{E}\sigma} \sum_{n=0}^{\infty} \mathbb{P}(T^* < (W_n + U - x)^+ ; \sigma > n)$$

$$= \mathbb{P}(T^* < (W + U - x)^+) = \mathbb{P}(T^* < W + U - x)$$

(using $T^* > 0$). □

Theorem 3.5 provides a new proof of the basic property $W \overset{\mathscr{D}}{=} V$ of the $M/G/1$ queue. In fact, when T is exponentially distributed, we have $T \overset{\mathscr{D}}{=} T^*$ and hence

$$V \overset{\mathscr{D}}{=} (W + U - T^*)^+ \overset{\mathscr{D}}{=} (W + U - T)^+ \overset{\mathscr{D}}{=} W.$$

Problems

3.1. Define R_t as the residual service time of the customer being served at time t ($R_t = 0$ if the system is empty). Show that in equilibrium $\mathbb{P}(R \leqslant x) = 1 - \rho + \rho B_0(x)$ and find a similar formula for the waiting time to the next arrival.

3.2. Let $\{B_t\}$ be the forwards recurrence time of the arrival process. Show that $\{B_t, V_t\}_{t \geqslant 0}$ is a strong Markov process.

3.3. Show that the assumption $\mathbb{E}\sigma < \infty$ of Proposition 3.2 is indispensable [hint: $T_n = 1 + \theta B_n$, where $\theta \in (0,1)$ is irrational, $B_n = \pm 1$ with probability $\frac{1}{2}$ and $\sigma = \inf\{n \geqslant 1 : B_0 + \cdots + B_{n-1} = 0\}$].

Notes

See next section.

4. QUEUE LENGTH PROCESSES

In the same way as the actual waiting-time process is obtained by observing virtual waiting time just before arrival instants, it is sometimes of considerable interest to look at the queue length at certain random times. In particular, seen from the point of view of the individual customer, the queue length at the time of arrival is a basic quantity and motivates the study of $\{Q_n^{\mathrm{A}}\}_{n \in \mathbb{N}}$, cf. Definition 1.4. To distinguish from $\{Q_n^{\mathrm{A}}\}$, $\{Q_n^{\mathrm{D}}\}$, we use the terminology 'queue length at an arbitrary point of time' when considering the continuous-time process $\{Q_t\}_{t \geqslant 0}$, and Q in Definition 1.4 refers to this case (i.e. $\mathbb{P}(Q = j) = \lim_{t \to \infty} \mathbb{P}(Q_t = j)$).

The pattern of analysis of the various queue length processes is much the same as for the virtual waiting-time process, namely to relate the paths within a busy cycle to the corresponding actual waiting times. Again, we start with some simple examples.

Theorem 4.1. (LITTLE'S FORMULA) *Suppose $\rho < 1$ and that A is non-lattice. Then the arrival rate $\lambda = 1/\mu_A$, the mean equilibrium queue length $l = \mathbb{E}Q$ at an arbitrary point of time and the mean equilibrium sojourn time $w = \mathbb{E}(W + U)$ are related by $l = \lambda w$.*

Proof. By reference to Proposition 3.1, regenerative processes apply to $\{Q_t\}_{t \geqslant 0}$ exactly as to the virtual waiting time to show that a limiting equilibrium distribution exists and is given by

$$\mathbb{E}f(Q) = \frac{1}{\mathbb{E}C}\mathbb{E}\int_0^C f(Q_t)\mathrm{d}t. \tag{4.1}$$

Letting $f(x) = x$, it is seen that each of the customers $n = 0, 1, \ldots, \sigma - 1$ provides exactly the contribution $W_n + U_n$ to $\int_0^C Q_t\,\mathrm{d}t$. Hence

$$l = \mathbb{E}Q = \frac{1}{\mathbb{E}C}\mathbb{E}\int_0^C Q_s\,\mathrm{d}s = \frac{1}{\mu_A \mathbb{E}\sigma}\mathbb{E}\sum_{n=0}^{\sigma-1}\{W_n + U_n\}$$

$$= \lambda\mathbb{E}(W + U) = \lambda w.$$

□

A heuristical proof of Little's formula has already been given in Problem III.1.5. It is clear that the above argument is valid for queueing systems far more general than the FIFO $GI/G/1$ model.

Theorem 4.2. *Suppose* $\rho < 1$. *Then* $Q^{\mathrm{A}}, Q^{\mathrm{D}}$ *are well defined and have the same distribution given by*

$$\mathbb{P}(Q^{\mathrm{A}} \geqslant k) = \mathbb{P}(Q^{\mathrm{D}} \geqslant k) = \mathbb{P}(W + U \geqslant T^{(k)}), \tag{4.2}$$

$k = 1, 2, \ldots$. *If either* A *or* B *is continuous, this may be rewritten as*

$$\mathbb{P}(Q^{\mathrm{A}} = 0) = \mathbb{P}(Q^{\mathrm{D}} = 0) = \mathbb{P}(W = 0) = H(0), \tag{4.3}$$

$$\mathbb{P}(Q^{\mathrm{A}} \geqslant k) = \mathbb{P}(Q^{\mathrm{D}} \geqslant k) = \mathbb{P}(W > T^{(k-1)}) \qquad k = 1, 2, \ldots, \tag{4.4}$$

$$\mathbb{P}(Q^{\mathrm{A}} = k) = \mathbb{P}(Q^{\mathrm{D}} = k) = \mathbb{P}(T^{(k-1)} < W < T^{(k)}), \tag{4.5}$$

i.e.

$$\mathbb{P}(Q^{\mathrm{A}} = k) = \mathbb{P}(Q^{\mathrm{D}} = k) = \int_{0+}^{\infty} [A^{*(k-1)}(x) - A^{*k}(x)]\mathrm{d}H(x) \qquad k = 1, 2, \ldots. \tag{4.6}$$

Proof. Here $\{Q_n^{\mathrm{A}}\}$, $\{Q_n^{\mathrm{D}}\}$ are clearly regenerative with respect to the renewal sequence $\{\sigma(k)\}$ and thus the existence of the limiting distributions is immediate. Letting $N_+^{(k)}$, $N_-^{(k)}$ be the number of upcrossings from k to $k + 1$, and downcrossings from $k + 1$ to k, respectively, of $\{Q_t\}_{0 \leqslant t < C}$, we get

$$\mathbb{P}(Q^{\mathrm{A}} = k) = \frac{1}{\mathbb{E}\sigma}\mathbb{E}\sum_{n=0}^{\sigma-1} I(Q_{T_0 + \cdots + T_{n-1}-} = k)$$

$$= \frac{1}{\mathbb{E}\sigma}\mathbb{E}\sum_{n=0}^{\sigma-1} I(Q_{T_0 + \cdots + T_{n-1}-} = k, Q_{T_0 + \cdots + T_{n-1}} = k + 1) = \mathbb{E}N_+^{(k)}/\mathbb{E}\sigma.$$

Similarly, $\mathbb{P}(Q^{\mathrm{D}} = k) = \mathbb{E}N_-^{(k)}/\mathbb{E}\sigma$ and since clearly $N_+^{(k)} = N_-^{(k)}$, it is clear that $Q^{\mathrm{A}} = Q^{\mathrm{D}}$. Now look at customer $n + k$ arriving $t = T_n + \cdots + T_{n+k-1}$ time units after n. The queue length just before his arrival will be $\geqslant k$ precisely if n is still present, i.e. if $W_n + U_n \geqslant t$. Thus for $k \geqslant 1$

$$\mathbb{P}(Q^{\mathrm{A}} \geqslant k) = \lim_{n \to \infty} \mathbb{P}(W_n + U_n \geqslant T_n + \cdots + T_{n+k-1})$$

$$= \lim_{n \to \infty} \mathbb{P}(W_n + U \geqslant T^{(k)}) = \mathbb{P}(W + U \geqslant T^{(k)}) \qquad (4.7)$$

(the last step uses $W_n \to W$ in total variation and not only weakly). Now if either A or B is continuous, then so is the distribution F of $U - T$, hence also that of $W + U - T$. Hence since $(W + U - T)^+ \stackrel{\mathscr{D}}{=} W$, (4.7) becomes

$$\mathbb{P}(W + U > T^{(k)}) = \mathbb{P}(W + U - T > T^{(k-1)}) = \mathbb{P}(W > T^{(k-1)})$$

$$= \int_{0+}^{\infty} A^{*(k-1)}(x) \mathrm{d}H(x)$$

[the point $x = 0$ must be excluded from the domain of integration to deal with the case $k = 1$, $T^{(k-1)} = 0$]. From this (4.3)–(4.6) follow by easy manipulations. □

Theorem 4.3. *Suppose $\rho < 1$ and that A is non-lattice. Then a limiting equilibrium distribution of the queue length Q at an arbitrary point of time exists and is given by* $\mathbb{P}(Q = 0) = 1 - \rho$,

$$\mathbb{P}(Q \geqslant k) = \mathbb{P}(V > T^{(k-1)}) = \rho \mathbb{P}(W + U^* > T^{(k-1)}), \qquad (4.8)$$

i.e.

$$\mathbb{P}(Q = k) = \int_0^{\infty} [A^{*(k-1)}(x) - A^{*k}(x)] \mathrm{d}H * B_0(x) \qquad k = 1, 2, \ldots \qquad (4.9)$$

If A is spread out, then even $Q_t \to Q$ in total variation.

Proof. The existence was noted above and we shall apply (4.1) to evaluate the distribution of Q. First, $Q_t = 0$ is equivalent to the system being empty, i.e. $V_t = 0$ and thus by (3.2), $\mathbb{P}(Q = 0) = 1 - \rho$. Now for $n = 0, \ldots, \sigma - 1$ let

$$J_n = [U_0 + \cdots + U_{n-1}, U_0 + \cdots + U_n)$$

be the interval between the departure of $n - 1$ and n (with the convention $J_0 = [0, U_0)$). If $t \in J_n$, we will then have $Q_t \geqslant k$ if and only if customers $0, \ldots, n + k - 1$ have arrived, i.e. $T_0 + \cdots + T_{n+k-2} \leqslant t$. Hence by (4.1)

$$\mathbb{P}(Q \geqslant k) = \frac{1}{\mathbb{E}C} \mathbb{E} \int_0^C I(Q_t \geqslant k) \mathrm{d}t = \frac{1}{\mathbb{E}C} \mathbb{E} \sum_{n=0}^{\sigma-1} \int_{J_n} I(Q_t \geqslant k) \mathrm{d}t$$

$$= \frac{1}{\mathbb{E}C} \mathbb{E} \sum_{n=0}^{\sigma-1} |[U_0 + \cdots + U_{n-1}, U_0 + \cdots + U_n) \cap [T_0 + \cdots + T_{n+k-2}, \infty)|$$

$$= \frac{1}{\mathbb{E}C} \mathbb{E} \sum_{n=0}^{\sigma-1} \{(U_0 + \cdots + U_n - T_0 - \cdots - T_{n+k-2})^+ - (U_0 + \cdots + U_{n-1} - T_0 - \cdots - T_{n+k-2})^+\}$$

$$= \frac{1}{\mathbb{E}C} \mathbb{E} \sum_{n=0}^{\sigma-1} \{(W_n + U_n - T_n - \cdots - T_{n+k-2})^+ - (W_n - T_n - \cdots - T_{n+k-2})^+\} \tag{4.10}$$

$$= \frac{1}{\mathbb{E}C} \mathbb{E} \sum_{n=0}^{\sigma-1} \{(W_n + U - T^{(k-1)})^+ - (W_n - T^{(k-1)})^+\} \tag{4.11}$$

(evaluating the nth term in (4.10) by conditioning upon $\mathscr{F}_{n-1}$). But conditionally upon $T^{(k-1)} = x$, (4.11) is of the same form as (3.5) and hence becomes

$$\rho \mathbb{P}(U^* > T^{(k-1)} - W) = \rho \mathbb{P}(V > T^{(k-1)} \mid V > 0) = \mathbb{P}(V > T^{(k-1)}),$$

completing the proof. □

Problems

4.1. Consider the set-up of Theorem 4.2. Explain that if $\mathbb{P}(U - T = 0) > 0$ it may happen that $\mathbb{P}(Q^A = 0)$ is effectively smaller than $\mathbb{P}(W = 0)$.
4.2. Derive the distribution of Q^D by a direct argument similar to the one used for Q^A.

Notes

Arguments like the proofs of Theorems 3.5 and 4.3 are given in Harrison and Lemoine (1976) and Lemoine (1974). See also Cohen (1976). Little's formula has given rise to an extensive literature, see the survey by Ramalhoto *et al.* (1983).

5. THE ROBUSTNESS OF THE ACTUAL WAITING TIME

We consider here and in the next section for each $k = 0, 1, 2, \ldots$ a $GI/G/1$ queueing system with service-time distribution $B^{(k)}$, interarrival distribution $A^{(k)}$ and $U_n^{(k)}$, $T_n^{(k)}$, $X_n^{(k)}$, $S_n^{(k)}$, $W_n^{(k)}$, $W^{(k)}$, etc. defined the obvious way. The problem, stated in a rough form, is to study the limiting behaviour of $W^{(k)}$ as $k \to \infty$ under appropriate conditions, assuming that $\rho_k < 1$ for $k = 1, 2, \ldots$ and that $A^{(k)} \to A^{(0)}$, $B^{(k)} \to B^{(0)}$ in the sense of weak convergence. In Section 6, we consider the *heavy traffic* case where the limit has traffic intensity $\rho_0 = 1$, whereas the situation here is $\rho_0 < 1$. It is then reasonable to ask for conditions under which $W^{(k)} \xrightarrow{\mathscr{D}} W$. This is denoted as a *robustness* (or *stability*, *insensitivity* or *continuity*) property of the waiting time, and is of importance for example to justify the approximation of a queueing system with given $A^{(0)}$, $B^{(0)}$ by systems with $A^{(k)}$, $B^{(k)}$ of phase type (cf. III.6).

To facilitate notation, we suppress from now on indices n and $k = 1, 2, \ldots$ whenever convenient (thus, for example, $\mathbb{E}U \to \mathbb{E}U^{(0)}$ or $\lim_{k\to\infty} \mathbb{E}U = \mathbb{E}U^{(0)}$ means $\mathbb{E}U_n^{(k)} \to \mathbb{E}U_n^{(0)}$, $k \to \infty$).

We shall first state and prove the main result in random walk terms, and thereafter reformulate in terms more natural for queues.

Theorem 5.1. *Consider random walks* $\{S_n\}_{n\in\mathbb{N}}$, $\{S_n^{(0)}\}_{n\in\mathbb{N}}$ *with* $F \xrightarrow{w} F^{(0)}$, $k \to \infty$, $\mu_0 = \mathbb{E}X_n^{(0)} < 0$, $k = 0, 1, 2, \ldots$. *Then* $M \xrightarrow{\mathscr{D}} M^{(0)}$ *provided that the* X^+ *are uniformly integrable or equivalently that* $\mathbb{E}X^+ \to \mathbb{E}X^{(0)+}$.

The key step of the proof is

Lemma 5.2. *Define* $K_n = \max_{r \geq n} S_r$. *Then*

$$\overline{\lim_{k \to \infty}}\ \mathbb{P}(K_n > 0) \to 0 \quad \text{as } n \to \infty.$$

Proof. By general results on weak convergence, $\mathbb{E}X^+ \to \mathbb{E}X^{(0)+}$ is equivalent to the uniform integrability of the X^+ since $X^+ \xrightarrow{\mathscr{D}} X^{(0)+}$. Choose $c < 0$ such that $\mathbb{E}X^{(0)} \vee c < 0$ and define $\check{X}_n = X_n \vee c$. Then $\check{X}_n \xrightarrow{\mathscr{D}} X_n^{(0)}, \check{S}_n \geq S_n, \check{K}_n \geq K_n$. Hence for the proof it is no restriction to assume that the X are uniformly bounded below, say, by c. Then the X themselves are uniformly integrable, hence $\mu = \mathbb{E}X \to \mu_0 < 0$. Now for $\mu < 0$,

$$\mathbb{P}(K_n > 0) = \mathbb{P}\left(\max_{r \geq n} \frac{S_r}{r} > 0\right)$$

$$= \mathbb{P}\left(\max_{r \geq n}\left\{\frac{S_r}{r} - \mu\right\} > -\mu\right) \leq \frac{1}{|\mu|}\mathbb{E}\left|\frac{S_n}{n} - \mu\right|,$$

using the fact that $\{S_r/r - \mu\}_{r=n,n+1,\cdots}$ is a backwards martingale and Kolmogorov's inequality. Decompose $S_n - n\mu$ as $\tilde{S}_n + \tilde{\tilde{S}}_n$, where

$$\tilde{X}_n = X_n I(X_n \leq d) - \mathbb{E}X_n I(X_n \leq d),$$

$$\tilde{\tilde{X}}_n = X_n I(X_n > d) - \mathbb{E}X_n I(X_n > d)$$

with d satisfying $\mathbb{P}(X_n^{(0)} = d) = 0$, $\mathbb{E}X_n I(X_n > d) < \varepsilon$ for all k. Then $\tilde{\sigma}^2 = \mathbb{V}\mathrm{ar}\, \tilde{X}_n \to \tilde{\sigma}_0^2 = \mathbb{V}\mathrm{ar}\, \tilde{X}_n^{(0)} < \infty$ and

$$\mathbb{E}\left|\frac{S_n}{n} - \mu\right| \leq \mathbb{E}\left|\frac{\tilde{S}_n}{n}\right| + \mathbb{E}\left|\frac{\tilde{\tilde{S}}_n}{n}\right| \leq \frac{\tilde{\sigma}}{\sqrt{n}} + 2\varepsilon,$$

using the Cauchy–Schwarz inequality. Hence

$$\overline{\lim_{n \to \infty}}\ \overline{\lim_{k \to \infty}}\ \mathbb{P}(K_n > 0) \leq \overline{\lim_{n \to \infty}}\ \frac{1}{|\mu_0|}\left\{\frac{\tilde{\sigma}_0}{\sqrt{n}} + 2\varepsilon\right\} = \frac{2\varepsilon}{|\mu_0|}$$

and since ε is arbitrary, the proof is complete. □

Proof of Theorem 5.1. From $X \xrightarrow{\mathscr{D}} X^{(0)}$ it follows that $\{X_r\}_{r=0}^n \xrightarrow{\mathscr{D}} \{X_r^{(0)}\}_{r=0}^n$ and hence by the continuous mapping theorem $M_n \xrightarrow{\mathscr{D}} M_n^{(0)}$. Now let x satisfy $\mathbb{P}(M^{(0)} = x) = 0$. Then also $\mathbb{P}(M_n^{(0)} = x) = 0$ for each n and hence

$$\overline{\lim_{k \to \infty}}\ \mathbb{P}(M^{(k)} > x) \leq \overline{\lim_{k \to \infty}}\ \{\mathbb{P}(M_n^{(k)} > x) + \mathbb{P}(K_n^{(k)} > 0)\}$$

$$= \mathbb{P}(M_n^{(0)} > x) + \overline{\lim_{k \to \infty}}\ \mathbb{P}(K_n^{(k)} > 0),$$

$$\underline{\lim_{k \to \infty}}\ \mathbb{P}(M^{(k)} > x) \geq \underline{\lim_{k \to \infty}}\ \mathbb{P}(M_n^{(k)} > x) = \mathbb{P}(M_n^{(0)} > x).$$

Letting $n \to \infty$ yields $\mathbb{P}(M > x) \to \mathbb{P}(M^{(0)} > x)$ and hence $M \xrightarrow{\mathscr{D}} M^{(0)}$ □

Apparently the point mass of M at zero is of particular interest, but $\mathbb{P}(M=0)\to\mathbb{P}(M^{(0)}=0)$ does not follow alone from $M\xrightarrow{\mathcal{D}}M^{(0)}$. However:

Proposition 5.3. *If in addition to the assumptions of Theorem 5.1 the distribution $F^{(0)}$ of $X^{(0)}$ is continuous, then $\mathbb{P}(M=0)\to\mathbb{P}(M^{(0)}=0)$.*

Proof. The assumptions ensure that $\mathbb{P}(S_n^{(0)}=0)=0$ for each $n\geqslant 1$ and hence $\mathbb{P}(M_n=0)\to\mathbb{P}(M_n^{(0)}=0)$. Now argue exactly as above. □

Corollary 5.4. *Consider for $k=0,1,2,\ldots$ $GI/G/1$ queues with $A\xrightarrow{w}A^{(0)}$, $B\xrightarrow{w}B^{(0)}$, $\rho_0<1$, Then $W\xrightarrow{\mathcal{D}}W^{(0)}$ provided that the U are uniformly integrable, or equivalently, that $\mu_B\to\mu_B^{(0)}$. If in addition either $A^{(0)}$ or $B^{(0)}$ is continuous, then also $\mathbb{P}(W=0)\to\mathbb{P}(W^{(0)}=0)$.*

Proof. Appealing to the interpretation $X=U-T$, $W\overset{\mathcal{D}}{=}M$, it is straightforward to check the assumptions of Theorem 5.1 and Proposition 5.3 (the uniform integrability of the X^+ follows from $X^+\leqslant U$ and the uniform integrability of the U). □

Problems

5.1. Let $F^{(k)}$ be concentrated at -1, k with point masses $1-1/2k$, $1/2k$ and $F^{(0)}=\lim F^{(k)}$. Show that

$$\mathbb{P}(M^{(k)}\geqslant 1)\geqslant\mathbb{P}(X_n=k\quad\text{for some } n=1,\ldots,k)\to e^{-1/2}$$

and deduce that $M^{(k)}\to M^{(0)}$ does not hold.

Notes

Robustness problems are treated for example in Stoyan (1983) and Borovkov (1976), Of extensions, we mention for example the $GI/G/s$ case treated in Brandt and Lisek (1983) and robustness of the moments as derived in Asmussen and Johansen (1986).

6. HEAVY TRAFFIC LIMIT THEOREMS

If, in the set-up of Section 5, the limiting traffic intensity ρ_0 is $=1$ rather than <1, we are in the situation of *heavy traffic* where all queueing systems are heavily congested. We expect again $W\xrightarrow{\mathcal{D}}W^{(0)}$, but now $W^{(0)}=M^{(0)}=\infty$ a.s. It will turn out that a more precise result can be obtained, namely that under suitable conditions $|\mu|W$ is approximately exponentially distributed. We start again by formulating this for the random walk setting (σ^2 denotes $\mathbb{V}\text{ar}\,X$).

Theorem 6.1. *Consider random walks $\{S_n\}_{n\in\mathbb{N}}$, $\{S_n^{(0)}\}_{n\in\mathbb{N}}$ with $\mu<0$, $\mu\to 0$, $\underline{\lim}\,\sigma^2>0$, and the X^2 uniformly integrable. Then $Y=|\mu|M/\sigma^2$ is approximately exponentially distributed with intensity 2, $\mathbb{P}(Y>y)\to e^{-2y}$. Furthermore, $\mathbb{E}Y\to 1/2$.*

Remark 6.2. The conditions of Theorem 6.1 are not intrinsically different from the apparently stronger

$$X \xrightarrow{\mathscr{D}} X^{(0)}, \qquad \sigma^2 \to \sigma_0^2 > 0, \qquad \mu_0 = 0. \tag{6.1}$$

Indeed, the uniform integrability ensures that $\{F\}$ is tight. Thus every subsequence $\{k'\}$ has a weakly convergent subsequence $\{k''\}$, $X^{(k'')} \xrightarrow{\mathscr{D}} X^{(0)}$ for some $X^{(0)}$. But then by uniform integrability, $\mu_0 = \lim \mu_{k''} = 0$, $\sigma_0^2 = \lim \sigma_{k''}^2 > 0$. Furthermore, a standard analytical argument shows that if we can show the asymptotic exponentiality for $\{k''\}$, then it will hold for $\{k\}$ as well. Hence *for the proof we can* (and shall) *assume that* (6.1) *holds*. Also, by rescaling, we may take $\sigma^2 = 1$ and then define Y as $|\mu| M = -\mu M$ rather than as $|\mu| M/\sigma^2$. □

Two approaches to Theorem 6.1 will be considered, the first being based on characteristic functions $\varphi(y; Y) = \mathbb{E}e^{iyY}$. Thus we have to show $\varphi(y; Y) = \varphi(-\mu y; M) \to (1 - iy/2)^{-1}$. In the proof, we let $\mu_2 = \mathbb{E}X^2$ (thus $\mu_2 \to 1$ since $\mu \to 0$, $\sigma^2 \to 1$).

Lemma 6.3. *For each y, it holds as $k \to \infty$ that*

$$\varphi(-\mu y; X) = 1 - i\mu^2 y - \frac{\mu^2 y^2}{2} + o(\mu^2). \tag{6.2}$$

Proof. Define $g(z) = e^{iyz} - 1 - iyz + y^2 z^2/2$. Then for each $\varepsilon > 0$, we have

$$|g(z)| \leqslant c_\varepsilon |z|^3, \qquad |z| \leqslant \varepsilon, \qquad |g(z)| \leqslant d_\varepsilon z^2, \qquad |z| > \varepsilon,$$

and hence

$$\begin{aligned} |\mathbb{E}g(-\mu X)| &\leqslant c_\varepsilon \mathbb{E}[|-\mu X|^3; |-\mu X| \leqslant \varepsilon] + d_\varepsilon \mathbb{E}[(\mu X)^2; |-\mu X| > \varepsilon] \\ &\leqslant \mu^2 \{\varepsilon c_\varepsilon \mathbb{E}X^2 + d_\varepsilon \mathbb{E}[X^2; |-\mu X| > \varepsilon]\}, \\ &\overline{\lim_{k \to \infty}} |\mathbb{E}g(-\mu X)|/\mu^2 \leqslant \varepsilon c_\varepsilon, \end{aligned}$$

using the uniform integrability. Since c_ε remains bounded as $\varepsilon \downarrow 0$, it follows that

$$\varphi(-\mu y; X) - \left(1 - i\mu^2 y - \frac{\mu^2 y^2}{2} \mu_2\right) = \mathbb{E}g(-\mu X) = o(\mu^2)$$

and the lemma follows since $\mu_2 \to 1$. □

Proof of Theorem 6.1. We first note that as in Section 5 we have $M \xrightarrow{\mathscr{D}} \infty$. But

$$\mathbb{E}[(M + X)^-]^2 \leqslant \mathbb{E}(X^-)^2 \mathbb{P}(M \leqslant c) + \mathbb{E}[X^2; X < -c].$$

Letting first $k \to \infty$ and next $c \to \infty$ yields

$$\mathbb{E}[(M + X)^-]^2 \to 0 \quad (\text{hence } \mathbb{E}(M + X)^- \to 0). \tag{6.3}$$

From this $\mathbb{E}Y \to \frac{1}{2}$ is immediately apparent from (2.5). Now for each z

$e^{iyz^+} = e^{iyz} + 1 - e^{-iyz^-}$. Letting $Z = M + X$ and taking expectations we get

$$\varphi(y; M) = \varphi(y; M)\varphi(y; X) + 1 - \varphi(y; -(M+X)^-)$$
$$= \frac{1 - \varphi(y; -(M+X)^-)}{1 - \varphi(y; X)} \tag{6.4}$$

Since $e^{iz} - 1 - iz = z^2 O(1)$ for z real, we get

$$\varphi(-\mu y; -(M+X)^-) = 1 + i\mu y \mathbb{E}(M+X)^- + O(1)\mu^2 y^2 \mathbb{E}[(M+X)^-]^2$$
$$= 1 - i\mu^2 y + o(\mu^2),$$

using (1.8) and (6.3). Hence by Lemma 6.3 and (6.4),

$$\varphi(-\mu y; M) = \frac{i\mu^2 y + o(\mu^2)}{i\mu^2 y + \mu^2 y^2/2 + o(\mu^2)} \to \frac{1}{1 - iy/2}. \qquad \square$$

The second proof of Theorem 6.1 involves more advanced tools (weak convergence in function space) but is perhaps more illuminating and yields additional information, namely asymptotic relations for the $M_n^{(k)}, n < \infty$. We let $\{B_\xi(t)\}_{t \geq 0}$ denote Brownian motion with unit variance and drift ξ. The *inverse Gaussian distribution function* $G(t; \xi, c)$ corresponding to parameters $\xi, c > 0$ is the distribution of the first passage time $\tau(\xi, c) = \inf\{t \geq 0 : B_\xi(t) \geq c\}$,

$$G(T; \xi, c) = \mathbb{P}\left(\max_{0 \leq t \leq T} B_\xi(t) \geq c\right) = \mathbb{P}(\tau(\xi, c) \leq T). \tag{6.5}$$

This distribution (defective for $\xi < 0$) can in fact be found explicitly. We defer the derivation to XII.3 and here use only the formula

$$\| G(\cdot; \xi, c) \| = \mathbb{P}\left(\max_{0 \leq t < \infty} B_\xi(t \geq c)\right) = e^{2\xi c}, \qquad \xi < 0. \tag{6.6}$$

Proposition 6.4. *Under the conditions of Theorem 6.1, it holds for any $T < \infty$ that* $|\mu| M_{[T\sigma^2/\mu^2]}/\sigma^2 \xrightarrow{\mathcal{D}} \max_{0 \leq t \leq T} B_{-1}(t)$,

$$\mathbb{P}\left(\frac{|\mu|}{\sigma^2} M_{[T\sigma^2/\mu^2]} > y\right) \to G(T; -1, y)$$

(here $[\cdot]$ denotes integer part).

Proof. We may again assume that (6.1) holds with $\sigma_0^2 = 1$. Let $\{c\} = \{c^{(k)}\}$ be any sequence with $c^{(k)} \to \infty$ and define

$$B(t) = B^{(k)}(t) = \frac{1}{\sqrt{c}}\{S_{[ct]} - [ct]\mu\}.$$

It then follows from the invariance principle (Donsker's theorem) in its standard form (e.g. Billingsley, 1968, Ch. 3) that $B \xrightarrow{\mathcal{D}} B_0$ in $D[0, \infty)$. Taking $c = \mu^{-2}$ we have

$[ct]\mu/\sqrt{c} \to -t$, i.e.

$$\{|\mu| S_{[t/\mu^2]}\}_{0 \leqslant t < \infty} = \left\{ B(t) + \frac{[ct]\mu}{\sqrt{c}} \right\}_{0 \leqslant t < \infty} \xrightarrow{\mathcal{D}} \{B_0(t) - t\}_{0 \leqslant t < \infty} \stackrel{\mathcal{D}}{=} B_{-1}.$$

Hence, since $f \to \sup_{0 \leqslant t \leqslant T} f(t)$ is almost everywhere continuous on D with respect to any probability distribution concentrated on the continuous functions, it follows from the continuity of B_{-1} that

$$|\mu| M_{[T/\mu^2]} = \sup_{0 \leqslant t \leqslant T} |\mu| S_{[t/\mu^2]} \xrightarrow{\mathcal{D}} \max_{0 \leqslant t \leqslant T} B_{-1}(t)$$

which yields the desired conclusion in view of $\sigma^2 \to 1$. □

Proof 2 of Theorem 6.1. We assume again $\sigma^2 \to 1$ and write

$$Y = Y_1 \vee Y_2 = (|\mu| M_{[T/\mu^2]}) \vee \left(\sup_{n > T/\mu^2} (|\mu| S_n) \right).$$

Here by (6.6) and Proposition 6.4,

$$\lim_{T \to \infty} \lim_{k \to \infty} \mathbb{P}(Y_1 > y) = \lim_{T \to \infty} G(T; -1, y) = \mathrm{e}^{-2y}, \tag{6.7}$$

whereas $\{(S_n/n - \mu)^2\}$ is a backwards submartingale, hence

$$\mathbb{P}(Y_2 > 0) = \mathbb{P}\left(\max_{n > T/\mu^2} \left\{ \frac{S_n}{n} - \mu \right\} > -\mu \right)$$

$$\leqslant \frac{1}{\mu^2} \mathbb{E}(S_{[T/\mu^2]}/[T/\mu^2] - \mu)^2 = \frac{\sigma^2}{\mu^2 [T/\mu^2]},$$

$$\overline{\lim_{T \to \infty}}\ \overline{\lim_{k \to \infty}}\ \mathbb{P}(Y_2 > 0) \leqslant \overline{\lim_{T \to \infty}} \frac{1}{T} = 0. \tag{6.8}$$

Combining (6.7) and (6.8), the desired conclusion is obtained exactly as in the proof of Theorem 5.1. □

Corollary 6.5. *Consider $GI/G/1$ queueing systems with $A \xrightarrow{w} A^{(0)}$, $B \xrightarrow{w} B^{(0)}$, where $A^{(0)}, B^{(0)}$ are not both degenerate, $\rho < 1$, $\rho \to \rho_0 = 1$ and the U^2, T^2 uniformly integrable. Then $y = |\mu| W/\sigma^2$ is approximately exponentially distributed with intensity 2 and $\mathbb{E}Y \to \frac{1}{2}$. Here $\mu = \mathbb{E}X = \mathbb{E}U - \mathbb{E}T$, $\sigma^2 = \mathbb{V}\mathrm{ar}\, X$. Furthermore, for each T*

$$\mathbb{P}\left(\frac{|\mu|}{\sigma^2} W_{[T\sigma^2/\mu^2]} > y \right) \to G(T; -1, y)$$

The proof is a routine check and is omitted. Clearly, the conditions can be weakened corresponding to those of Theorem 6.1, but the case where $F \to F^{(0)}$ is satisfied while the condition $A \to A^{(0)}$, $B \to B^{(0)}$ fails rarely occurs in practice.

Results of the type in Corollary 6.5 are of high potential relevance, since the heavy traffic situation occurs widely in practice (when designing a service facility, one usually avoids for economical reasons keeping the server idle for a large proportion of the time). Given a queue with ρ smaller than but close to 1, we may imbed the system in the set-up of Corollary 6.5, writing $A = A^{(k)}$, $B = B^{(k)}$ for some large k. It is then suggested that the following approximations may be used:

$$\mathbb{E}W \cong \mathbb{V}\mathrm{ar}\, X/2\mathbb{E}(-X), \qquad \mathbb{P}(W > y) \cong \exp\{-2\mathbb{E}(-X)y/\mathbb{V}\mathrm{ar}\, X\} \tag{6.9}$$

Note that when $\mathbb{E}X \cong 0$, we have $\mathbb{E}X^2 \cong \mathbb{V}\mathrm{ar}\ X$ and one may thus replace $\mathbb{V}\mathrm{ar}\ X$ by $\mathbb{E}X^2$ in (6.9). However, inspection of (2.5) shows that $\mathbb{E}X^2/2\mathbb{E}(-X)$ and $\mathbb{V}\mathrm{ar}\ X/2\mathbb{E}(-X)$ are both upper bounds for $\mathbb{E}W$ and hence $\mathbb{V}\mathrm{ar}\ X/2\mathbb{E}(-X)$ is the best approximation.

We return to heavy traffic approximations in Chapter XII, but finally we mention that in view of the formulae of Sections 3 and 4, it is straightforward to derive analogues of Corollary 6.5 for virtual waiting time, queue lengths and so on (cf. Problem 6.1).

Problems

6.1. Show that under the conditions of Corollary 6.5 the equilibrium virtual waiting time V has the same limiting distribution as W. Show similarly, using the results of Section 4, that $|\mu|Q/\sigma^2$, $|\mu|Q^A/\sigma^2$ have limiting exponential distributions with intensities $2\mu_A$.

Notes

Heavy traffic limit theory was largely initiated by Kingman in the 1960s and is surveyed in a broad setting in Whitt (1974). The present characteristic function proof is essentially that of Blomqvist (1974), whereas the diffusion approximation point of view has been noticed by several authors. To the references of Whitt (1974) we add here among many Kleinrock (1976, Ch. 2), Grandell (1977, 1978), Siegmund (1979), Köllerström (1981), Asmussen (1984) and Harrison (1985).

CHAPTER IX

Explicit Examples in the Theory of Random Walks and the Single-Server Queues

1. ASCENDING LADDER HEIGHTS. $GI/M/1$

In this chapter, we shall look at random walk examples, where one of the ladder height distributions G_+, G_- can be evaluated explicitly. If this is possible, both distributions are then (at least in principle) available in view of the formulae

$$F = G_+ + G_- - G_+ * G_-, \tag{1.1}$$

$$\hat{G}_+ = \frac{\hat{F} - \hat{G}_-}{1 - \hat{G}_-}, \quad \hat{G}_- = \frac{\hat{F} - \hat{G}_+}{1 - \hat{G}_+}, \tag{1.2}$$

$$1 - \hat{F} = (1 - \hat{G}_+)(1 - \hat{G}_-), \tag{1.3}$$

(the transform relations (1.2) and (1.3) always make sense for characteristic functions $\hat{F}(s) = \int_{-\infty}^{\infty} e^{isx}\,\mathrm{d}F(x)$, etc. but depending on the context also sometimes for generating functions $\hat{F}(s) = \sum_{-\infty}^{\infty} s^n f_n$, etc., moment generating functions $\hat{F}(s) = \int_{-\infty}^{\infty} \mathrm{e}^{sx}\,\mathrm{d}F(x)$, etc. and so on),

$$G_-(A) = F * U_+(A) = \left(F * \sum_{n=0}^{\infty} G_+^{*n} \right)(A), \qquad A \subseteq (-\infty, 0] \tag{1.4}$$

$$G_+(A) = F * U_-(A), \qquad A \subseteq (0, \infty). \tag{1.5}$$

Having found both G_+ and G_-, now basically all interesting quantities can be found. In particular, in the case of the $GI/G/1$ queue, the distribution

$$H(x) = (1 - \| G_+ \|)U_+(x) = (1 - \| G_+ \|) \sum_{n=0}^{\infty} G_+^{*n}(x) \tag{1.6}$$

of the maximum is now simply that of the equilibrium actual waiting time W. Equivalently in transform formulation,

$$\hat{H}(s) = \frac{1 - \hat{G}_+(0)}{1 - \hat{G}_+(s)} = \frac{\hat{F}'(0)}{\hat{G}'_-(0)} \frac{1 - \hat{G}_-(s)}{1 - \hat{F}(s)}. \tag{1.7}$$

Using the formulae of VIII.3–4 we can then also find the equilibrium distributions of the virtual waiting time, queue lengths and so on.

Ascending and descending ladder heights are, of course, symmetrically defined. However, in the $GI/G/1$ case there is no symmetry between positive and negative values since for the steady state to exist we need the condition $\mu = \mathbb{E}X_n = \mathbb{E}U_n - \mathbb{E}T_n < 0$. Therefore ascending and descending ladder heights need to be treated separately, and for the purpose of evaluating (1.6) it is therefore somewhat simpler if G_+ rather than G_- can be found. The $GI/M/1$ queue provides the simplest example where this is possible, and $M/G/1$ the simplest example where G_- can be found. For this reason $GI/M/1$ can be argued to be of slightly more elementary nature than $M/G/1$ (however, presumably $M/G/1$ is most often a more realistic model than $GI/M/1$).

We give first the general random walk formulation and specialize next to queues. Consider first the lattice case, say F is concentrated on $\mathbb{Z}$ with point probabilities $\{f_n\}_{n\in\mathbb{Z}}$. We call the random walk *right-continuous* or *skip-free to the right* if $f_2 = f_3 = \cdots = 0$ or equivalently $\mathbb{P}(X_n > 1) = 0$. Similarly, the random walk is *left-continuous* or *skip-free to the left* if $f_{-2} = f_{-3} = \cdots 0$ or equivalently $\mathbb{P}(X_n < -1) = 0$.

Theorem 1.1. *For a right-continuous lattice random walk on* $\mathbb{Z}$ *with* $-\infty \leqslant \mu \leqslant 0$ *it holds that*:

(a) G_+ *is concentrated at* 1, *say with mass* $\theta = g_1^{(+)} = \| G_+ \|$. *Here* $\theta = 1$ *if* $\mu = 0$, *whereas for* $\mu < 0$ θ *can be evaluated as the unique solution* < 1 *of*

$$1 = \mathbb{E}\theta^{-X_n} = \frac{1}{\theta} f_1 + f_0 + \theta f_{-1} + \theta^2 f_{-2} + \cdots. \tag{1.8}$$

(b) *For* $\mu < 0$, *the distribution of* M *is geometric with parameter* θ, $\mathbb{P}(M = n) = (1-\theta)\theta^n$.

(c) G_- *is given by the point probabilities*

$$g_n^{(-)} = \sum_{k=-\infty}^{n} \theta^{n-k} f_k, \qquad n = 0, -1, -2, \ldots.$$

Proof. That $S_{\tau_+} = 1$ on $\{\tau_+ < \infty\}$ is an obvious consequence of right-continuity. Also it is clear that G_+^{*n} is concentrated at n with point mass θ^n, and since $1 - \| G_+ \| = 1 - \theta$ (b) is therefore immediately apparent from (1.6). Similarly, (c) follows by letting $A = \{n\}$ in (1.4). Also, from VII.2.4 we have $\theta = 1$ for $\mu = 0$ and $\theta < 1$ for $\mu < 0$. In the last case also

$$1 = \| G_- \| = \sum_{n=-\infty}^{0} g_n^{(-)} = \sum_{k=-\infty}^{0} f_k \sum_{n=k}^{0} \theta^{n-k} = \sum_{k=-\infty}^{0} f_k \frac{1-\theta^{1-k}}{1-\theta},$$

$$1 - \theta = 1 - f_1 - \sum_{k=-\infty}^{0} f_k \theta^{1-k}$$

and (1.8) follows (for an alternative proof using transforms, see Problem 1.1). Furthermore, $\varphi(\theta) = \mathbb{E}\theta^{-X_n}$ is strictly convex, $\varphi(1) = 1$, $\varphi'(1) = -\mu > 0$ and $\lim_{\theta\downarrow 0}\varphi(\theta) = \infty$. This ensures that the solution in $(0, 1)$ of $\varphi(\theta) = 1$ is unique.

□

The case of F having an exponential density on $(0, \infty)$ is very similar:

Theorem 1.2. *Suppose that* $-\infty \leqslant \mu \leqslant 0$ *and that* $\mathrm{d}F(x)/\mathrm{d}x$ *exists for* $x > 0$ *and is of the form* $\alpha\,\mathrm{e}^{-\delta x}$ (no assumptions are made on the form of $\mathrm{d}F(x)$ for $x \leqslant 0$). *Then*:

(a) G_+ *is* (possibly defective) *exponential with intensity* δ, $\mathrm{d}G_+(x)/\mathrm{d}x = \theta\delta\,\mathrm{e}^{-dx}$, $x > 0$. *Here* $\theta = 1$ *if* $\mu = 0$, *whereas for* $\mu < 0$ θ *can be evaluated as the unique solution* < 1 *of*

$$1 = \mathbb{E}\mathrm{e}^{\eta X_n} = \int_{-\infty}^{\infty} \mathrm{e}^{\eta x}\,\mathrm{d}F(x) \tag{1.9}$$

where $\eta = \delta(1-\theta)$;

(b) *If* $\mu < 0$, *then the distribution of* M *is a mixture with weights* $1-\theta$, θ *of the distribution degenerate at zero and the exponential distribution with intensity* $\eta = \delta(1-\theta)$. *That is,* $\mathbb{P}(M \leqslant m) = 1 - \theta\,\mathrm{e}^{-\eta m}$;

(c) $$G_-(x) = \begin{cases} \dfrac{1}{1-\theta}\left\{F(x) - \theta\,\mathrm{e}^{-\eta x}\displaystyle\int_{-\infty}^{x} \mathrm{e}^{\eta y}\,\mathrm{d}F(y)\right\}, & x \leqslant 0, \mu < 0 \\ (1+\delta x)F(x) - \delta\displaystyle\int_{-\infty}^{x} y\,\mathrm{d}F(y), & x \leqslant 0, \mu = 0. \end{cases}$$

Proof. The conditional distribution $F^{(+)}$ of X_n given $\{X_n > 0\}$ is exponential with intensity δ. Hence the distribution of S_{τ_+} is so, S_{τ_+} being an overshot corresponding to $F^{(+)}$. More formally, for $x > 0$

$$\begin{aligned}\mathbb{P}(\tau_+ = n, S_{\tau_+} > x) &= \mathbb{P}(X_n > x - S_{n-1}, S_k \leqslant 0, k = 1, \ldots, n-1) \\ &= \mathrm{e}^{-\delta x}\mathbb{P}(X_n > -S_{n-1}, S_k \leqslant 0, k = 1, \ldots, n-1) = \mathrm{e}^{-\delta x}\mathbb{P}(\tau_+ = n),\end{aligned}$$

and $\mathrm{d}G_+(x)/\mathrm{d}x = \theta\delta\,\mathrm{e}^{-\delta x}$ follows by summing over n and letting $\theta = \sum_1^\infty \mathbb{P}(\tau_+ = n) = \|G_+\|$. Hence if $\mu < 0$, $\sum_1^\infty G_+^{*n}$ has density

$$\sum_{n=1}^{\infty} \theta^n \delta^n \frac{x^{n-1}}{(n-1)!}\mathrm{e}^{-\delta x} = \theta\delta\,\mathrm{e}^{-\eta x} = \frac{\theta}{1-\theta}\eta\mathrm{e}^{-\eta x}$$

and

$$U_+(x) = 1 + \frac{\theta}{1-\theta}(1 - \mathrm{e}^{-\eta x}) = \frac{1}{1-\theta} - \frac{\theta}{1-\theta}\mathrm{e}^{-\eta x}.$$

Combined with (1.6) and (1.4), the formulae in (b) and (c) follow (for $\mu = 0$ we get instead $U_+(x) = 1 + \delta x$). It remains to evaluate θ. For $\mu = 0$ we have $\theta = \|G_+\| = 1$. For $\mu < 0$, we again use

$$1 = \|G_-\| = \frac{1}{1-\theta}\left\{F(0) - \theta\int_{-\infty}^{0} \mathrm{e}^{\eta y}\,\mathrm{d}F(y)\right\}. \tag{1.10}$$

Now necessarily $\alpha = 1 - F(0)$ and thus

$$\int_0^\infty \mathrm{e}^{\eta x}\,\mathrm{d}F(x) = (1 - F(0))\int_0^\infty \delta\,\mathrm{e}^{(\eta-\delta)x}\,\mathrm{d}x = (1 - F(0))\frac{1}{\theta}. \tag{1.11}$$

Solving (1.11) for $F(0)$ and inserting in (1.10) shows that (1.9) holds. The uniqueness of a solution $\eta > 0$ to (1.9) is proved along similar lines as in Theorem 1.1. □

Using Theorem 1.2, we can now evaluate the steady-state properties of $GI/M/1$. An alternative derivation relating also to Theorem 1.1 will be presented in Section 3.

Theorem 1.3. *Consider the equilibrium* $GI/M/1$ *queue* ($\mathrm{d}B(x)/\mathrm{d}x = \delta\,\mathrm{e}^{-\delta x}$, $\rho = 1/\delta\mu_A < 1$). *Then*:

(a) *The equation*

$$1 = \mathbb{E}\,\mathrm{e}^{\eta X_n} = \mathbb{E}\,\mathrm{e}^{\eta(U_n - T_n)} = \frac{\delta}{\delta - \eta} \cdot \int_0^\infty \mathrm{e}^{-\eta y}\,\mathrm{d}A(y) \tag{1.12}$$

has a unique solution $\eta > 0$;

(b) *The actual waiting-time distribution is given by* $\mathbb{P}(W \leqslant x) = 1 - \theta\,\mathrm{e}^{-\eta x}$, *where* $\theta = 1 - \eta/\delta = \int_0^\infty \mathrm{e}^{-\eta y}\,\mathrm{d}A(y)$;

(c) *If* A *is non-lattice, then the virtual waiting-time distribution is given by* $\mathbb{P}(V \leqslant x) = 1 - \rho\,\mathrm{e}^{-\eta x}$;

(d) *If* A *is non-lattice, then the distribution of the queue length at an arbitrary point of time is modified geometric given by* $\mathbb{P}(Q = 0) = 1 - \rho$, $\mathbb{P}(Q \geqslant k) = \rho\theta^{k-1}$, $k = 1, 2, \ldots$.

(e) *The distribution of the queue length just before an arrival or just after a departure is geometric with parameter* θ, *i.e. with point probabilities* $\pi_n = (1 - \theta)\theta^n$.

Proof. The distribution of $X_n = U_n - T_n$ is of the form considered in Theorem 1.2, since for $x > 0$

$$\mathbb{P}(X_n > x \mid X_n > 0) = \mathbb{P}(U_n > T_n + x \mid U_n > T_n) = \mathrm{e}^{-\delta x}$$

Furthermore, the assumption $\mu < 0$ is equivalent to $\rho < 1$ and hence Theorem 1.2 immediately yields (a) and (b), the last expression for θ being a consequence of (1.12). For (c), we use VIII.3.5. The distribution of U^* and U is the same when U is exponential, and we get the Laplace transform of $W + U^*$ as

$$\left[1 - \theta + \theta\frac{\eta}{\eta + s}\right] \cdot \frac{\delta}{\delta + s} = \frac{\eta}{\eta + s},$$

proving (c). For (d) and (e), we note first that in the notation of VIII.4 it follows by conditioning upon $T^{(k-1)}$ that for $k = 1, 2, \ldots$

$$\mathbb{P}(W > T^{(k-1)}) = \theta\mathbb{E}\exp\{-\eta T^{(k-1)}\} = \theta(\mathbb{E}\exp\{-\eta T\})^{k-1} = \theta^k.$$

Hence (e) is immediately apparent from VIII.4.2; (d) follows similarly from VIII.4.3 and $\mathbb{P}(V > T^{(k-1)}) = \rho\theta^{k-1}$. □

Problems

1.1. Show (1.8) and (1.9) by means of (1.3).
1.2. Find the parameter θ in (1.8) in the case $f_{-1}+f_1=1$ of a Bernoulli random walk.
1.3. Derive a lattice analogue of Theorem 1.2 for the case $f_n=p\delta^n$, $n\geqslant 1$.
1.4. Analyse the $GI/G/1$ queue with $\mathrm{d}B(x)/\mathrm{d}x=0$ $x<1$, $=\delta \mathrm{e}^{-\delta(x-1)}$ $x\geqslant 1$, and A concentrated on $(1,\infty)$.

2. DESCENDING LADDER HEIGHTS. $M/G/1$

We start again with the lattice case:

Theorem 2.1. *For a left-continuous lattice random walk on* $\mathbb{Z}$ $(f_{-2}=f_{-3}=\cdots=0)$ *with* $f_{-1}+f_0<1$, $-\infty\leqslant\mu\leqslant 0$ *it holds that:*

(a) G_- *is concentrated on* $\{-1,0\}$ *with point probabilities* $g_{-1}^{(-)}=f_{-1}$, $\zeta=g_0^{(-)}=1-f_{-1}$;
(b) G_+ *has point masses* $g_n^{(+)}=r_n/f_{-1}$, $n=1,2,\ldots$, *where* $r_n=f_n+f_{n+1}+\cdots$ *Furthermore* $\|G_+\|=(f_1+2f_2+\cdots)/f_{-1}=1+\mu/f_{-1}$;
(c) *For* $\mu<0$, *the point probabilities* $v_n=\mathbb{P}(M=n)$ *are given by* $v_0=-\mu/f_{-1}$,

$$v_n=\frac{r_n}{f_{-1}}v_0+\frac{r_{n-1}}{f_{-1}}v_1+\cdots+\frac{r_1}{f_{-1}}v_{n-1},\qquad n\geqslant 1.$$

Proof. That $g_n^{(-)}=0$ for $n=-2,-3,\ldots$ is clear by left-continuity, which also implies that $S_{\tau_-}=-1$ can only occur if $X_1=-1$. Hence $g_{-1}^{(-)}=f_{-1}$, and $g_0^{(-)}=1-f_{-1}$ because of $\|G_-\|=1$. To evaluate G_+, we use (1.5). Now G_-^s is degenerate at -1, hence U_-^s is counting measure on $\{0,-1,-2,\ldots\}$ and U_- has point masses $(1-\zeta)^{-1}=f_{-1}^{-1}$ at $0,-1,-2$, cf. the proof of VII.3.2. Thus the asserted expression for $g_n^{(+)}$ follows from (1.5) and $\|G_+\|$ comes out by just summing the $g_n^{(+)}$. For (c), it is now clear that $v_0=1-\|G_+\|=-\mu/f_{-1}$ and also the formula for the $v_n, n\geqslant 1$, is clear since $\{v_n\}$ is proportional to the renewal sequence governed by $\{g_n^{(+)}\}$, cf. I.(2.1). □

Theorem 2.2. *Suppose that* $-\infty<\mu\leqslant 0$ *and that* $\mathrm{d}F(x)/\mathrm{d}x$ *exists for* $x\leqslant 0$ *and is of the form* $\alpha\,\mathrm{e}^{\beta x}$, $\beta>0$. (No assumptions are made on the form of $\mathrm{d}F(x)$ for $x>0$.) *Then*

(a) G_- *is exponential with intensity* β, $\mathrm{d}G_-(x)/\mathrm{d}x=\beta\,\mathrm{e}^{\beta x}, x\leqslant 0$;
(b) $G_+(x)=F(x)+\beta\{\mu-\int_x^\infty(1-F(y))\mathrm{d}y\}$, $x>0$, $\|G_+\|=1+\mu\beta$

[the expression (1.6) for the distribution of M does not substantially simplify. See, however, the discussion for $M/G/1$ below.]

Proof. The assumption on F ensures that the distribution of X_n given $X_n\leqslant 0$ is exponential with intensity β [up to the sign]. Hence (a) follows exactly as in Theorem 1.2(a), using $\|G_-\|=1$, and by VII.2.3(c)

$$1-\|G_+\|=\frac{1}{\mathbb{E}\tau_-}=\frac{\mu}{\mathbb{E}S_{\tau_-}}=-\mu\beta.$$

Now by properties of the Poisson process, $U_-[y,0] = 1 - \beta y$, $y \leqslant 0$. Hence for $x > 0$ (1.5) yields

$$\| G_+ \| - G_+(x) = G_+(x, \infty) = 1 - F(x) + \int_x^\infty \beta(y - x)\mathrm{d}F(y)$$

$$= 1 - F(x) + \beta \int_x^\infty (1 - F(y))\mathrm{d}y$$

and (b) follows. □

For the $M/G/1$ case, G_+ takes a somewhat simpler form:

Theorem 2.3. *Consider the equilibrium $M/G/1$ queue* ($\mathrm{d}A(x)/\mathrm{d}x = \beta \mathrm{e}^{-\beta x}$, $\rho = \beta\mu_B < 1$). *Then:*

(a) *The ascending ladder height distribution G_+ corresponding to $X_n = U_n - T_n$ is absolutely continuous with density $g_+(x) = \beta(1 - B(x)) = \rho\, \mathrm{d}B_0(x)/\mathrm{d}x$;*

(b) *The actual waiting time W and virtual waiting time V have a common distribution given by*

$$H(x) = \mathbb{P}(W \leqslant x) = \mathbb{P}(V \leqslant x) = (1 - \rho) \sum_{n=0}^{\infty} \rho^n B_0^{*n}(x). \tag{2.1}$$

In particular, the two first moments are

$$\mathbb{E}W = \mathbb{E}V = \frac{\rho\mu_B^{(2)}}{2(1 - \rho)\mu_B} \tag{2.2}$$

$$\mathbb{E}W^2 = \mathbb{E}V^2 = \frac{\rho\mu_B^{(3)}}{3(1 - \rho)\mu_B} + \frac{\beta^2 \mu_B^{(2)^2}}{2(1 - \rho)^2} \tag{2.3}$$

where $\mu_B^{(k)} = \mathbb{E}U_n^k$, $k = 2, 3$;

(c) *The queue lengths just before an arrival time, just after a departure time and at an arbitrary point of time have all the same distribution which can be expressed in terms of the distribution (2.1) and the Poisson distribution by $\mathbb{P}(Q = 0) = 1 - \rho$,*

$$\mathbb{P}(Q = k) = \int_{0+}^\infty \mathrm{e}^{-\beta t} \frac{(\beta t)^{k-1}}{(k-1)!} \mathrm{d}H(t) = \rho \int_0^\infty \mathrm{e}^{-\beta t} \frac{(\beta t)^{k-1}}{(k-1)!} \mathrm{d}H * B_0(t), \tag{2.4}$$

$k = 1, 2, \ldots.$ *In particular,*

$$\mathbb{E}Q = \rho\{1 + \beta(\mathbb{E}W + \mathbb{E}U^*)\} = \rho + \beta\mathbb{E}W = \rho + \frac{\rho^2 \mu_B^{(2)}}{2(1 - \rho)\mu_B^2}. \tag{2.5}$$

The formula (2.1) (or its transform analogue, Problem 2.1, or sometimes just (2.2)) is often referred to as the *Pollaczek–Khintchine formula.*

Proof. We first note that $F(x) = \mathbb{P}(X_n \leqslant x)$ has a density, which for $x < 0$ is of the form $\alpha \mathrm{e}^{\beta x}$ and for $x > 0$ is given by

$$\int_x^\infty \beta e^{-\beta(y-x)}\, dB(y).$$

Also for $x > 0$,

$$1 - F(x) = \int_x^\infty (1 - e^{-\beta(y-x)}) dB(y) = 1 - B(x) - e^{\beta x} \int_x^\infty e^{-\beta y}\, dB(y).$$

Hence, using Theorem 2.2(b), g_+ exists and is given by

$$\frac{dF(x)}{dx} + \beta(1 - F(x)) = \beta(1 - B(x)) = \rho \frac{dB_0(x)}{dx}.$$

That $W \stackrel{\mathscr{D}}{=} V$ was previously found in VIII.3.5 (cf. also III.10(a) and V.3.5) and thus (2.1) reduces to (1.6). Then

$$\begin{aligned} \mathbb{E}W &= (1-\rho) \sum_{n=0}^\infty \rho^n \int_0^\infty y\, dB_0^{*n}(y) = (1-\rho) \sum_{n=0}^\infty \rho^n n \int_0^\infty y\, dB_0(y) \\ &= (1-\rho) \sum_{n=0}^\infty \rho^n n \frac{\mu_B^{(2)}}{2\mu_B} = \frac{\rho \mu_B^{(2)}}{2(1-\rho)\mu_B}. \end{aligned}$$

The proof of (2.3) is similar (though a little more lengthy) and left to the Problems. The equality in distribution in (c) follows from $W \stackrel{\mathscr{D}}{=} V$ and VIII.4.2–3. Also VIII.4.3 yields $\mathbb{P}(Q = 0) = 1 - \rho$ and

$$\mathbb{P}(Q \geq k) = \mathbb{P}(W > T^{(k-1)}) = \int_{0+}^\infty \sum_{l=k-1}^\infty e^{-\beta t} \frac{(\beta t)^l}{l!} dH(t)$$

from which (2.4) follows. The proof of (2.5) is then easy. □

Problems

2.1. Find $\hat{G}(s) = \int_0^\infty e^{-sx}\, dG_+(s)$ in $M/G/1$, both directly from Theorem 2.3(a) and by Wiener–Hopf factorization. Show hereby that

$$\hat{H}(s) = \mathbb{E}\, e^{-sW} = \frac{(1-\rho)s}{s - \beta + \beta \hat{B}(s)}, \qquad s > 0.$$

2.2. Check the formula (2.3) for $\mathbb{E}W^2$ in $M/G/1$, both by proceeding as in the proof of Theorem 2.3(b) and by finding $\hat{H}''(0)$ [hint: L'Hospital's rule].
2.3. Check that the expression for $\mathbb{E}W$ and $\mathbb{E}Q$ in $M/G/1$ are compatible with Little's formula and that $\mathbb{E}W = \mathbb{E}V$ is compatible with VIII.3.3.
2.4. Show that $\mathbb{E}W, \mathbb{E}Q$ in $M/G/1$ are minimized for $M/D/1$ if one imposes the constraints that β and μ_B are fixed.

3. IMBEDDED MARKOV CHAIN ANALYSIS FOR $GI/M/1$

The method of the imbedded Markov chain is the traditional approach to $GI/M/1$ and $M/G/1$ in much of the literature, and also applicable in some situations where the random walk methods used in Sections 1 and 2 do not apply (examples are $GI/M/m$, cf. XI.3, and some systems with arrivals or services in groups). We shall exemplify the method here by giving an alternative derivation of the steady-state characteristics of $GI/M/1$ and $M/G/1$.

The idea is to observe that the queue lengths observed just before arrival times in $GI/M/1$ and just after departure times in $M/G/1$ are Markov chains $\{Y_n\}$, the steady-state distribution π of which can be calculated. The remaining characteristics can then either be found immediately from π, or related to π by using regenerative processes in a somewhat similar manner as when relating the steady-state properties to the distribution of W, cf. VIII.3–4.

Starting with $GI/M/1$, the Markov property of $\{Y_n\}$ was previously derived in III.7.2 and the transition matrix found to be

$$P = \begin{pmatrix} r_0 & q_0 & 0 & 0 & \\ r_1 & q_1 & q_0 & 0 & \cdots \\ r_2 & q_2 & q_1 & q_0 & \\ \vdots & & & & \ddots \end{pmatrix},$$

where

$$q_k = \int_0^\infty e^{-\delta t} \frac{(\delta t)^k}{k!} \, dA(t), \qquad r_n = q_{n+1} + q_{n+2} + \cdots.$$

By direct insertion it is now seen that $\pi_n = (1-\theta)\theta^n$ solves $\pi P = \pi$, provided that θ satisfies $\sum_0^\infty r_n \theta^n = 1$, $\sum_0^\infty q_n \theta^n = \theta$. An elementary calculation shows the first of these equations to be a consequence of the second. If $\eta \in (0, \delta)$, $\theta \in (0, 1)$ are connected by $\eta = \delta(1-\theta)$, $\theta = 1 - \eta/\delta$, we may rewrite as

$$1 - \frac{\eta}{\delta} = \sum_0^\infty q_n \theta^n = \int_0^\infty e^{-\eta t} \, dA(t).$$

This is the same as (1.12), and arguing as in Section 1 it is seen that a solution exists and is unique if $\rho < 1$. Alternatively, π can be derived by remarking that $\{Y_n\}$ is a Lindley process governed by the f given by $f_1 = q_0$, $f_0 = q_1$, $f_{-1} = q_2, \ldots$, hence the stationary distribution is that of M and Theorem 1.1 immediately applies.

To study the queue length in continuous time, we first note that $Y_0 = 0$, $\sigma = \inf\{n \geqslant 1 : Y_n = 0\}$. Hence

$$\mathbb{E}\sigma = 1/\pi_0 = 1/(1-\theta) \quad \text{and} \quad \mathbb{E} \sum_0^{\sigma-1} I(Y_n = k) = \pi_k/\pi_0 = \theta^k,$$

cf. I.(3.4). By Wald's identity, $\mathbb{E}C = \mu_A \mathbb{E}\sigma = \mu_A/(1-\theta)$, cf. VIII.3.1. Now let $j \geqslant 1$ and note that

$$\mathbb{P}(Q = j) = \frac{1}{\mathbb{E}C} \mathbb{E} \int_0^C I(Q_t = j) \, dt = \frac{1}{\mathbb{E}C} \mathbb{E} \sum_{n=0}^{\sigma-1} I_n,$$

$$I_n = \int_{T_0 + \cdots + T_{n-1}}^{T_0 + \cdots + T_n} I(Q_t = j) \, dt.$$

To evaluate $\mathbb{E}(I_n; \sigma > n)$, we condition upon $\{Y_n = k, \sigma > n\}$. Just after the arrival of customer n, a total of $k+1$ customers are present. Thus for the queue length to be j, s time units later, and customer $n+1$ having not yet arrived, we need to have $T_n > s$ and $k+1-j$ service events in a time interval of length s. Thus for $k = j-1, j, \ldots$

$$\mathbb{E}(I_n; \sigma > n \mid Y_n = k, \sigma > n) = \int_0^\infty \mathrm{e}^{-\delta s}\frac{(\delta s)^{k+1-j}}{(k+1-j)!}(1 - A(s))\mathrm{d}s$$

$$= \int_0^\infty \delta^{-1} \sum_{l=k+2-j}^{\infty} \mathrm{e}^{-\delta t}\frac{(\delta t)^l}{l!}\mathrm{d}A(t) = \delta^{-1} r_{k+1-j}, \quad (3.1)$$

$$\mathbb{P}(Q = j) = \frac{1}{\mathbb{E}C}\mathbb{E}\sum_{n=0}^{\infty}\sum_{k=j-1}^{\infty} I(Y_n = k, \sigma > n)\delta^{-1} r_{k+1-j}$$

$$= \frac{\rho}{\mathbb{E}\sigma}\sum_{k=j-1}^{\infty} r_{k+1-j}\mathbb{E}\sum_{n=0}^{\infty} I(Y_n = k, \sigma > n)$$

$$= \rho(1-\theta)\sum_{k=j-1}^{\infty} r_{k+1-j}\frac{\pi_k}{\pi_0}$$

$$= \rho(1-\theta)\theta^{j-1}\sum_{l=0}^{\infty} r_l \theta^l = \rho(1-\theta)\theta^{j-1},$$

$$\mathbb{P}(Q = 0) = 1 - \sum_{j=1}^{\infty}\mathbb{P}(Q = j) = 1 - \rho.$$

The above arguments are typical for standard imbedded Markov chain analysis and have therefore been given in some detail. One may note that (given the general tools of Chapters VII and VIII) the calculations of Section 1 are at least as short and also make the particular form of the solution more transparent. However, for simple queues like $GI/M/1$ and $M/G/1$ a variety of further more or less related special methods are available and may sometimes lead to quite elegant arguments. Here is such an example of a further derivation of the relation between π_k and $\mathbb{P}(Q = k)$. Let N_1 be the number of times $\{Q_t\}_{0 \leqslant t < C}$ has a downcrossing from $j+1$ to j and let N_2 be the number of upcrossings. Then obviously $N_1 = N_2$. But an upcrossing corresponds to the queue length just before an arrival being j, i.e. a visit of $\{Y_n\}_{0 \leqslant n < \sigma}$ to j. Since downcrossings occur at rate δ from state $j+1$, $\mathbb{E}N_1 = \mathbb{E}N_2$ yields

$$\delta\mathbb{E}\int_0^C I(Q_t = j+1)\mathrm{d}t = \mathbb{E}\sum_{n=0}^{\sigma-1} I(Y_n = j),$$

$$\delta\mathbb{E}C\mathbb{P}(Q = j+1) = \pi_j/\pi_0 = \theta^j, \; \mathbb{P}(Q = j+1) = \frac{\theta^j}{\delta\mathbb{E}C} = \rho(1-\theta)\theta^j.$$

Having found π_k and $\mathbb{P}(Q = k)$, the remaining steady-state characteristics now follow easily: for the queue length just after departure times, the up- and downcrossing argument in the proof of VIII.4.2 applies, and for the actual and virtual waiting time we can condition upon the queue length and use the exponential distributions of the residual service time of the customer being served and the service time of the customer in line in just the same way as for the $M/M/1$ case in III.10(a). We omit the details.

Notes

Imbedded Markov chain analysis was largely initiated by Kendall (1953) and has since then become an extremely popular tool. Of further approaches to queues like $GI/M/1$ and $M/G/1$, we mention in particular Markov renewal theory and semi-regenerative processes, which are related to imbedded Markov chain analysis and which we shall study in Chapter X. A survey of up- and downcrossing arguments is given in Hordijk and Tijms (1976) and Cohen (1976).

4. IMBEDDED MARKOV CHAIN ANALYSIS FOR $M/G/1$

We consider the $M/G/1$ queue with $\rho < 1$ and let Y_n denote the queue length just after the departure of customer $n-1$, $Y_0 = 0$. If K_n is the number of customers arriving while n is being served,

$$q_k = \mathbb{P}(K_n = k) = \int_0^\infty \mathrm{e}^{-\beta t}\frac{(\beta t)^k}{k!}\,\mathrm{d}B(t),$$

we then clearly have

$$Y_{n+1} = (Y_n - 1)^+ + K_n \tag{4.1}$$

and it is seen that $\{Y_n\}$ is a Markov chain with transition matrix

$$P = \begin{pmatrix} q_0 & q_1 & q_2 & q_3 & \cdots \\ q_0 & q_1 & q_2 & q_3 & \cdots \\ 0 & q_0 & q_1 & q_2 & \cdots \\ 0 & 0 & q_0 & q_1 & \cdots \\ 0 & 0 & 0 & q_0 & \cdots \\ & & & & \vdots \end{pmatrix}.$$

Irreducibility is obvious since all $q_k > 0$, and also

$$\mathbb{E}K_n = \sum_{k=0}^\infty k q_k = \int_0^\infty \beta t\,\mathrm{d}B(t) = \beta\mu_B = \rho$$

so that $\mathbb{E}(Y_{n+1} | Y_n = i) = \rho + i - 1$ for $i = 1, 2, \ldots$ and it is a matter of routine to check from Foster's criteria I.5 that we have recurrence when $\rho \leqslant 1$ and ergodicity when $\rho < 1$ [when $\rho = 1$, there is in fact null recurrence and when $\rho > 1$, there is transience, cf. Problem 4.1].

Assume in the following that $\rho < 1$. Then the equation $\pi P = \pi$ becomes

$$\left.\begin{aligned} \pi_0 &= \pi_0 q_0 + \pi_1 q_0, \\ \pi_1 &= \pi_0 q_1 + \pi_1 q_1 + \pi_2 q_0, \\ \pi_2 &= \pi_0 q_2 + \pi_1 q_2 + \pi_2 q_1 + \pi_3 q_0. \\ &\vdots \end{aligned}\right\} \tag{4.2}$$

Letting $r_n = q_{n+1} + q_{n+2} + \cdots$, it follows by adding the first n equations and solving for $\pi_{n+1}q_0$ that

$$\left.\begin{aligned} \pi_1 q_0 &= \pi_0 r_0, \\ \pi_2 q_0 &= \pi_0 r_1 + \pi_1 r_1, \\ \pi_3 q_0 &= \pi_0 r_2 + \pi_1 r_2 + \pi_2 r_1. \\ &\vdots \end{aligned}\right\} \tag{4.3}$$

If we sum these equations and note that $\sum r_n = \rho$, we get

$$(1-\pi_0)q_0 = \pi_0\rho + (1-\pi_0)(\rho - r_0)$$

from which it easily follows that $\pi_0 = 1-\rho$. The remaining π_n are then recursively determined by (4.3), but cannot be found in closed formulae. The equations (4.1)–(4.3) contain quite a lot of information on π, however. Let us look at (4.1), which in the limit becomes

$$Y \stackrel{\mathscr{D}}{=} (Y-1)^+ + K = Y - I(Y>0) + K \tag{4.4}$$

(with the obvious notation). Taking squared expectations we get

$$\mathbb{E}Y^2 = \mathbb{E}Y^2 + \mathbb{P}(Y>0) + \mathbb{E}K^2 - 2\mathbb{E}Y + 2\mathbb{E}Y\mathbb{E}K - 2\mathbb{P}(Y>0)\mathbb{E}K,$$

Now $\mathbb{E}K = \rho, \mathbb{P}(Y>0) = 1-\pi_0 = \rho$,

$$\mathbb{E}K^2 = \int_0^\infty \sum_{k=0}^\infty k^2 \mathrm{e}^{-\beta t}\frac{(\beta t)^k}{k!}\mathrm{d}B(t) = \int_0^\infty \{\beta t + \beta^2 t^2\}\mathrm{d}B(t) = \rho + \beta^2\mu_B^{(2)}.$$

Thus

$$\begin{aligned} \mathbb{E}Y &= \frac{1}{2(1-\mathbb{E}K)}\{\mathbb{P}(Y>0)(1-2\mathbb{E}K) + \mathbb{E}K^2\} \\ &= \frac{1}{2(1-\rho)}\{\rho - 2\rho^2 + \rho + \beta^2\mu_B^{(2)}\} = \rho + \frac{\rho^2\mu_B^{(2)}}{2(1-\rho)\mu_B^2} \end{aligned}$$

in accordance with Theorem 2.3(c). Also the generating function $\hat{\pi}(s) = \sum_0^\infty s^n\pi_n = \mathbb{E}s^Y$ can be found in the same way. In fact, (4.4) yields

$$\hat{\pi}(s) = \mathbb{E}s^Y = \mathbb{E}s^{Y-I(Y>0)}\mathbb{E}s^K = (\pi_0 + \pi_1 + s\pi_2 + s^2\pi_3 + \cdots)\sum_{n=0}^\infty s^n q_n,$$

$$s\hat{\pi}(s) = (\hat{\pi}(s) + \pi_0(s-1))\hat{q}(s) = (\hat{\pi}(s) + (1-\rho)(s-1))\hat{q}(s),$$

$$\hat{\pi}(s) = \frac{(1-\rho)(1-s)\hat{q}(s)}{\hat{q}(s)-s} \tag{4.5}$$

where

$$\hat{q}(s) = \int_0^\infty \mathrm{e}^{-\beta t}\sum_{k=0}^\infty \frac{(s\beta t)^k}{k!}\mathrm{d}B(t) = \int_0^\infty \mathrm{e}^{-\beta t(1-s)}\mathrm{d}B(t) = \hat{B}(\beta(1-s)).$$

The equation (4.1) also permits an alternative characterization of π. To this end, let $Z_n = Y_{n+1} - K_n$, $n = 0, 1, 2, \ldots$. Then $Z_0 = 0$,

$$Z_{n+1} = Y_{n+2} - K_{n+1} = (Y_{n+1} - 1)^+ = (Z_n + K_n - 1)^+$$

so that $\{Z_n\}$ is a Lindley process corresponding to the random walk with increments distributed as $K_n - 1$, i.e. with point probabilities $f_{-1} = q_0$, $f_0 = q_1, \ldots$. This random walk is left-continuous with $\mu = \rho - 1 < 0$, so that Theorem 2.1 applies to evaluate the distribution of its maximum M and thereby the limiting distribution of Z_n. But then clearly in the limit $Z \overset{\mathcal{D}}{=} (Y-1)^+$ so that $\pi_{k+1} = \mathbb{P}(M = k)$, $k = 1, 2, \ldots$, and in fact (4.3) and Theorem 2.1(c) are seen to provide essentially the same formulae. Compare also Problem III.7.2.

By a similar method as for $GI/M/1$, we shall next relate π to the distribution of the queue length at an arbitrary point of time and thereby show that in fact they are equal. We first note that the definition of Y_n ensures that $\sigma = \inf\{n \geqslant 1: W_n = 0\}$ coincides with $\inf\{n \geqslant 1: Y_n = 0\}$, and thus $\mathbb{E}\sigma = 1/\pi_0 = 1/(1-\rho)$, $\mathbb{E}\sum_0^{\sigma-1} I(Y_n = k) = \pi_k/\pi_0$. Now let $j \geqslant 1$ be fixed and let I_n be the time Q_t spends in state j while n is being served. Then by the usual formula for the regenerative process $\{Q_t\}$, we have

$$\mathbb{P}(Q = j) = \frac{1}{\mathbb{E}C}\mathbb{E}\int_0^C I(Q_t = j)\mathrm{d}t = \frac{1}{\mathbb{E}C}\mathbb{E}\sum_{n=0}^{\sigma-1} I_n.$$

Now the service period of customer n starts with Y_n customers present for $n = 1, \ldots, \sigma - 1$ and with 1 for $n = 0$. Consider first $n = 1, \ldots, \sigma - 1$. For the queue length to be j, s time units later, and n still being under service, we must have $Y_n = k \in \{1, \ldots, j\}$ (since $n = 1, \ldots, \sigma - 1$ excludes $Y_n = 0$), $U_n > s$ and $j - k$ arrivals in a time interval of length s. Thus for $k = 1, \ldots, j$ we get exactly as in (3.1)

$$\begin{aligned}\mathbb{E}(I_n; \sigma > n \mid Y_n = k, \sigma > n) &= \int_0^\infty \mathrm{e}^{-\beta s}\frac{(\beta s)^{j-k}}{(j-k)!}(1 - B(s))\mathrm{d}s \\ &= \beta^{-1}\int_0^\infty \sum_{l=j-k+1}^{\infty} \mathrm{e}^{-\beta s}\frac{(\beta s)^l}{l!}\mathrm{d}B(s) = \beta^{-1} r_{j-k}.\end{aligned}$$

The similar expression for $n = 0$ becomes $\beta^{-1} r_{j-1}$ and hence

$$\begin{aligned}\mathbb{P}(Q = j) &= \frac{1}{\mathbb{E}C}\left\{\mathbb{E}I_0 + \mathbb{E}\sum_{k=1}^{j}\sum_{n=1}^{\infty} I(Y_n = k, \sigma > n)\beta^{-1} r_{j-k}\right\} \\ &= \frac{1}{\mathbb{E}\sigma}\left\{r_{j-1} + \sum_{k=1}^{j} r_{j-k}\mathbb{E}\sum_{n=1}^{\infty} I(Y_n = k, \sigma > n)\right\} \\ &= \pi_0 r_{j-1} + \pi_1 r_{j-1} + \pi_2 r_{j-2} + \cdots + \pi_j r_0 = \pi_j, \end{aligned} \tag{4.6}$$

the last equality being a consequence of (4.3).

Again, there is a more elegant proof of $\pi_j = \mathbb{P}(Q = j)$ based on up- and downcrossings. The number N_1 of upcrossings of $\{Q_t\}_{0 \leqslant t < C}$ from j to $j+1$ is the same as the number N_2 of downcrossings from $j+1$ to j. But a downcrossing to j

means a departure leaving j customers behind, i.e. a visit of $\{Y_n\}$ to j, $N_2 = \sum_0^{\sigma-1} I(Y_n = j)$, and upcrossings correspond to arrivals in state j which occur at rate β. Hence $\mathbb{E}N_2 = \mathbb{E}N_1$ yields

$$\mathbb{E}\sigma\pi_j = \mathbb{E}\int_0^C \beta I(Q_t = j)\mathrm{d}t = \beta\mathbb{E}C\mathbb{P}(Q = j) = \mathbb{E}\sigma\mathbb{P}(Q = j).$$

The method used for $GI/M/1$ to find the waiting-time distribution does not apply here since we do not know the residual service of the customer facing the server [at least this requires some work, cf. Problem 4.3]. However, there is a direct argument relating the distributions of W and Q: simply observe that the number of customers left behind when n departs are the ones which arrived during his sojourn time $W_n + U_n$. Hence if N is a Poisson process with intensity β which is independent of W, U, we get in the limit

$$Y \overset{\mathscr{D}}{=} N_{W+U} \tag{4.7}$$

From this we get, for example, $\mathbb{E}Y = \beta(\mathbb{E}W + \mathbb{E}U) = \rho + \beta\mathbb{E}W$ in accordance with (2.5) and also the Laplace transform of W can be found. We leave the details to the Problems.

Problems

4.1. Show by a modification of the argument leading to $\pi_0 = 1 - \rho$ from (4.3) that the stationary measure π has $\|\pi\| = \infty$ for $\rho = 1$ and thereby that $\{Y_n\}$ is null recurrent for $\rho = 1$. Show also that there is transience for $\rho > 1$ [hint: $Y_{n+1} \geqslant Y_n - 1 + K_n$].

4.2. Derive $\hat{\pi}(s)$ directly from (4.3) [hint: multiply the nth equation by s^{n+1} and sum over $n = 0, 1, 2, \ldots$]. Check the formula for the mean by differentiation.

4.3. Let R_t denote the service being attained by the customer facing the server at time t and $D^{(n)}(x) = \mathbb{P}(R \leqslant x,\ Q = n)$, $n = 1, 2, \ldots$ (thus $\|D^{(n)}\| = \pi_n$). Show, using regenerative processes, that

$$\frac{\mathrm{d}D^{(n)}(x)}{\mathrm{d}x} = \beta(1 - B(x))\mathrm{e}^{-\beta t}\left\{(\pi_0 + \pi_1)\frac{(\beta t)^{n-1}}{(n-1)!} + \sum_{k=0}^{n-2} \pi_{n-k}\frac{(\beta t)^k}{k!}\right\}.$$

4.4. Use (4.7) and (4.5) to rederive the Pollaczek–Khintchine formula for the Laplace transform of W in Problem 2.1.

5. MORE ON LATTICE DISTRIBUTIONS

We proceed here and in Section 6 to present some further cases (close to being all known ones) where the random walk problems can be solved quite explicitly.

We consider throughout the typical case in queueing theory, $\mu < 0$ (the case $\mu > 0$ is essentially symmetric, whereas some modifications may be needed for $\mu = 0$). We shall in some cases be satisfied with evaluating either G_+ or G_- since it is then obvious how to proceed for, say, the distribution of the maximum.

The idea is as in Sections 1 and 2 to recognize the form of one of the ladder height distributions by a probabilistic argument. The remaining unknowns are then constants, which can be found analytically (though in a rather more sophisticated manner than in Sections 1 and 2) from the transform version (1.3) of the Wiener–Hopf factorization identity. More precisely the arguments proceed by relations between the roots of the equations $\hat{F}(s)=1$, $\hat{G}_+(s)=1$, $\hat{G}_-(s)=1$ provided by (1.3). Here we shall say that the equation $\varphi(s)=0$ (where $\varphi: D\to\mathbb{C}$ is a continuous function on a complex domain $D\subseteq\mathbb{C}$) has the roots $\alpha_1,\ldots,\alpha_r$ in $D_1\subseteq D$ if $\varphi(z)=(z-\alpha_1)\cdots(z-\alpha_r)\psi(z)$ with $\alpha_1,\ldots,\alpha_r\in D_1$ and $\psi(z)$ continuous and non-zero for $z\in D_1$. Note that some α_i may coincide, corresponding to multiple roots.

In the present section, F is taken to be lattice, without loss of generality aperiodic on $\mathbb{Z}$ with point probabilities $f_k=\mathbb{P}(X=k)$. Then also G_+, G_- are concentrated on $\mathbb{Z}$ and the relevant transforms are the generating functions $\hat{F}(s)=\sum_{-\infty}^{\infty}s^k f_k$, etc. These are always defined for $|s|=1$, but some may have larger domains. For example, $G_+(s)$ is also defined for s in the closed unit disc $\Delta=\{s\in\mathbb{C}:|s|\leqslant 1\}$ and $\hat{G}_-(s)$ for $s^{-1}\in\Delta$. The situations that we are able to deal with extend the right- and left-continuous case by assuming that the support of F is unbounded at most in one direction. The simplest case is boundedness to the right:

Theorem 5.1. *If for some $r>0$, $f_r>0$ and $f_{r+s}=0$, $s=1,2,\ldots$, then the equation $\hat{F}(s)=1$ has exactly r roots $\alpha_1,\ldots,\alpha_r\in\mathbb{C}\backslash\Delta$ outside the unit circle, which determine G_+ by means of*

$$1-\hat{G}_+(s)=\left(1-\frac{s}{\alpha_1}\right)\cdots\left(1-\frac{s}{\alpha_r}\right). \tag{5.1}$$

Proof. Clearly G_+ is concentrated on $\{1,\ldots,r\}$ with $G_+(\{r\})\geqslant f_r>0$. Thus $G_+(s)$ is a polynomial of degree r with $\hat{G}_+(0)=0$ and we may write

$$1-\hat{G}_+(s)=\left(1-\frac{s}{\alpha_1}\right)\cdots\left(1-\frac{s}{\alpha_r}\right).$$

From $\|G_+\|<1$ it follows that $|\hat{G}_+(s)|<1$ for $s\in\Delta$ and thus $\alpha_i\notin\Delta$, $i=1,\ldots,r$. Also for $s\notin\Delta$ we have $|\hat{G}_-(s)|<1$, and (1.3) then shows that $\alpha_1,\ldots,\alpha_r$ are exactly the roots of $\hat{F}(s)=1$ in $\mathbb{C}\backslash\Delta$. □

The case of boundedness to the left is only slightly more complicated:

Theorem 5.2. *Suppose that for some $r>0$, $f_{-r}>0$ and $f_{-r-s}=0$, $s=1,2,\ldots$ Then the equation $\hat{F}(s)=1$ has exactly r roots $\alpha_1,\ldots,\alpha_r$ in Δ, one of which is 1 and the rest are within the unit circle, say $\alpha_r=1$, $\alpha_1,\ldots,\alpha_{r-1}\in\operatorname{int}\Delta$. These roots determine $\hat{G}_-$ by means of*

$$1-\hat{G}_-(s)=C\left(1-\frac{1}{s}\right)\left(1-\frac{\alpha_1}{s}\right)\cdots\left(1-\frac{\alpha_{r-1}}{s}\right),\qquad s\neq 0, \tag{5.2}$$

$$C=\frac{(-1)^{r+1}f_{-r}}{\alpha_1\cdots\alpha_{r-1}} \tag{5.3}$$

Proof. Clearly G_- is concentrated on $\{0, -1, \ldots, -r\}$ with $G_-(\{-r\}) = f_{-r}$. Thus $\hat{G}_-(s)$ is a polynomial of degree r in $1/s$ with coefficient f_{-r} to s^{-r} and we may write

$$1 - \hat{G}_-(s) = \frac{(-1)^{r+1} f_{-r}}{\alpha_1 \cdots \alpha_r}\left(1 - \frac{\alpha_1}{s}\right)\cdots\left(1 - \frac{\alpha_r}{s}\right).$$

Again, $|\hat{G}_-(s)| < 1$ for $s \notin \Delta$ and hence $\alpha_1, \ldots, \alpha_r \in \Delta$, and similarly $\hat{G}_+(s) \neq 1$, $s \in \Delta$, because of $\| G_+ \| < 1$. Hence $\alpha_1, \ldots, \alpha_r$ are indeed the roots in Δ of $\hat{F}(s) = 1$, and clearly one root is $s = 1$. Here $s = 1$ cannot be a double root, since then $1 - \hat{F}(s) = (1-s)^2 \psi(s)$ with ψ continuous at $s = 1$ would imply $\hat{F}'(1) = 0$, contradicting $\mu < 0$. Also $|\alpha_i| = 1$, $\alpha_i \neq 1$, is impossible since then $\hat{G}_-(\alpha_i) = 1$ would imply G_- to be periodic which is impossible unless F is so, cf. VII.1.3. □

Example 5.3. Suppose $X = Y - 2$, where Y is geometric with parameter $\frac{1}{2}$, i.e. $f_j = 2^{-j-3}$, $j = -2, -1, 0, \ldots$, $f_j = 0$, $j < -2$,

$$\hat{F}(s) = s^{-2}\mathbb{E}s^Y = \frac{1}{s^2(2-s)}.$$

Here $\mu = -1 < 0$ and $r = -2$ in Theorem 5.2, and the equation $\hat{F}(s) = 1$ becomes $s^3 - 2s^2 + 1 = 0$. Obviously $s = 1$ is root and since $s^3 - 2s^2 + 1 = (s-1)(s^2 - s - 1)$, the remaining ones are $\alpha_\pm = (1 \pm \sqrt{5})/2$, where $\alpha_+ \notin \Delta$, $\alpha_- \in \Delta$. Thus $\alpha_1 = \alpha_-$,

$$1 - \hat{G}_-(s) = \frac{(-1)^3 \frac{1}{2}}{1 \cdot \alpha_-}\left(1 - \frac{1}{s}\right)\left(1 - \frac{\alpha_-}{s}\right),$$

$$\hat{G}_-(s) = \frac{1}{2s^2} + \frac{3 - \sqrt{5}}{2(\sqrt{5}-1)s} + \frac{\sqrt{5}-2}{\sqrt{5}-1},$$

corresponding to

$$\hat{G}_-(\{-2\}) = \tfrac{1}{2} (= f_{-2}), \qquad \hat{G}_-(\{-1\}) = \frac{3-\sqrt{5}}{2(\sqrt{5}-1)}, \qquad \hat{G}_-(\{0\}) = \frac{\sqrt{5}-2}{\sqrt{5}-1}.$$

Also

$$1 - \hat{G}_+(s) = \frac{1 - \hat{F}(s)}{1 - \hat{G}_-(s)} = \frac{-(s^3 - 2s^2 + 1)/s^2(2-s)}{C(1 - 1/s)(1 - \alpha_-/s)}$$

$$= \frac{-(s-1)(s-\alpha_+)(s-\alpha_-)}{C(2-s)(s-1)(s-\alpha_-)} = \frac{\alpha_+ - s}{C(2-s)},$$

$$\hat{G}_+(s) = \frac{2C - Cs - \alpha_+ + s}{C(2-s)} = \frac{1-C}{C}\frac{s}{2-s} = (\sqrt{5}-2)\frac{s}{2-s}$$

which means that G_+ is defectively geometric with parameter $\frac{1}{2}$ on $\{1, 2, \ldots\}$ and $\| G_+ \| = \sqrt{5} - 2$. It is illustrative to check with the alternative solution provided by Problem 1.3, which yields the same form of G_+ but with $\|G_+\| = (1-\gamma/2)/\gamma(1-\frac{1}{2})$, where γ is the unique solution >1 of $\hat{F}(s)=1$, i.e. $\gamma = \alpha_+$, so that indeed

$$\|G_+\| = \frac{1 - \dfrac{1+\sqrt{5}}{4}}{\dfrac{1+\sqrt{5}}{2}(1-\frac{1}{2})} = \frac{3-\sqrt{5}}{1+\sqrt{5}} = \sqrt{5}-2. \qquad \square$$

Problems

5.1. Express in Theorem 5.1 the point probabilities of G_+ in terms of the elementary symmetric functions of the α_i^{-1}.

5.2. Show by direct inspection of the shape of $\hat{F}(s)$ that for $r=2$ in Theorem 5.1 α_1, α_2 are both real and of opposite sign.

Notes

See next section.

6. PHASE-TYPE DISTRIBUTIONS

We shall next generalize the results of Sections 1 and 2 for $GI/M/1$ and $M/G/1$ to the queueing situation $X = U - T$, where either the service time U or the interarrival time T is of phase type. More precisely, we consider the case of mixtures of Erlang distributions, but start with some more general discussion of closedness properties of the class $\mathscr{PH}_{\text{AT}}$ under overshot formation (these observations are of key importance not only here but also for, say, finding explicit examples in renewal theory, cf. Problems 6.1 and 6.2).

Consider the distribution $H(t) = H_\pi(t) = \mathbb{P}_\pi(\tau \leqslant t)$ of the time $\tau = \inf\{t \geqslant 0: Y_t = \Delta\}$ to absorption in a Markov jump process $\{Y_t\}$ which moves on $E \cup \{\Delta\}$ and has initial distribution π concentrated on E (if $\pi_i = 1$, we write $H = H_i$ and thus $H_\pi = \sum_{i\in E}\pi_i H_i$).

Lemma 6.1. (i) *The overshot distribution* $H^{(t)}(s) = H(t+s)/(1-H(t))$ *is obtained by changing* π *to* v *given by* $v_i = b_i(t) = P_\pi(Y_t = i \mid \tau > t)$; (ii) *if* $X = U - T$ *with* U *having distribution* H, *then the conditional distribution of* X *given* $X > 0$ *is of the form* H_λ *for some* λ, *i.e.*

$$\frac{\mathrm{d}F(x)}{\mathrm{d}x} = c\frac{\mathrm{d}H_\lambda(x)}{\mathrm{d}x}, \qquad x>0 \qquad (c = \mathbb{P}(X>0)); \tag{6.1}$$

(iii) *if in a random walk problem* (6.1) *holds, then the ascending ladder height distribution* G_+ *is of the form* $G_+ = \|G_+\| H_\omega$ *for some* ω.

Proof. (i) is immediately apparent from

$$\begin{aligned} H^{(t)}(s) &= \mathbb{P}_\pi(\tau \leqslant t+s \mid \tau > t) \\ &= \sum_{i\in E} b_i(t)\mathbb{P}_\pi(\tau \leqslant t+s \mid \tau > t, Y_t = i) = \sum_{i\in E} b_i(t) H_i(s) = H_v(s), \end{aligned}$$

and (ii) follows with $\lambda_i = \int_0^\infty b_i(t)\mathrm{d}A(t)$ from

$$\begin{aligned}\mathbb{P}(X \leqslant s | X > 0) &= \mathbb{P}(U \leqslant T + s | U > T)\\ &= \int_0^\infty \mathbb{P}(U \leqslant t + s | U > t)\mathrm{d}A(t)\\ &= \sum_{i\in E} \int_0^\infty b_i(t)H_i(s)\mathrm{d}A(t) = H_\lambda(s).\end{aligned}$$

The explanation for (iii) is that for getting from $(-\infty, 0]$ to $(0, \infty)$, the random variable S_{τ_+} has to be drawn from some $H^{(t)}$ and thus by (i) G_+ is a mixture of the H_i. More precisely, with $c_i(t) = \mathbb{P}_\lambda(Y_t = i | \tau > t)$ we get

$$\begin{aligned}\mathbb{P}(\tau_+ = n, S_{\tau_+} \leqslant x) &= \mathbb{P}(-S_{n-1} < X_n \leqslant x - S_{n-1}, S_1 \leqslant 0, \ldots, S_{n-1} \leqslant 0)\\ &= \mathbb{E}\left[\sum_{i\in E} c_i(-S_{n-1})H_i(x); S_1 \leqslant 0, \ldots, S_{n-1} \leqslant 0, S_n > 0\right]\\ &= \sum_{i\in E} H_i(x)\mathbb{E}[c_i(-S_{n-1}); \tau_+ = n],\end{aligned}$$

$$G_+(x) = \mathbb{P}(S_{\tau_+} \leqslant x; \tau_+ < \infty) = \| G_+ \| H_\omega(x)$$

where

$$\omega_i = \mathbb{E}[c_i(-S_{\tau_+-1}) | \tau_+ < \infty]. \tag{6.2}$$

□

We now specialize to mixtures of Erlang distributions, say K parallel channels with $j \leqslant n(i)$ phases and intensity α_i at the ijth channel. Thus in the above notation, $H_{ij}, H_\pi, H_\lambda, H_\omega, G_+$, etc. are all within the class of distributions with a moment generating function (m.g.f.) of the form

$$\hat{H}(s) = \sum_{i=1}^{k} \sum_{j=1}^{n(i)} \beta_{ij}(H)\left(\frac{\alpha_i}{\alpha_i - s}\right)^j, \qquad \sum_{i,j} \beta_{ij}(H) = \| H \|. \tag{6.3}$$

It is obviously no restriction in the following to assume that no two α_i are equal. Though the m.g.f. $\int_0^\infty \mathrm{e}^{sx}\,\mathrm{d}H(x)$ is defined only for $\mathrm{Re}\, s < \alpha_0 = \min(\alpha_1, \ldots, \alpha_k)$, the r.h.s. of (6.3) is then well defined and analytic whenever $s \notin \{\alpha_1, \ldots, \alpha_k\}$. We also denote this extension by $\hat{H}$. Similarly, for $\mathrm{Re}\, s > 0$ and $s \notin \{\alpha_1, \ldots, \alpha_k\}$ we write

$$\hat{F}(s) = \int_{-\infty}^0 \mathrm{e}^{sx}\,\mathrm{d}F(x) + c\hat{H}_\lambda(s).$$

Lemma 6.1(iii) determines G_+ as a mixture of Erlang distribution, except that the weights (the $\beta_{ij}(G_+)$ in (6.3)) seem exceedingly difficult to evaluate by means of (6.2). However, the weights are determined by the form of $\hat{G}_+$. Letting $R(s) = \prod_1^k (\alpha_i - s)^{n(i)}$, it follows from (6.3) that we may write $1 - \hat{G}_+(s) = P(s)/R(s)$, where $P(s)$ is a polynomial, and we have:

Theorem 6.2. *Assume that the right tail of F is a mixture of Erlang distribution in the sense of* (6.1), *that all $\beta_{in(i)}(H_\lambda) > 0$ and that $\mu < 0$. Then G_+ is also a* (defective) *mixture of Erlang distributions. Furthermore, the equation* $\hat{F}(s) = 1$ *has exactly* $n = n(1) + \cdots + n(k)$ *roots* $\rho_1, \ldots, \rho_n$ *in* $\Omega = \{s \in \mathbb{C} \setminus \{\alpha_1, \ldots, \alpha_k\} : \mathrm{Re}\, s > 0\}$, *and these determine* G_+ *by means of*

$$1 - \hat{G}_+(s) = \frac{(\rho_1 - s) \cdots (\rho_n - s)}{(\alpha_1 - s)^{n(1)} \cdots (\alpha_k - s)^{n(k)}} \tag{6.4}$$

Proof. As $s \to -\infty$, we have $\hat{G}_+(s) \to 0$ and $R(s) \approx (-s)^n$. Therefore (in the above notation) $P(s)$ is of degree n with coefficient $(-1)^n$ to s^n. That is, $P(s) = (\rho_1 - s) \cdots (\rho_n - s)$ for some $\rho_1, \ldots, \rho_n$. Then by (1.3),

$$1 - \hat{F}(s) = \frac{(\rho_1 - s) \cdots (\rho_n - s)}{R(s)} (1 - \hat{G}_-(s)) \tag{6.5}$$

for $0 \leqslant \mathrm{Re}\, s < \alpha_0$ and hence by analytic continuation also for $\mathrm{Re}\, s > 0$, $s \notin \{\alpha_1, \ldots, \alpha_k\}$.

Now $\mathrm{Re}\, \rho_l \leqslant 0$ would imply $\int_0^\infty e^{\rho_l s} \mathrm{d}G_+(s) = 1$ which is impossible because of $\| G_+ \| < 1$. Also we cannot have $\rho_l \in \{\alpha_1, \ldots, \alpha_k\}$ since then for some i the pole of $1 - \hat{G}_+(s)$ at $s - \alpha_i$ is of order $\leqslant n(i) - 1$ which is excluded by $\beta_{in(i)}(G_+) > 0$. Hence indeed $\rho_l \in \Omega$. Furthermore, $G_-(s) \neq 1$, $s \in \Omega$, and hence (6.5) shows that $\rho_1, \ldots, \rho_n$ are exactly the roots of $\hat{F}(s) = 1$ in Ω. □

Corollary 6.3. *Consider a stable GI/PH/1 queue,* $X = U - T$ *with* $\mathbb{E}U < \mathbb{E}T$ *and the distribution B of the service time U a mixture of Erlang distributions such that* (6.3) *holds for B with all* $\beta_{in(i)}(B) > 0$. *Then the equation* $\hat{A}(-s)\hat{B}(s) = 1$ *has exactly* $n = n(1) + \cdots + n(k)$ *roots* $\rho_1, \ldots, \rho_n$ *in* $\Omega = \{s \in \mathbb{C} \setminus \{\alpha_1, \ldots, \alpha_k\}$: $\mathrm{Re}\ s > 0\}$ *and these determine the actual waiting-time distribution by means of*

$$\mathbb{E}\, e^{sW} = \frac{\rho_1 \cdots \rho_n}{\alpha_1^{n(1)} \cdots {}_k^{n(k)}} \frac{(\alpha_1 - s)^{n(1)} \cdots (\alpha_k - s)^{n(k)}}{(\rho_1 - s) \cdots (\rho_n - s)},$$

$\mathrm{Re}\ s < \alpha_0$. *In particular,*

$$\mathbb{E}W = \sum_{l=1}^{n} \frac{1}{\rho_l} - \sum_{i=1}^{k} \frac{n(i)}{\alpha_i}$$

Proof. By Lemma 6.1(ii), we may apply Theorem 6.2. Interpreting again $\hat{B}(s)$ as the analytic continuation of the m.g.f. of B, it is clear that $\hat{A}(-s)\hat{B}(s)$ is an analytic continuation of the m.g.f. of F to Ω so that the equations $\hat{F}(s) = 1$ and $\hat{A}(-s)\hat{B}(s) = 1$ are the same. Thus G_+ is given by Theorem 6.2. In particular,

$$1 - \| G_+ \| = 1 - \hat{G}_+(0) = \frac{\rho_1 \cdots \rho_n}{\alpha_1^{n(1)} \cdots \alpha_k^{n(k)}},$$

$$\mathbb{E}\, e^{sW} = (1 - \| G_+ \|) \sum_{n=0}^{\infty} \hat{G}_+(s)^n = \frac{1 - \| G_+ \|}{1 - \hat{G}_+(s)}$$

which is the same as the asserted expression. The expression for $\mathbb{E}W$ follows easily by differentiating and letting $s = 0$. □

Example 6.4. $GI/H_k/1$. This corresponds to $n(1) = \cdots = n(k) = 1$, $n = k$, $\hat{B}(s) = \sum_1^k \beta_i(B)\alpha_i/(\alpha_i - s)$. We can assume $\alpha_1 < \cdots < \alpha_k$, and considered as functions of a real argument $s > 0$, $\hat{A}(-s)^{-1}$ and $\hat{B}(s)$ then have the shape illustrated in Fig. 6.1. The slopes at $s = 0$ are μ_A and μ_B. In the stable case, $\mu_A > \mu_B$ and since $\hat{B}(s) \to \infty$, $s \uparrow \alpha_1$, the equation $\hat{A}(-s)\hat{B}(s) = 1$ has a root $\rho_1 \in (0, \alpha_1)$. Also $\hat{B}(s) \to \infty$, $s \downarrow \alpha_{i-1}$, and $\hat{B}(s) \to \infty$, $s \uparrow \alpha_i$, and hence there is at least one root $\rho_i \in (\alpha_{i-1}, \alpha_i)$. This gives a total of $k = n$ roots and Corollary 6.3 tells us that we need not look n either for further real roots n or for complex roots. □

Example 6.5. $GI/E_n/1$. This corresponds to $k = 1$, $\hat{B}(s) = \alpha^n/(\alpha - s)^n$. The shape of the figure corresponding to Fig. 6.1 depends on whether n is even or uneven, the two possibilities being depicted in Figs. 6.2(a) and (b) respectively. It is seen that for n even there is one real root $\rho_1 \in (0, \alpha)$, one real root $\rho_2 > \alpha$ and $n - 2$ complex roots $\rho_3, \ldots, \rho_n$ (which falls into a pair of complex conjugates). For n uneven, there is only one real root $\rho_1 \in (0, \alpha)$ and $n - 1$ complex roots $\rho_2, \ldots, \rho_n$. □

The case of the left tail of F being of phase type is very similar. We omit the random walk formulation and pass right on to queues:

Theorem 6.6. *Consider a stable $PH/G/1$ queue with the distribution A of the interarrival time T being a mixture of Erlang distributions such that* (6.3) *holds for A with all $\beta_{in(i)}(A) > 0$. Then the equation*

$$1 = \hat{F}(s) = \hat{A}(-s)\hat{B}(s) = \sum_{i=1}^{k} \sum_{j=1}^{n(i)} \beta_{ij}(A)\left(\frac{\alpha_i}{\alpha_i + s}\right)^j \cdot \mathbb{E}\, e^{sU}$$

has exactly $n - 1$ roots $\rho_1, \ldots, \rho_{n-1}$ in the domain $\Omega = \{s \in \mathbb{C} \setminus \{-\alpha_1, \ldots, -\alpha_k\}$:

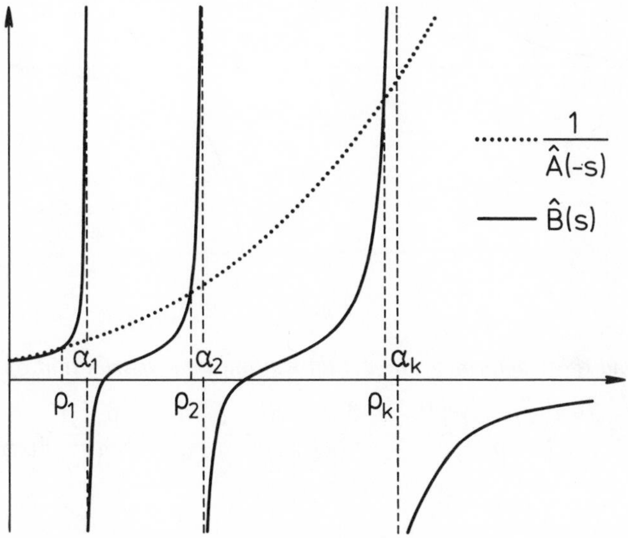

Fig. 6.1

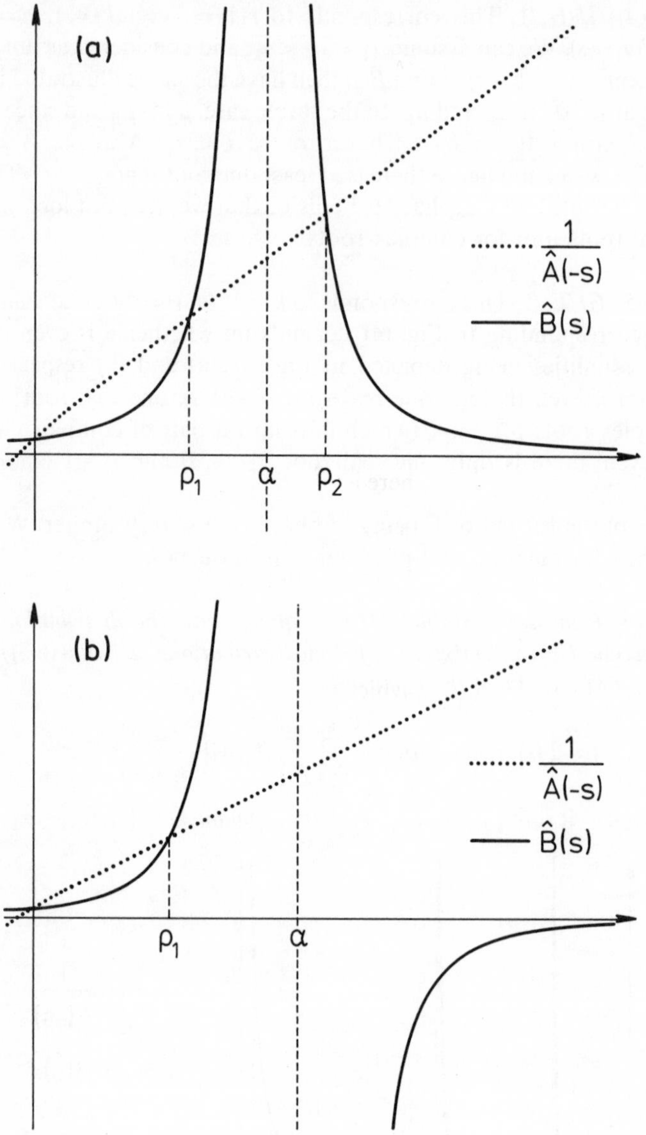

Fig. 6.2

Re $s<0\}$, *and these determine the actual waiting-time distribution by means of*

$$\mathbb{E}\,e^{sW} = \frac{(\mu_A-\mu_B)\alpha_1^{n(1)}\cdots\alpha_k^{n(k)}}{(-1)^{n-1}\rho_1\cdots\rho_{n-1}}\frac{s(s-\rho_1)\cdots(s-\rho_{n-1})}{(\alpha_1+s)^{n(1)}\cdots(\alpha_k+s)^{n(k)}(1-\hat{F}(s))},$$

Re $s<0$.

Proof. Exactly as above, we may conclude that the distribution of $-S_{\tau_-}$ is of the form (6.3) with all $\beta_{ij}>0$. Letting $R(s)=\Pi_1^k(\alpha_i+s)^{n(i)}$, this implies that for the m.g.f. we have $1-G_-(s)=P(s)/R(s)$, where P is a polynomial of degree exactly n with $P(-\alpha_i)\neq 0$, $i=1,\ldots,k$. Because of $\|G_-\|=1$, one root is $s=0$ and because of $\hat{G}_-(s)\to 0$, Re $s\to\infty$, $P(s)\approx R(s)\approx s^n$ so that we may write $P(s)=s(s-\rho_1)\cdots(s-\rho_{n-1})$, where $\rho_1,\ldots,\rho_{n-1}\in\mathbb{C}\backslash\{-\alpha_1,\ldots,-\alpha_n\}$. Also $\operatorname{Re}\rho_i\geqslant 0$ is excluded since then $\int_{-\infty}^0 e^{\rho_i x}\,dG_-(x)=1$ yields a contradiction. Hence $\rho_i\in\Omega$, and by (1.3) and analytic continuation,

$$1-\hat{F}(s)=(1-\hat{G}_+(s))\frac{s(s-\rho_1)\cdots(s-\rho_{n-1})}{R(s)}.$$

For $s\in\Omega, |\hat{G}_+(s)|<1$ and hence $\rho_1,\ldots,\rho_{n-1}$ are exactly the roots of $\hat{F}(s)=1$ in Ω.

It follows that for $\operatorname{Re} s<0$

$$\mathbb{E}\,e^{sW}=(1-\|G_+\|)\frac{1}{1-\hat{G}_+(s)}=(1-\|G_+\|)\frac{s(s-\rho_1)\cdots(s-\rho_{n-1})}{R(s)(1-\hat{F}(s))}.$$

Here $1-\|G_+\|$ can be evaluated (Problem 6.5), but is more easily determined by $\mathbb{E}\,e^{sW}\to 1$, $s\uparrow 0$ which yields

$$(1-\|G\|_+)\frac{(-\rho_1)\cdots(-\rho_{n-1})}{R(0)(\mu_A-\mu_B)}=1$$

and the proof is complete. □

Example 6.7. $H_k/GI/1$ Here $n(1)=\cdots=n(k)=1$, $n=k$, $\hat{A}(-s)=\sum_1^k\beta_i(A)\alpha_i/(\alpha_i+s)$. The situation is depicted in Fig. 6.3. The slopes of $\hat{A}(-s)$, $\hat{B}(s)^{-1}$ at $s=0$ are

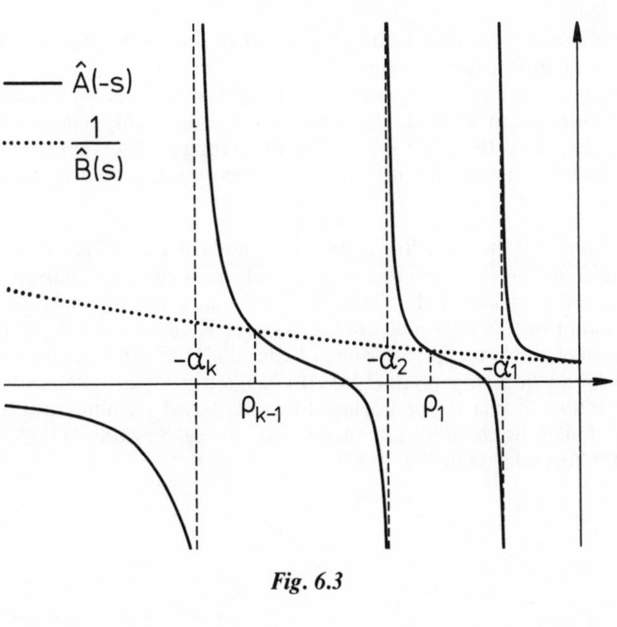

Fig. 6.3

$-\mu_A$ and $-\mu_B$ respectively, and the roots are $\rho_i \in (-\alpha_{i+1}, -\alpha_i)$, $i = 1, \dots n-1$. □

Problems

6.1. Consider a renewal process with interarrival distribution of H_2 type,

$$\frac{\mathrm{d}F(x)}{\mathrm{d}x} = \beta(F)\alpha_1 \mathrm{e}^{-\alpha_1 x} + (1-\beta(F))\alpha_2 \mathrm{e}^{-\alpha_2 x}.$$

Show that the distribution G_t of the forwards recurrence time B_t is again of the same type with $f(t) = \beta(G_t)$ satisfying the differential equation

$$f'(t) = \varepsilon - (\varepsilon + \eta) f(t), \qquad \varepsilon = \beta(F)\alpha_2, \qquad \eta = (1-\beta(F))\alpha_1.$$

Solve for f using $f(0) = \beta(F)$ and check that $\lim_{t\to\infty} f(t)$ is in agreement with $\mathrm{d}G_\infty/\mathrm{d}x = (1-F(x))/\mu$. Show finally that the renewal density is

$$u(t) = \frac{\partial}{\partial x} G_t(x)|_{x=0} = \alpha_1 f(t) + \alpha_2(1-f(t))$$

$$= \frac{1}{\mu} + \frac{\beta(F)(1-\beta(F))(\alpha_1-\alpha_2)^2}{\varepsilon+\eta} \mathrm{e}^{-(\varepsilon+\eta)t}.$$

6.2. Consider a renewal process with E_2 interarrival distribution, $\mathrm{d}F/\mathrm{d}x = \alpha^2 x \mathrm{e}^{-\alpha x}$. Show that G_t is of the form

$$\frac{\mathrm{d}G_t}{\mathrm{d}x} = f(t)\alpha \mathrm{e}^{-\alpha x} + (1-f(t))\alpha^2 x \mathrm{e}^{-\alpha x}$$

with f satisfying the differential equation $f'(t) = \alpha - 2\alpha f(t)$. Solve for f using $f(0) = 0$ and check that $\lim_{t\to\infty} f(t)$ is in agreement with $\mathrm{d}G_\infty/\mathrm{d}x = (1-F(x))/\mu$. Show finally that the renewal density is

$$u(t) = \alpha f(t) = \frac{\alpha}{2}(1-\mathrm{e}^{-2\alpha x}) = \frac{1}{\mu} - \frac{\alpha}{2}\mathrm{e}^{-2\alpha x}$$

6.3. Show in Corollary 6.3 that the distribution of W is in $\mathscr{PH}_{\mathrm{AT}}$ [hint: $\mathscr{PH}_{\mathrm{AT}}$ is closed under the formation of geometric sums].

6.4. In Example 6.5, it was tacitly assumed that the real roots ρ_1, ρ_2 cannot be multiple. Show this in more detail. Explain also why there is indeed only one root in $(0, \alpha_1)$.

6.5. Evaluate $\mathbb{E}S_{\tau_-}$ and thereby $1 - \|G_+\|$ in the setting of Theorem 6.6.

6.6. Explain in more detail why in Example 6.7 there cannot be real roots in $(-\alpha_1, 0)$.

Notes

The results of this and the preceding section are standard, but the proofs avoid the usual key step of reference to Rouché's theorem, a result from complex analysis which allows the counting and location of the roots of $\hat{F}(s) = 1$ in a purely analytical way. For a thorough account of this approach see for example Kleinrock (1975) or Takács (1962) (there is also an example on pp. 407–408 of Feller, 1971). Further results along the same lines can be found in Kemperman (1961). In X.5, we study an alternative method for numerical solution of phase-type queues. Methods based on numerical integration of characteristic functions could also be mentioned, see e.g. Siegmund (1985, 8.51, X.4) (an example of the formulae is in XII. (6.12)).

CHAPTER X

Multidimensional Methods

1. NON-NEGATIVE MATRICES

Finite square matrices with non-negative elements occur in a variety of contexts in applied probability. The so-called *Perron–Frobenius theory* of such matrices describes in quite some detail their spectral properties and therefore also the asymptotic properties of their powers, and is therefore a powerful and indispensable tool for many applications. We shall develop this theory here by exploiting the intimate connection to Markov chains with a finite number of states. Applications will be touched upon in the next sections.

We start by recalling some facts from linear algebra. Let A be any $p \times p$ matrix and define for $\lambda \in \mathbb{C}$ $E_\lambda = \{x \in \mathbb{C}^p : x \neq 0,\ Ax = \lambda x\}$. Thus $\mathrm{sp}(A) = \{\lambda : E_\lambda \neq \varnothing\}$ is the set of eigenvalues of A or the *spectrum* of A, and $\mathrm{spr}(A) = \sup\{|\lambda| : \lambda \in \mathrm{sp}(A)\}$ is the *spectral radius* of A. If $\lambda \in \mathrm{sp}(A)$, then λ is a root in the characteristic polynomial $\det(A - \lambda I)$, and if the multiplicity is 1, we call λ *simple*. Then also the geometric multiplicity $\dim(E_\lambda \cup \{0\})$ is 1, i.e. the eigenvector is unique up to a constant. If $\lambda \in \mathrm{sp}(A)$, then λ is also eigenvalue for the transposed matrix A^{T}. The existence of an eigenvector for A^{T} may then be written as $vA = \lambda v$ for some row vector v, called a *left eigenvector* for A ($x \in E_\lambda$ is a *right eigenvector*). The following lemma is standard (all statements are easy to verify if one writes A on the Jordan canonical form):

Lemma 1.1. (i) $\mathrm{sp}(A^m) = \{\lambda^m : \lambda \in \mathrm{sp}(A)\}$; (ii) *the A^m-multiplicity of $\lambda \in \mathrm{sp}(A^m)$ is the sum of the A-multiplicities of the $\lambda_i \in \mathrm{sp}(A)$ with $\lambda_i^m = \lambda$;* (iii) *if $\lambda \in \mathrm{sp}(A)$ is not simple, then either* $\dim E_\lambda \cup \{0\} > 1$ *or for any $h \in E_\lambda$ we can find k with $Ak = h + \lambda k$;* (iv) $A^n = \mathrm{O}(n^k[\mathrm{spr}(A)]^n)$ *for some* $k = 0, 1, 2, \ldots$.

We start by examining the spectral properties of ergodic transition matrices:

Proposition 1.2. *Let $P = (p_{ij})_{i,j=1,\ldots,p}$ be an ergodic $p \times p$ transition matrix with stationary distribution π. Then $\mathrm{spr}(P) = 1$ and 1 is a simple eigenvalue of P with π and $e = (1 \cdots 1)^{\mathrm{T}}$ as corresponding left and right eigenvectors. Furthermore for $\lambda \in \mathrm{sp}(P)$, $\lambda \neq 1$, we have $|\lambda| < 1$ and with $\lambda_1 = \max\{|\lambda| : \lambda \in \mathrm{sp}(P),\ \lambda \neq 1\}$ it holds that for some k the powers $P^n = (p_{ij}^n)$ satisfy*

$$p_{ij}^n = \pi_j + \mathrm{O}(n^k \lambda_1^n), \qquad n \to \infty. \tag{1.1}$$

Proof. It is clear that $\pi P = \pi, Pe = e$ and hence $1 \in \mathrm{sp}(P)$. Also $h \in E_1$ means that h is harmonic and thus $h = ce$ (cf. I.5.1; the extension to the complex case is easy). Thus if 1 is not simple, Lemma 1.1(iii) shows that we can find k with $Pk = e + k$. But then $P^n k = ne + k$ which in Markov chain terms means that $\mathbb{E}_i k_{X(n)} = n + k_i$. This is impossible since k has only a finite number of coordinates. Similarly, the ergodic theorem means that $P^n \to e\pi$ and hence if $\lambda \in \mathrm{sp}(P)$, $k \in E_\lambda$, we have $\lambda^n k = P^n k \to e\pi k$. But $\lambda^n k$ can only converge if $|\lambda| < 1$ or $\lambda = 1$.

It only remains to prove (1.1). Write $P = P_1 + P_2$ with $P_1 = e\pi$, $P_2 = P - e\pi$. It is then readily checked that $P_1 P_2 = P_2 P_1 = 0$ and hence $P^n = P_1^n + P_2^n$. It is also easily seen that $P_1^n = P_1 = e\pi$. Hence by Lemma 1.1(iv) it suffices to show that if $\lambda \in \mathrm{sp}(P_2)\backslash\{0\}$, then $|\lambda| \leqslant \lambda_1$. But from $P_2 k = \lambda k$ we get $Pk = (P_1 + P_2)\lambda^{-1} P_2 k = \lambda k$, i.e. $\lambda \in \mathrm{sp}(P)$. If $\lambda = 1$, we would have $k = ce$ and hence $P_2 k = 0$ which is impossible. Hence $|\lambda| \leqslant \lambda_1$. □

If $\lambda_1 < \delta < 1$, (1.1) may be rewritten as $p_{ij}^n = \pi_j + \mathrm{O}(\delta^n)$. Hence the ergodic steady state is approached at a geometric rate, and we have obtained a second proof of a result of VI.2,

Corollary 1.3. *Any irreducible aperiodic Markov chain with a finite number of states is geometrically ergodic.*

We shall now derive a close analogue of Proposition 1.2 for matrices A which are *non-negative*, i.e. $a_{ij} \geqslant 0$ for all i, j. We shall adopt the definitions of irreducibility and the period from transition matrices to non-negative matrices by noting that they depend only on the pattern of entries i, j with $a_{ij} = 0$. Thus A is *irreducible* if for any i, j we can find m such that $a_{ij}^m > 0$, and we have:

Lemma 1.4. *If A is an irreducible non-negative matrix, then the greatest common divisor d of the m with $a_{ii}^m > 0$ does not depend on i. If $d = 1$, then it holds for all sufficiently large m that $a_{ij}^m > 0$ for all i, j.*

Proof. Choose a transition matrix $P = (p_{ij})$ with $p_{ij} > 0$ for exactly the same ij as for which $a_{ij} > 0$. Then $a_{ij}^m > 0$ precisely when $p_{ij}^m > 0$ and results from Chapter I complete the proof. □

The d in Lemma 1.4 is called the *period* of A, and A is *aperiodic* if $d = 1$.

Theorem 1.5 (PERRON–FROBENIUS) *Let A be an irreducible $p \times p$ matrix. Then:*

(a) *The spectral radius λ_0 of A is strictly positive and a simple eigenvalue of A with the corresponding left and right eigenvectors v, h satisfying $v_i > 0$, $h_i > 0$ for $i = 1, \ldots, p$;*

(b) *If A is also aperiodic, then $\lambda_1 = \sup\{|\lambda| : \lambda \in \mathrm{sp}(A)\backslash\{\lambda_0\}\} < \lambda_0$. Furthermore, if we normalize v, h by $vh = \sum_1^p v_i h_i = 1$, then for some k*

$$A^n = \lambda^n hv + \mathrm{O}(n^k \lambda_1^n), \qquad n \to \infty; \tag{1.2}$$

(c) *If A has period $d > 1$, then $|\lambda| \leqslant \lambda_0$ for any $\lambda \in \mathrm{sp}(A)$. Furthermore, $\lambda \in \mathrm{sp}(A)$, $|\lambda| = \lambda_0$ holds exactly when λ is of the form $\lambda_0 \theta^k$, $k = 0, \ldots, d-1$, with $\theta^k = \mathrm{e}^{2\pi i k/d}$ the roots of unity.*

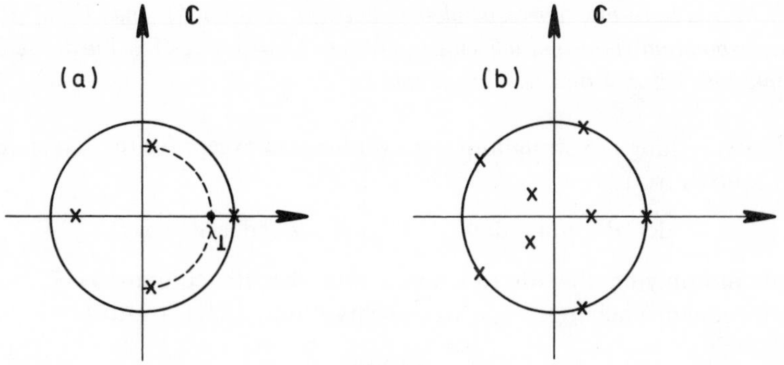

Fig. 1.1

Figure 1.1 depicts sp(A) for the aperiodic case in (a) and for the periodic case in (b). The eigenvalues fall in pairs of complex conjugates since A is real. We shall refer to λ_0 as the *Perron–Frobenius root* of A.

The proof of the Perron–Frobenius theorem will be reduced to the Markov case in Proposition 1.2. We need some lemmas.

Lemma 1.6. *If A has all $a_{ij} > 0$, then there exists $\lambda \in \mathrm{sp}(A)$, $x \in E_\lambda$ with $\lambda > 0$, $x_i > 0$, $i = 1, \ldots, p$.*

Proof. The basic observation is that all $a_{ij} > 0$ implies

$$x_i \geqslant 0, \qquad \sum_{i=1}^{p} x_i > 0 \quad \Rightarrow \quad \text{all components of } Ax \text{ are} > 0. \tag{1.3}$$

Define

$$K = \left\{ x \in \mathbb{R}^p : 0 \leqslant x_i \leqslant 1, \sum_{1}^{p} x_i = 1 \right\},$$
$$S = \{\mu \geqslant 0 : Ax \geqslant \mu x \quad \text{for some } x \in K\},$$
$$\lambda = \sup\{\mu : \mu \in S\}.$$

For a given $x \in K$, (1.3) implies $Ax \geqslant \varepsilon x$ for small enough ε. Hence $\lambda > 0$. Also, for $x \in K$ we can bound the components of Ax say by $m = \max_{ij} a_{ij}$, and since $x_i \geqslant 1/p$ for at least one i, $Ax \geqslant \mu x$ then implies that $\mu \leqslant m$. Hence $\lambda < \infty$ and we can find $\lambda_n \in S$, $x_n \in S$ with $\lambda_n \uparrow \lambda$, $Ax_n \geqslant \lambda_n x_n$. Passing to a subsequence if necessary, we may assume that $x = \lim x_n$ exists. Then $Ax \geqslant \lambda x$ and we shall complete the proof by showing that indeed $Ax = \lambda x$ ($x_i > 0$ is then ensured by (1.3)). Otherwise let $y = cAx$ with $c > 0$ chosen so that $y \in K$. Then $Ay - \lambda y = cA(Ax - \lambda x)$ has all components > 0 by (1.3). Hence $Ay \geqslant (\lambda + \varepsilon)y$ for some $\varepsilon > 0$, a contradiction. □

Lemma 1.7. *Suppose that $Ak = \lambda k$ with $\lambda > 0$ and all $k_i > 0$. Then the matrix P with elements $a_{ij}k_j/\lambda k_i$ is a transition matrix, $Pe = e$, and the formulae*

$$\lambda^A = \lambda\lambda^P, \qquad h_i^A = k_i h_i^P, \qquad \pi_i^A = \pi_i^P / k_i$$

establish a one-to-one correspondence between $\lambda^A \in \mathrm{sp}(A)$ *and* $\lambda^P \in \mathrm{sp}(P)$ *and the corresponding right and left eigenvectors* ($\pi^A A = \lambda^A \pi^A$, etc.) *Furthermore,* λ^A *is simple for* A *if and only if* λ^P *is simple for* P.

Proof. Everything is a straightforward verification except for the last statement which follows from

$$\det(P - \mu I) = \det(\lambda^{-1}A - \mu I) = \lambda^{-p}\det(A - \mu\lambda I).$$

Indeed, multiplying the ith row by k_i and the jth column by k_j^{-1} leaves the determinant unchanged and transforms P into $\lambda^{-1}A$, I into I. □

Proof of Theorem 1.3 in the aperiodic case. Choose first m with all $a_{ij}^m > 0$, cf. Lemma 1.4, and next λ, k with $A^m k = \lambda k$, $\lambda > 0$, all $k_i > 0$, cf. Lemma 1.6. Then by Lemma 1.7, 1 is simple for $P^m = (a_{ij}^m k_j/\lambda k_i)$ and hence λ simple for A^m. If $\lambda_0 \in \mathrm{sp}(A)$ satisfies $\lambda_0^m = \lambda$, then by Lemma 1.1(ii) λ_0 is simple for A. Choose $h \in E_{\lambda_0}$. Then $A^m h = \lambda_0^m h = \lambda h$, and since λ is simple for A^m, it follows that we may take $h = k$. Then by non-negativity, $Ah = \lambda_0 h$ implies $\lambda_0 > 0$ and $P = (a_{ij}h_j/\lambda_0 h_i)$ is a transition matrix. Applying Proposition 1.2 and Lemma 1.7 everything then comes out in a straightforward manner. For (1.2), note that if $\pi P = \pi$, $\pi e = 1$ and we let $v_i = \pi_i/h_i$, then $vA = \lambda_0 v$, $vh = 1$ and

$$\begin{aligned} a_{ij}^n &= \lambda_0^n \frac{p_{ij}^n h_i}{h_j} = \lambda_0^n \frac{h_i}{h_j}\left\{\pi_j + \mathrm{O}\left(n^k\left(\frac{\lambda_1}{\lambda_0}\right)^n\right)\right\} \\ &= \lambda_0^n h_i v_j + \mathrm{O}(n^k \lambda_1^n). \end{aligned}$$

□

Proof of Theorem 1.5 in the periodic case. We can reorder the coordinates by a cyclic class argument so that A has the form

$$\begin{pmatrix} 0 & A_1 & 0 & & 0 \\ 0 & 0 & A_2 & & 0 \\ & & & \vdots & \vdots \\ 0 & 0 & 0 & \cdots & A_{d-1} \\ A_d & 0 & 0 & \cdots & 0 \end{pmatrix}.$$

Then A^d is block-diagonal with diagonal elements $B_k = A_k A_{k+1} \cdots A_d A_1 \cdots A_{k-1}$ which are irreducible aperiodic. Let μ_k be the Perron–Frobenius root of B_k and $B_k h^{(k)} = \mu_k h^{(k)}$ with $h_i^{(k)} > 0$. Now

$$B_k A_k h^{(k+1)} = A_k B_{k+1} h^{(k+1)} = \mu_{k+1} A_k h^{(k+1)}$$

(identifying $d+1$ with 1). Since $A_k h^{(k+1)} \neq 0$, it follows that $\mu_{k+1} \in \mathrm{sp}(B_k)$ and hence $\mu_{k+1} \leqslant \mu_k$. Hence all μ_k are equal, say μ, and we may take $h^{(k)} = A_k h^{(k+1)} = A_k \cdots A_{d-1} h^{(d)}$. Now

$$\det(A^d - \eta I) = \prod_{k=1}^d \det(B_k - \eta I).$$

This shows that if $\lambda \in \mathrm{sp}(A)$, then $\eta = \lambda^d$ is in $\mathrm{sp}(B_k)$ for some k. Hence $|\lambda| = |\eta|^{1/d} \leqslant |\mu|^{1/d} = \lambda_0$ (say) and $|\lambda| = \lambda_0$ can only occur if $\lambda^d = \mu$, i.e. λ is of the form $\lambda_0 \theta^k$ for some k. Also the A^d-multiplicity of μ is exactly d. By Lemma 1.1(ii) the proof is now complete if we can show that each $\lambda_0\theta^k$ is an eigenvalue and that $z^{(0)} \in E_{\lambda_0}$ may be taken with all $z_i > 0$. But an easy calculation shows that

$$z^{(k)} = ((\lambda_0\theta^k)^0 h^{(1)} \cdots (\lambda_0\theta^k)^{d-1} h^{(d)})$$

satisfies $Az^{(k)} = \lambda_0\theta^k z^{(k)}$. □

Problems

1.1. Is it true that if P is an *infinite* ergodic transition matrix, then all $p_{ij}^n > 0$ for some n?
1.2. Suppose that A is an irreducible aperiodic non-negative matrix such that A^m is a transition matrix for some $m = 1, 2, \ldots$. Show then that A is itself a transition matrix. Show also that the result fails in the periodic case.
1.3. Let A be irreducible and non-negative, and assume that $Ax \leqslant \lambda x$ with $x_i \geqslant 0$, some $x_i > 0$ and $\lambda \geqslant 0$. Show that $\mathrm{spr}(A) \leqslant \lambda$ provided that either (i) A is irreducible, or (ii) all $x_i > 0$. Show also in case (i) that $\mathrm{spr}(A) < \lambda$ if in addition $Ax \neq \lambda x$.
1.4. Use the Perron–Frobenius theory to give an alternative proof of III.5.1 on the uniqueness of the solution β to the equation $\beta = \alpha + \beta\Gamma$ in queueing networks, e.g. in the following steps: (1) Problem 1.3 with $x = e$ yields $\mathrm{spr}(\Gamma) \leqslant 1$; (2) $\mathrm{spr}(\Gamma) = 1$ is impossible because of the assumption (ii) preceding III.5.1; (3) $I - \Gamma$ is invertible. Note that assumption (i) is not needed.
1.5. Let A be an irreducible intensity matrix, $a_{ii} \leqslant 0$, $a_{ij} \geqslant 0$, $i \neq j$, $Ae = 0$. Show that 0 is a simple eigenvalue and that $\mathrm{Re}\,\lambda < 0$ for any $\lambda \in \mathrm{sp}(A)$ with $\lambda \neq 0$ [hint: let $P = e^A$ and use $\mathrm{sp}(P) = \{e^\lambda : \lambda \in \mathrm{sp}(A)\}$].

Notes

Non-negative matrices are treated in a number of places, e.g. Gantmacher (1959), Seneta (1981) and the appendices in Karlin and Taylor (1975) and Çinlar (1975). Of extensions of the Perron–Frobenius theorem, we mention in particular operator versions such as the Krein–Rutman theorem, e.g. Schaefer (1966), and the more probabilistic inspired discussion of Athreya and Ney (1982) and Nummelin (1984).

2. MARKOV RENEWAL THEORY

By a *Markov renewal process* we understand a point process where the inter-arrival times $T_0, T_1, \ldots$ are not independent identically distributed but governed by a Markov chain $\{J_n\}$ with (finite or countable) state space E. This Markov dependence of the T_n may be formulated in various equivalent ways. One formulation is that T_n is sampled according to the current values of J_n, J_{n+1}. With $\mathscr{H} = \sigma(J_0, J_1, \ldots)$ this means that $T_0, T_1, \ldots$ are conditionally independent given $\mathscr{H}$ with

$$\mathbb{P}(T_n \leqslant t | \mathscr{H}) = \mathbb{P}(T_n \leqslant t | J_n, J_{n+1}) = G_{ij}(t) \qquad \text{on } \{J_n = i, J_{n+1} = j\} \quad (2.1)$$

for a suitable family of distributions $(G_{ij})_{i,j \in E}$ on $(0, \infty)$. Equivalently, one may think of (J_{n+1}, T_n) being sampled simultaneously according to the current value of J_n. That is, $\{J_{n+1}, T_n\}$ is a Markov chain on $E \times (0, \infty)$ with the transition function depending only on the first coordinate. In particular, letting $Q=(q_{ij})$ denote the transition matrix of $\{J_n\}$ and $F_{ij}=q_{ij}G_{ij}$, we have

$$F_{ij}(t)=\mathbb{P}(J_{n+1}=j,\ T_n\leqslant t\,|\,J_n=i)=\mathbb{P}_i(J_1=j,\ T_0\leqslant t) \tag{2.2}$$

where $\mathbb{P}_i$, $\mathbb{E}_i$ refer to the case $J_0=i$. The matrix F whose elements are the measures F_{ij} is called the *semi-Markov kernel*, and we define the associated *semi-Markov process* $\{W_t\}_{t\geqslant 0}$ by $W_t=J_0$ for $t<T_0$, $W_t=J_1$ for $T_0\leqslant t<T_0+T_1$ and so on. Obviously, the semi-Markov process and the Markov renewal process are in one-to-one correspondence (at least subject to regularity conditions like $q_{ii}=0$, $T_0+T_1+\cdots=\infty$), and we shall not keep a formal distinction between them.

We note that the process reduces to a renewal process if E consists of one point, to a Markov chain with state space E if all G_{ij} are degenerate at 1 and to a continuous-time Markov process with state space E if all G_{ij} are exponential with intensities depending only on i, $1-G_{ij}(t)=\mathrm{e}^{-\lambda(i)t}$. Thus, the Markov renewal process may be said to extend the continuous-time Markov jump process in the same way as the renewal process extends the Poisson process. We take these remarks as sufficient motivation for developing the theory and give just one practical example.

Example 2.1. Suppose in the traffic theory example IV.1.2 that two types of vehicles are possible, for example cars and trucks. Then clearly the distribution of the distance between two vehicles depends in an essential way on their types. One could also model clumping by letting the type of a vehicle be its number in a clump. Suppose that a clump consists of n cars with probability q_n ($q_1+q_2+\cdots=1$) and that the sizes of the clumps are independent. Then the Markov chain goes from state n to state 1 with probability $q_n/(q_n+q_{n+1}+\cdots)$ and to state $n+1$ otherwise, and one could take all $G_{n1}=H_1$, all $G_{n(n+1)}=H_2$ (with H_1 stochastically larger than H_2). □

The following observation is the key to Markov renewal theory:

Proposition 2.2. *The instants t of returns to i ($W_t=i, W_{t-}\neq i$) from a renewal process which is non-terminating if $\{J_n\}$ is recurrent. In that case, the mean interarrival time is μ/ν_i where ν is the stationary measure for $\{J_n\}$ and $\mu=\sum_{i,j\in E}\nu_i\mu_{ij}$ with $\mu_{ij}=\int_0^\infty t\,\mathrm{d}F_{ij}(t)$.*

Proof. The first statement is obvious and letting $\tau=\inf\{n\geqslant 1: J_n=i\}$, it is seen that if $J_0=i$, then $T_0+\cdots+T_{\tau-1}$ is the first interarrival interal. Hence, letting $m(i,j)=\int_0^\infty t\,\mathrm{d}G_{ij}(t)$ (so that $\mu_{ij}=q_{ij}m(i,j)$), the mean interarrival time is

$$\begin{aligned}\mathbb{E}_i(T_0+\cdots+T_{\tau-1})&=\mathbb{E}_i\sum_{n=0}^{\infty}\mathbb{E}_i[T_n;\tau>n\,|\,\mathscr{H}]\\&=\mathbb{E}_i\sum_{n=0}^{\infty}I(\tau>n)m(J_n,J_{n+1})\\&=\sum_{j,k\in E}m(j,k)\mathbb{E}_i\sum_{n=0}^{\infty}I(J_n=j,\ J_{n+1}=k,\tau>n)\end{aligned}$$

$$= \sum_{j,k\in E} m(j,k) q_{jk} \mathbb{E}_i \sum_{n=0}^{\infty} I(J_n = j, \tau > n)$$

$$= \sum_{j,k\in E} \mu_{jk} \frac{v_j}{v_i} = \frac{\mu}{v_i}. \qquad \square$$

In a standard manner, one can now prove that if $\{J_n\}$ is irreducible recurrent, then the interarrival distributions corresponding to Proposition 2.2 are either non-lattice for all i, or all lattice with the same span. We call the Markov renewal process (or equivalently the semi-Markov kernel F) *non-lattice* in the first case and *lattice* in the second.

For two kernels F, G we define $H = G * F$ as the kernel with elements

$$H_{ij}(t) = \sum_{k\in E} G_{ik} * F_{kj}(t) = \sum_{k\in E} \int_0^t F_{kj}(t-u)\,\mathrm{d}G_{ik}(u)$$

and the convolution powers F^{*k} the obvious way (with $F_{ij}^{*0}(t) = I(i=j)$). Using (2.2) and induction, one can immediately check that the interpretation in terms of Markov renewal processes is

$$F_{ij}^{*k}(t) = \mathbb{P}_i(J_k = j, T_0 + \cdots + T_{k-1} \leqslant t). \tag{2.3}$$

Define finally the Markov renewal kernel U as $U = \sum_0^\infty F^{*n}$. We then have the following generalization of Blackwell's theorem:

Theorem 2.3. *Consider a Markov renewal process with semi-Markov kernel F and $J_0 = W_0 = i$. Then $U_{ij}(t)$ is the expected number of returns to j before t,*

$$U_{ij}(t) = \sum_{n=0}^{\infty} \mathbb{E}_i(J_n = j, T_0 + \cdots + T_{n-1} \leqslant t). \tag{2.4}$$

In particular, $U_{ij}(t) < \infty$ and in the non-lattice case it holds in the notation of Proposition 2.2 that $U_{ij}(t+a) - U_{ij}(t) \to a v_j/\mu$.

Proof. Here (2.4) is clear from (2.3), and the rest of the theorem is an immediate consequence of Proposition 2.2 and one-dimensional renewal theory. $\square$

The *Markov renewal equation* (or *multivariate renewal equation*, system of *coupled renewal equations*, etc.) has the form

$$Z_i(t) = z_i(t) + \sum_{j\in E} \int_0^t Z_j(t-u)\,\mathrm{d}F_{ij}(u), \qquad i \in E, \tag{2.5}$$

where the Z_i are unknown functions on $[0, \infty)$, the z_i known functions on $[0, \infty)$ and the F_{ij} known bounded measures on $[0, \infty)$. Equation (2.5) can be rewritten in matrix form as $Z = z + F * Z$, with the convolution defined in the manner consistent with (2.5), and in a similar manner as in one dimension we have:

Proposition 2.4. *Suppose that F is a semi-Markov kernel* (i.e. $Q = (\|F_{ij}\|)$ a transition matrix) *and $\{J_n\}$ is irreducible recurrent. Then if $z_i \geqslant 0$ and the z_i are* *bounded on finite intervals uniformly in i, it holds that $Z = U * z$ is the unique solution to* (2.5) *with the Z_i uniformly bounded on finite intervals.*

Proof. Since $U = F^{*0} + F * U$, $F^{*0} * z = z$, it is clear that $Z = U * z$ is well defined and solves (2.5). Given two solutions of the type considered, their difference D satisfies $D = F * D = \cdots = F^{*k} * D$, hence $|D| \leqslant F^{*k} * |D|$ and $\sup_i |D_i(t)| \leqslant K(\tau) < \infty$ when $0 \leqslant t \leqslant \tau$. To prove that $D \equiv 0$, let $0 \leqslant t \leqslant \tau$ and assume without loss of generality that $K(\tau) = 1$. Then $|D| \leqslant F^{*k} * 1$ on $[0, \tau]$ and hence by (2.4),

$$|D_i(t)| \leqslant \sum_{j \in E} F_{ij}^{*k}(t) = \mathbb{P}_i(T_0 + \cdots + T_{k-1} \leqslant t).$$

We claim that $T_0 + T_1 + \cdots = \infty$ a.s. so that indeed $D_i(t) = 0$ follows as $k \to \infty$. To see this, note simply that the T_n with $J_n = i$, $J_{n+1} = j$ are independent identically distributed given $\mathscr{H}$ and not degenerate at 0, and that $\mathbb{P}(J_n = i, J_{n+1} = j$ infinitely often$) = 1$ for some i, j. □

Using Theorem 2.3, one can now deduce exactly as in one dimension that if z_j is directly Riemann integrable (d.R.i.), then

$$U_{ij} * z_j(t) \to \frac{v_j}{\mu} \int_0^\infty z_j(x)\,\mathrm{d}x. \tag{2.6}$$

From this one expects the generalization

$$Z_i(t) = \sum_{j \in E} U_{ij} * z_j(t) \to \frac{1}{\mu} \sum_{j \in E} v_j \int_0^\infty z_j(x)\,\mathrm{d}x. \tag{2.7}$$

of the key renewal theorem. However, if E is infinite one cannot deduce (2.7) from (2.6) without imposing some further conditions. We shall not go into this but be satisfied by noting:

Corollary 2.5. *Suppose in addition to the assumptions of Proposition* 2.4 *that F is non-lattice, that E is finite or, more generally, that $z_i = 0$ except for a finite number of i, and that the z_i are d.R.i. Then* (2.7) *holds.*

We shall also derive an analogue of the asymptotic estimates of VI.5 for the case where the rows of Q not necessarily sum to 1:

Theorem 2.6. *Consider the Markov renewal equation* (2.5) *with $E = \{1, \ldots, p\}$ and $Q = (\| F_{ij} \|)$ irreducible. Define, for some real β, $a_{ij} = \int_0^\infty e^{\beta u}\,\mathrm{d}F_{ij}(u)$, suppose that $A = (a_{ij})$ has spectral radius* 1 *and choose v, h with $vA = v$, $Ah = h$, $v_i > 0$, $h_i > 0$ $i = 1, \ldots, p$. Then*

$$\mathrm{d}\tilde{F}_{ij}(u) = \frac{h_j}{h_i} e^{\beta u}\,\mathrm{d}F_{ij}(u)$$

defines a semi-Markov kernel $\tilde{F}$ with $\tilde{Q} = (\| \tilde{F}_{ij} \|)$ irreducible recurrent. Let further

$$\tilde{Z}_i(t) = e^{\beta t} Z_i(t)/h_i, \qquad \tilde{z}_i(t) = e^{\beta t} z_i(t)/h_i$$

Then $Z = z + F * Z$ implies that $\tilde{Z} = \tilde{z} + \tilde{F} * \tilde{Z}$. Hence if $\tilde{F}$ is non-lattice and the $\tilde{z}_i$

d.R.i., then

$$\lim_{t\to\infty} \tilde{Z}_i(t) = \frac{1}{\tilde{\mu}} \sum_{j=1}^{p} \tilde{v}_j \int_0^\infty \tilde{z}_j(x)\,dx,$$

where $\tilde{v}$ is stationary for $\tilde{Q}$, hence

$$\lim_{t\to\infty} e^{\beta t} Z_i(t) = \frac{h_i \sum_{j=1}^{p} v_j \int_0^\infty e^{\beta x} z_j(x)\,dx}{\sum_{k,j=1}^{p} v_k h_j \int_0^\infty x e^{\beta x}\,dF_{kj}(x)}. \tag{2.8}$$

Proof. That $\tilde{Q} = (h_j a_{ij}/h_i)$ is a transition matrix is immediate and has in fact already been noted in Section 1 where it was also found that $\tilde{v}_j = v_j h_j$ (the existence of $v, h, \tilde{v}$ is ensured by Theorem 1.5). The rest of the proof is trivial manipulations. □

For conditions for β to exist, see Problem 2.3.

Example 2.7. Consider the Lotka–Sharpe population model from IV.2.2, but assume now that each woman has one of p types. The type can change during life and could, for example, be one of p social groups, one of p geographical regions in which the woman lives, or the parity of the woman, i.e. the number of children already born (then group p comprises all women having $p-1$ or more children). In such situations, it is highly relevant to assume that the birth-rates and survival rates depend on types, say type i women aged a give birth to type j daughters at rate $\lambda_{ij}(a)$ and can be found in group j, t time units later in an average proportion of ${}_tp_a(i,j)$. Then if $Z_i(t)$ is the rate of birth of type i girls at time t, $f_0^{(j)}(a)$ the density of type j women aged a in the initial population, we get exactly as for $p=1$

$$Z_i(t) = z_i(t) + \sum_{j,k=1}^{p} \int_0^t Z_j(t-s)\,{}_sp_0(j,k)\lambda_{ki}(s)\,ds,$$

where

$$z_i(t) = \sum_{j,k=1}^{p} \int_0^\infty f_0^{(j)}(a)\,{}_tp_a(j,k)\lambda_{ki}(a+t)\,da.$$

This is of the form (2.5) with

$$\frac{dF_{ij}(x)}{dx} = \sum_{k=1}^{p} {}_xp_0(j,k)\lambda_{ki}(x)$$

and the programme of Theorem 2.6 may be carried through to obtain asymptotic estimates of the form $Z_i(t) \cong e^{\beta t} c_i$. □

Problems

2.1. Suppose that the Markov chain $J_0, J_1, \ldots$ imbedded in a Markov renewal process is transient and define the lifetime L as $L = T_0 + T_1 + \cdots$. Explain that, in contrast to the recurrent case, it is possible that $L < \infty$ a.s. Define $Z_i(t) = \mathbb{P}_i(L \leqslant t)$. Show that $Z = U * Z$

and that the solution of the Markov renewal equation need not be unique in the transient case.

2.2. Consider a Markov renewal equation of the form

$$\begin{aligned} Z_1(t) &= z_1(t) + Z_1 * F_{11}(t) + Z_2 * F_{12}(t) \\ Z_2(t) &= z_2(t) + \qquad\qquad\quad Z_2 * F_{22}(t) \end{aligned}$$

where

$$0 < \| F_{11} \| < 1, \qquad \| F_{11} \| + \| F_{12} \| = \| F_{22} \| = 1.$$

Show that the solution is unique and find its limiting behaviour.

2.3. Consider the set-up of Theorem 2.6 and suppose that Q is irreducible with $1 < \operatorname{spr} Q < \infty$. Show that β always exists [hint: let $A = A_\beta$ be as in Theorem 2.6, $\rho(\beta) = \operatorname{spr}(A_\beta)$, $p(A, \lambda) = \det(A - \lambda I)$. Show that $\partial p / \partial \lambda$ evaluated at $A = A_\beta$, $\lambda = \rho(\beta)$ is non-zero and thereby, using the implicit function theorem, that $\rho(\beta)$ is continuous in β. Show finally $\rho(\beta) \to 0$, $\beta \to \infty$.]

Notes

See the next section for Markov renewal theory in general. Theorem 2.6 has a long history in the theory of branching processes, the present version and proof being noted independently by Asmussen and Hering (1977) and Athreya and Ney (1978b). Further demographic applications along the lines of Example 2.7 are in Braun (1978).

3. SEMI-REGENERATIVE PROCESSES

The concept of semi-regenerativity generalizes regenerative processes by allowing the regeneration points to be of several types, indexed by $i \in E$, where E is finite or countable. Thus instead of an imbedded renewal process we have an imbedded Markov renewal process specified say by $\{(J_n, T_n)\}_0^\infty$. Each time state i is entered, the semi-regenerative process $\{X_t\}$ is restarted subject to the ith set of initial conditions and independent of the Markov renewal process up to that time.

More formally, let $T = \mathbb{N}$ or $T = [0, \infty)$ and let $(\mathbb{P}_i)_{i \in E}$ be a governing set of probabilities for $\{X_t\}_{t \in T}$. We then call $\{X_t\}$ *semi-regenerative* if we can find a Markov renewal process (possibly defined on an enlarged probability space), such that for any n the conditional distribution of $\{X_{t+T_0+\cdots+T_{n-1}}\}_{t \in T}$ given $T_0, \ldots, T_{n-1}, J_0, \ldots, J_{n-1}, J_n = i$ is the same as the $\mathbb{P}_i$-distribution of $\{X_t\}_{t \in T}$ itself. Thus if E consists of one point, the concept reduces to regenerative processes. Even in the general case, we have:

Proposition 3.1. *Any semi-regenerative process with $\{J_n\}$ recurrent is regenerative, with the instants of returns of the Markov renewal process to $i \in E$ as an imbedded renewal process.*

This is an immediate consequence of the definitions. From the point of view of proving the existence of limiting distributions, semi-regenerative processes are therefore not a much more powerful tool than regenerative processes. Rather, the formulae derived for the limits may be somewhat more convenient than the expression $(\mathbb{E}C)^{-1} \mathbb{E} \int_0^C f(X_t)\,dt$ for regenerative processes, and at least serve the purpose of doing some reduction once and for all:

Proposition 3.2. *Consider a semi-regenerative process with $\{J_n\}$ irreducible recurr-*

ent, say with stationary measure ν. *Define* $C = T_0$ *and suppose that* $\mu = \sum_{i \in E} \nu_i \mathbb{E}_i C$ *is finite. Then:*

(i) *If* $T = [0, \infty)$, *the imbedded Markov renewal process is non-lattice and* $\{X_t\}$ *has metric state space and right-continuous paths, then the limiting distribution exists and is given by*

$$\mathbb{E}f(X) = \frac{1}{\mu} \sum_{j \in E} \nu_j \mathbb{E}_j \int_0^C f(X_t) \mathrm{d}t; \tag{3.1}$$

(ii) *If* $T = \mathbb{N}$ *and the imbedded Markov renewal process is aperiodic on* $\mathbb{N}$, *then the limiting distribution exists and is given by* (3.1) *with* $\int_0^C$ *replaced by* $\sum_0^{C-1}$.

Proof. First check that the expression given for μ is the same as in Proposition 2.2. Hence if $\mu < \infty$, we may appeal to Proposition 3.1 and use V.1.3 to get the existence of the limit as well as the expression

$$\mathbb{E}f(X) = \frac{1}{\mathbb{E}_i \tilde{C}} \mathbb{E}_i \int_0^{\tilde{C}} f(X_t) \mathrm{d}t = \frac{\nu_i}{\mu} \mathbb{E}_i \int_0^{\tilde{C}} f(X_t) \mathrm{d}t \tag{3.2}$$

where

$$\tilde{C} = T_0 + \cdots + T_{\tau-1}, \qquad \tau = \inf\{n \geq 1 : J_n = i\}.$$

Now the semi-regenerative property implies

$$\mathbb{E}_i\left[\int_{T_0 + \cdots + T_{k-1}}^{T_0 + \cdots + T_k} f(X_t) \mathrm{d}t \mid T_1, \ldots, T_{k-1}, J_0, \ldots, J_k\right] = \mathbb{E}_j \int_0^C f(X_t) \mathrm{d}t$$

on $\{J_k = j\}$. Hence (3.2) can be written as

$$\frac{\nu_i}{\mu} \mathbb{E}_i \sum_{n=0}^{\tau-1} \mathbb{E}_{J_n} \int_0^C f(X_t) \mathrm{d}t = \frac{\nu_i}{\mu} \sum_{j \in E} \mathbb{E}_j \int_0^C f(X_t) \mathrm{d}t \, \mathbb{E}_i \sum_{n=0}^{\tau-1} I(J_n = j)$$

and since the last factor is just ν_j/ν_i, (3.1) follows. The proof of (ii) is entirely similar. □

Corollary 3.3. *Let* $\{X_n\}$ *be an irreducible recurrent Markov chain with discrete state space E and let* X_k^F *be the value of* X_n *at the* k*th visit to* $F \subseteq E$. *Then a measure* ν *is stationary for* $\{X_n\}$ *if and only if* $(\nu_i)_{i \in F}$ *is stationary for* $\{X_n^F\}$ *and*

$$\nu_j = \sum_{k \in F} \nu_k \mathbb{E}_k \sum_{n=0}^{\tau(F)-1} I(X_n = j), \tag{3.3}$$

where $\tau(F) = \inf\{n \geq 1 : X_n \in F\}$.

Proof. Let $i \in F$ be fixed and define

$$\tau = \inf\{n \geq 1 : X_n = i\}, \qquad \nu_j = \mathbb{E}_i \sum_{n=0}^{\tau-1} I(X_n = j).$$

Then according to I.3.8 (cf. also VI.3.4), it only remains to show that (3.3) holds.

But letting

$$J_n = X_n^F, \qquad C = \tau(F), \qquad \tau^F = \inf\{n \geqslant 1 : X_n^F = i\},$$

the expectation in the definition of v_j may then be evaluated exactly as in the proof of Proposition 3.2 and we get

$$\begin{aligned} v_j &= \sum_{k \in F} \mathbb{E}_i \sum_{n=0}^{\tau^F - 1} I(X_n^F = k) \mathbb{E}_k \sum_{n=0}^{C-1} I(X_n = j) \\ &= \sum_{k \in F} v_k^{(F)} \mathbb{E}_k \sum_{n=0}^{C-1} I(X_n = j) = \sum_{k \in F} v_k \mathbb{E}_k \sum_{n=0}^{C-1} I(X_n = j). \end{aligned}$$

□

Example 3.4. We consider the $M/G/1$ queue length process $\{Q_t\}$ and its imbedded Markov chain $\{J_n\}$ ($J_n = Y_n$ in the notation of IX.4). Letting $T_0 + \cdots + T_{k-1}$ be the time of the kth departure, it is clear that the development of $\{Q_t\}$ after that time depends solely on the number J_k of customers left behind. Hence we have a semi-regenerative process for $\rho < 1$ and we shall illustrate the content of (3.1) by evaluating the expressions in question. First, $C = T_0$ is a service interval if $J_0 > 0$ so that $\mathbb{E}_i C = \mu_B$, $i = 1, 2, \ldots$. For $i = 0$, C also comprises the exponential waiting time to the arrival of a customer and hence

$$\mu = \pi_0\left(\frac{1}{\beta} + \mu_B\right) + \sum_{i=1}^{\infty} \pi_i \mu_B = \frac{1-\rho}{\beta} + \mu_B = \frac{1}{\beta}.$$

Now for $j \geqslant 1$ fixed $\alpha_k = \mathbb{E}_k \int_0^C I(Q_t = j)\,dt$ has essentially already been evaluated in IX.4. Indeed, it is immediately seen that $\alpha_0 = \mathbb{E} I_0 = \beta^{-1} r_{j-1}$, whereas $\alpha_k = 0$ for $k > j$ and otherwise

$$\alpha_k = \mathbb{E}(I_n; \sigma > n \mid J_n = k, \sigma > n) = \beta^{-1} r_{j-k}$$

in the notation of IX.4. Hence (3.1) becomes

$$\mathbb{P}(Q = j) = \beta \sum_{k=0}^{\infty} \pi_k \alpha_k = \pi_0 r_{j-1} + \sum_{k=1}^{j} \pi_k r_{j-k}$$

which is the same as IX.(4.6). Thus for this case semi-regenerative processes lead to essentially the same calculations as imbedded Markov chain analysis. □

Notes

A standard source for Markov renewal theory and semi-regenerative processes (in the case of a discrete E) is Çinlar (1975). As argued in the text, the extension from renewal theory and regenerative processes does not present intrinsically new mathematical difficulties, but nevertheless, the versatility of the set-up makes it highly useful and popular in applications.

If $\{J_n\}$ has a general state space, one may use the regeneration scheme developed for Harris recurrent Markov chains in VI.3 to develop a version of Proposition 3.2 for the case where $\{J_n\}$ has a general state space. One needs, however, then to restrict attention to the case where the distribution of T_n depends solely on J_n and not in addition J_{n+1} (this is because the regeneration for $\{J_n\}$, say at τ, occurs not in a fixed state $J_\tau = i$ but according to λ, and thus the cycle length $T_0 + \cdots + T_{\tau-1}$ for the semi-regenerative process would otherwise depend on the initial conditions for the next cycle which invalidates the basic renewal equation). The details are given in Athreya *et al.* (1978), whereas the problem in its full generality is treated in Berbee (1979). A further main reference in the area is Kesten (1974).

4. RANDOM WALKS ON A MARKOV CHAIN

The process that we are concerned with may simply be thought of as a Markov renewal process where the increments can be both positive and negative. That is, we have given a Markov chain $((J_n, X_n))$ with state space $E \times \mathbb{R}$ (E discrete) and transition function depending only on the first coordinate, and consider the $E \times \mathbb{R}$-valued Markov chain $((J_n, S_n))$, where $S_n = X_0 + \cdots + X_n$. Thus (S_n) is a random walk with the increment X_n chosen according to the current value of (J_{n-1}, J_n), and for this reason use the expression a *random walk on a Markov chain* (different terms are *Markov additive process* (MAP), *Markov-modulated random walk* (MMRW), *random walk in a random environment*, etc.). Note that compared to Markov renewal processes X_n corresponds to T_{n-1} and that we have allowed an initial 'delay' X_0.

Example 4.1. Suppose that the X_n are integer-valued and define

$$G_{ij}(k) = \mathbb{P}_i(J_1 = j, X_1 = k) = \mathbb{P}(J_1 = j, X_1 = k | J_0 = i) \tag{4.1}$$

Then $((J_n, S_n))$ is a Markov chain on $E \times \mathbb{Z}$, and the transition matrix is of block form with $G(l-k)$ as the (k, l)th block. Conversely, given a transition matrix of this form, (4.1) defines a random walk on a Markov chain. □

In the following, $\mathbb{P}_{ix}$ refers to the case $J_0 = i$, $X_0 = S_0 = x$, and we assume that (J_n) is irreducible positive recurrent with stationary distribution ν, and that $\mathbb{E}_i|X_1| < \infty$ for all i. We denote by μ the column vector with elements $\mu(i) = \mathbb{E}_i X_1$. Thus $\nu\mu$ is the mean drift of (S_n), and we have:

Proposition 4.2. *We have* (a′) $S_n \to -\infty$, (b′) $\overline{\lim}\, S_n = \infty$, $\underline{\lim}\, S_n = -\infty$, (c′) $S_n \to \infty$ $\mathbb{P}_{ix}$ *a.s. for all ix according as* (a) $\nu\mu < 0$, (b) $\nu\mu = 0$, (c) $\nu\mu > 0$. *Defining* $\sigma = \inf(n \geqslant 1 : S_n \leqslant 0)$, *it holds for each fixed i in case* (a) *that* $\overline{\lim}_{x\to\infty} \mathbb{E}_{ix}\sigma/x \leqslant -(\nu\mu)^{-1}$, *and in case* (b) *with E finite that* $\mathbb{E}_{ix}\sigma = \infty$ *for all sufficiently large x.*

Proof. Let $\tau(i, k)$ be the time of the kth visit of J_n to state i and $\tau(i) = \tau(i, 1)$. Then the $\tau(i, k) + 1$ are regeneration points for $\{X_n\}$ and hence by V.3.1, S_n/n has the limit

$$\frac{\mathbb{E}_i S_{\tau(i)}}{\mathbb{E}_i \tau(i)} = \frac{1}{\mathbb{E}_i \tau(i)} \mathbb{E}_i \sum_{n=1}^{\tau(i)} \mu(X_{n-1}) = \mathbb{E}_\nu \mu(X) = \nu\mu$$

(the same conclusion can also be obtained in a slightly more direct way by applying VI.3.7 to $\{(J_n, X_n)\}$). From $S_n/n \to \nu\mu$ it is immediately apparent that (a)⇒(a′), (c)⇒(c′). In case (b), $(S_{\tau(i,k)})$ has independent identically distributed increments with mean zero and hence by VII.2.4 $\overline{\lim}\, S_{\tau(i,k)} = \infty$, $\underline{\lim} = -\infty$, implying (b′). In case (a), let $\kappa(x) = \inf\{k : S_{\tau(i,k)} \leqslant -x)$. Then since the $\tau(i,k) - \tau(i,k-1)$ are independent identically distributed with respect to $\mathbb{P}_{i0}$, we get from Problem VII.2.3 that $\mathbb{E}_{i0}\kappa(x) \cong -x/\mathbb{E}_{i0}S_{\tau(i)}$ and hence by Wald's identity

$$\mathbb{E}_{ix}\sigma \leqslant \mathbb{E}_{i0}\tau(i,\kappa(x)) = \mathbb{E}_{i0}\tau(i)\mathbb{E}_{i0}\kappa(x) \cong -\frac{x}{\nu\mu}.$$

In case (b), let $\tau_-(k)$ be the kth strict ascending ladder epochs for S_n. Then $(J_{\tau_-(k)})$ is a Markov chain and letting $w(j) = \inf(k \geqslant 1 : J_{\tau_-(k)} = j)$, there exists in the finite case an irreducible absorbing class $E_1 \subseteq E$ such that $\mathbb{P}_{i0}(w(j) < \infty) > 0$ for all $j \in E_1$. Clearly, $w(j) < \infty$ $\mathbb{P}_{j0}$ a.s. for $j \in E_1$, and we may write $\tau_-(w(j)) = \tau(j;\delta)$ for a suitable δ which is a stopping time w.r.t. $\{\mathscr{F}_{\tau(j,k)}\}$. In fact, $\mathbb{E}_{j0}\delta = \infty$ since otherwise

$$0 > \mathbb{E}_{j0}S_{\tau(j;\delta)} = \mathbb{E}_{j0}\delta\mathbb{E}_{j0}S_{\tau(j)} = \mathbb{E}_{j0}\delta \cdot 0,$$

and therefore also

$$\mathbb{E}_{j0}\tau_-(w(j)) = \mathbb{E}_{j0}\delta\mathbb{E}_{j0}\tau(j) = \infty.$$

But letting $m(j) = \mathbb{E}_{j0}\tau_-(1)$ and noting that $\{J_{\tau_-(k)}\}$ is positive recurrent on E_1, say with stationary distribution $\pi_j = (\mathbb{E}_{j0}w(j))^{-1}$, we get

$$\mathbb{E}_{j0}\tau_-(w(j)) = \mathbb{E}_{j0}\sum_{n=0}^{w(j)-1}(\tau_-(n+1) - \tau_-(n))$$

$$= \mathbb{E}_{j0}\sum_{n=0}^{w(j)-1} m(J_{\tau_-(n)}) = \sum_{l \in E_1}\frac{\pi_l}{\pi_j}m(l).$$

Hence $m(l) = \infty$ for some $l \in E_1$. But if x is so large that $\mathbb{P}_{i0}(S_{\tau_-(w(l))} < -x) > 0$, then also $\mathbb{E}_{ix}\sigma = \infty$. □

Now define a *Lindley process on a Markov chain* as $W_{n+1} = (W_n + X_n)^+$, where $W_0 \geqslant 0$ is some arbitrary value.

Theorem 4.3. If $\nu\mu \geqslant 0$, *then* $W_n \xrightarrow{\mathscr{D}} \infty$ *irrespective of initial conditions. If* $\nu\mu < 0$, *then the Markov chain* $\{(J_n, X_n, W_n)\}$ *has a stationary distribution* π. *More precisely, a stationary doubly infinite version* $\{(J_n^*, X_n^*, W_n^*)\}_{n\in\mathbb{Z}}$ *may be constructed by letting*

$$W_n^* = \sup_{-\infty < k \leqslant n}\{X_k^* + \cdots + X_{n-1}^*\} = \sup_{-\infty < k \leqslant n-1}\{X_k^* + \cdots + X_{n-1}^*\}^+,$$

where $\{(J_n^*, X_n^*)\}_{n\in\mathbb{Z}}$ *is a stationary doubly infinite version of* $\{(J_n, X_n)\}$. *If, furthermore,* $\{J_n\}$ *is aperiodic, then (irrespective of the initial conditions)* $(J_n, X_n, W_n) \to \pi$ *in total variation.*

Proof. In just the same way as in III.7 we have $W_n = (W_0 + S_n) \wedge K_n$, where K_n is the maximum of the $S_n - S_k$ with $1 \leqslant k \leqslant n$. But considering the cases $\nu\mu = 0$, $\nu\mu > 0$ separately, Proposition 4.2(b′), (c′) imply $K_n \to \infty$ a.s. if $\nu\mu \geqslant 0$, hence $W_n \xrightarrow{\mathscr{D}} \infty$. If $\nu\mu < 0$, it follows similarly as in the proof of Proposition 4.2 that $X_k^* + \cdots + X_{n-1}^* \to -\infty$ as $k \to -\infty$, and hence $W_n^* < \infty$ a.s. The stationarity of the chain $\{(J_n^*, X_n^*, W_n^*)\}$ is obvious, and since obviously

$$W_{n+1}^* = \sup_{-\infty<k\leqslant n} \{X_k^* + \cdots + X_n^*\}^+ = (W_n^* + X_n^*)^+,$$

it is indeed a version of $\{(J_n, X_n, W_n)\}$. To show total variation convergence to π, we shall then show that the stationary version and a version with some arbitrary starting values $J_0 = i$, $X_0 = x$, $W_0 = w$ can be coupled (cf. VI.2). Since $\{J_n\}$ is ergodic, $\{J_n\}$ and $\{J_n^*\}$ can be coupled, i.e. we may assume $J_n = J_n^*$ for $n \geqslant T-1$ (say) and hence also $X_n = X_n^*$, $n \geqslant T$. With $S_n^{(T)} = S_{n+T} - S_T$ we then have from III.7 that

$$W_{n+T} = \max\{W_T + S_n^{(T)}, S_n^{(T)} - S_1^{(T)}, \ldots, S_n^{(T)} - S_{n-1}^{(T)}, 0\},$$
$$W_{n+T}^* = \max\{W_T^* + S_n^{(T)}, S_1^{(T)} - S_1^{(T)}, \ldots, S_n^{(T)} - S_{n-1}^{(T)}, 0\}$$

But for n sufficiently large, say $n \geqslant T_1$, we have $S_n^{(T)} < 0$ so that $W_{n+T} = W_{n+T}^*$ and $T + T_1$ is the desired coupling epoch. □

Example 4.4. Suppose that the arrival instants $T_0 + \cdots + T_{n-1}$ of a single-server FIFO queue form a Markov renewal process governed by $\{J_{n-1}\}$ and that the service times $U_0, U_1, \ldots$ are independent identically distributed and independent of $\{(J_n, T_n)\}$. This system, denoted by $SM/G/1$ (the extention to $SM/SM/1$ is straightforward) may frequently be more appealing than $GI/G/1$, since it may be easier to imagine how arrivals can be Markov-modulated than how they can form a renewal process. The waiting-time process $\{W_n\}$ is then of the above form with $X_n = U_n - T_n$, and Theorem 4.3 yields the desired ergodicity results. Also, virtual waiting times and queue lengths in continuous time can easily be treated, see Problem 4.1. □

Example 4.5. Let $\{Z_t\}$ be the Markov jump process denoting the number of the K lines which are busy in Erlang's loss system III.3e, let $\{\theta_n\}$ be the jump chain and let $T_0, T_1, \ldots$ be the corresponding exponential holding times (thus Z_t jumps from θ_0 to θ_1 at time T_0, to θ_2 at time $T_0 + T_1$ and so on). Suppose now that in addition to ordinary telephone traffic the system also carries data messages of lower priority. Modelling these as a deterministic fluid flow, it holds for suitable constants σ, τ independent of Z_t that data is received at rate σ and is processed at rate $\tau(K-i)$ if $Z_t = i$. Let V_t be the amount of unprocessed data at time t, let $W_n = V_{T_0} + \cdots + T_{n-1}-$ be $\{V_t\}$ observed at the nth jump of $\{Z_t\}$ and define $X_n = (\sigma - \tau(K-i))T_n$. It is then readily seen that $W_{n+1} = (W_n + X_n)^+$ so that $\{W_n\}$ is a Lindley process on the Markov chain $\{\theta_n\}$, and Theorem 4.3 applies in a routine manner to find the ergodic behaviour, cf. Problem 4.5. Similarly, $\{V_t\}$ could be modelled as a continuous time Lindley process governed by $\{Z_t\}$. Presumably, this is the more interesting point in the present example, but we shall not develop the formalism here. □

The Lindley process forms one of several possible modifications of the Markov-modulated random walk at zero. Such modifications are highly adaptable from the point of view of specific applications, and here we shall consider a general class given by transition matrices of the form

$$P=\begin{pmatrix} M & H(1) & H(2) & H(3) & \cdots \\ K(1) & G(0) & G(1) & G(2) & \\ K(2) & G(-1) & G(0) & G(1) & \\ K(3) & G(-2) & G(-1) & G(0) & \ddots \end{pmatrix}, \tag{4.2}$$

cf. Example 4.1. Particular important forms are obtained by letting $\{S_n\}$ be right- or left-continuous (skip-free) for levels, i.e. of one of the forms

$$\underset{\text{(a)}}{\begin{pmatrix} B(0) & A(0) & 0 & 0 & \cdots \\ B(1) & A(1) & A(0) & 0 & \\ B(2) & A(2) & A(1) & A(0) & \\ B(3) & A(3) & A(2) & A(1) & \ddots \\ \vdots & & & & \end{pmatrix}}, \quad \underset{\text{(b)}}{\begin{pmatrix} B(0) & B(1) & B(2) & B(3) & \cdots \\ A(0) & A(1) & A(2) & A(3) & \\ 0 & A(0) & A(1) & A(2) & \\ 0 & 0 & A(0) & A(1) & \ddots \\ \vdots & & & & \end{pmatrix}}. \tag{4.3}$$

We say that matrices of the form (4.3a) are of the *GI/M/1 type* and those of the form (4.3b) of the *M/G/1 type*, the obvious motivation being the imbedded Markov chains in IX.3–4 (corresponding to E being a single point, i.e. blocks reducing to numbers).

We return to matrices of the $GI/M/1$ type in more depth in the next section, and give here a simple ergodicity criterion covering most practical situations. Let e denote the vector with all components equal to 1.

Proposition 4.6. *Suppose in* (4.2) *that E is finite, that both P and* $G=\sum_{-\infty}^{\infty}G(k)$ *are irreducible and stochastic, and let* ν *be the stationary distribution of G,* $\mu=\sum_{-\infty}^{\infty}kG(k)e$. *Then P is recurrent if and only if* $\nu\mu\leqslant 0$, *and positive recurrent if and only if* (a) $\sum_1^{\infty}kH(k)e<\infty$, (b) $\nu\mu<0$.

Proof. We consider a Markov-modulated random walk $\{(J_n,S_n)\}$ constructed from the $G(k)$ as in Example 4.1 and a $E\times\mathbb{N}$-valued Markov chain $\{(I_n,L_n)\}$ governed by P (L for level). Then the two chains evolve in the same way in levels $\geqslant 1$. More precisely, if we let $\sigma=\inf\{n\geqslant 1: S_n\leqslant 0\}$, $\tau=\inf\{n\geqslant 1: L_n=0)$ and start the chains in such a way that $I_0=J_0$, $L_0=S_0\geqslant 1$, then $\tau=\sigma$ and $I_n=J_n$, $L_n=S_n$ for $n<\tau$. Hence, recurrence of $\{(I_n,L_n)\}$ is equivalent to $\mathbb{P}_{jl}(\sigma<\infty)$ for all $j\in E$, $l\geqslant 0$, i.e. by Proposition 4.2 to $\nu\mu\leqslant 0$, and (by I.3.9) positive recurrence is equivalent to $\mathbb{E}_{i0}\tau<\infty$ for all $i\in E$. Now clearly

$$\mathbb{E}_{i0}\tau=\sum_{j\in E}m_{ij}+\sum_{l=1}^{\infty}\sum_{j\in E}h_{ij}(l)\{\mathbb{E}_{jl}\sigma+1\}. \tag{4.4}$$

If (a) fails, then Proposition 4.2 shows that (irrespective of $\nu\mu<0$ or $\nu\mu=0$) this expression cannot be finite. If conversely (a) and (b) hold, then Proposition 4.2 shows immediately that (4.4) is finite. Finally, if (a) holds but not (b), we can choose l_0 such that $\mathbb{E}_{jl}\sigma=\infty$ for all $l\geqslant l_0$ and all $j\in E$, and (by irreducibility)

$i \in E$ such that $\mathbb{P}_{i0}(L_n \geqslant l_0, \tau > n) > 0$ for some n. Then $\mathbb{E}_{i0}\tau = \infty$, and positive recurrence fails. $\qquad \square$

Problems

4.1. Which form of (4.2) corresponds to a Lindley process on $E \times \mathbb{N}$?

4.2. Consider the $SM/G/1$ or $SM/SM/1$ queue with $0 < \rho = \mathbb{E}_\nu U/\mathbb{E}_\nu T < 1$. Show that if $i \in E$ is a state such that $\mathbb{P}(J^*_{n-1} = i, W^*_n = 0) > 0$, then the arrival instants of customers with $J_{n-1} = i$, $W_n = 0$ form regeneration points for $\{V_t\}_{t \geqslant 0}$ and $\{Q_t\}_{t \geqslant 0}$. Use Wald's identity to show that the mean cycle length is finite and show (under a suitable non-lattice condition) that V_t, Q_t have weak limits.

4.3. State and prove more general versions of Proposition 4.6, allowing G to be reducible.

4.4. Show that if G is irreducible and substochastic ($Ge \leqslant e$, $Ge \neq e$), then a matrix of the form (4.1) is always positive recurrent [hint: use Problem 4.3].

4.5. Show in Example 4.5 that $\{W_n\}$ is ergodic if and only if $(\varepsilon + \eta(1 - E_k(\eta))) < 1$ where $\varepsilon = \sigma/\tau$ and $\eta, E_k(\eta)$ are as in III.3e.

Notes

The literature on random walks on Markov chains is extensive. For further aspects of the topic, we mention for example Arjas and Speed (1973) and Arndt (1984). The present proof of Theorem 4.3 is from Asmussen and Thorisson (1987), whereas a more restrictive version of Proposition 4.6 is given by Keilson and Wishart (1965) (the $GI/M/1$ case is also covered by Neuts, 1981). The model in Example 4.5 is from Gaver and Lehoczky (1982).

5. MATRIX–GEOMETRIC STATIONARY DISTRIBUTIONS

The ergodicity criterion for transition matrices of the form (4.3) has been found in Proposition 4.6 and Problem 4.3, but the non-trivial problem of finding the stationary distribution π remains. We shall consider here the (easiest) $GI/M/1$ case (4.3a), start by the mathematical theory and proceed to discuss examples and applications below.

We thus consider throughout a Markov chain $\{(J_n, L_n)\}$ governed by a transition matrix P of the form (4.3a) with E finite. The typical state is of the form $il \in E \times \mathbb{N}$ with i denoted as the *phase* and l as the *level*. The Markov chain (J_n, L_n) goes from il to jk with probability $a_{ij}(l - k + 1)$, $1 \leqslant k \leqslant l + 1$, and to $j0$ with probability $b_{ij}(l)$. Levels $k > l + 1$ cannot be reached in one step from level l, i.e. the chain is *skip-free to the right for levels*.

The key to the analysis of the ergodicity problem is to fix l and think of $\{(J_n, L_n)\}$ as a semi-regenerative process with the semi-regeneration points as the returns to level l (which may be of any of the types $i \in E$). To this end, in the notation of Section 3, we let $C = \inf\{n \geqslant 1 : L_n = l\}$. The role of the $\{J_n\}$-chain of Section 3 is then taken by $\{J_n^{(l)}\}$ where $J_n^{(l)}$ is the phase at the kth return of $\{(J_n, L_n)\}$ to level l (in the transient case, $J_n^{(l)}$ is defined only for a finite number of n). Define $R(k)$ as the matrix with elements

$$r_{ij}(k) = \mathbb{E}_{il} \sum_{n=0}^{C-1} I(J_n = j, L_n = l + k) \qquad (5.1)$$

(this is finite also in the transient case) and write $R = R(1) = (r_{ij})$. For measures π on $E \times \mathbb{N}$, we use block notation like $\pi = (\pi_l)_{l \in \mathbb{N}} = (\pi_0 \pi_1 \ldots)$, where $\pi_l = (\pi_{il})_{i \in E}$.

Lemma 5.1. (i) *The matrices $R(k)$ do not depend on the choice of $l = 0, 1, 2, \ldots$;* (ii) *in the recurrent case, $\pi = (\pi_0 \pi_1 \cdots)$ is a stationary measure for $\{(J_n, L_n)\}$ if and only if π_0 is stationary for $\{J_n^{(0)}\}$ and $\pi_k = \pi_0 R(k)$, $k = 1, 2, \ldots$.*

Proof. Here (i) is a consequence of the structure of P which shows that $r_{ij}(k)$ can only be positive if the chain moves upwards from l, and also that the excursions above l are governed by the same probabilities. For (ii), let $l = 0$ and combine the definition of $R(k)$ with Corollary 3.3 to see that π_0 is necessarily stationary for $\{J_n^{(0)}\}$ and connected to π_k by

$$\pi_{jk} = \sum_{i \in E} \pi_{i0} \mathbb{E}_{i0} \sum_{n=0}^{C-1} I(J_n = j, L_n = k) = \sum_{i \in E} \pi_{i0} r_{ij}(k). \qquad \square$$

Now define $D(k; n)$, $R(k; N)$ by

$$d_{ij}(k; n) = \mathbb{P}_{il}(J_n = j, L_n = l + k, C > n),$$

$$r_{ij}(k; N) = \sum_{n=0}^{N} d_{ij}(k; n) = \mathbb{E}_{il} \sum_{n=0}^{(C-1) \wedge N} I(J_n = j, L_n = l + k).$$

Then

$$R(k; N) \uparrow R(k) = \sum_{n=0}^{\infty} D(k; n), \qquad D(k; 0) = \delta_{0k} I \tag{5.2}$$

Lemma 5.2. $R(k) = R^k$, $R(k; N) \leqslant R(1; N)^k$.

Proof. Consider an excursion from il and upwards. The visits to $t(l + k + 1)$ can be divided into segments separated by visits to level $l + k$, the expected number of visits to $t(l + k + 1)$ following $s(l + k)$ before levels $\leqslant l + k$ are reached again begin r_{st}. Hence

$$r_{it}(k + 1) = \sum_{s=1}^{p} r_{is}(k) r_{st}, \qquad R(k + 1) = R(k)R = \cdots = R^{k+1}.$$

Similarly, a visit to $t(l + k + 1)$ contributes only to $r_{it}(k; N)$ if the preceding visit to $s(l + k)$ occurs not later than N and not more than time N has passed since then. This yields $R(k + 1; N) \leqslant R(k; N)R(1; N)$ and the truth of the second statement. $\square$

For any $E \times E$ matrix X, define

$$\hat{A}[X] = \sum_{k=0}^{\infty} X^k A(k), \qquad \hat{B}[X] = \sum_{k=0}^{\infty} X^k B(k)$$

Lemma 5.3. *The Markov chain $\{J_n^{(0)}\}$ has transition matrix $\hat{B}[R] = \sum_{l=0}^{\infty} R^l B(l)$.*

Proof. Let C be defined as above, corresponding to $l = 0$. Then the transition matrix Q (say) of $\{J_n^{(0)}\}$ has elements

$$\begin{aligned} q_{ij} &= \mathbb{P}_{i0}(C < \infty, J_C = j) \\ &= \sum_{n=1}^{\infty} \sum_{k=0}^{\infty} \sum_{s \in E} \mathbb{P}_{i0}(C = n, J_{n-1} = s, L_{n-1} = k, J_n = j) \\ &= \sum_{n=1}^{\infty} \sum_{k=0}^{\infty} \sum_{s \in E} d_{is}(k; n-1) b_{sj}(k) \end{aligned}$$

so that

$$Q = \sum_{n=1}^{\infty} \sum_{k=0}^{\infty} D(k; n-1) B(k) = \sum_{k=0}^{\infty} R^k B(k) = \hat{B}[R]. \qquad \square$$

The following main result is now easily derived:

Theorem 5.4. *Consider an irreducible positive recurrent transition matrix P of the $GI/M/1$ type (4.3a). Then $\{J_n^{(0)}\}$ is positive recurrent as well, i.e. $\pi_0 \hat{B}[R] = \pi_0$ admits a solution π_0 with all $\pi_{i0} > 0$ which is unique up to a constant, and if we normalize π_0 by $\pi_0 \sum_0^\infty R^l e = 1$, then the stationary distribution for P is $\pi = (\pi_0 \pi_1 \pi_2 \cdots)$ with $\pi_l = \pi_0 R^l$, $l = 1, 2, \ldots$.*

Proof. That $\{J_n^{(0)}\}$ is positive recurrent follows from I.3.8, and the remaining statements are immediate from Lemmas 5.1–5.3. $\square$

Two problems remain, to discuss conditions for positive recurrence and to evaluate R.

Corollary 5.5. *Suppose that both P and $A = A(0) + A(1) + \cdots$ are irreducible and stochastic, and let v be the stationary distribution of A and $\lambda = \sum_0^\infty kA(k)e$. Then recurrence of P is equivalent to $v\lambda \geqslant 1$ and positive recurrence to $v\lambda > 1$.*

Proof. Just note that in the notation of Proposition 4.6 we have $G(k) = A(1-k)$, $k = 1, 0, -1, \ldots,$

$$\mu = \sum_{k=-\infty}^{\infty} kG(k)e = \sum_{k=-\infty}^{1} kA(1-k)e = e - \lambda$$

so that $v\mu = 1 - v\lambda$. Also, since $H(k) = 0$, $k = 2, 3, \ldots$, condition (a) is vacuous. $\square$

Proposition 5.6. *R is solution to $R = \hat{A}[R]$ and the minimal non-negative solution to this equation. Furthermore, R can be evaluated by successive iterations, say as limit of the non-decreasing sequence $\{X(n)\}$ given by $X(0) = 0$, $X(n+1) = \hat{A}[X(n)]$.*

Proof. For $n = 0, 1, 2, \ldots$, it follows easily from the definition of the $D(k; n)$ that

$$d_{ij}(1; n+1) = \sum_{k=0}^{\infty} \sum_{s \in E} d_{is}(k; n) a_{sj}(k). \tag{5.3}$$

Using (5.2) and summing over n yields $R = \hat{A}[R]$. Now clearly $X(1) \geqslant X(0)$,

$R \geqslant X(0)$ and hence by induction

$$X(n+1) - X(n) = \sum_{k=0}^{\infty} (X(n)^k - X(n-1)^k)A(k) \geqslant 0,$$

$$R - X(n) = \sum_{k=0}^{\infty} (R^k - X(n-1)^k)A(k) \geqslant 0.$$

Hence, for some X we have $X(n) \uparrow X$, $X \leqslant R$, and also clearly $X = \hat{A}[X]$. It remains to show $R \leqslant X$. Now summing (5.3) for $n = 0, \ldots, N$ and using Lemma 5.2 and $D(1;0) = 0$, we get

$$R(1; N+1) = \sum_{k=0}^{\infty} R(k; N)A(k) \leqslant \sum_{k=0}^{\infty} R(1; N)^k A(k) = \hat{A}[R(1; N)].$$

But $R(1;1) = A(0) = X(1)$ and hence $R(1;2) \leqslant \hat{A}[X(1)] = X(2)$. Continuing, we get by induction that $R(1;N) \leqslant X(N)$. Letting $N \to \infty$ yields $R \leqslant X$. □

Having evaluated R, we have the following alternative criterion for positive recurrence (valid also without conditions on A):

Corollary 5.7. *Suppose that P is irreducible. Then positive recurrence is equivalent to* $\operatorname{spr}(R) < 1$.

Proof. Suppose first that $\{(J_n, L_n)\}$ is recurrent. Then also $\{J_n^{(0)}\}$ is recurrent and therefore (since the state space is finite and the chain irreducible) has a stationary distribution π_0 with all $\pi_{il} > 0$, i.e. $\pi_0 \hat{B}[R] = \pi_0$ (cf. Lemma 5.3). Then π as defined by $\pi_l = \pi_0 R^l$ is a stationary measure for P according to Lemma 5.1(ii) and thus positive recurrence is equivalent to $|\pi| = \pi_0 \sum R^l e < \infty$, or equivalently to the convergence of $\sum_0^\infty R^l$. This always holds if $\lambda_0 = \operatorname{spr}(R) < 1$ since then $R^l = \mathrm{O}(\delta^l)$ for $\lambda_0 < \delta < 1$. If conversely $\lambda_0 \geqslant 1$ and $Rx = \lambda_0 x$, $x \neq 0$, then $R^l x = \lambda_0^l x$ does not tend to zero and $\sum_0^\infty R^l$ cannot converge.

It only remains to prove that in the transient case we always have $\lambda_0 \geqslant 1$. But let i satisfy $\mathbb{P}_{i0}(C = \infty) > 0$. Then by the skip-free property,

$$\sum_{j \in E} r_{ij}(k) \geqslant \mathbb{P}_{i0}(L_n = k \text{ for some } n < C)$$

$$\geqslant \mathbb{P}_{i0}(C = \infty, L_n \to \infty) = \mathbb{P}_{i0}(C = \infty).$$

Therefore $R(k) \to 0$ cannot hold so that we must have $\lambda_0 \geqslant 1$. □

Before proceeding to examples and applications, we state and prove the analogous main results in continuous time:

Theorem 5.8. *Consider an irreducible Markov jump process $\{(J_t, L_t)\}_{t>0}$ with intensity matrix P of the form (4.3a). Then the equation $\hat{A}[R] = 0$ admits a minimal non-negative solution, and the process is positive recurrent if and only if* $\operatorname{spr}(R) < 1$. *In that case, the stationary distribution $(\pi_0 \pi_1 \cdots)$ is given by $\pi_k = \pi_0 R^k$, where $\pi_0 \geqslant 0$ solves $\pi_0 \hat{B}[R] = 0$ and is normalized by $\pi_0 \sum_0^\infty R^l e = 1$. If, furthermore,*

$A=\sum_0^\infty A(k)$ is an irreducible intensity matrix, then a stationary distribution v for A exists, and positive recurrence is equivalent to $v\lambda>0$, where $\lambda=\sum_0^\infty kA(k)e$.

Proof. The form (4.3a) implies that the elemens of P are bounded by

$$\tau_0=\max_{i\in E}|a_{ii}(1)|\vee|b_{ii}(0)|$$

and hence we are in a position to use uniformization (cf. Problem II.4.1). Thus let $\tau>\tau_0$ and define $P^*=\tau^{-1}P+I$. Then P^* is a transition matrix, P is a positive recurrent intensity matrix if and only if P^* is a positive recurrent transition matrix, and in that case the stationary distributions are the same, see again Problem II.4.1. But P^* is of the $GI/M/1$ type corresponding to

$$A^*(k)=\begin{cases}\tau^{-1}A(k) & k\neq 1\\ \tau^{-1}A(k)+I & k=1\end{cases},\qquad B^*(k)=\begin{cases}\tau^{-1}B(k) & k\neq 0\\ \tau^{-1}B(k)+I & k=0\end{cases}$$

Here

$$\hat{A}^*[X]=\tau^{-1}\hat{A}[X]+X,\qquad \hat{B}^*[X]=\tau^{-1}\hat{B}[X]+I.$$

Thus $\hat{A}^*[R]=R$ if and only if $\hat{A}[R]=0$, and $\pi_0\hat{B}^*[R]=\pi_0$ if and only if $\pi_0\hat{B}[R]=0$. This shows the first part of the theorem, and the last one concerning the role of the condition $v\lambda>0$ is seen similarly or by a continuous-time version of Proposition 4.5. □

The crux of the theory developed so far is the amenability of the form of π for numerical implementation on a computer, and also the versatility of the set-up, which includes a variety of models for rather different situations. Here are three examples:

Example 5.9. Consider the $GI/\mathscr{PH}_{\mathrm{AT}}/1$ queue where the service-time distribution has representation (E,Q,π), cf. III.6, and let the level L_n be the number of customers just *before* the nth arrival and the phase I_n the state of the server just *after* the nth arrival. Let $p_{ij}^t(k)$ be the probability that the server, when facing an infinitely long queue and started at time zero in state i, will at time t have completed service of k customers and be in state j. With A the interarrival distribution, it is then seen that the transition matrix is of the form (4.3a) corresponding to

$$a_{ij}(l)=\int_0^\infty p_{ij}^t(l)\,\mathrm{d}A(t),$$

$$b_{ij}(l)=\pi_j\sum_{k=l+1}^{\infty}\sum_{r\in E}a_{ir}(k).$$

Indeed, the expression for $a_{ij}(l)$ is clear and so is the one for $b_{ij}(l)$ once one observes that the double sum is the probability that a server in state i will dispose of at least l customers within a service period. □

Example 5.10. Let M be the maximum of a lattice random walk with point probabilities f_k satisfying $f_r>0$, $f_{r+s}=0$ for $s=1,2,\dots$ and $\mu=\mathbb{E}X=$

$\sum_{-\infty}^{r} k f_k < 0$, and define $\eta_k = \mathbb{P}(M = k)$. An algorithm for determining the ladder height distribution G_+ (and thereby η) is given in IX.5.1 and involves the roots $\alpha_1, \ldots, \alpha_r \in \mathbb{C} \backslash \Delta$ of $\hat{F}(s) = 1$. Alternatively, we may note that η is also the stationary distribution of the corresponding Lindley process and that this has a transition matrix P of the $GI/M/1$ type. In fact (taking for simplicity $r = 2$ here and in the following), we have

$$
P = \left(
\begin{array}{cc|cc|cc}
h_1 & f_1 & f_2 & 0 & 0 & 0 \\
h_0 & f_0 & f_1 & f_2 & 0 & 0 \\
\hline
h_{-1} & f_{-1} & f_0 & f_1 & f_2 & 0 \\
h_{-2} & f_{-2} & f_{-1} & f_0 & f_1 & f_2
\end{array}
\right),
$$

where $h_k = 1 - f_k - f_{k+1} - \cdots - f_2$, and η may therefore be expressed in matrix–geometric form $(\pi_0 \;\; \pi_0 R \;\; \pi_0 R^2 \cdots)$, where π_0 has two components and R is 2×2, say $E = \{0, 1\}$. Not surprisingly π_0, R and α_1, α_2 are intimately connected. In fact, from X.5.1 it can be seen after some calculations that

$$
\eta_j = \frac{1}{\alpha_1 - \alpha_2} \left\{ \frac{\alpha_1}{\alpha_2^j} - \frac{\alpha_2}{\alpha_1^j} \right\} \eta_0, \qquad j = 0, 1, 2, \ldots.
$$

Recalling that $\eta_{2k} = \pi_{0k}$, $\eta_{2k+1} = \pi_{1k}$, one can then check that $\pi_k = \pi_{k-1} R$ for all k determines R uniquely and that

$$
R = \frac{1}{\alpha_1 \alpha_2} \left(
\begin{array}{cc}
-1 & \alpha_1 + \alpha_2 \\
-\alpha_1^{-1} - \alpha_2^{-1} & 1 + \alpha_1/\alpha_2 + \alpha_2/\alpha_1
\end{array}
\right).
$$

In particular, the eigenvalues of R are $\alpha_1^{-2}, \alpha_2^{-2}$. $\qquad \square$

Example 5.11. We let L_t be the queue length at time t in a single-server queue where the rates of arrivals, and service completions, respectively, is of the form $\beta(J_t), \delta(J_t)$, where $\{J_t\}_{t \geqslant 0}$ is an ergodic Markov jump process on $\{1, \ldots, p\}$, say with intensity matrix Q and stationary distribution ν. For obvious reasons, this system is denoted as *the M/M/1 queue in a Markovian environment* or the *Markov-modulated M/M/1 queue (MM/MM/1)* and is useful in situations where, say, the arrivals do not show the regularity of a Poisson process. Obviously $\{(J_t, L_t)\}$ is a Markov jump process with intensity of the form (4.3a) which, letting $\Delta(\beta)$ be the diagonal matrix with diagonal elements $\beta_1, \ldots, \beta_p$, etc. corresponds to

$$
A(0) = \Delta(\beta), \qquad A(2) = \Delta(\delta), \qquad A(1) = Q - \Delta(\beta + \delta),
$$
$$
B(0) = Q - \Delta(\beta), \qquad B(1) = \Delta(\delta),
$$

all other $B(i)$, $A(j)$ equal to zero. In the notation of Theorem 5.8, we have $A = Q$ (hence A is irreducible) and $\lambda = \delta - \beta$. Hence ergodicity is equivalent to $\nu\beta < \nu\delta$ (scalar product!) which is intuitive since $\nu\beta$ represents the average arrival rate and $\nu\delta$ the average service rate. The stationary distribution π as given by

Theorem 5.8 is matrix–geometric. The rate matrix R can not be found explicitly, but at least $\hat{A}[R]=0$ reduces to a quadratic,

$$\Delta(\beta)+R[Q-\Delta(\beta+\delta)]+R^2\Delta(\delta)=0.$$

Also, π_0 can be evaluated as $\pi_0=v(I-R)$. In fact,

$$\begin{aligned} v(I-R)\hat{B}[R] &= v(I-R)(Q-\Delta(\beta)+R\Delta(\delta)) \\ &= v(-\Delta(\beta)+R\Delta(\delta)-RQ+R\Delta(\beta)-R^2\Delta(\delta)) = v0 = 0, \end{aligned}$$

$$v(I-R)\sum_{k=0}^{\infty} R^k e = vIe = 1. \qquad \square$$

Problems

5.1. Show that $\operatorname{spr}(R)=1$ if A, P are irreducible and P is transient or null recurrent [hint: Problem 1.3].

Notes

The theory was initiated by M. F. Neuts and a main source in his 1981 book. The present approach is somewhat different from the mathematical point of view (one key step, Lemma 5.3, was observed by Tweedie, 1982). The set-up has by now become widespread in applied work and the literature contains an abundance of examples, the particular features of which are not always fully caught in the present condensed survey. For instance in Example 5.9, it is messy just to evaluate the matrices $A(k)$, $B(k)$ and there exist ways to compute R which are more efficient than the iteration scheme of Proposition 5.6 (see Lucantoni and Ramaswami, 1985).

CHAPTER XI

Many-Server Queues

1. COMPARISONS WITH $GI/G/1$

Many-server queues present some of the most intricate problems in queueing theory and provide examples of how models, which are simple and well motivated from practical situations, may lead to substantial mathematical difficulties. Not only are the steady-state characteristics far more difficult to evaluate than in the single-server case, but also even just to show existence of (unique) limits presents major difficulties when pursuing the model in its greatest generality.

We consider the standard $GI/G/s$ queue with customers $n = 0, 1, 2, \ldots$, service times $U_0, U_1, \ldots$ (governed by $B(\mathrm{d}x)$), interarrival times $T_0, T_1, \ldots$ (governed by $A(\mathrm{d}x)$) and FCFS queue discipline, meaning that the customers join service in the order they arrive (this does *not* as for $s = 1$ imply the FIFO property of a similar ordering of the departures). The model may be represented in various ways, one of the most obvious being that the customers form one line in the order of arrival and the customer in front joins the first server to become idle. However, most often we think of each server having his own waiting line and the arriving customer joining the line which is the first to become available, i.e. which has the least residual work. For mathematical purposes, we order the residual work in the various lines at time t and thus obtain a vector $V_t = (V_t^{(1)} \cdots V_t^{(s)})$ satisfying $V_t^{(1)} \leqslant V_t^{(2)} \leqslant \cdots \leqslant V_t^{(s)}$. It is of particular interest to observe V_t just before the arrival instants $\tau(n) = T_0 + \cdots + T_{n-1}$ and we write $W_n = V_{\tau(n)-}$. Thus the FCFS discipline implies that $W_n^{(1)}$ is the waiting time of the nth customer. As generalization of $W_{n+1} = (W_n + U_n - T_n)^+$ we also have the celebrated recurrence relation

$$(W_{n+1}^{(1)}, \ldots, W_{n+1}^{(s)}) = R((W_n^{(1)} + U_n - T_n)^+, (W_n^{(2)} - T_n)^+, \ldots, (W_n^{(s)} - T_n)^+), \tag{1.1}$$

where R is the operator on $\mathbb{R}^s$ which orders the coordinates in ascending order (one immediate implication is that $\{W_n\}$ is a Markov chain). Finally, we let Q_t denote the queue length (number of customers in the system) at time t and write

$$|v| = |(v^{(1)} \cdots v^{(s)})| = v^{(1)} + \cdots + v^{(s)} \quad \text{when } v^{(i)} \geqslant 0.$$

Thus, for example, $|V_t|$ is the residual work in system at time t.

By good luck, many problems in the theory of the $GI/G/s$ queue which are difficult to approach directly may be reduced to the case $s=1$ by obtaining suitable bounds in terms of single-server systems with the same traffic intensity. For the present applications, it suffices to consider an initial empty queue, and this will be done for the sake of simplicity. Starting with the lower (and easier) bound, let $\{W_n^*\}$, $\{V_t^*\}$, etc. refer to a $GI/G/1$ queue with the same interarrival times $T_0, T_1, \ldots$ and service times $U_0/s, U_1/s, \ldots$. Loosely speaking, the server in this system works the same way as when all s servers are busy in the $GI/G/s$ system, i.e. as when this system is working at its highest capacity, and in fact we have:

Theorem 1.1. *For initially empty systems, it holds that* $sV_t^* \leqslant |V_t|$ *for all* $t \geqslant 0$.

Proof. We first show, more generally, that $sV_{0-}^* \leqslant |V_{0-}|$ implies $sV_t^* \leqslant |V_t|$ for $0 \leqslant t < T_0$. This follows simply by using the inequality $(x+y)^+ \leqslant x^+ + y^+$ s times to obtain

$$sV_t^* = s(V_{0-}^* + U_0/s - t)^+ \leqslant \left(V_{0-}^{(1)} + U_0 - t + \sum_{i=2}^{s} (V_{0-}^{(i)} - t) \right)^+$$

$$\leqslant (V_{0-}^{(1)} + U_0 - t)^+ + \sum_{i=2}^{s} (V_{0-}^{(i)} - t)^+ = |V_t|$$

for $t < T_0 = \tau(1)$. In particular, $sV_{\tau(1)-}^* \leqslant |V_{\tau(1)-}|$ so that repeating the argument yields $sV_t^* \leqslant |V_t|$, $\tau(1) \leqslant t < \tau(2)$ and the desired conclusion follows by iteration. □

Note that we cannot infer similar bounds for the waiting times themselves: If say $U_0 > sT_0$, then $W_1^{(1)} = 0$ but $W_1^* = (U_0/s - T_0)^+ > 0$.

As upper bound, we shall consider a s-server queue with the same interarrival times and service times, but with a different allocation of the customers to the servers (by an allocation we mean simply a $\{1, \ldots, s\}$ valued function $\sigma(n)$ telling which server customer n joins). In addition to the $GI/G/s$ FCFS allocation rule, corresponding to a customer joining a server with the lowest workload, a main example is the *cyclic discipline* $\sigma(ks+i) = i$, where every sth customer goes to server i. Then the queue in front of server i is simply a $GI/G/1$ queue with interarrival time distribution A^{*s} and service-time distribution B (these s queues are highly dependent due to the dependence between their interarrival times).

Intuitively, one feels that the given $GI/G/s$ rule should be optimal, and indeed this will now in a certain sense be shown to be the case. We let $0 \leqslant J_0 \leqslant J_1 \leqslant \cdots$ be the ordered epochs of initiation of service in the given $GI/G/s$ FCFS system and similarly let $0 \leqslant \tilde{J}_0 \leqslant \tilde{J}_1 \leqslant \cdots$ refer to the modified system (the FCFS rule then simply means that J_n is the instant where customer n initiates service). The crux is now that in the modified system we allocate the service times $U_0, U_1, \ldots$ not according to their order of arrival, but rather according to the order in which they join service. Thus the service time of a particular customer is chosen from the sequence $U_0, U_1, \ldots$ in a way that depends on $\{T_n\}$ and the service times of other customers. However, by independence, distributional properties remain the same (for example in the cyclical case, any server still faces a $GI/G/1$ system). Similar remarks apply to modifications as in the proof of Theorem 1.3 below.

Letting $\min^{(k)}$ be the kth-order statistics ($k = 0, 1, 2, \ldots$), it follows that the departure times from the two systems are

$$D_k = \min^{(k)}\{J_n + U_n : n \geqslant 0\}, \qquad \tilde{D}_k = \min^{(k)}\{\tilde{J}_n + U_n : n \geqslant 0\} \tag{1.2}$$

($k = 0, 1, 2, \ldots$) and we shall show:

Theorem 1.2. *For initially empty sytems, it holds that* $D_k \leqslant \tilde{D}_k$ *for all* k. *In particular,* $Q_t \leqslant \tilde{Q}_t$ *for all* $t \geqslant 0$.

Proof. The crux is to establish the relations

$$\tilde{J}_n \geqslant \max\{\tau(n), \tilde{D}_{n-s}\}, \qquad J_n = \max\{\tau(n), D_{n-s}\}, \tag{1.3}$$

$$\tilde{D}_k = \min^{(k)}\{\tilde{J}_n + U_n : 0 \leqslant n < k + s\}, \tag{1.4}$$

$$D_k = \min^{(k)}\{J_n + U_n : 0 \leqslant n < k + s\}, \tag{1.5}$$

where in (1.3) we let $D_k = \tilde{D}_k = 0$ for $k < 0$. In the first half of (1.3) we have obviously $\tilde{J}_n \geqslant \tau(n)$ and thus the assertion is true for $n = 0, \ldots, s-1$. Also, for $n \geqslant s$ it would follow from $\tilde{J}_n < \tilde{D}_{n-s}$ that at time $\tilde{J}_n$ at least $n + 1 - (n - s) = s + 1$ customers were receiving service, which is impossible. This shows the first half of (1.3). For $n \geqslant k + s$ we then get $\tilde{D}_k \leqslant \tilde{J}_{k+s} \leqslant \tilde{J}_n < \tilde{J}_n + U_n$, and (1.4) follows by combining with (1.2) (of course, (1.5) is just a special case of (1.4)).

For the second half of (1.3), note that if $W_n^{(1)} = 0$ (i.e. customer n does not wait), then $J_n = \tau(n)$, whereas otherwise $J_n = D_{n-s}$. The claim thus follows by noting that $W_n^{(1)} = 0$ if and only if $D_{n-s} \leqslant \tau(n)$.

It now follows from (1.3)–(1.5) that $\tilde{J}_n \geqslant J_n$ for all n. Indeed, this is obvious for $n \leqslant s - 1$, and if $\tilde{J}_n \geqslant J_n$ for all $n \leqslant N$, then

$$\begin{aligned} D_{N+1-s} &= \min^{(N+1-s)}\{J_n + U_n : n \leqslant N\} \\ &\leqslant \min^{(N+1-s)}\{\tilde{J}_n + U_n : n \leqslant N\} = \tilde{D}_{N+1-s}, \\ \tilde{J}_{N+1} &\geqslant \max\{\tau(N+1), \tilde{D}_{N+1-s}\} \geqslant \max\{\tau(N+1), D_{N+1-s}\} = J_{N+1}. \end{aligned}$$

It has also been proved that $\tilde{D}_n \geqslant D_n$ for all n, and since the arrivals in the modified system and the $GI/G/s$ queue are the same, it is immediate that $\tilde{Q}_t \geqslant Q_t$ for all $t \geqslant 0$. □

It is tempting to assert that bounds similar to Theorem 1.2 hold also for the total work in system. This is, however, false in the sense of sample paths (Problem 1.1) but will now be shown to hold in the sense of stochastical ordering:

Theorem 1.3. *For all* t, *it holds that* $|V_t| \overset{\mathscr{D}}{\leqslant} |\tilde{V}_t|$. *Similarly* $|W_n| \overset{\mathscr{D}}{\leqslant} |\tilde{W}_n|$ *for all* n.

Proof. We proceed by a modification of the construction used so far. Fixing t, the idea is to treat the $m = \inf\{n: J_n > t\}$, respectively $\tilde{m} = \inf\{n: \tilde{J}_n > t\}$, customers joining service before t in the same way as before, but to allocate service times to the $M - m$, respectively $M - \tilde{m}$ (M = number of arrivals before t), customers awaiting service in the natural order. Thus the service times of customers arriving before time t are a permutation (different for the $GI/G/1$ system and the alternative rule for allocation to servers) of $U_0, \ldots, U_{M-1}$. The modified systems obtained this way are denoted by superscripts$^{\#}$, and obviously $V_t^{\#} \stackrel{\mathscr{D}}{=} V_t$, $\tilde{V}_t^{\#} \stackrel{\mathscr{D}}{=} \tilde{V}_t$ so that it suffices to show $V_t^{\#} \leqslant \tilde{V}_t^{\#}$. But because of $J_n \leqslant \tilde{J}_n$, we have $m \geqslant \tilde{m}$, and hence

$$V_t^{\#} = \sum_{n=0}^{m-1} (J_n + U_n - t)^+ + \sum_{n=m}^{M-1} U_n \leqslant \sum_{n=0}^{\tilde{m}-1} (\tilde{J}_n + U_n - t)^+$$

$$+ \sum_{n=\tilde{m}}^{m-1} U_n + \sum_{n=m}^{M-1} U_n = \tilde{V}_t^{\#}.$$

The similar inequality for the W_n follows by just the same argument. □

Problems

1.1. Let $s = 2$, $T_0 = 1/2$, $T_1 = 3/4$, $U_0 = 3/2$, $U_1 = 100$, $U_2 = 1/2$, and suppose that in the modified system customers 0, 1 are allocated to server 1, customer 2 to server 2. Show that $|V_1| = 100$, $|\tilde{V}_1| = 1$.

Notes

The area is surveyed in Stoyan (1982), to the references of which we add here Foss (1980). The literature on the subject is notorious for many erroneous or incomplete arguments! Theorem 1.1 is standard, Theorem 1.2 is from Wolff (1977) and Theorem 1.3 from Wolff (1987).

2. REGENERATION AND EXISTENCE OF LIMITS

Motivated from the single-server case, our aim is to show that waiting-time vectors, queue length processes and so on have limits for $\rho = \mathbb{E}U/s\mathbb{E}T < 1$ but not for $\rho \geqslant 1$.

The case $\rho \geqslant 1$ is by far the easiest. In fact, letting W_n^*, V_t^*, etc. refer to the $GI/G/1$ system in Theorem 1.1 (which has the same traffic intensity as the given $GI/G/s$ system), we have $V_t^* \xrightarrow{\mathscr{D}} \infty$, $W_n^* \xrightarrow{\mathscr{D}} \infty$ and thus it is immediately apparent from Theorem 1.1 that $|V_t| \xrightarrow{\mathscr{D}} \infty$, $|W_n| \xrightarrow{\mathscr{D}} \infty$. We give a slightly stronger result below (Corollary 2.6) and pass right on to the more interesting and difficult case $\rho < 1$.

The straightforward generalization of the $GI/G/1$ methodology would be to base the analysis on the sequence $\{\sigma(k)\}$ of customers arriving at an empty system. These are obviously regeneration points for say $\{W_n\}$, but (perhaps somewhat unexpectedly), it turns out that $\rho < 1$ alone is not enough to ensure that the renewal process $\{\sigma(k)\}$ will be non-terminating. Define

$$\alpha_+ = \operatorname{ess\,sup} A = \sup\{x: \mathbb{P}(T > x) > 0\},$$
$$\beta_- = \operatorname{ess\,inf} B = \inf\{x: \mathbb{P}(U < x) > 0\}.$$

Example 2.1. Suppose that $\beta_- > \alpha_+$ (e.g. for $s=2$, T may be uniformly distributed on $(\frac{1}{2}, 1)$ and U degenerate at $\frac{5}{4}$ so that $\rho = \frac{5}{6} < 1$). Then $U_n > T_n$ which means that customer n is still present in the system when $n+1$ arrives. That is, *the system never becomes empty.* □

It turns out, in fact, that $\beta_- < \alpha_+$ ensures that $\{\sigma(k)\}$ will be non-terminating. This assumption does not appear terribly restrictive from the point of view of applications (where typically either A or B has support on the whole of $(0, \infty)$ so that $\beta_- < \alpha_+$ is automatic). Nevertheless, we shall pursue the general case, which presents a classical problem and for which many ingenious and interesting ideas have been developed.

It was noted in Section 1 that $\{W_n\}$ is a Markov chain on $E = \{w: 0 \leqslant w^{(1)} \leqslant \cdots \leqslant w^{(s)}\}$, and we shall show:

Theorem 2.2. *If $\rho < 1$, then $\{W_n\}$ is Harris ergodic on E. In particular, there exists an E-valued random variable W such that $W_n \to W$ in total variation. In particular, the waiting times converge, $W_n^{(1)} \to W^{(1)}$ in total variation.*

The proof rests on two lemmas:

Lemma 2.3. *If $\rho < 1$, then $\{W_n\}$ (or equivalently $\{|W_n|\}$) is tight.*

Proof. This follows simply by comparison with single-server queues: If the modified system of Section 1 corresponds to the cyclical allocation rule, then $\{|\tilde{W}_n|\}$ is the sum of s (dependent) waiting time sequences in $GI/G/1$ queues with $\rho < 1$. Hence $\{|\tilde{W}_n|\}$ is tight, and therefore $\{|W_n|\}$ is so according to Theorem 1.3. □

Lemma 2.4. *If $\rho < 1$, then for all sufficiently large K the set $R_K = \{w \in E : w^{(s)} \leqslant K\}$ is a regeneration set for $\{W_n\}$ in the sense of* VI.3.

Proof. Recalling that $\alpha_+ = \operatorname{ess\,sup} A, \beta_- = \operatorname{ess\,inf} B$, it follows from $\rho < 1$ that $\beta_- < s\alpha_+$. Hence we can find η and ε such that the event $F_k = \{U_k < s\eta - \varepsilon, T_k > \eta\}$ has positive probability, say δ. Let r be an integer so large that $r > sK/\varepsilon + 2s$ and define $F = F_0 \ldots F_{r-1}$. Loosely speaking, a main idea of the proof is that each occurrence of a F_k decreases residual work (at least when $W_k^{(i)} > \eta$ for all i) and that r has been chosen so large that the dependence on the particular value of $W_0 = w \in R_K$ becomes unimportant after r steps. To make this more precise, consider again the cyclical system of Theorem 1.2 and let $W_0 = w \in R_K$. It is then easy to check (say from the expression III.7.3 for the $GI/G/1$ waiting time) that customer n does not have to wait provided that $r - 2s \leqslant n \leqslant r$ and that F occurs. Hence the queue length at the $(r-s)$th arrival is at most $s-1$ in the cyclical system, therefore also in the $GI/G/s$ system. This means that customers $r-s, \ldots, r-1$ enter service immediately. Thus with

$$\lambda(A) = \mathbb{P}_0(W_s \in A \mid F_0 \cdots F_{s-1}),$$

we have

$$\mathbb{P}_w(W_r \in A) \geqslant \delta^r \lambda(A), \qquad w \in R_K \tag{2.1}$$

and it only remains to show that R_L is recurrent for some (and then necessarily all larger) L. But let $L = s\eta$, $G_K = \{W_n \in R_K \text{ infinitely often}\}$, $G = \{\underline{\lim}\, W_n^{(s)} < \infty\}$. Then λ in (2.1) is concentrated on R_L and hence (say by the conditional Borel–Cantelli lemma) $\mathbb{P}G_L \geqslant \mathbb{P}G_K$ for all K. Thus also $\mathbb{P}G_L \geqslant \mathbb{P}G$ since $G_K \uparrow G$. But $\mathbb{P}G = 1$ by tightness, hence $\mathbb{P}G_L = 1$. □

Proof of Theorem 2.2. It follows from Lemma 2.4 that $\{W_n\}$ is Harris recurrent. Also, aperiodicity follows since (2.1) holds for all sufficiently large r (cf. Problem VI.3.2) and we only have to show $|\pi| < \infty$, where π is the stationary measure. Now a perusal of the construction of π in VI.3 easily shows that R_K being a regeneration set implies $\pi(R_K) < \infty$. Hence if $|\pi| = \infty$, VI.3.9 would yield $\mathbb{P}_w(W_n \in R_K) \to 0$ for all K, i.e. $W_n^{(s)} \xrightarrow{\mathscr{D}} \infty$. But this contradicts Lemma 2.3. □

In continuation of Example 2.1, we also get:

Corollary 2.5. *If $\rho < 1$ and $\beta_- < \alpha_+$, then the sequence $\{\sigma(k)\}$ of customers entering an empty system* (i.e. satisfying $W_{\sigma(k)} = (0 \cdots 0)$) *is an aperiodic non-terminating renewal process with finite mean interarrival time.*

Proof. The condition $\beta_- < \alpha_+$ ensures $\mathbb{P}F'_k = \delta' > 0$, where $F'_k = \{U_k < \eta' - \varepsilon < \eta' < T_k\}$. Just as in the proof of Lemma 2.5 (even easier!) it then follows that for all sufficiently large r

$$\mathbb{P}_w(W_r = (0, \ldots, 0)) \geqslant \delta'^r, \qquad w \in R_K.$$

Since R_K is recurrent, a geometrical trial argument then shows that $(0, \ldots, 0)$ is so, and the rest of the argument is much the same as before. □

Returning to the case $\rho \geqslant 1$ for a brief remark, we shall show:

Corollary 2.6. *If $\rho \geqslant 1$, then the waiting time process $\{W_n^{(1)}\}$ satisfies $W_n^{(1)} \xrightarrow{\mathscr{D}} \infty$.*

This follows simply by combining the estimate $|W_n| \xrightarrow{\mathscr{D}} \infty$ observed earlier with the following bound on the dispersion of the servers:

Lemma 2.7. *Define $Z_n^{(i)} = W_n^{(s)} - W_n^{(i)}$. Then the sequence $\{Z_n\}$ is tight.*

Proof. Since all $Z_n^{(i)} \geqslant 0$, it suffices to show that $\{|Z_n|\}$ is tight. Now

$$|Z_{n+1}| = W_{n+1}^{(s)} - (W_n^{(1)} + U_n - T_n)^+ + \sum_{i=2}^{s} \{(W_{n+1}^{(s)} - (W_n^{(i)} - T_n)^+\}.$$

Define $H = (W_n^{(1)} + U_n - T_n)^+$, $K = (W_n^{(s)} - T_n)^+$. If $H \geqslant K$, we have $W_{n+1}^{(s)} = H$ and get

$$|Z_{n+1}| = \sum_{i=2}^{s} \{H - (W_n^{(i)} - T_n)^+\} \leqslant \sum_{i=2}^{s} \{W_n^{(1)} + U_n - W_n^{(i)}\} \leqslant (s-1)U_n,$$

using the inequality $h^+ - j^+ \leqslant h - j$ valid for $h \geqslant j$. If $H < K$, we have $W_{n+1}^{(s)} = K$

and get similarly

$$|Z_{n+1}| = K - H + \sum_{i=2}^{s} \{K - (W_n^{(i)} - T_n)^+\}$$
$$\leqslant W_n^{(s)} - (W_n^{(1)} + U_n) + \sum_{i=2}^{s} \{W_n^{(s)} - W_n^{(i)}\} = |Z_n| - U_n.$$

Thus

$$|Z_{n+1}| \leqslant \max\{|Z_n| - U_n, (s-1)U_n\} \leqslant \cdots$$
$$\leqslant \max\left\{|Z_0| - \sum_{i=0}^{n} U_i, (s-1)U_k - \sum_{i=k+1}^{n} U_i : 0 \leqslant k \leqslant n\right\}$$
$$\stackrel{\mathscr{D}}{=} \max\left\{|Z_0| - \sum_{i=0}^{n} U_i, (s-1)U_k - \sum_{i=0}^{k-1} U_i : 0 \leqslant k \leqslant n\right\}$$
$$\stackrel{\mathscr{D}}{\rightarrow} \max_{0 \leqslant k < \infty}\left\{(s-1)U_k - \sum_{i=0}^{k-1} U_i\right\}.$$

This limit is finite a.s. since $\mathbb{E}U < \infty$ implies $U_k/k \to 0$ and $\sum_0^{k-1} U_i \approx k\mathbb{E}U$. This shows that $\{|Z_n|\}$ is tight. □

We pass on to continuous time.

Corollary 2.8. *Suppose $\rho < 1$. Then $Q = \lim_{t\to\infty} Q_t$ and $V = \lim_{t\to\infty} V_t$ exist* (a) *in the sense of weak convergence provided A is non-lattice and* $\mathbb{P}(U < T) > 0$, (b) *in the sense of total variation convergence provided A is spread out.*

Proof. In case (a), the system regenerates in the usual sense at the instants

$$\tau(k) = T_0 + \cdots + T_{\sigma(k)-1}$$

of arrivals at an empty system. The cycle length distribution is the $\mathbb{P}_0$-distribution of $\tau(1)$. This is non-lattice by VIII.3.2 and the mean $\mathbb{E}_0\tau(1)$ is $\mathbb{E}_0\sigma(1)\mathbb{E}T < \infty$. Thus case (a) is just a standard application of regenerative processes and Corollary 2.5.

In case (b), we define

$$Q_t^* = (Q_t, A_t, R_t^{(1)}, \ldots, R_t^{(s)}) \in \mathbb{N} \times [0, \infty)^{s+1},$$

where A_t is the backwards recurrence time of the arrival process and the $R_t^{(i)}$ are the ordered residual service times just before time t (the residual service time at an empty channel is defined as zero). Then $\{Q_t^*\}$ is a Markov process, and we shall carry out the proof by a slightly tricky application of ideas from VI.2–3. We let ψ be the distribution of $Q^*_{T_0+\cdots+T_{s-1}}$ conditionally upon an initially empty queue and the events $F_0, \ldots, F_{s-1}$ of the proof of Lemma 2.4. Arguing as in the proof of Lemma 2.4, we can find a stopping time τ with $\mathbb{E}\tau < \infty$ such that Q_τ^* is distributed according to ψ. Hence, along similar lines as in VI.3.2, it follows that $\{Q_t^*\}$ has a stationary version $\{\tilde{Q}_t^*\}$, and we shall complete the proof by constructing a coupling of $\{Q_t^*\}$ to $\{\tilde{Q}_t^*\}$. First, it follows from VI.2.3 that we can construct a coupling epoch S for $\{A_t\}$, $\{\tilde{A}_t\}$. This means that there exists $r, \tilde{r}$ such that $S = T_0 + \cdots + T_{r-1} = \tilde{T}_0 + \cdots + \tilde{T}_{\tilde{r}-1}$ and $T_{r+k} = \tilde{T}_{\tilde{r}+k}$, $k = 0, 1, \ldots$ Then

$$\{V_{S+T_r+\cdots+T_{r+k-1}-}\}_{k\in\mathbb{N}}, \qquad \{\tilde{V}_{S+T_r+\cdots+T_{r+k-1}-}\}_{k\in\mathbb{N}} \tag{2.2}$$

are both versions of $\{W_n\}$, and by Harris ergodicity, there exists (VI.3.13) a coupling epoch K such that the chains (2.2) at time K have at least one component equal to zero. But this implies that $\{Q_t^*\}$ and $\{\tilde{Q}_t^*\}$ agree at $S + T_r + \cdots + T_{r+K-1}$ which hence may be taken as the desired coupling epoch. □

Notes

The theory (in particular in discrete time) goes back to a remarkable *tour de force* paper by Kiefer and Wolfowitz (1955). Several (not always correct) variants have been suggested. The key step here, the construction of regeneration points in Lemma 2.4, can be found in Gnedenko and Kovalenko (1968; their upper bound, ensuring positive recurrence, is incorrect). Harris ergodicity was proved by Charlot, Chidouche and Hamami (1978) as an extension of the Kiefer–Wolfowitz argument (Lemma 2.7 is from that paper). Another frequency cited paper in the area is Loynes (1962), dealing with non-independent input, and Lisek (1982) develops a general approach to the ergodic theory of processes governed by evolution equations such as (1.1). Corollary 2.5 was observed by Whitt (1972). For the existence of limits in continuous time under minimal conditions, see Miyazawa (1977) and references therein.

3. THE $GI/M/s$ QUEUE

The $GI/M/s$ case corresponds to B being exponential, $1 - B(x) = e^{-\delta x}$. Within the theory of the $GI/G/s$ queue, this is not only the simplest explicit example (as for $s = 1$) but also more or less the only one (for example, even the main $M/G/s$ case of Poisson arrivals does not seem to simplify substantially).

We shall base the analysis on imbedded Markov chains, cf. IX.3–4, and let Y_n denote the queue length just before the arrival of customer n. Then:

Proposition 3.1. *$\{Y_n\}$ is a Markov chain on $\mathbb{N}$. The transition matrix P is of the form*

$$\begin{array}{c} \\ 0 \\ 1 \\ \vdots \\ s-2 \\ s-1 \\ s \\ s+1 \\ \vdots \end{array}
\begin{array}{c} \begin{array}{cccccccc} 0 & 1 & 2 & & s-1 & s & s+1 & \end{array} \\
\begin{pmatrix}
p_{00} & p_{01} & 0 & 0 & 0 & 0 & 0 & 0 \\
p_{10} & p_{11} & p_{12} & 0 & 0 & 0 & 0 & 0 \\
\vdots & & & & \vdots & & & \\
p_{(s-2)0} & p_{(s-2)1} & p_{(s-2)2} & p_{(s-2)3} \cdots & p_{(s-2)(s-1)} & 0 & 0 & 0 \\
p_{(s-1)0} & p_{(s+1)1} & p_{(s-1)2} & p_{(s-1)3} \cdots & p_{(s-1)(s-1)} & q_0 & 0 & 0 \\
p_{s0} & p_{s1} & p_{s2} & p_{s3} \cdots & p_{s(s-1)} & q_1 & q_0 & 0 \\
p_{(s+1)0} & p_{(s+1)1} & p_{(s+1)2} & p_{(s+1)3} \cdots & p_{(s+1)(s-1)} & q_2 & q_1 & q_0
\end{pmatrix} \\ \hfill \ddots \end{array}$$

with the elements given by

$$q_k = \int_0^\infty e^{-\delta st}\frac{(\delta st)^k}{k!}A(dt), \qquad k = 0, 1, 2, \ldots, \tag{3.1}$$

$$p_{ij} = \int_0^\infty b_{i+1-j}(i+1, 1-e^{-\delta t})A(dt), \qquad j \leqslant i+1 \leqslant s, \tag{3.2}$$

$$p_{ij} = \int_0^\infty A(dt)\int_0^t b_{s-j}(s, 1-e^{-\delta(t-y)})E_{i+1-s}(dy), \qquad i+1 > s > j, \tag{3.3}$$

where E_k denotes the Erlang distribution with k stages and intensity δs, and $b_k(n, p)$ is the binomial probability $\binom{n}{k}p^k(1-p)^{n-k}$.

Proof. This is seen by arguments which are similar to the case $s = 1$ in III.7.2, but also somewhat more elaborate. With K_n the number of customers being served between the arrivals of customers n and $n+1$, we have $Y_{n+1} = (Y_n + 1 - K_n)^+$ so that obviously $p_{ij} = \mathbb{P}_i(K_0 = i+1-j)$. In the following let $Y_0 = i$. Then from $K_0 \geqslant 0$ it is clear that $p_{ij} = 0$ for $j > i+1$. Also, if $i \geqslant s-1$, $j \geqslant s$, then $Y_1 = j$ means that all servers are busy and perform a total of $k = i+1-j$ service events before $T_0 = t$. The probability of this being obviously q_k, it follows that $p_{ij} = q_{i+1-j}$. For (3.2), note that if $i \leqslant s-1$, then all $i+1$ customers present at time zero receive service immediately. Thus conditionally upon $T_0 = t$, the distribution of K_0 is binomial with parameters $(i+1, 1-e^{-\delta t})$ and (3.2) follows. Finally for (3.3), note first that if $i+1 > s$, then in order for $Y_1 = j < s$, the waiting line must disappear at time $y < T_0 = t$ (say). This is equivalent to $S = y < T_0 = t$, where S has the distribution E_{i+1-s} and is independent of T_0. After time y, the s servers need then complete $s-j$ services in $[y, t]$. The probability of this being obviously $b_{s-j}(s, 1-e^{-\delta(t-y)})$, (3.3) follows. □

The next step in the further analysis of the system is to derive the stationary distribution of $\{Y_n\}$:

Theorem 3.2. *The Markov chain $\{Y_n\}$ has a stationary distribution π if and only if $\rho = (s\delta\mathbb{E}T)^{-1} < 1$. In that case, π may then be computed as $\pi_i = Cv_i$ where:*

(i) $v_i = \theta^i, i \geqslant s-1$, *with θ the unique solution in (0, 1) of the transcendental equation*

$$\theta = \int_0^\infty \exp\{-\delta s(1-\theta)y\}A(dy); \tag{3.4}$$

(ii) $v_{s-2}, \ldots, v_0$ *are recursively determined by*

$$v_j = \frac{1}{p_{j(j+1)}}\left\{v_{j+1}(1-p_{(j+1)(j+1)}) - \sum_{i=j+2}^{\infty} v_i p_{i(j+1)}\right\}, \qquad j = s-2, \ldots, 0; \tag{3.5}$$

(iii) $1/C = v_0 + \cdots + v_{s-2} + \theta^{s-1}/(1-\theta)$.
Equivalently, (i) may be formulated as the length of the queue being geometrically distributed given that a queue exists.

Proof of Theorem 3.2. Define $E = \{s-1, s, s+1, \ldots\}$. Then once $\{0, 1, \ldots, s-2\}$ is entered, the next visit of $\{Y_n\}$ to E occurs necessarily at state $s-1$. Letting Q be the transition matrix of the Markov chain obtained by restricting $\{Y_n\}$ to E and $r_n = 1 - q_0 - \cdots - q_n$, it follows that Q is given as

$$\begin{array}{c} \\ s-1 \\ s-2 \\ \\ \\ \end{array} \begin{array}{c} \begin{array}{cc} s-1 & s-2 \end{array} \\ \begin{pmatrix} r_0 & q_0 & 0\cdots \\ r_1 & q_1 & q_0 \\ r_2 & q_2 & q_1 \\ \vdots & & \ddots \end{pmatrix} \end{array}.$$

This is of the same form as in III.7.2 (replacing the δ there by δs), and we may infer immediately that Q is not positive recurrent when $\rho \geqslant 1$. Hence P cannot be so either. Conversely, when $\rho < 1$ a stationary measure v for Q with $|v| < \infty$ exists according to III.7.2, and combining with IX.1 and IX.3 it is seen that v indeed may be taken to be of the form in (i) for $i \geqslant s-1$. By I.3.8, v has a unique extension to a stationary measure for P, and considering the $(j+1)$th coordinate of the equation $vP = v$ yields

$$v_{j+1} = \sum_{i=0}^{\infty} v_i p_{i(j+1)} = \sum_{i=j}^{\infty} v_i p_{i(j+1)}$$

which implies (3.5). Since $v_{s-1}, v_s, \ldots$ are known, (3.5) can then be solved for v_{s-2} and we may repeat the argument to get $v_{s-3}, \ldots, v_0$. Finally, it is clear that for $\pi = Cv$ to be a stationary distribution we simply have to let $C^{-1} = |v|$, and this is equivalent to (iii). □

The remaining steady-state characteristics can now easily be found. Consider first the queue length at an arbitrary point of time:

Corollary 3.3. *If $\rho < 1$ and A is non-lattice, then $\pi_k^* = \lim_{t\to\infty} \mathbb{P}(Q_t = k)$ exists and is given by*

$$\pi_0^* = 1 - \pi_1^* - \pi_2^* - \cdots = 1 - \rho - \frac{1}{\delta \mathbb{E}T} \sum_{k=1}^{s-1} \pi_{k-1}\left(\frac{1}{k} - \frac{1}{s}\right),$$

$$\pi_k^* = \begin{cases} \pi_{k-1}/k\delta\mathbb{E}T & k = 1, \ldots, s \\ \rho\pi_{k-1} & k = s, s+1, \ldots. \end{cases}$$

Proof. Existence follows immediately from Corollary 2.8(a). To derive the form of π_k^*, exactly the same up- and downcrossing argument as for the single-server case in IX.3 applies. Using the regeneration at idle times, the cycle length for $\{Y_n\}$, and $\{Q_t\}$ is $\sigma = \inf\{n \geqslant 1 : Y_n = 0\}$ and $C = T_0 + \cdots + T_{\sigma-1}$ respectively. In particular, $\mathbb{E}_0 C = \mathbb{E}_0 \sigma \mathbb{E}T$. Let $k \geqslant 1$ be fixed and note that the number N_1 of downcrossings of $\{Q_t\}_{0 \leqslant t < C}$ from k to $k-1$ is the same as the number N_2

of upcrossings. But such an upcrossing of $\{Q_t\}$ corresponds to the queue length just before an arrival being $k-1$, i.e. a visit of $\{Y_n\}_{0\leqslant n<\sigma}$ to $k-1$. Letting δ_k be the rate of downcrossings from state k, we have $\delta_k = s\delta$, $k \geqslant s$ and $\delta_k = k\delta$, $1 \leqslant k \leqslant s$. But $\mathbb{E}N_1 = \mathbb{E}N_2$ means that

$$\delta_k \mathbb{E}_0 \int_0^C I(Q_t = k)\,\mathrm{d}t = \mathbb{E}_0 \sum_{n=0}^{\sigma-1} I(Y_n = k-1)$$

i.e.

$$\delta_k \mathbb{E}_0 C \pi_k^* = \pi_{k-1}/\pi_0 = \pi_{k-1}\mathbb{E}_0\sigma$$

From this the result follows by easy manipulations. □

Corollary 3.4. *If $\rho < 1$, then the waiting-time process $\{W_n^{(1)}\}$ has a total variation limit which is a mixture with weights $\zeta = \pi_0 + \cdots + \pi_{s-1}$, $1-\zeta$ of an atom at zero and the exponential distribution with intensity $\eta = s\delta(1-\theta)$.*

Proof. An arriving customer has to wait if and only if he meets $Y = s$ or more customers in the system. Thus the atom at zero has obviously weight $\mathbb{P}(Y \leqslant s-1) = \pi_0 + \cdots + \pi_{s-1}$ in equilibrium. If $Y \geqslant s$, the customer has to wait until $Y - s + 1$ services have been completed, i.e. until $Y - s + 1$ events have occurred in a Poisson process with intensity $s\delta$. Since the distribution of $Y - s$ given $Y \geqslant s$ is geometric with parameter θ, the distribution of $W^{(1)}$ given $W^{(1)} > 0$ can therefore be evaluated as for the case $s = 1$ as a geometric mixture of Erlang distributions, and this leads immediately to the conclusion of the corollary. □

Problems

3.1. Explain that consideration of queue lengths just after departure times in $M/G/s$ does not lead to a Markov chain when $s > 1$.

Notes

The $GI/M/s$ queue is a favourite topic of many textbooks, of which we mention for example Takács (1962) and Gross and Harris (1974). For further aspects of the system, a basic reference is de Smit (1972). The reader interested in further explicit examples may consult Ovuworie (1980) for a survey and references, and Hillier and Yu (1981) and Seelen *et al.* (1985) for numerical tables.

CHAPTER XII

Conjugate Processes

1. CONJUGATE RANDOM WALKS

A family $(F_\theta)_{\theta\in\Theta}$ of distribution on $\mathbb{R}$ is called a *conjugate family* if the F_θ are mutually equivalent with densities of the form

$$\frac{\mathrm{d}F_\theta}{\mathrm{d}F_{\theta_0}}(x) = \exp\{(\theta-\theta_0)x - c(\theta_0;\theta)\} \tag{1.1}$$

and if for some fixed $\theta_0\in\Theta$ the parameter set Θ contains all $\theta\in\mathbb{R}$ for which (1.1) defines a distribution for some $c(\theta_0;\theta)$. Since $1 = \int \mathrm{d}F_\theta/\mathrm{d}F_{\theta_0}\cdot \mathrm{d}F_{\theta_0}$, (1.1) implies that $c(\theta_0;\theta)$ is given in terms of the cumulant generating function $\kappa(\theta_0;\cdot)$ of F_{θ_0} by

$$c(\theta_0;\theta) = \kappa(\theta_0;\theta-\theta_0) = \log\int_{-\infty}^{\infty} \exp\{(\theta-\theta_0)x\}\,\mathrm{d}F_{\theta_0}(x). \tag{1.2}$$

Therefore (1.1) also implies

$$\kappa(\theta;\beta) = \kappa(\theta_0;\theta-\theta_0+\beta) - \kappa(\theta_0;\theta-\theta_0) \tag{1.3}$$

and in conjunction with $\Theta = \{\theta:\kappa(\theta_0;\theta-\theta_0)<\infty\}$ for some (and then all) $\theta_0\in\Theta$, (1.3) is easily seen to be equivalent to $(F_\theta)_{\theta\in\Theta}$ being a conjugate family. The term *conjugate* stems from two distributions satisfying a relation of the form (1.1) being called conjugate.

Example 1.1. Let (F_θ) be the family of normal distribution on $\mathbb{R}$ with fixed variance, say T, and arbitrary mean $\mu = \mu_\theta$. Writing μ_θ on the form θT, $\mathrm{d}F_\theta/\mathrm{d}F_{\theta_0}$ becomes

$$\exp\left\{-\frac{1}{2T}[(x-\theta T)^2 - (x-\theta_0 T)^2]\right\} = \exp\{(\theta-\theta_0)x + [\theta_0^2-\theta^2]T/2\}$$

and since $\Theta = \mathbb{R}$, we have a conjugate family with $\kappa(\theta;\beta) = \frac{1}{2}T(\beta^2+2\theta\beta)$. □

For each θ, we can now consider a random walk with increments distributed according to F_θ. Letting the governing probability measure be, say, $\mathbb{P}_\theta$, it follows from (1.1) and (1.2) that we have the relation

$$\frac{\mathrm{d}\mathbb{P}_{\theta_0}}{\mathrm{d}\mathbb{P}_\theta}(x_1,\ldots,x_n) = \exp\{(\theta_0-\theta)s_n - n\kappa(\theta;\theta_0-\theta)\}$$

($s_n = x_1 + \cdots + x_n$) for the set of joint distributions of $(X_1,\ldots,X_n)$. In particular, if $G \in \mathscr{F}_n = \sigma(X_1,\ldots,X_n)$, then we can evaluate $\mathbb{P}_{\theta_0}G$ by integrating $\mathrm{d}\mathbb{P}_{\theta_0}/\mathrm{d}\mathbb{P}_\theta$ with respect to $\mathrm{d}\mathbb{P}_\theta$ over G and get the basic relation

$$\mathbb{P}_{\theta_0}G = \mathbb{E}_\theta[\exp\{(\theta_0-\theta)S_n - n\kappa(\theta;\theta_0-\theta)\};G]. \tag{1.4}$$

By standard arguments, this formula has an extention to random variables Z which are $\mathscr{F}_n$-measurable and (say) bounded and non-negative

$$\mathbb{E}_{\theta_0}Z = \mathbb{E}_\theta[\exp\{(\theta_0-\theta)s_n - n\kappa(\theta;\theta_0-\theta)\};Z]$$

Similar remarks apply many times in the following.

In applications, we are not given a conjugate family but rather a single F. We can then, however, form the conjugate family generated by F, i.e. choose θ_0 arbitrarily and define the family by (1.1) and (1.2). Then F is imbedded as $F = F_{\theta_0}$ in (F_θ) and (1.4) applies to express $\mathbb{P}G = \mathbb{P}_{\theta_0}G$, $G \in \mathscr{F}_n$, as an expectation corresponding to a different distribution (namely F_θ) of the increments of the random walk. At a first sight this looks to be a complicated way of evaluate $\mathbb{P}G$. However, in many cases the F_θ-random walk has more convenient properties than the given one, as will be demonstrated by a number of applications.

Example 1.2. In queueing theory, we have $F(x) = \mathbb{P}(U - T \leqslant x)$ with U, T having the service-time distribution B, and the interarrival distribution A, respectively. If we define $A_\theta, B_\theta, F_\theta$ by

$$\frac{\mathrm{d}A_\theta}{\mathrm{d}A}(x) = \exp\{-(\theta-\theta_0)x - \kappa_A(\theta_0-\theta)\},$$

$$\frac{\mathrm{d}B_\theta}{\mathrm{d}B}(x) = \exp\{(\theta-\theta_0)x - \kappa_B(\theta-\theta_0)\}$$

$$F_\theta(x) = \mathbb{P}_\theta(U - T \leqslant x) = \int_0^\infty B_\theta(t+x)\,\mathrm{d}A_\theta(t),$$

then F_θ corresponds precisely to U, T having distributions B_θ, A_θ, and the conjugate family therefore corresponds to a family of queues with parameters A_θ, B_θ conjugate to the given A, B. Furthermore, we can extend $\mathscr{F}_n$ to $\sigma(U_1,\ldots,U_n;T_1,\ldots,T_n)$ and it is then readily seen that (1.4) still remains valid. □

The basic case for queueing theory is F having negative mean μ, but not being concentrated on $(-\infty, 0]$. The typical shape of the cumulant generating function (c.g.f) $\kappa(\cdot)$ of F is illustrated in Fig. 1.1(a). The slope at zero is $\mu < 0$ and κ is strictly convex. Since $\operatorname{supp} F \cap (0,\infty) \neq \varnothing$, $\kappa(\beta) \to \infty$ as $\beta \to \infty$. In fact $\kappa(\beta)$ may be $+\infty$ for large but finite β, and in the case of heavy tails we may even have $\kappa(\beta) = \infty$ for all $\beta \neq 0$ (this situation is therefore excluded from the following discussion). However, typically the constant $\gamma_0 > 0$ defined uniquely by $\kappa'(\gamma_0) = 0$ will exist [e.g. this is always the case if $\operatorname{supp} F$ is bounded to the right]. We shall then take $\theta_0 = -\gamma_0$ and have $0 \in \Theta$. By (1.3),

$$\mu_\theta = \mathbb{E}_\theta X = \kappa'(\theta;0) = \kappa'(\theta-\theta_0).$$

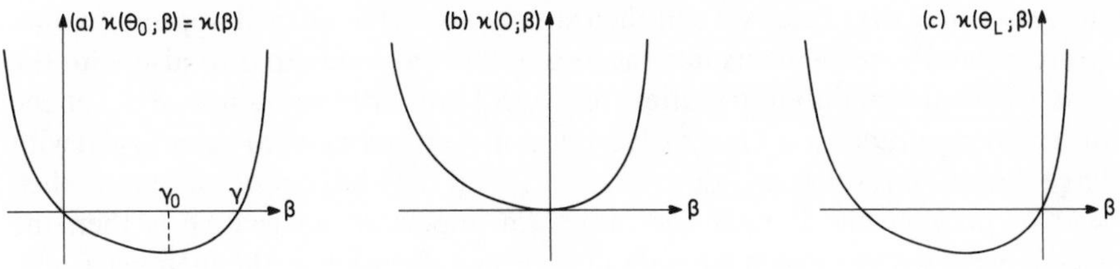

Fig. 1.1

In particular, $\mu_0 = \kappa'(-\theta_0) = \kappa'(\gamma_0) = 0$ (and $\mu_\theta < 0$, $\theta < 0$, $\mu_\theta > 0$, $\theta > 0$). Figure 1.1(b) depicts the c.g.f. $\kappa(0;\beta)$.

In addition to $\theta = 0$, a further parameter value $\theta = \theta_L$ is of particular importance. This is defined as $\theta_L = \theta_0 + \gamma$, where $\gamma > \gamma_0$ is the solution of the *Lundberg equation* $\kappa(\gamma) = 0$ (similar remarks as above apply to the existence). Since $\theta_L > 0$, we have $\mu_{\theta_L} > 0$ and the c.g.f. $\kappa(\theta_L;\beta)$ is depicted in Fig. 1.1(c). Note how both $\kappa(0;\beta)$ and $\kappa(\theta_L;\beta)$ emerge from $\kappa(\beta)$ by simple geometric operations.

Example 1.3. Consider $M/M/1$ with A, B having intensities $\beta < \delta$. Then as in Example 1.2, A_θ, B_θ have intensities $\beta_\theta = \beta + \theta - \theta_0$ and $\delta_\theta = \delta - \theta + \theta_0$, respectively, where θ_0 is determined by the requirement $\mathbb{E}_0 X = 0$, i.e. $\beta_0 = \delta_0$ or $\theta_0 = (\beta - \delta)/2$. The Lundberg equation $\kappa(\gamma) = 0$ is

$$\log\frac{\delta}{\delta-\gamma} + \log\frac{\beta}{\beta+\gamma} = 0$$

and the solution $\gamma > 0$ is $\gamma = \delta - \beta$. Thus $\beta_{\theta_L} = \delta$, $\delta_{\theta_L} = \beta$, and $\mathbb{P}_{\theta_L}$ correspond to interchanging the given intensities β, δ. □

Problems

1.1. Show that F_θ is lattice if and only if F is lattice, spread out if and only if F is spread out and satisfies Cramér's condition (C) if and only if F does so [hint: if $|\hat{F}(t_k)| \to 1$, then $\cos t_k x \to 1$ for a.a. x with respect to the symmetrized distribution H with characteristic function $\hat{H}(t) = \hat{F}(t)\hat{F}(-t)$. Hence, $\hat{H}_\theta(t_k) \to 1$ and $|\hat{F}_\theta(t_k)| \to 1$.]

Notes

Conjugate distribution were introduced independently by Khintchine in statistical mechanics and by Lundberg in risk theory. The terms 'Lundberg equation', 'Lundberg parameter' and so on are therefore standard in risk theory, but not for random walks and queues where the terminology is rather diverse (e.g. Feller, 1971, talks of the 'associated random walk').

What we have called here a conjugate family of distributions is in statistical terms a particular simple case of an exponential family and corresponds for example in the terminology of Barndorff–Nielsen (1978) to a one-parameter full exponential family in canonical representation.

2. THE SADDLE-POINT METHOD. RELAXATION TIME APPROXIMATIONS

Suppose that we are interested in studying the behaviour of $\mathbb{E}g_n(S_n) = \mathbb{E}_{\theta_0}g_n(S_n)$ for a suitable sequence of functions. If g_n is of the form $g_n(x) = g((x - n\mu)/\sqrt{n})$, the problem is easy since we can then simply apply the normal approximation. More generally, refinements like the local CLT may apply to deal also with the case where the main contribution to $\mathbb{E}g_n(S_n)$ comes from values of S_n of the order of magnitude $n\mu + O(n^{1/2})$. However, in many cases we are concerned with large deviation results, say $g_n(x) = I(x > x_n)$ and the above procedure may then lead to poor results if x_n is too large. An important alternative is then the *saddle-point method* which consists in choosing, for each n, the member $\theta = \theta_n$ of the conjugate family satisfying $\mathbb{E}_\theta S_n = x_n$. Then by (1.4), we can write $\mathbb{P}(S_n > x_n) = \mathbb{P}_{\theta_0}(S_n > x_n)$ as

$$\exp\{(\theta_0 - \theta)x_n - n\kappa(\theta; \theta_0 - \theta)\}\mathbb{E}_\theta[\exp\{(\theta_0 - \theta)(S_n - x_n)\}; S_n - x_n > 0] \quad (2.1)$$

and CLT expansions for the $\mathbb{P}_\theta$-distribution of $S_n - x_n$ then hopefully permit an estimation of the expectation. For example, let $x_n = n(\mu + \varepsilon)$, $\varepsilon > 0$. By a translation, we can achieve $\mu + \varepsilon = 0$ (i.e. $\mu < 0$) and have then:

Theorem 2.1 (BAHADUR AND RAO). *Consider a random walk where F has negative mean and satisfies Cramér's condition (C). Suppose that the solution $\gamma_0 > 0$ to $\kappa'(\gamma_0) = 0$ exists and satisfies $|\kappa'''(\gamma_0)| < \infty$, and define $\delta = e^{\kappa(\gamma_0)}$, $\sigma_0^2 = \kappa''(\gamma_0)$. Then*

$$\mathbb{P}(S_n > 0) = \frac{\delta^n}{\gamma_0\sqrt{2\pi\sigma_0^2 n}}\{1 + o(1)\}.$$

Proof. Since $x_n = 0$ and $\mathbb{E}_0 S_n = 0$, the appropriate choice is $\theta = 0$ for all n. By (1.3), $\kappa(0; \theta_0) = -\kappa(\gamma_0)$ and hence according to (2.1) $\mathbb{P}(S_n > 0) = \delta^n I_n$ where $I_n = \mathbb{E}_0[e^{-\gamma_0 S_n}; S_n > 0]$. Define $F_n(x) = \mathbb{P}_0(S_n \leqslant x\sigma_0\sqrt{n})$. Then F_1 satisfies (C), cf. Problem 1.1, and also the third moment exists because of $|\kappa'''(\gamma_0)| < \infty$. Hence (Bhattacharya and Rao, 1976, Theorem 20.1) we may write

$$F_n(dx) = \Phi(dx) + n^{-1/2} f'(x)\,dx + G_n(dx), \quad (2.2)$$

where $f(x) = \eta(1 - x^2)e^{-x^2/2}$ for some constant η and G_n is a (possibly signed) measure with $\|G_n\| = o(n^{-1/2})$. Thus with $g(x) = \exp\{-\gamma_0\sigma_0 x\}$ we have

$$\begin{aligned} I_n &= \int_0^\infty g(xn^{1/2})F_n(dx) \\ &= \int_0^\infty g(xn^{1/2})\Phi(dx) + n^{-1/2}\int_0^\infty g(xn^{1/2})f'(x)\,dx + O(\|G_n\|). \end{aligned}$$

Here $g(xn^{1/2}) \to 0$ as $n \to \infty$ and hence by dominated convergence $\int gf' = o(1)$. Hence, using dominated convergence once more, we get

$$\begin{aligned} I_n &= \int_0^\infty g(xn^{1/2})\frac{1}{\sqrt{2\pi}}e^{-x^2/2}\,dx + o(n^{-1/2}) \\ &= \frac{1}{\sqrt{2\pi n}}\int_0^\infty e^{-\gamma_0\sigma_0 y}e^{-y^2/2n}\,dy + o(n^{-1/2}) \end{aligned}$$

$$= \frac{1}{\sqrt{2\pi n}} \left\{ \int_0^\infty e^{-\gamma_0 \sigma_0 y} \, dy + o(1) \right\} = \frac{(\gamma_0 \sigma_0)^{-1}}{\sqrt{2\pi n}} \{1 + o(1)\}$$

and the proof is complete. □

The above result is classical and well suited to demonstrate the method, but of course somewhat out of the mainstream of the rest of this book. However, a slight variant produces a highly relevant queueing relation:

Theorem 2.2. *Consider the waiting-time process $W_0, W_1, \ldots$ of a stable ($\rho < 1$) GI/G/1 queue. Then if $F(x) = \mathbb{P}(U - T \leqslant x)$ satisfies the assumptions of Theorem 2.1,*

$$\mathbb{E}W - \mathbb{E}W_{N-1} = \frac{\delta^N}{N^{3/2}} \frac{1}{\gamma_0^2 (1-\delta)\sqrt{2\pi\sigma_0^2}} \{1 + o(1)\}. \tag{2.3}$$

Proof. It follows from VII.4.5 that $\mathbb{E}W - \mathbb{E}W_{N-1} = \sum_N^\infty \mathbb{E}S_n^+/n$. Here exactly as above we have $\mathbb{E}S_n^+ = \delta^n J_n$, where

$$J_n = \mathbb{E}_0[S_n e^{-\gamma_0 S_n}; S_n > 0] = \int_0^\infty h(xn^{1/2}) F_n(dx)$$

where $h(x) = \sigma_0 x \exp\{-\gamma_0 \sigma_0 x\}$ is bounded with $h(xn^{1/2}) \to 0$, $n \to \infty$. Thus exactly as above

$$\begin{aligned} J_n &= \int_0^\infty h(xn^{1/2}) \frac{1}{\sqrt{2\pi}} e^{-x^2/2} \, dx + o(n^{-1/2}) \\ &= \frac{\sigma_0}{\sqrt{2\pi n}} \int_0^\infty y e^{-\gamma_0 \sigma_0 y} e^{-y^2/2n} \, dy + o(n^{-1/2}) \\ &= \frac{\sigma_0}{\sqrt{2\pi n}} \int_0^\infty y e^{-\gamma_0 \sigma_0 y} \, dy + o(n^{-1/2}) = \frac{1}{\gamma_0^2 \sqrt{2\pi\sigma_0^2 n}} + o(n^{-1/2}), \end{aligned}$$

$$\begin{aligned} \mathbb{E}W - \mathbb{E}W_{N-1} &= \sum_{n=N}^\infty \frac{\delta^n J_n}{n} = \frac{1}{\gamma_0^2 \sqrt{2\pi\sigma_0^2}} \sum_{n=N}^\infty \frac{\delta^n}{n^{3/2}} \{1 + o(1)\} \\ &= \frac{\delta^N}{N^{3/2}} \left\{ \frac{1}{\gamma_0^2 (1-\delta)\sqrt{2\pi\sigma_0^2}} + o(1) \right\} \end{aligned}$$ □

Obviously, (2.3) is of both the same form and the same spirit as the relaxation-time approximation for $M/M/1$ in III.9 (also, the proof in III.9 is based upon the same method, cf. in particular III.9.10). It is natural to ask whether there is an analogue for the distribution of W_n rather than its mean, and indeed:

Theorem 2.3. *Under the assumption of Theorem 2.2,*

$$\mathbb{P}(W_N \leqslant x) - \mathbb{P}(W \leqslant x) = \frac{\delta^N}{N^{3/2}} e^{-\gamma_0 x} U_+^{(0)}(x) c \{1 + o(1)\}, \tag{2.4}$$

where $U_+^{(0)}$ is the ascending ladder height renewal measure of the mean zero walk and

$$c = \frac{\delta^{3/2}[1 - \mathbb{E}_0 e^{-\gamma_0 S_{\tau_+}}]}{(1-\delta)\gamma_0\sqrt{2\pi\sigma_0^2}}.$$

The proof is substantially more involved and will not be given here.

Problems

2.1. Show that under the conditions of Theorem 2.1 the limiting distribution of S_n given $S_n > 0$ is exponential [hint: Laplace transforms].

Notes

A survey of the saddle-point method is given in Barndorff–Nielsen and Cox (1979). Theorem 2.1 is from a classical paper by Bahadur and Rao (1960), whereas Theorem 2.2 is from Heathcote (1967) and Heathcote and Winer (1969). Theorem 2.3 is from Veraverbeke and Teugels (1975/76), with the constants rewritten somewhat here. For continuous-time analogues, see Teugels (1977, 1982).

3. CONTINUOUS TIME. THE INVERSE GAUSSIAN DISTRIBUTION

We consider as in III.8 a continuous-time random walk $\{S_t\}_{t\geq 0}$ with D-paths and (for the sake of simplicity) jumps of locally bounded variation. Then $\{S_t\}$ is specified as in III.(8.1) by a drift ξ, the variance σ^2 of the Brownian component and a jump measure ν satisfying III.(8.2), and from this it follows readily that the c.g.f. is

$$\kappa(\beta) = \frac{1}{t}\log \mathbb{E}e^{\beta S_t} = \xi\beta + \frac{\sigma^2\beta^2}{2} + \int_{-\infty}^{\infty}(e^{\beta y} - 1)\nu(dy) \tag{3.1}$$

(this is finite exactly when $\int e^{\beta y} I(|y| > 1)\nu(dy) < \infty$.

We now fix θ_0, and for any θ with $\kappa(\theta - \theta_0) < \infty$ in analogy with the discrete time case define

$$\kappa(\theta;\beta) = \kappa(\beta + \theta - \theta_0) - \kappa(\theta - \theta_0). \tag{3.2}$$

A simple calculation shows that this is again of the form (3.1) with ξ, $\nu(dy)$ replaced by $\xi_\theta = \xi + \sigma^2(\theta - \theta_0)$ and $\nu_\theta(dy) = e^{(\theta-\theta_0)y}\nu(dy)$, respectively, and the same σ^2, i.e.

$$\kappa(\theta;\beta) = [\xi + \sigma^2(\theta - \theta_0)]\beta + \frac{\sigma^2\beta^2}{2} + \int_{-\infty}^{\infty}(e^{\beta y} - 1)e^{(\theta-\theta_0)y}\nu(dy) \tag{3.3}$$

Also $\int |y| \wedge 1\nu_\theta(dy) < \infty$ follows easily from the same relation for ν and the ν-integrability of $e^{(\theta-\theta_0)y}$, and hence κ_θ *is the c.g.f. corresponding to a different continuous-time random walk with paths of locally bounded variation.*

Example 3.1. Consider the case of a compound Poisson process, possibly with a drift term, i.e. $\sigma^2 = 0$, $0 < \alpha = \|\nu\| < \infty$, and let $B = \nu/\alpha$. It follows immediately that κ_θ again corresponds to a compound Poisson process with the same drift ξ and intensity

$$\alpha_\theta = \| \nu_\theta \| = \int_{-\infty}^{\infty} e^{(\theta-\theta_0)y}\,\nu(dy),$$

$$B_\theta(dy) = \frac{\nu_\theta(dy)}{\alpha_\theta} = \frac{e^{(\theta-\theta_0)y}B(dy)}{\int_{-\infty}^{\infty} e^{(\theta-\theta_0)x}B(dx)}. \qquad \square$$

The conjugate family can be constructed as follows:

Theorem 3.2. *Let for each θ with $\kappa(\theta-\theta_0)<\infty$ $\mathbb{P}_\theta$ denote the probability measure on $D[0,\infty)$ corresponding to a continuous-time random walk $\{S_t\}$ with c.g.f. $\kappa(\theta;\cdot)$ and write $\mathbb{P}=\mathbb{P}_{\theta_0}$. Then for each $T<\infty$ the restrictions $\mathbb{P}_\theta^{(T)}$ of the $\mathbb{P}_\theta$ to $\mathscr{F}_T = \sigma(S_t; t\leqslant T)$ are mutually equivalent with densities given by*

$$\frac{d\mathbb{P}_\theta^{(T)}}{d\mathbb{P}_{\theta_0}^{(T)}} = \exp\{(\theta-\theta_0)S_T - T\kappa(\theta_0;\theta-\theta_0)\} \tag{3.4}$$

That is, for $G\in\mathscr{F}_T$

$$\mathbb{P}G = \mathbb{P}_{\theta_0}G = \mathbb{E}_\theta[\exp\{(\theta_0-\theta)S_T - T\kappa(\theta;\theta_0-\theta)\}; G]. \tag{3.5}$$

Proof. Define $\mathscr{F}_T^{(n)} = \sigma(S_{kT/n}; k=1,\ldots,n)$. Then, since $\{S_{kT/n}\}$ is a discrete-time random walk corresponding to c.g.f.'s $\kappa(\theta;\cdot)/n$ of the increments, it follows from (1.4) that (3.5) holds whenever $G\in\mathscr{F}_T^{(n)}$. The truth of this for all n implies (3.5) for all $G\in\mathscr{F}_T$, and (3.4) is just a reformulation of (3.5). $\square$

As example of the use of the identity (3.5), we shall now derive the inverse Gaussian distribution $G(\cdot;\xi,c)$ which has already been met in VIII.6 and is of basic importance in the present chapter as well. We recall that $G(T;\xi,c) = \mathbb{P}(\tau(\xi,c)\leqslant T)$, where $\tau(\xi,c)$ is the time of first passage of Brownian motion with drift ξ and unit variance to level $c>0$.

Theorem 3.3. *$G(T;\xi,c)=0$, $T<0$, whereas for $T>0$*

$$G(T;\xi,c) = 1 - \Phi\left(\frac{c}{\sqrt{T}} - \xi\sqrt{T}\right) + e^{2\xi c}\Phi\left(-\frac{c}{\sqrt{T}} - \xi\sqrt{T}\right). \tag{3.6}$$

In particular, $\| G(\cdot;\xi,c)\| = e^{2\xi c}$, $\xi<0$, $=1$, $\xi\geqslant 0$. The density $g(T;\xi,c)$ and the c.g.f. $\lambda(\xi,c;\alpha)$ are

$$\frac{c}{\sqrt{2\pi}}T^{-3/2}\exp\left\{\xi c - \frac{1}{2}\left(\frac{c^2}{T}+\xi^2 T\right)\right\}, \qquad T>0, \tag{3.7}$$

$$\log\int_0^\infty e^{\alpha t}g(t;\xi,c)\,dt = \xi c - c\sqrt{\xi^2-2\alpha}, \qquad \alpha\leqslant\frac{\xi^2}{2}, \tag{3.8}$$

and for $\xi\geqslant 0$ the mean and variance are

$$\mathbb{E}\tau(\xi,c) = \frac{c}{\xi}, \qquad \mathbb{V}\mathrm{ar}\,\tau(\xi,c) = \frac{c}{\xi^3}. \tag{3.9}$$

In the proof, we represent Brownian motion with drift ξ as a probability measure $\mathbb{P}_\xi$ on $D[0,\infty)$ and then have a conjugate family $(\mathbb{P}_\xi)$, cf. (3.3) and Theorem 3.2. Furthermore, if $M_T=\sup_{0\leqslant t\leqslant T}S_t$, then $G(T;\xi,c)=\mathbb{P}_\xi(M_T\geqslant c)$. We shall need:

Lemma 3.4. *For $x,y\geqslant 0$, the joint $\mathbb{P}_0$-density of (M_T,S_T) at $(x,x-y)$ is*

$$\sqrt{\frac{2}{\pi}}\frac{x+y}{T^{3/2}}\exp\left\{-\frac{1}{2T}(x+y)^2\right\}. \tag{3.10}$$

Proof. The joint distribution of (M_T,S_T) was found in III.8.5 by means of the reflection principle,

$$\mathbb{P}_0(M_T\geqslant x,S_T\leqslant z)=\mathbb{P}_0(S_T\geqslant 2x-z)=\int_{2x-z}^{\infty}\varphi_T(u)\,\mathrm{d}u,$$

where $\varphi_T(u)=\mathrm{e}^{-u^2/2T}/\sqrt{2\pi T}$. Differentiating first with respect to x and next to z, the density at x,z becomes

$$2\frac{\partial}{\partial z}\varphi_T(2x-z)=\frac{2(2x-z)}{T}\varphi_T(2x-z).$$

Substituting $z=x-y$, (3.10) follows. □

Proof of Theorem 3.3. Using (3.5) and Lemma 3.4, we get

$$\begin{aligned}G(T;\xi,c)&=\mathbb{P}(\tau(\xi,c)\leqslant T)=\mathbb{P}_\xi(M_T\geqslant c)\\&=\mathbb{E}_0[\exp\{\xi S_T-\xi^2T/2\};M_T\geqslant c]\\&=\int_c^\infty\mathrm{d}x\int_0^\infty\mathrm{d}y\exp\{\xi(x-y)-\xi^2T/2\}\frac{2(x+y)}{T}\varphi_T(x+y).\end{aligned}$$

Substituting $u=x+y$, $v=x-y$, the Jacobian cancels the factor 2 and we get

$$\begin{aligned}&\mathrm{e}^{-\xi^2T/2}\int_c^\infty-\varphi_T'(u)\int_{2c-u}^{u}\mathrm{e}^{\xi v}\,\mathrm{d}v\,\mathrm{d}u\\&=\mathrm{e}^{-\xi^2T/2}\left\{\left[-\varphi_T(u)\int_{2c-u}^{u}\mathrm{e}^{\xi v}\,\mathrm{d}v\right]_c^\infty+\int_c^\infty(\mathrm{e}^{\xi u}+\mathrm{e}^{\xi(2c-u)})\varphi_T(u)\,\mathrm{d}u\right\}\\&=0-0+\int_c^\infty\varphi_T(u-\xi T)\,\mathrm{d}u+\mathrm{e}^{2\xi c}\int_c^\infty\varphi_T(u+\xi T)\,\mathrm{d}u\end{aligned}$$

which is the same as (3.6). The density is found by differentiating (3.6) with respect to T and becomes

$$\begin{aligned}&\frac{1}{2}\left\{(cT^{-3/2}+\xi T^{-1/2})\varphi\left(\frac{c}{\sqrt{T}}-\xi\sqrt{T}\right)\right.\\&\qquad\left.+\mathrm{e}^{2\xi c}(cT^{-3/2}-\xi T^{1/2})\varphi\left(-\frac{c}{\sqrt{T}}-\xi\sqrt{T}\right)\right\}\end{aligned}$$

$$=\frac{1}{2}\left\{(cT^{-3/2}+\xi T^{-1/2})\varphi\left(\frac{c}{\sqrt{T}}-\xi\sqrt{T}\right)\right.$$

$$\left.+(cT^{-3/2}-\xi T^{1/2})\varphi\left(\frac{c}{\sqrt{T}}-\xi\sqrt{T}\right)\right\}$$

$$=cT^{-3/2}\varphi\left(\frac{c}{\sqrt{T}}-\xi\sqrt{T}\right)$$

which is the same as (3.7). The expression for $\|G(\cdot;\xi,c)\|$ follows letting $T\to\infty$ in (3.6). For (3.8), suppose first $\xi\geqslant 0$ and note that if we substitute $\theta=-\xi^2/2$, then (3.7) shows that $\{G(\cdot;\xi(\theta),c)\}_{\theta\leqslant 0}$ is a conjugate family with

$$\kappa(0;\theta)=c(0;\theta)=c(\xi(0)-\xi(\theta))=-c\sqrt{-2\theta},$$

cf. (1.2). Thus by (1.3)

$$\lambda(\xi,c;\alpha)=\kappa(-\xi^2/2;\alpha)=\kappa(0;-\xi^2/2+\alpha)-\kappa(0;-\xi^2/2)$$
$$=\xi c-c\sqrt{\xi^2-2\alpha}.$$

This shows (3.8) for $\xi\geqslant 0$ and we get the mean and variance as

$$\left.\frac{\partial\lambda}{\partial\alpha}\right|_{\alpha=0}=\left.\frac{c}{\sqrt{\xi^2-2\alpha}}\right|_{\alpha=0}=\frac{c}{\xi},$$

$$\left.\frac{\partial^2\lambda}{\partial\alpha^2}\right|_{\alpha=0}=\left.c(\xi^2-2\alpha)^{-3/2}\right|_{\alpha=0}=\frac{c}{\xi^3}.$$

It remains to show that (3.8) holds also for $\xi<0$. But then by (3.7)

$$\int_0^\infty e^{\alpha t}g(t;\xi,c)\,dt=e^{2\xi c}\int_0^\infty e^{\alpha t}g(t;-\xi,c)\,dt$$
$$=\exp\{2\xi c+(-\xi c-\sqrt{\xi^2-2\alpha})\}=\exp\{\xi c-\sqrt{\xi^2-2\alpha}\}.\qquad\square$$

We formulate the last step of the proof as

Corollary 3.5. *For* $\xi<0$, $\mathbb{P}(\tau(\xi,c)<\infty)=e^{2\xi c}$, *and the distribution of* $\tau(\xi,c)$ *given* $\{\tau(\xi,c)<\infty\}$ *is the same as the distribution of* $\tau(-\xi,c)$.

Some further noteworthy properties of the distribution are in

Proposition 3.6. (i) $G(T;\xi,c)=G(T/u^2;\xi u;c/u)$; (ii) *for* $c_1,c_2>0$ $G(\cdot;\xi,c_1+c_2)=G(\cdot;\xi,c_1)*G(\cdot;\xi,c_2)$; (iii) *if* $\xi>0$, *then* $\tau(\xi,c)$ *is asymptotically normal as* $c\to\infty$ *with mean* c/ξ *and variance* c/ξ^3.

Proof. Case (i) is clear from (3.6). For (ii), note just that $\tau(\xi,c_1+c_2)=\tau(\xi,c_1)+V$, where V is the time of first passage from level c_1 to level c_1+c_2. But by the strong Markov property, V is independent of $\tau(\xi,c_1)$ and distributed as $\tau(\xi,c_2)$. Finally, (iii) is an immediate consequence of (ii) and (3.9). $\square$

It should be noted that the proof of (ii) more generally tells that for fixed $\xi \geqslant 0$ the process $\{\tau(\xi, c)\}_{c \geqslant 0}$ has stationary independent increments. The paths being non-decreasing and $\tau(\xi, c)$ being supported by the whole of $(0, \infty)$ (excluding a drift term), the general representation theorem for such processes therefore tells us that $\{\tau(\xi, c)\}_{c \geqslant 0}$ is a pure jump process with jump measure v_ξ concentrated on $(0, \infty)$ and satisfying $\int_0^\infty y \wedge 1 v_\xi(\mathrm{d}y) < \infty$,

$$\int_0^\infty (\mathrm{e}^{\beta y} - 1) v_\xi(\mathrm{d}y) = \log \mathbb{E} \mathrm{e}^{\beta \tau(\xi, 1)} = \lambda(\xi, 1; \beta) = \xi - \sqrt{\xi^2 - 2\beta}.$$

An elementary calculation shows that the solution is

$$v_\xi(\mathrm{d}y) = \frac{\sqrt{\pi} \mathrm{e}^{-y\xi^2/2}}{2y^{3/2}} \mathrm{d}y.$$

For $\xi = 0$, $\{\tau(0, c)\}$ *is called the positive stable process with exponent* $\frac{1}{2}$.

Notes

The (standard) likelihood ratio identity (3.4) and its stopping time extension in the next section is related to *Girsanov's formula* which occurs in the general theory of stochastic processes, see e.g. Liptser and Shiryaev (1977).

A number of variants of the derivation of the inverse Gaussian distribution are around, see for example Skorohod (1965), pp. 171ff., Harrison (1985), Ch. I) and Siegmund (1985, III.3). Further properties of the distribution can be found in Johnson and Kotz (1969).

4. THE FUNDAMENTAL IDENTITY OF SEQUENTIAL ANALYSIS

Consider as in Sections 1 and 3 a conjugate family $(\mathbb{P}_\theta)_{\theta \in \Theta}$ governing a random walk $\{S_t\}_{t \in T}$ in discrete or continuous time. Define $\mathscr{F}_T = \sigma(S_t; t \leqslant T)$, with the usual extension to stopping times.

Theorem 4.1 *Let τ be any stopping time and let $G \in \mathscr{F}_\tau$, $G \subseteq \{\tau < \infty\}$. Then for each θ_0, $\theta \in \Theta$*

$$\mathbb{P}_{\theta_0} G = \mathbb{E}_\theta[\exp\{(\theta_0 - \theta)S_\tau - \tau\kappa(\theta; \theta_0 - \theta)\}; G]. \tag{4.1}$$

Proof. By monotone convergence, it is sufficient to consider the case where $G \subseteq \{\tau \leqslant T\}$. Then $G \in \mathscr{F}_T$ and by (1.4) and (3.5)

$$\mathbb{P}_{\theta_0} G = \mathbb{E}_\theta[\exp\{(\theta_0 - \theta)S_T - T\kappa(\theta; \theta_0 - \theta)\}; G].$$

Writing $T = (T - \tau) + \tau$, $S_T = (S_T - S_\tau) + S_\tau$ this becomes $\mathbb{E}_\theta R_1 R_2$, where

$$R_1 = \mathbb{E}_\theta[\exp\{(\theta_0 - \theta)(S_T - S_\tau) - (T - \tau)\kappa(\theta; \theta_0 - \theta)\} | \mathscr{F}_\tau]$$
$$R_2 = \exp\{(\theta_0 - \theta)S_\tau - \tau\kappa(\theta; \theta_0 - \theta)\} I(G).$$

But any process of the type considered is easily seen to be strong Markov, cf. I.6.2, III.8.1. Thus on $\{\tau \leqslant T\}$, it follows from $\mathbb{E}_\theta \exp\{\alpha S_u\} = \exp\{u\kappa(\theta; \alpha)\}$ that $R_1 = 1$, completing the proof. □

The formula (4.1) is obviously just obtained by allowing N in (1.4) and T in (3.5) to be a stopping time. It is known in the literature (sometimes in variants like (4.2) below) as *Wald's fundamental identity* or the *fundamental identity of sequential analysis*. It will play a crucial role in the rest of the chapter and we shall start by some simple illustrations, in particular the relation of (4.1) to some standard martingale identities.

Consider a given random walk $\{S_t\}_{t\in T}$ in discrete or continuous time, and let the c.g.f. of S_t be $t\kappa(\alpha)$. It is then a matter of routine (cf. e.g. the proof of III.9.8) to check that $\{\exp\{\alpha S_t - t\kappa(\alpha)\}\}_{t\in T}$ is a martingale for each α with $\kappa(\alpha) < \infty$. It is thus reasonable to ask for conditions on an a.s. finite stopping time τ for

$$\mathbb{E}\exp\{\alpha S_\tau - \tau\kappa(\alpha)\} = 1 \tag{4.2}$$

to hold. The answer is provided by Theorem 4.1: we imbed the given process corresponding to $\theta = 0$ in a conjugate family $(\mathbb{P}_\theta)_{\theta\in\Theta}$, $\kappa(0;\theta) = \kappa(\theta)$. Then $\kappa(\alpha) < \infty$ implies $\alpha \in \Theta$ and if we let $\theta_0 = \alpha$, $\theta = 0$ in (4.1), $G = \{\tau < \infty\}$ we get

$$\mathbb{P}_\alpha(\tau < \infty) = \mathbb{E}\exp\{\alpha S_\tau - \tau\kappa(\alpha)\}$$

since $\mathbb{P}(\tau < \infty) = 1$ implies that $I(\tau < \infty)$ is vacuous on the r.h.s. of (4.1). That is, (4.2) *holds if and only if* $\mathbb{P}_\alpha(\tau < \infty) = 1$. Here are some main examples:

(a) $\tau = \inf\{t : S_t \notin [-a, b]\}$ with $a, b > 0$. Here τ is finite for any random walk in discrete or continuous time (except for the trivial case $S_t \equiv 0$). Hence $\mathbb{P}_\alpha(\tau < \infty) = 1$ for all α and (4.2) *holds always*.

(b) $\tau = \inf\{t : S_t > u\}$ with $u > 0$. Here $\mathbb{P}_\alpha(\tau < \infty) = 1$ if and only if the $\mathbb{P}_\alpha$-process has drift $\geqslant 0$. That is, (4.2) *holds if and only if* $\mathbb{E}_\alpha S_t/t = \kappa'(\alpha; 0) = \kappa'(\alpha) \geqslant 0$.

Example 4.2. Let $\{S_t\}_{t\geqslant 0}$ be Brownian motion with drift $\xi \geqslant 0$ and let $\tau = \tau(\xi, c) = \inf\{t : S_t \geqslant c\}$, cf. Section 3. Here $\kappa(\alpha) = \alpha^2/2 + \xi\alpha$ so that $\kappa'(\alpha) \geqslant 0$ when $\alpha \geqslant -\xi$. Since $S_\tau = c$, (4.2) becomes

$$1 = \mathbb{E}\exp\{\alpha c - \tau(\xi, c)(\alpha^2/2 + \xi\alpha)\}.$$

For each $\theta \leqslant \xi^2/2$, there is a unique $\alpha \geqslant -\xi$ satisfying $-\alpha^2/2 - \xi\alpha = \theta$, namely $\alpha = \sqrt{\xi^2 - 2\theta} - \xi$. Thus

$$\mathbb{E}e^{\theta\tau(\xi,c)} = e^{-\alpha c} = \exp\{\xi c - c\sqrt{\xi^2 - 2\theta}\}$$

which gives another proof of the formula (3.8) for the c.g.f. of the inverse Gaussian distribution (for another application of this method, see III.9.8). □

We finally remark that if in a formal manner we differentiate (4.2) with respect to α, put $\alpha = 0$ and note that $\kappa'(0) = \mathbb{E}X = \mu$, we obtain Wald's identity $\mathbb{E}[S_\tau - \tau\mu] = 0$ (of course, this derivation requires unnecessarily strong conditions on the existence of exponential moments). Similarly, a further differentiation produces Wald's second moment identity, cf. Appendix A8.

Example 4.3. Many of the techniques studied in the present chapter originate from sequential analysis. As a digression, we shall present some of the ideas in that setting, stressing the probability calculations rather than going deep into the statistical aspects (which are practical as well as touching upon questions in the foundations of statistical inference).

Suppose we are given a family of conjugate random walks specified by $(F_\theta)_{\theta\in\Theta}$ with $\mathbb{E}_0 X = \kappa'(0;0) = 0$ and want to test the hypothesis $H_0:\theta \geqslant 0$ versus the alternative $H_1:\theta < 0$. The traditional likelihood ratio test based on a fixed sample size N then rejects for small values of S_N, say $S_N < -a_N$. In the sequential setting, one instead prescribes two constants $a, b > 0$ and proceeds by drawing $X_1, X_2, \ldots$ one after another. If $S_N < -a$ at stage N, we reject H_0. Similarly, H_0 is accepted if $S_N > b$, whereas if $-a \leqslant S_N \leqslant b$, the sampling is continued by drawing X_{N+1}. That is, the sampling stops at time $\tau = \inf\{N \geqslant 1 : S_N \notin [-a, b]\}$ by rejecting H_0 if $S_\tau < -a$ and accepting H_0 if $S_\tau > b$.

The probability problems are to compute the so-called operation characteristic $k(\theta) = \mathbb{P}_\theta(\text{accept } H_0) = \mathbb{P}_\theta(S_\tau > b)$, one purpose being to tell how a, b should be chosen to meet certain prescribed requirements on $k(\theta)$. Also, we want to say something about the distribution of the sample size τ, in particular to evaluate $\mathbb{E}_\theta \tau$. The idea is to observe that if, as will typically be the case, both a and b are large compared to the individual sizes of the X_n, then we may neglect the excess (or overshot) of S_τ over the boundaries-a, b, i.e. use the approximations $S_\tau \approx -a$ on $\{S_\tau < -a\}$, $S_\tau \approx b$ on $\{S_\tau > b\}$. Now the assumption $0 = \mathbb{E}_0 X_n = \kappa'(0;0)$ ensures that to each $\theta \neq 0$ we can find θ_L of opposite sign satisfying $\kappa(0;\theta) = \kappa(0;\theta_L)$, i.e. $\kappa(\theta;\theta_L - \theta) = 0$. Then by (4.2), $1 = \mathbb{E}_\theta \exp\{(\theta_L - \theta)S_\tau\}$ which neglecting the excess over the boundary yields

$$\begin{aligned} 1 &\approx e^{a(\theta-\theta_L)}\mathbb{P}_\theta(S_\tau < -a) + e^{b(\theta_L-\theta)}\mathbb{P}_\theta(S_\tau > b) \\ &= k(\theta)\{e^{b(\theta_L-\theta)} - e^{a(\theta-\theta_L)}\} + e^{a(\theta-\theta_L)}, \end{aligned}$$

i.e.

$$k(\theta) \approx \frac{1 - e^{a(\theta-\theta_L)}}{e^{b(\theta_L-\theta)} - e^{a(\theta-\theta_L)}} \tag{4.3}$$

Similarly, Wald's identity $\mu_\theta \mathbb{E}_\theta \tau = \mathbb{E}_\theta S_\tau$ yields

$$\mu_\theta \mathbb{E}_\theta \tau \approx -a(1 - k(\theta)) + bk(\theta)$$

from which an approximation to $\mathbb{E}_\theta \tau$ follows by inserting (4.3). The case $\theta = 0$ requires a separate treatment. Letting $\theta \uparrow 0$ in (4.3), we have $\theta_L \downarrow 0$ and get

$$k(0) = \lim k(\theta) \approx \lim \frac{(\theta_L - \theta)a}{(\theta_L - \theta)(b + a)} = \frac{a}{a+b}$$

whereas for $\mathbb{E}_0 \tau$, Wald's second moment identity yields

$$\mathbb{E}_0 \tau \approx \frac{a^2(1 - k(0)) + b^2 k(0)}{\kappa''(0;0)} \approx \frac{ab}{\kappa''(0;0)}. \qquad \square$$

Notes

A major recent reference for sequential analysis is Siegmund (1985). It should be noted that the setting of conjugate families is not required for all parts of Example 4.3. Instead, pairs of equivalent distributions suffice.

5. THE CRAMÉR–LUNDBERG APPROXIMATION

We consider in the rest of this chapter a given random walk $\{S_t\}_{t\in T}$ in discrete or continuous time, $\kappa'(0)=\mathbb{E}S_1<0$, and the corresponding conjugate family $(\mathbb{P}_\theta)_{\theta\in\Theta}$ with $\theta_0<0$ defined as in Section 1, i.e. as solution of $\kappa'(-\theta_0)=0$ and $\kappa(\theta_0;\cdot)=\kappa(\cdot)$.

We assume here that the solution $\gamma>0$ of the Lundberg equation $\kappa(\gamma)=0$ exists, let $\theta_L=\theta_0+\gamma$ and refer to $\mathbb{P}_{\theta_L}$ as the *Lundberg process*. Its role is largely due to the fact that the fundamental identity (4.1) takes a particular simple form for this case. In fact, by (1.3)

$$\begin{aligned}\kappa(\theta_L;\theta_0-\theta_L)&=\kappa((\theta_L-\theta_0)+(\theta_0-\theta_L))-\kappa(\theta_L-\theta_0)\\&=\kappa(0)-\kappa(\gamma)=0\end{aligned}$$

and hence for $G\in\mathscr{F}_\tau$, $G\subseteq\{\tau<\infty\}$

$$\mathbb{P}G=\mathbb{E}_{\theta_L}[\exp\{-\gamma S_\tau\};G]. \tag{5.1}$$

As the main example on the use of this formula, we shall look at the distribution of the maximum $M=\sup_{t\geqslant 0}S_t$. To this end, take $\tau=\inf\{t\geqslant 0:S_t>u\}$ and let $B(u)=S_\tau-u$ be the overshot.

Theorem 5.1. $\mathbb{P}(M>u)\leqslant e^{-\gamma u}$ *for any* $u\geqslant 0$.

Proof. Let $G=\{\tau<\infty\}$. Then $\mathbb{P}_{\theta_L}G=1$ because of $\mathbb{E}_{\theta_L}S_t>0$, and hence because of $B(u)\geqslant 0$

$$\mathbb{P}(M>u)=\mathbb{P}G=\mathbb{E}_{\theta_L}\exp\{-\gamma S_\tau\}=e^{-\gamma u}\mathbb{E}_{\theta_L}e^{-\gamma B(u)}\leqslant e^{-\gamma u}. \tag{5.2}$$

□

The result of Theorem 5.1 is known as *Lundberg's inequality* in risk theory (see further, Ch. XIII), and has been reproved and refined in queueing theory (where $M\overset{\mathscr{D}}{=}W$) by several authors, see the Notes. The argument in (5.2) is apparently just to neglect the excess $B(u)$ of S_τ over the boundary u. A minor refinement produces the celebrated *Cramér–Lundberg approximation*:

Theorem 5.2. *If* $B(u)$ *converges in* $\mathbb{P}_{\theta_L}$*-distribution as* $u\to\infty$, *say to* $B(\infty)$, *then* $\mathbb{P}(M>u)\approx Ce^{-\gamma u}$, *where* $C=\mathbb{E}_{\theta_L}e^{-\gamma B(\infty)}$.

Proof. Since $e^{-\gamma x}$ is bounded and continuous on $[0,\infty)$, we just have to note that in (5.2) $\mathbb{E}_{\theta_L}e^{-\gamma B(u)}\to C$ by general results on weak convergence. □

The assumption on the existence of $B(\infty)$ will typically hold if just $\kappa'(\gamma)<\infty$. We shall not go into the complications of general Lévy processes but only note:

Theorem 5.3. *For a discrete-time non-lattice random walk* $B(\infty)$ *exists with respect to* $\mathbb{P}_{\theta_L}$ *if* $\kappa'(\gamma)<\infty$. *In that case, C is given in terms of the ladder height distributions* *by*

$$C = \frac{1 - \|G_+\|}{\gamma \int_0^\infty x e^{\gamma x} \mathrm{d}G_+(x)} = \frac{\mathbb{E}X}{\gamma\kappa'(\gamma)\mathbb{E}S_{\tau_-}}\left\{1 - \int_{-\infty}^0 e^{\gamma x} \mathrm{d}G_-(x)\right\}. \tag{5.3}$$

Proof. Since $\kappa'(\gamma) = \kappa'(\theta_L; 0) = \mathbb{E}_{\theta_L} X$, the existence was noted in VII.2.1, and we also find that the $\mathbb{P}_{\theta_L}$-distribution of $B(\infty)$ has density $(1 - G_+^{(L)}(x))/\mu_+^{(L)}$, where $G_+^{(L)}$ is the ascending ladder height distribution for $\mathbb{P}_{\theta_L}$ and $\mu_+^{(L)}$ its mean. Now if we put $\tau = \tau_+$, $G = \{S_{\tau_+} \in A, \tau_+ < \infty\}$ in (5.1), we get

$$G_+(A) = \mathbb{P}G = \mathbb{E}_{\theta_L}[\exp\{-\gamma S_{\tau_+}\}; S_{\tau_+} \in A] = \int_A e^{-\gamma x} \mathrm{d}G_+^{(L)}(x) \tag{5.4}$$

which shows that $G_+(\mathrm{d}x) = e^{-\gamma x} \mathrm{d}G_+^{(L)}(x)$. Hence

$$C = \mathbb{E}_{\theta_L} e^{-\gamma B(\infty)} = \int_0^\infty e^{-\gamma x}(1 - G_+^{(L)}(x))/\mu_+^{(L)} \mathrm{d}x$$

$$= \frac{1}{\mu_+^{(L)}} \int_0^\infty \frac{1}{\gamma}(1 - e^{-\gamma y})\mathrm{d}G_+^{(L)}(y) = \frac{1}{\gamma\mu_+^{(L)}}(1 - \|G_+\|)$$

and since $\mu_+^{(L)} = \int_0^\infty x e^{\gamma x} \mathrm{d}G_+(x)$, the first identity in (5.3) follows. For the second, note first that as in (5.4), $G_-(\mathrm{d}x) = e^{-\gamma x} G_-^{(L)}(\mathrm{d}x)$, and thus $\{\ldots\}$ is just $1 - \|G_-^{(L)}\|$. Now just note that

$$1 - \|G_+\| = \frac{1}{\mathbb{E}\tau_-} = \frac{\mathbb{E}X}{\mathbb{E}S_{\tau_-}}, \qquad \frac{1}{\mu_+^{(L)}} = \frac{1}{\mathbb{E}_{\theta_L} X \mathbb{E}_{\theta_L} \tau_+^{(L)}} = \frac{1 - \|G_-^{(L)}\|}{\kappa'(\gamma)}. \qquad \square$$

In view of $W \stackrel{\mathcal{D}}{=} M$, the Cramér–Lundberg approximation states that under appropriate conditions the tail of the waiting-time distribution of a $GI/G/1$ queue is asymptotically exponential. This results is superficially similar to the heavy traffic approximation of VIII.6, but of course the range of parameters for which these two results give an exponential approximation to $\mathbb{P}(W > u)$ is not the same (neither are the constants equal).

Example 5.4. For $GI/M/1$, we have in the notation of IX.1.3 that $\mathbb{P}(W > u) = \theta e^{-\eta u}$. It follows that the Cramér–Lundberg approximation is exact in this case and that $\theta = C$, $\eta = \gamma$ (it is straightforward to check directly from the expression for G_+ in IX.1.2(a) that indeed (5.3) reduces to θ). For $M/G/1$, G_- is exponential with intensity β and (5.3) becomes

$$\frac{\mu_B - 1/\beta}{\gamma\kappa'(\gamma)(-1/\beta)}\left\{1 - \beta\int_{-\infty}^0 e^{(\gamma+\beta)x} \mathrm{d}x\right\} = \frac{1-\rho}{\gamma\kappa'(\gamma)}\left\{1 - \frac{\beta}{\gamma+\beta}\right\} = \frac{1-\rho}{\kappa'(\gamma)(\gamma+\beta)} \qquad \square$$

Just as the heavy traffic approximation has a time-dependent version VIII.6.4 (in terms of the inverse Gaussian distribution), so is the case for the Cramér–Lundberg approximation. This time the correction factor is normal:

Theorem 5.5. *Suppose in addition to the conditions of Theorem 5.2 that* $\sigma_L^2 = \kappa''(\gamma) = \mathbb{V}\mathrm{ar}_{\theta_L} X < \infty$ *and define*

$$\mu_L = \mathbb{E}_{\theta_L} X = \kappa'(\gamma), \qquad \lambda = 1/\mu_L, \qquad \omega^2 = \sigma_L^2/\mu_L^3.$$

Then

$$\mathbb{P}(M_N > u) \approx C\mathrm{e}^{-\gamma u}\Phi\left(\frac{N-\lambda u}{\omega u^{1/2}}\right), \qquad u \to \infty \tag{5.5}$$

in the sense that if N varies with u in such a way that $x = \lim (N - \lambda u)/\omega u^{1/2}$ *exists, then* (5.5) *is* $C\mathrm{e}^{-\gamma u}\Phi(x) + \mathrm{o}(\mathrm{e}^{-\gamma u})$.

The proof rests on two lemmas:

Lemma 5.6. *As* $u \to \infty$, (i) $\tau(u)/u \xrightarrow{\mathbb{P}_{\theta_L}} \mu_L^{-1}$; (ii) $\mathbb{E}_{\theta_L}\tau(u)/u \to \mu_L^{-1}$; (iii) $\tau(u)$ *is asymptotically normal with mean* λu *and variance* $\omega^2 u$ *with respect to* $\mathbb{P}_{\theta_L}$.

Proof. First note the standard LLN and CLT,

$$Y_t = \frac{S_t}{t} \to \mu_L \text{ a.s.}, \qquad Z_t = \frac{S_t - t\mu_L}{t^{1/2}} \xrightarrow{\mathscr{D}} N(0, \sigma_L^2)$$

with respect to $\mathbb{P}_{\theta_L}$. Let $t = \tau$ and write $S_\tau = u + B(u)$. Since $\tau \to \infty$ and $B(u) \xrightarrow{\mathscr{D}} B(\infty)$, we have $B(u)/\tau \xrightarrow{\mathbb{P}_{\theta_L}} 0$ and hence $Y_\tau \to \mu_L$ implies $u/\tau \xrightarrow{\mathbb{P}_{\theta_L}} \mu_L$ and (i). The proof of (ii) is the same as for the elementary renewal theorem in IV.1 (or as in Problem VII.2.3). By (i) and Anscombe's theorem, $Z_\tau \xrightarrow{\mathscr{D}} N(0, \sigma_L^2)$, and (iii) follows since $B(u)/\tau^{1/2} \xrightarrow{\mathscr{D}} 0$ implies

$$Z_\tau \overset{\mathscr{D}}{\approx} \frac{u - \tau\mu_L}{\tau^{1/2}} \overset{\mathscr{D}}{\approx} \mu_L^{3/2}\frac{\tau - u/\mu_L}{u^{1/2}}.$$

□

Lemma 5.7. $B(u)$ *and* τ *are asymptotically independent as* $u \to \infty$. *That is, for* f, g *bounded and continuous*

$$\mathbb{E}_{\theta_L} f(B(u)) g\left(\frac{\tau - \lambda u}{\omega u^{1/2}}\right) \to \mathbb{E}_{\theta_L} f(B(\infty))\mathbb{E}g(U), \tag{5.6}$$

where U is standard normal.

Proof. Define $u' = u - u^{1/4}$. Then the distribution of $\tau(u) - \tau(u')$ given $\mathscr{F}_{\tau(u')}$ is readily seen to be degenerate at zero if $S_{\tau(u')} > u$ and otherwise that of $\tau(v)$ with $v = u - S_{\tau(u')} = u^{1/4} - B(u')$. Hence

$$\begin{aligned}\mathbb{E}[\tau(u) - \tau(u')] &= \mathbb{E}[\tau(u^{1/4} - B(u')); B(u') \leqslant u^{1/4}] \\ &\leqslant \mathbb{E}\tau(u^{1/4}) = \mathrm{O}(u^{1/4})\end{aligned}$$

and thus in (5.6), we can replace $\tau(u)$ by $\tau(u')$. Defining $h(u) = \mathbb{E}_{\theta_L} f(B(u))$ so that $h(u) \to h(\infty) = \mathbb{E}_{\theta_L} f(B(\infty))$ it follows similarly that

$$\begin{aligned}\mathbb{E}[f(B(u)) \mid \mathscr{F}_{\tau(u')}] &= h(u^{1/4} - B(u'))I(B(u') \leqslant u^{1/4}) \\ &\quad + f(B(u') - u^{1/4})I(B(u') > u^{1/4}) \xrightarrow{\mathbb{P}_{\theta_L}} h(\infty)\cdot 1 + 0\end{aligned}$$

using the result $u^{1/4} - B(u') \xrightarrow{\mathbb{P}_{\theta_L}} \infty$ of $B(u') \xrightarrow{\mathscr{D}} B(\infty)$. Hence

$$\mathbb{E}f(B(u))g\left(\frac{\tau(u') - \lambda u}{\omega u^{1/2}}\right) = \mathbb{E}(\mathbb{E}[f(B(u))|\mathscr{F}_{\tau(u')}]g\left(\frac{\tau(u') - \lambda u}{\omega u^{1/2}}\right)$$

$$\approx h(\infty)\mathbb{E}g\left(\frac{\tau(u') - \lambda u}{\omega u^{1/2}}\right) \approx h(\infty)\mathbb{E}g(U). \qquad \square$$

Proof of Theorem 5.5 By the fundamental identity and (5.6),

$$\begin{aligned}\mathbb{P}(M_N > u) = \mathbb{P}(\tau(u) \leqslant N) &= \mathrm{e}^{-\gamma u}\mathbb{E}_{\theta_L}[\mathrm{e}^{-\gamma B(u)}; \tau(u) \leqslant N] \\ &= \mathrm{e}^{-\gamma u}\mathbb{E}_{\theta_L}[\mathrm{e}^{-\gamma B(u)}; \tau(u) \leqslant \lambda u + x\omega u^{1/2} + \mathrm{o}(u^{1/2})] \\ &= \mathrm{e}^{-\gamma u}\{\mathbb{E}_{\theta_L}\mathrm{e}^{-\gamma B(\infty)}\Phi(x) + \mathrm{o}(1)\}. \qquad \square\end{aligned}$$

Notes

General references on risk theory are given in XIII.1. In the queueing setting, Lundberg's inequality was first proved by Kingman, see Stoyan (1983, Ch. 5.3) and Asmussen (1982) for a survey and references. The Lundberg parameter occurs in a variety of settings in the literature, for example Feller (1971), Segerdahl (1955), von Bahr (1974), Asmussen (1981, 1982), Siegmund (1985) and Neuts (1986). Lemma 5.7 is known in the literature as *Stam's lemma.*

6. SIEGMUND'S CORRECTED HEAVY TRAFFIC APPROXIMATIONS

The heavy traffic condition of the mean $\mathbb{E}X$ being smaller than but close to zero corresponds in the setting of the imbedding $F = F_{\theta_0}$ in a conjugate family with $\mu_0 = 0$ to also $\theta_0 = -\gamma_0$ being smaller than but close to zero. To derive heavy traffic approximations, for a given conjugate family $(\mathbb{P}_\theta)_{\theta\in\Theta}$ we therefore consider the limiting behaviour of $\mathbb{P}_{\theta_0}$ in terms of $\mathbb{P}_0$ subject to the limit $\theta_0 \uparrow 0$.

We shall only consider the discrete random walk case. It is assumed that Θ contains a neighbourhood of zero and, for the ease of notation, that the scale is chosen such that $\mathbb{V}\mathrm{ar}_0 X = \kappa''(0, 0) = 1$. Then for small θ

$$\kappa(0; \theta) = \frac{\theta^2}{2!} + \frac{\theta^3}{3!}\mathbb{E}_0 X^3 + \cdots, \tag{6.1}$$

$$\mu_\theta = \kappa'(0; \theta) = \theta + \mathrm{O}(\theta^2) \tag{6.2}$$

$$\mathbb{V}\mathrm{ar}_\theta X = \kappa''(0; \theta) = 1 + \mathrm{O}(\theta), \qquad \mathbb{E}_\theta X^2 = 1 + \mathrm{O}(\theta). \tag{6.3}$$

Also, $\theta_L > 0$ connected to $\theta_0 < 0$ by means of $\kappa(0; \theta_0) = \kappa(0; \theta_L)$ is well defined, and by (6.1) we have in the limit $\theta_0 \uparrow 0$ that

$$\frac{\theta_L}{-\theta_0} \to 1, \qquad \frac{\gamma}{-\theta_0} = \frac{\theta_L - \theta_0}{-\theta_0} \to 2. \tag{6.4}$$

We shall let u vary with θ_0 in such a way that any of the equivalent relations

$$u\theta_0 \to -\xi, \qquad u\theta_L \to \xi \tag{6.5}$$

hold for some $\xi \geqslant 0$ and as previously let $\tau(u) = \inf\{n \geqslant 1 : S_n > u\}$. Some preliminary estimates are immediately apparent from VIII.6:

Proposition 6.1. *As* $\theta_0 \uparrow 0$, $\gamma \mathbb{E}_{\theta_0} M \to 1$ *and subject to* (6.5),

$$\mathbb{P}_{\theta_0}(M > u) = \mathbb{P}_{\theta_0}(\tau(u) < \infty) \to \mathrm{e}^{-2\xi}, \tag{6.6}$$

$$\mathbb{P}_{\theta_0}\left(\frac{\tau(u)}{u^2} \leqslant T\right) \to G(T; -\xi, 1). \tag{6.7}$$

Proof. The condition VIII.(6.1) is immediately apparent since clearly $F_{\theta_0} \xrightarrow{w} F_0$, and by (6.2) and (6.3) $\mu_{\theta_0} \to \mu_0 = 0$, $\mathbb{E}_{\theta_0} X^2 \to 1 = \mathbb{E}_0 X^2$. Hence, for example, VIII.6.1 implies $\mathbb{E}_{\theta_0}[-\mu_{\theta_0} M / \mathbb{E}_{\theta_0} X^2] \to \frac{1}{2}$ which is equivalent to $\gamma \mathbb{E}_{\theta_0} M \to 1$ by (6.2)–(6.4). For (6.7), apply VIII.6.4 to get

$$\mathbb{P}_{\theta_0}\left(\tau\left(\frac{-y\mathbb{E}_{\theta_0} X^2}{\mu_{\theta_0}}\right) \leqslant \frac{t}{\mu_{\theta_0}^2}\right) \to G(t; -1, y).$$

Letting $y = -u\theta_0$, $t = \xi^2 T$ and using (6.2), (6.3) and (6.5) this implies

$$\mathbb{P}_{\theta_0}(\tau(u) \leqslant Tu^2) \to G(\xi^2 T; -1; \xi) = G(T; -\xi, 1),$$

cf. Proposition 3.6(i). Similar estimates yield (6.6). □

We shall now study improvements of these estimates, obtained in a 1979 paper by D. Siegmund. The idea is to estimate the excess over the boundary more carefully and thereby obtain correction terms of lower magnitude $\mathrm{O}(\gamma)$, $\mathrm{O}(\gamma^2), \ldots$ (in view of (6.2) or (6.4), we might as well have replaced γ by $-\theta_0$ or $-\mu_{\theta_0}$). Only the form of the first-order correction term will be derived rigorously, but the second term is included in the statement of the results because of their importance as approximations. Considering first (6.6), we have $\gamma u \cong 2\xi$. Hence $\mathbb{P}_{\theta_0}(M > u) \cong \mathrm{e}^{-\gamma u}$ and we have:

Theorem 6.2. *Suppose that the* $\mathbb{P}_0$*-distribution* F_0 *of* X *is spread out. Then as* $\theta_0 \uparrow 0$, $u\theta_0 \to -\xi$

$$\mathbb{P}_{\theta_0}(M > u) = \mathrm{e}^{-\gamma(u+\beta)} + \mathrm{o}(\gamma^2) \tag{6.8}$$

where $\beta = \mathbb{E}_0 S_{\tau_+}^2 / 2\mathbb{E}_0 S_{\tau_+} = \mathbb{E}_0 B(\infty)$.

The proof is based on the relation $\mathbb{P}_{\theta_0}(M > u) = \mathrm{e}^{-\gamma u} C(u)$ with $C(u) = \mathbb{E}_{\theta_L} \mathrm{e}^{-\gamma B(u)}$, which was used in the proof of the Cramér–Lundberg approximation. However, $C(u)$ must now be estimated in a different manner since θ_L is no longer fixed. We shall need some lemmas, in particular a variant (Lemma 6.4) of Lemma 5.7:

Lemma 6.3. $\mathbb{E}_0 \exp\{\varepsilon S_{\tau_+}\} < \infty$ *for any* $\varepsilon > 0$ *with* $\kappa(0; \varepsilon) < \infty$.

Proof. This can be obtained either by an easy variant of the proof of VIII.2.1 or by Wiener–Hopf factorization, cf. Problem 6.1. □

Lemma 6.4. *$B(u)$ and $\tau(u)/u^2$ are asymptotically independent with respect to $\mathbb{P}_{\theta_L}$ as $\theta_0 u \to -\xi$, with the limiting distribution of $\tau(u)/u^2$ being $G(\cdot;\xi,1)$ and that of $B(u)$ the $\mathbb{P}_0$-distribution of $B(\infty)$. That is, for f, g bounded and continuous*

$$\mathbb{E}_{\theta_L} f(B(u)) g\left(\frac{\tau(u)}{u^2}\right) \to \mathbb{E}_0 f(B(\infty)) \int_0^\infty g(x) G(\mathrm{d}x;\xi,1). \tag{6.9}$$

More generally, there is an $\varepsilon > 0$ such that (6.9) holds when f is continuous with $f(x) = \mathrm{O}(\mathrm{e}^{\varepsilon x})$.

Proof. By the fundamental identity, the l.h.s. of (6.9) is

$$\mathbb{E}_0 f(B(u)) g\left(\frac{\tau(u)}{u^2}\right) \exp\{\theta_L(u + B(u)) - \tau\kappa(0;\theta_L)\} \tag{6.10}$$

and thus we have to inspect the $\mathbb{P}_0$-limit of $(B(u), \tau(u)/u^2)$. Clearly, $B(u) \xrightarrow{\mathscr{D}} B(\infty)$ and by Proposition 6.1, $\tau(u)/u^2 \xrightarrow{\mathscr{D}} G(\cdot;0,1)$. By a variant of the proof of Lemma 5.7, it is seen that we also have asymptotic independence: with $u' = u - u^{1/4}$ the only new estimate needed is $\tau(u) - \tau(u') = \mathrm{o}(u^2)$ which follows from the stochastical domination by $\tau(u^{1/4})$ and $\tau(u^{1/4})/u^{1/2} \xrightarrow{\mathscr{D}} G(\cdot;0,1)$. Also by Lemma 6.3 and renewal theory, we have $\mathbb{E}_0 \mathrm{e}^{\delta B(u)} \to \mathbb{E}_0 \mathrm{e}^{\delta B(\infty)} < \infty$ for $\delta < \delta_0$, in particular $\mathbb{E}_0 \mathrm{e}^{\delta B(u)} \leqslant c$ for all u. For $\varepsilon < \delta$, we then have that $\varepsilon + \theta_L < \delta$ eventually, and using uniform integrability and $\kappa(0;\theta_L) \cong \xi^2/2u^2$, it follows that the limit of (6.10) exists for $f(x) = \mathrm{O}(\mathrm{e}^{\varepsilon x})$ and is

$$\mathbb{E}_0 f(B(\infty)) \int_0^\infty g(x) \exp\{\xi - \xi^2 x/2\} G(\mathrm{d}x;0,1) = \mathbb{E}_0 f(B(\infty)) \int_0^\infty g(x) G(\mathrm{d}x;\xi,1).$$

□

To obtain the second-order correction terms, the following two lemmas are needed:

Lemma 6.5. *As $\theta_0 u \to -\xi$, it holds for some $\varepsilon > 0$ that $\mathbb{E}_{\theta_L} B(u) = \mathbb{E}_{\theta_L} B(\infty) + \mathrm{O}(\mathrm{e}^{-\varepsilon u})$.*

Lemma 6.6. $\mathbb{E}_{\theta_L} S_{\tau_+}^k = \mathbb{E}_0 S_{\tau_+}^k + k\theta_L/(k+1)\mathbb{E}_0 S_{\tau_+}^{k+1} + \mathrm{o}(\gamma), \qquad k = 1, 2, \ldots.$

It is seen that Lemma 6.6 is a uniform version of VI.2.6, and the proof may proceed by first showing that the $\mathbb{P}_{\theta_L}$-distributions of $B(0) = S_{\tau_+}$ have a common absolute continuous component, and next to check that the estimates in VI.2 hold uniformly in small θ_L. In addition, the proof of Lemma 6.6 is not unreasonably complicated, but the details of both proofs have been omitted.

Proof of Theorem 6.2. By Taylor expansion,

$$C(u) = \mathbb{E}_{\theta_L} e^{-\gamma B(u)} = \mathbb{E}_{\theta_L}\left[1 - \gamma B(u) + \frac{\gamma^2}{2} B(u)^2 + \gamma^3 O(B(u)^3 e^{\gamma B(u)})\right].$$

Here the last term is $O(\gamma^3)$ by Lemma 6.4, while $\mathbb{E}_{\theta_L} B(u) \to \mathbb{E}_0 B(\infty) = \beta$. This is more than sufficient for $C(u) = 1 - \gamma\beta + o(\gamma)$ and thus that the remainder term in (6.8) is $o(\gamma)$. To see that it is actually $o(\gamma^2)$, note that by Lemmas 6.5 and 6.6

$$\mathbb{E}_{\theta_L} B(u) = \mathbb{E}_{\theta_L} B(\infty) + O(e^{-\varepsilon u}) = \frac{\mathbb{E}_{\theta_L} S_{\tau_+}^2}{2\mathbb{E}_{\theta_L} S_{\tau_+}} + O(e^{-\varepsilon u})$$

$$= \frac{\mathbb{E}_0 S_{\tau_+}^2 + 2\theta_L \mathbb{E}_0 S_{\tau_+}^3/3}{2\mathbb{E}_0 S_{\tau_+} + \theta_L \mathbb{E}_0 S_{\tau_+}^2} + o(\gamma)$$

$$= \beta + \theta_L\left(\frac{\mathbb{E}_0 S_{\tau_+}^3}{3\mathbb{E}_0 S_{\tau_+}} - \beta^2\right) + o(\gamma) = \beta + \frac{\gamma}{2}\left(\frac{\mathbb{E}_0 S_{\tau_+}^3}{3\mathbb{E}_0 S_{\tau_+}} - \beta^2\right) + o(\gamma),$$

$$\mathbb{E}_{\theta_L} B(u)^2 = \mathbb{E}_0 B(\infty)^2 + o(\gamma) = \frac{\mathbb{E}_0 S_{\tau_+}^3}{3\mathbb{E}_0 S_{\tau_+}} + o(\gamma).$$

Combining these estimates the terms involving $\mathbb{E}_0 S_{\tau_+}^3$ cancel and we get

$$C(u) = 1 - \gamma\beta + \frac{\gamma^2\beta^2}{2} + o(\gamma^2) = e^{-\gamma\beta} + o(\gamma^2),$$

$$\mathbb{P}(M > u) = e^{-\gamma u} C(u) = e^{-\gamma(u+\beta)} + o(\gamma^2). \qquad \square$$

There is also a similar refinement of $\mathbb{E}_{\theta_0} M \cong \gamma^{-1}$:

Theorem 6.7. *As* $\theta_0 \uparrow 0$,

$$\mathbb{E}_{\theta_0} M = \gamma^{-1} - \beta + \frac{\gamma}{2}\left[\frac{\mathbb{E}_0 S_{\tau_+}^3}{3\mathbb{E}_0 S_{\tau_+}} - \beta^2\right] + o(\gamma).$$

Proof. Using VIII.(2.3) and the fundamental identity,

$$\mathbb{E}_{\theta_0} M = \frac{\mathbb{E}_{\theta_0}[S_{\tau_+}; \tau_+ < \infty]}{1 - \mathbb{P}_{\theta_0}(\tau_+ < \infty)} = \frac{\mathbb{E}_{\theta_L} S_{\tau_+} \exp\{-\gamma S_{\tau_+}\}}{1 - \mathbb{E}_{\theta_L} \exp\{-\gamma S_{\tau_+}\}}$$

$$= \frac{\mathbb{E}_{\theta_L} S_{\tau_+} - \gamma \mathbb{E}_{\theta_L} S_{\tau_+}^2 + \gamma^2 \mathbb{E}\theta_L S_{\tau_+}^3/2 + O(\gamma^3)}{\gamma \mathbb{E}_{\theta_L} S_{\tau_+} - \gamma^2 \mathbb{E}_{\theta_L} S_{\tau_+}^3/2 + O(\gamma^3)}$$

from which the result follows by Lemma 6.6 after some elementary calculus. Again, Lemma 6.6 is not needed to prove rigorously that $\mathbb{E}_{\theta_0} M = \gamma^{-1} - \beta + o(1)$. $\square$

We mention also without proof the refinement of (6.7), i.e. a time-dependent version of the expansion (6.8) for $\mathbb{P}(M > u) = \mathbb{P}(\tau(u) < \infty)$,

$$\mathbb{P}(\tau(u) \leqslant tu^2) \cong G\left(tu^2 + \frac{u\mathbb{E}_0 X^3}{3}; -\frac{\gamma}{2}, u + \beta\right) \qquad (6.11)$$

$$\text{(by Proposition 3.6(i))} = G\left(t + \frac{\mathbb{E}_0 X^3}{3u}; -\frac{\gamma u}{2}, 1 + \frac{\beta}{u}\right).$$

Numerical studies (some examples of which are in XIII.2) indicate that the above approximations are superior to all others known, not only for θ_0 close to zero but in fact in a remarkably wide range. A deficit is that the constants such as β can be cumbersome to evaluate. We mention in this connecton the formula

$$\beta = \tfrac{1}{6}\mathbb{E}_0 X^3 - \frac{1}{\pi}\int_0^\infty t^{-2}\,\mathrm{Re}\log\{2[1-\phi(t)]/t^2\}\,\mathrm{d}t, \tag{6.12}$$

where $\varphi(t) = \mathbb{E}_0 \mathrm{e}^{itX}$, which can be implemented by numerical integraton. The proof is based upon Fourier inversion but omitted.

Problems

6.1. Let $F^{(k)}, S^{(k)}_{\tau_+}$, etc. correspond to $X^{(k)}_n = X_n \wedge k$. Show by Wiener–Hopf factorization that for $\alpha > 0$, $\kappa(0;\alpha) < \infty$

$$\mathbb{E}_0 \exp\{\alpha S^{(k)}_{\tau_+}\} \leqslant \frac{\mathrm{e}^{\kappa(0,\alpha)} - \mathbb{E}_0 \exp\{\alpha S^{(k)}_{\tau_-}\}}{1 - \mathbb{E}_0 \exp\{\alpha S^{(k)}_{\tau_-}\}}$$

and deduce for $k \to \infty$ that $\mathbb{E}_0 \exp\{\alpha S_{\tau_+}\} < \infty$.

6.2. Check that under the assumptions of Theorem 6.2 the constant C of the Cramér–Lundberg approximation satisfies $C = 1 - \gamma\beta + \mathrm{o}(\gamma)$.

Notes

The results are from Siegmund (1979). See also Siegmund (1985) where in particular the approach to time-dependent formulae such as (6.11) is somewhat different.

7. SIMULATING A CONJUGATE PROCESS

Instead of deriving approximations for quantities like $\mathbb{P}(M > u)$, it is tempting just to simulate on a computer. An obvious difficulty is that M cannot be simulated in finite time. One alternative, regenerative simulation, has already been discussed in V.2, but seems likely to be inefficient under heavy traffic conditions: the cycles (determined by the distribution of τ_-) will be long and hence few replicates will be obtained within a reasonable amount of computer time (in addition the variance on the estimates of say $\mathbb{E}M$ can be seen to be large under heavy traffic conditions).

Here we shall discuss a further alternative based upon the imbedding $F = F_{\theta_0}$ in a conjugate family. The idea is simply to note that for $\theta \geqslant 0$ we have $\mathbb{P}_\theta(\tau(u) < \infty)$ and hence by the fundamental identity with $G = \{\tau(u) < \infty\}$,

$$\mathbb{P}(M > u) = \mathbb{P}_{\theta_0}(\tau(u) < \infty) = \mathbb{E}_\theta[R_\theta; G] = \mathbb{E}_\theta R_\theta,$$

where

$$R_\theta = \exp\{(\theta_0 - \theta)S_{\tau(u)} - \tau(u)\kappa(\theta; \theta_0 - \theta)\}.$$

Thus we may simply simulate R_θ from $\mathbb{P}_\theta$. Not only can this be achieved in finite time, but also $v_\theta = \mathbb{V}\mathrm{ar}_\theta R_\theta$ can be expected to be small, at least for some $\theta \geqslant 0$. For example, consider the Lundberg case $\theta=\theta_L$. Then R_θ becomes $\mathrm{e}^{-\gamma u}\mathrm{e}^{-\gamma B(u)}$ and under heavy traffic conditions $\mathrm{e}^{-\gamma B(u)}\cong 1$. In particular, $\mathbb{V}\mathrm{ar}_{\theta_L}\mathrm{e}^{-\gamma B(u)}$ should be small and hence

$$v_{\theta_L}=\mathrm{e}^{-2\gamma u}\mathbb{V}\mathrm{ar}_{\theta_L}\mathrm{e}^{-\gamma B(u)}$$

even smaller.

Example 7.1. Consider, as in Example 1.3, $M/M/1$ with $\beta=0.85$, $\delta=1$, i.e. $\rho=0.85$, and $\mathbb{P}(M>u)=\rho\,\mathrm{e}^{-(1-\rho)u}=0.05$, i.e. $u=18.9$. Then $\gamma=0.15$, $\beta_{\theta_L}=1$, $\delta_{\theta_L}=0.85$ and since $\mathbb{P}_{\theta_L}(B(u)>y)=\exp\{-\delta_{\theta_L}y\}$, we get

$$\begin{aligned}v_{\theta_L}&=\mathrm{e}^{-2\gamma u}\{\mathbb{E}_{\theta_L}\mathrm{e}^{-2\gamma B(u)}-(\mathbb{E}_{\theta_L}\mathrm{e}^{-\gamma B(u)})^2\}\\&=\mathrm{e}^{-2\gamma u}\left\{\frac{\delta_{\theta_L}}{\delta_{\theta_L}+2\gamma}-\left(\frac{\delta_{\theta_L}}{\delta_{\theta_L}+\gamma}\right)^2\right\}\\&=0.00346\,[0.7391-0.7225]=0.0000575.\end{aligned}$$

This should be compared to the variance $p(1-p)=0.0475$ in the binomial distribution with parameter $p=0.05$, which approximately determines the variance if one uses crude simulation with an appropriate stopping criterion. Furthermore, Lundberg simulation compares favourably to regenerative simulation: in a test run with the same parameter values, each method being allowed 1 second CPU time, the empirical variance on the estimates of $\mathbb{P}(M>u)$ produced by the two methods resulted in ratios of 1:436. □

Now clearly it is not only important that v_θ should be small: i_θ, the expected computer time to create one replicate of R_θ, should also be so, since this yields many replicates within a fixed amount of computer time.

Example 7.2. With the same parameters for $M/M/1$, computer simulations of the $\mathbb{P}_\theta$-processes, each in 1 second CPU time, was performed for $\theta=\theta_L$ as well as other choices of θ. The point estimates, confidence intervals and number N_θ of replicates are depicted in Fig. 7.1. □

It is suggested from Example 7.2 that i_θ is decreasing with θ. To see this more precisely, note that R_θ is produced in $\tau(u)$ steps, the time needed for which should not substantially depend on θ. Hence *in the following we shall replace i_θ by $\mathbb{E}_\theta\tau(u)$*. Then, since F_θ is stochastically increasing with θ, $\tau(u)$ is decreasing and hence i_θ decreasing. On the other hand, it is also suggested by Example 7.2 that too large θ are not appropriate, i.e. that v_θ increases more than to balance the decrease in i_θ. For a theoretical study, we shall now *look for a θ minimizing $i_\theta v_\theta$*. That this is a reasonable single performance measure follows since within T units of time we get approximately $N_\theta=T/i_\theta$ replicates and hence an empirical variance v_θ/N_θ proportional to $i_\theta v_\theta$.

It will be proved that indeed, as suggested by Example 7.2, the Lundberg value θ_L is approximately optimal. To see this, we shall write θ in the form $\theta=\theta_L(1+\Delta)$ so that we are concerned with minimizing $f(\Delta)=i_{\theta_L(1+\Delta)}v_{\theta_L(1+\Delta)}$.

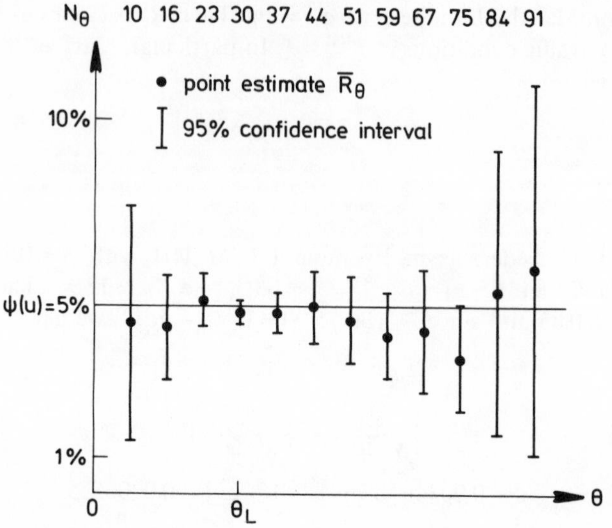

Fig. 7.1 Reproduced by permission of Elsevier Science Publishers from Asmussen (1985)

Theorem 7.3. *As $\theta_0 u \to -\xi$, we have $f(\Delta) = \infty$ ultimately if $\Delta > \sqrt{2} - 1$, whereas if $\Delta < \sqrt{2} - 1$.*

$$f(\Delta) = \frac{u^2 e^{-4\xi}}{\xi} \frac{1}{1+\Delta} [\exp\{-\Delta\xi + \xi - \xi(1 - \Delta^2 - 2\Delta)^{1/2}\} - 1]. \tag{7.1}$$

More precisely for $\Delta = 0$

$$f(0) \to e^{-4\xi} 4\xi \, \mathbb{V}\mathrm{ar}_0 B(\infty). \tag{7.2}$$

It is seen that, up to the factor u^2, the r.h.s. of (7.1) is a function of Δ which is zero for $\Delta = 0$ and is > 0 for $\Delta \neq 0$. Thus the minimum of $f(\Delta)$ is approximately at $\Delta = 0$ and also $f(\Delta)/f(0)$ goes to infinity for $\Delta \neq 0$ at rate u^2 so that in the limit the optimality of the Lundberg value is very marked.

The proof requires uniform integrability properties in the limiting relation $\tau(u)/u^2 \to G(\cdot; \xi, 1)$ in $\mathbb{P}_{\theta_L}$-distribution, cf. Lemma 6.4, more precisely the following result which we state without proof:

Lemma 7.4. *As $\theta_0 u \to -\xi$,*

$$\mathbb{E}_{\theta_L} \exp\{\lambda\tau(u)/u^2\} \begin{cases} = \infty \text{ ultimately} & \text{if } \lambda > \xi^2/2 \\ \to \exp\{\xi - \sqrt{\xi^2 - 2\lambda}\} & \text{if } \lambda < \xi^2/2. \end{cases}$$

Proof of Theorem 7.3. Since $\theta_L(1+\Delta)u \to \xi(1+\Delta)$, Lemma 7.4 and uniform integrability yield

$$\frac{1}{u^2} i_\theta \to \int_0^\infty t G(\mathrm{d}t; \xi(1+\Delta), 1) = \frac{1}{\xi(1+\Delta)}. \tag{7.3}$$

Also, by the fundamental identity and (1.3),

$$\begin{aligned} v_{\theta_L(1+\Delta)} &= \mathbb{E}_{\theta_L(1+\Delta)} R^2_{\theta_L(1+\Delta)} - (\mathbb{E}_{\theta_L(1+\Delta)} R_{\theta_L(1+\Delta)})^2 \\ &= \mathbb{E}_{\theta_L} \exp\{-(2\gamma + \theta_L \Delta) S_{\tau(u)} + \tau\kappa(\theta_L; \theta_L\Delta)\} - (\mathbb{P}_{\theta_0}(M > u))^2. \end{aligned} \tag{7.4}$$

Here the last term is just $(\mathrm{e}^{-2\xi})^2 + \mathrm{o}(1)$, whereas

$$\begin{aligned} (2\gamma + \theta_L\Delta) S_{\tau(u)} &= 2\gamma u + \theta_L \Delta u + \mathrm{o}(1) B(u) = (4 + \Delta)\xi + \mathrm{o}(1), \\ \kappa(\theta_L; \theta_L\Delta) &= \kappa(0; \theta_L(1+\Delta)) - \kappa(0; \theta_L) \\ &\cong \theta_L^2(1+\Delta)^2/2 - \theta_L^2/2 \cong \frac{(\Delta^2 + 2\Delta)\xi^2}{2u^2}. \end{aligned}$$

Therefore the integrand in the first term in (7.4) tends to

$$\exp\{-(4+\Delta)\xi + \tau(\xi, 1)(\Delta^2 + 2\Delta)\xi^2/2\}$$

in distribution (here $\tau(\xi, 1) \stackrel{\mathscr{D}}{=} G(\cdot; \xi, 1)$). Therefore using Lemma 7.4 and the form $\xi - \sqrt{\xi^2 - 2\lambda}$ of the cumulant generating function of $\tau(\xi, 1)$, (7.4) is ∞ ultimately if $\Delta^2 + 2\Delta > 1$, i.e. $\Delta > \sqrt{2} - 1$, whereas otherwise (7.4) becomes

$$\begin{aligned} &\exp\{-(4+\Delta)\xi\} \mathbb{E} \exp\{\tau(\xi, 1)(\Delta^2 + 2\Delta)\xi^2/2\} - \mathrm{e}^{-4\xi} + \mathrm{o}(1) \\ &= \exp\{-(4+\Delta)\xi + \xi - \xi\sqrt{1 - \Delta^2 - 2\Delta}\} - \exp\{-4\xi\} + \mathrm{o}(1) \end{aligned}$$

and combining with (7.3), (7.1) follows. For (7.2), we must inspect $v_{\theta_L} = \mathbb{V}\mathrm{ar}_{\theta_L} \exp\{-\gamma S_{\tau(u)}\}$ more carefully. But

$$|\mathrm{e}^{-\gamma B(u)} - 1 + \gamma B(u)| \leqslant \gamma^2 B(u)^2 \mathrm{e}^{\gamma B(u)},$$

hence

$$v_{\theta_L} \cong \mathrm{e}^{-2\gamma u} \mathbb{V}\mathrm{ar}_{\theta_L} \gamma B(u) \cong \mathrm{e}^{-4\xi} \frac{4\xi^2}{u^2} \mathbb{V}\mathrm{ar}_0 B(\infty). \qquad \square$$

Notes

The material is from Asmussen (1985), which in turn is related to Siegmund (1976).

CHAPTER XIII

Insurance Risk, Dam and Storage Models

1. INSURANCE RISK MODELS

We start by recalling the classical compound Poisson model for the risk reserve of an insurance company at time t,

$$R_t = u + pt - \sum_{n=1}^{N_t} U_n, \tag{1.1}$$

cf. IV.2.3. Thus $p > 0$ is the premium rate, u is the initial risk reserve, $\beta = \mathbb{E}N_t/t$ the arrival rate and $B(x) = \mathbb{P}(U \leqslant x)$ the claim size distribution, $\mu_B = \mathbb{E}U = \int_0^\infty x\,\mathrm{d}B(x)$.

A main problem in risk theory is the evaluation of the ruin probabilities

$$\psi(u) = \mathbb{P}(\inf_{0\leqslant t<\infty} R_t < 0 \mid R_0 = u), \qquad \psi(u, T) = \mathbb{P}(\inf_{0\leqslant t\leqslant T} R_t < 0 \mid R_0 = u). \tag{1.2}$$

A number of relevant observations and results can be found scattered around in the preceding chapters, but will now be put together.

We let

$$S_t = \sum_{n=1}^{N_t} U_n - pt, \qquad \tau(u) = \inf\{t > 0 : S_t > u\},$$

$$M_T = \sup_{0\leqslant t\leqslant T} S_t, \qquad M = \sup_{0\leqslant t<\infty} S_t.$$

Then $\tau(u)$ is the time to ruin and

$$\psi(u) = \mathbb{P}(\tau(u) < \infty) = \mathbb{P}(M > u), \qquad \psi(u, T) = \mathbb{P}(\tau(u) \leqslant T) = \mathbb{P}(M_T > u). \tag{1.3}$$

The interesting case is $p > \beta\mu_B$, meaning that on the average the claims in a unit time are strictly less than the premiums received, or equivalently that a surplus $p - \beta\mu_B$ of premiums is charged. The traditional measure of this effect is $\eta = p/\beta\mu_B - 1$, the so-called *safety loading*. It will be convenient in the following to assume $p = 1$. This can always be achieved by a time transformation (which leaves the safety loading unchanged).

Proposition 1.1. *If $\{V_t\}_{t\geq 0}$ is the virtual waiting-time process of an initially empty ($V_0 = 0$) $M/G/1$ queue with arrival intensity β and service-time distribution B, then $\psi(u, T) = \mathbb{P}(V_T > u)$, $\psi(u) = \mathbb{P}(V > u) = \mathbb{P}(W > u)$ with V, W the virtual and actual waiting times, respectively, in the steady state. Furthermore, the safety loading η of the risk process and the traffic intensity ρ of the queue are connected by $\eta = 1/\rho - 1$, $\rho = 1/(1+\eta)$.*

Proof. The last statement is immediate since $\rho = \beta\mu_B$. Thus if $\eta > 0$, then $\rho < 1$ and the steady state is well defined. Furthermore, it was shown in III.8.4 that $\{V_t\}_{t\geq 0}$ is simply the Lindley process generated by $\{S_t\}$. Hence $V_T \stackrel{\mathscr{D}}{=} M_T$, $W \stackrel{\mathscr{D}}{=} V \stackrel{\mathscr{D}}{=} M$, and (1.3) completes the proof. □

The relation $\psi(u) = \mathbb{P}(W > u)$ is also easily proved without using $W \stackrel{\mathscr{D}}{=} V$: if T_1, $T_2, \ldots$ are the interclaim times, then M is attained either at $t = 0$ or at a claim time (S_t decreases in between) so that

$$M = \max\{S_{T_1+\cdots+T_N}: N = 0, 1, 2\ldots\} = \max\left\{\sum_{n=1}^{N}(U_n - T_n): N = 0, 1, 2\ldots\right\},$$

and the last expression can be identified with the usual maximum representation of W in a discrete-time random walk.

It follows for $\rho < 1$ that the probability of ultimate ruin is given by the Pollaczek–Khintchine formula of IX.2,

$$1 - \psi(u) = \mathbb{P}(W \leqslant u) = (1-\rho)\sum_{n=0}^{\infty}\rho^n B_0^{*n}(u) \tag{1.4}$$

where $B_0(\mathrm{d}x) = (1 - B(x))/\mu_B$. The interpretation as a (normalized) transient renewal measure implies in particular that the survival probability $Z = 1 - \psi$ satisfies the renewal equation

$$Z(u) = 1 - \rho + \rho B_0 * Z(u) = 1 - \rho + \rho\int_0^u Z(u-x)(1 - B(x))/\mu_B \mathrm{d}x \tag{1.5}$$

which was found in IV.2.3 in a rather indirect way. An even more direct derivation of (1.5) proceeds from the identification of the $M/G/1$ ladder height distributions in IX.2. In fact, if $G_+(x) = \mathbb{P}(S_{\tau(0)} \leqslant x)$ is the ascending ladder height distribution for $\{S_t\}$, we obviously have

$$Z(u) = \mathbb{P}(\tau(u) = \infty) = \mathbb{P}(\tau(0) = \infty) + \int_0^u \mathbb{P}(\tau(u-x) = \infty)\mathrm{d}G_+(x)$$

$$= 1 - \|G_+\| + \int_0^u Z(u-x)\mathrm{d}G_+(x).$$

But it was found in IX.2 that $\|G_+\| = \rho$, $G_+ = \rho B_0$, and we are back to (1.5).

The mathematics of the compound Poisson risk model will be further discussed in the next section, but here we shall give a few examples of the modifications which have been suggested.

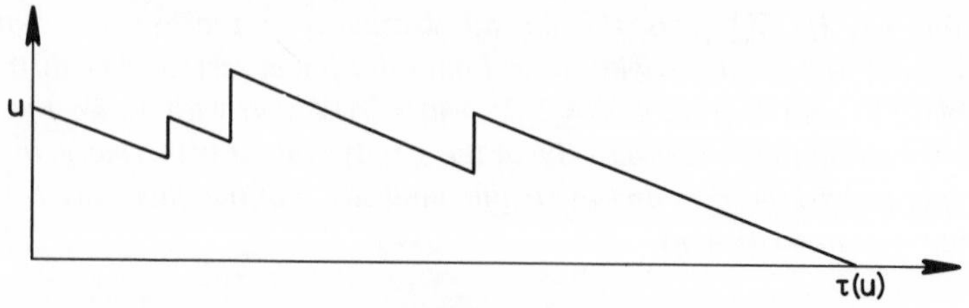

Fig. 1.1

Example 1.2. In pension fund arrangements, one may (in a rather idealized manner) think of a large number N of pension holders, each owing a certain part of the fund which is then paid out at a current rate. At the death of a pension holder, his remaining capital is then divided between the survivors. This leads to a process as in Fig. 1.1, depicting the remaining capital of some fixed pension holder,

$$R_t = u - pt + \sum_{n=1}^{N_t} T_n,$$

where $\{N_t\}$ is a Poisson process, say with intensity δ, which represents the times of death of the other pension holders and $T_1, T_2, \ldots$ are independent identically distributed, say with distribution A. The ruin probabilities are defined as in (1.2). This model is also related to queues and random walks. Take again $p = 1$ and let $U_0, U_1, \ldots$ denote the times between jumps. For ultimate ruin to occur, we must have $R_t < 0$, i.e. $u - R_t > u$, just before a jump time $t = U_0 + \cdots + U_{n-1}$. But here the values of $u - R_t$ are $U_0, U_0 + U_1 - T_1, U_0 + U_1 + U_2 - T_1 - T_2, \ldots$ so that $\psi(u) = \mathbb{P}(U_0 + W > u)$, where

$$W = \max\{0, U_1 - T_1, U_1 + U_2 - T_1 - T_2, \ldots\}$$

can be identified with the waiting time of a $GI/M/1$ queue with interarrival distribution A and service intensity δ. □

Example 1.3. A number of models have the feature that the development of the risk reserve process $\{R_t\}$ in between jumps is deterministic but not linear, i.e. governed by a differential equation $\dot{R} = p(R)$. For example, a company could pay out all its earnings as dividend once the risk reserve exceeds a certain level, say v, which corresponds to $p(r) = pI(r \leqslant v)$. Another possibility would be that the capital is invested at an interest rate say ε per unit time, meaning $p(r) = p + r\varepsilon$. Figure 1.2 depicts the combination $p(r) = (p + r\varepsilon)I(r \leqslant v)$ of the two cases, and a number of other variants are possible. The behaviour of the process up to the time $\tau(u)$ of ruin can be summarized in the so-called *storage equation*

$$R_t = u - \sum_{n=1}^{N_t} U_n + \int_0^t p(R_s)\,ds$$

to which we return within the context of dams. □

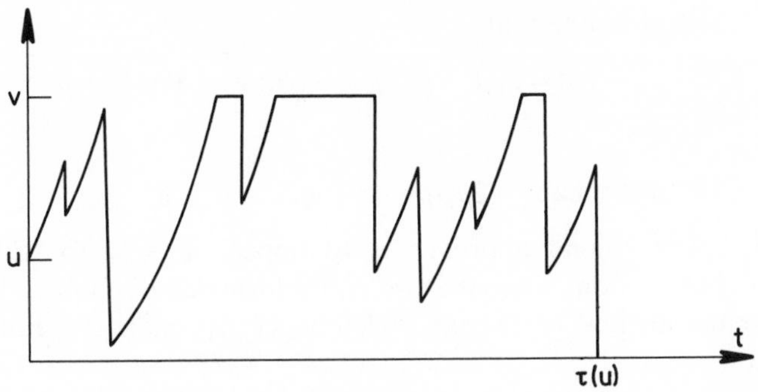

Fig. 1.2

Notes

Standard textbooks for risk theory are Seal (1969, 1978), Bühlmann (1970), Gerber (1979) and Beard *et al.* (1984). Much of the research of the area is published in specific journals like the *Scandinavian Actuarial Journal*, *ASTIN Bulletin*, *Insurance: Mathematics and Economics* and *Mitteilungen der Verein der Schweizerischen Versicherungsmathematiker*. The connection between risk theory and queueing theory was not always realized in the older literature, and there has been much parallel work. An example is Lundberg's inequality, see Notes to XII.5, and in fact many main results from queueing theory and random walks were pioneered in risk theory in the first half of the century.

2. RUIN PROBABILITY APPROXIMATIONS

We consider throughout the compound Poisson model (1.1) with $p=1$, $U>0$ and $\eta = 1/\rho - 1 > 0$.

It was found in Section 1 that the most explicit expression for $\psi(u)$ is given by the Pollaczek–Khintchine formula (1.4). For $\psi(u, T)$, the known expressions are even less explicit. However, a variety of approximations derived in earlier chapters can be used in (or adapted to) the present case. We shall give a brief survey and present some numerical comparisons, which are also of interest in a more general context for evaluating the merits of various methods of approximations.

The body of relevant formulae can be found in Chapter XII. The imbedding $\mathbb{P} = \mathbb{P}_{\theta_0}$ in a conjugate family $(\mathbb{P}_\theta)_{\theta\in\Theta}$ has been given in XII.3.1. We recall that the c.g.f. is

$$\kappa(\alpha) = \log \mathbb{E}\, e^{\alpha S_t}/t = \beta(\hat{\psi}(\alpha) - 1) - \alpha,$$

where $\hat{\psi}(\alpha) = \mathbb{E}\, e^{\alpha U}$, and $\gamma_0 = -\theta_0 > 0$ is given by $\kappa'(\gamma_0) = 0$, the constant $\gamma > 0$ determining the Lundberg parameter $\theta_L = \theta_0 + \gamma$ by $\kappa(\gamma) = 0$. Furthermore, $\{R_t\}$ evolves in the same way as in (1.1) under $\mathbb{P}_\theta$, the changed parameters being $\beta_\theta = \beta\hat{\psi}(\theta - \theta_0)$ and $\hat{\psi}_\theta(\alpha) = \hat{\psi}(\alpha + \theta - \theta_0)/\hat{\psi}(\theta - \theta_0)$.

For the probability of ruin $\psi(u)$ in infinite time two approaches are now possible. The first is to appeal to the discrete-time random walk interpretation $\psi(u) = \mathbb{P}(W > u)$. For example, for the heavy traffic approximation in VIII.6 we let $X = U - T$, and noting that

$$\mathbb{E}X = \mu_B - 1/\beta = \mu_B(1 - \rho^{-1}), \qquad \mathbb{V}\mathrm{ar}\, X = \mathbb{V}\mathrm{ar}\, U + \beta^{-2},$$

we get

$$\psi(u) \cong \exp\{-2\mu_B(\rho^{-1} - 1)u/(\mathbb{V}\mathrm{ar}\, U + \beta^{-2})\}, \tag{2.1}$$

cf. VIII.(6.9). The second approach is to appeal directly to $\{S_t\}$ being a continuous-time random walk and use either formulae developed directly for this case or the method of discrete skeletons, cf. Appendix A9. For example, noting that

$$\mathbb{E}S_1 = \beta\mu_B - 1 = \rho - 1, \qquad \mathbb{V}\mathrm{ar}\, S_1 = \beta\mathbb{E}U^2,$$

VIII.(6.9) leads to

$$\psi(u) \cong \exp\{-2(1 - \rho)u/\beta\mathbb{E}U^2\} \tag{2.2}$$

(since the heavy traffic conditions essentially require $\mathbb{V}\mathrm{ar}\, U \cong \sigma^2 > 0$, $\mu_B \cong 1/\beta$, cf. VIII.6.2, it is easy to see that, as expected, the coefficients to u in (2.1) and (2.2) are asymptotically equivalent).

For the ruin probability $\psi(u, T)$ in finite time the discrete-time random walk approach does, however, not apply (it leads instead to estimates for the probability of ruin after a finite number of claims). We shall therefore in the following consider only the direct continuous-time approach. The random walk imbedded at the times of claims and generating the actual waiting-time process of the corresponding $M/G/1$ queue is, however, still useful since we can identify its overshot process with that of $\{S_t\}$. We shall not spell out the details in the adaptation of earlier results to the present case, but only quote the resulting formulae. Considering first $\psi(u)$, the heavy traffic approximations in XII.(6.6) and XII.(6.8) become

$$\psi(u) \cong \mathrm{e}^{-2u\gamma_0}, \tag{2.3}$$

$$\psi(u) \cong \mathrm{e}^{-\gamma(u+b)}, \qquad b = \frac{\mathbb{E}_0 U^3}{3\mathbb{E}_0 U^2} \tag{2.4}$$

whereas the Cramér–Lundberg approximation in XII.5.2 becomes

$$\psi(u) \cong C\,\mathrm{e}^{-\gamma u}, \qquad C = \frac{1 - \rho}{(\gamma + \beta)\mathbb{E}_{\theta_L} U - 1}. \tag{2.5}$$

For $\psi(u, T)$, the analogues of (2.1), (2.3), (2.4) and (2.5) are given in VIII.6.4, XII.(6.7), XII.(6.11), XII.(5.5) respectively, and become

$$\psi(u, T) \cong G\left(\frac{T\beta\mathbb{E}U^2}{u^2}; -\frac{u(1-\rho)}{\beta\mathbb{E}U^2}, 1\right), \tag{2.6}$$

$$\psi(u, T) \cong G\left(\frac{T\beta_0\mathbb{E}_0 U^2}{u^2}; -u\gamma_0, 1\right), \tag{2.7}$$

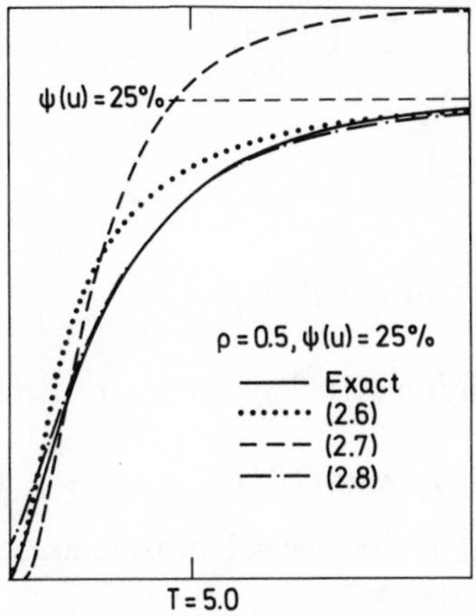

Fig. 2.1 Reproduced by permission of the Almqvist & Wiksell Periodical Company, Stockholm, from Asmussen (1984)

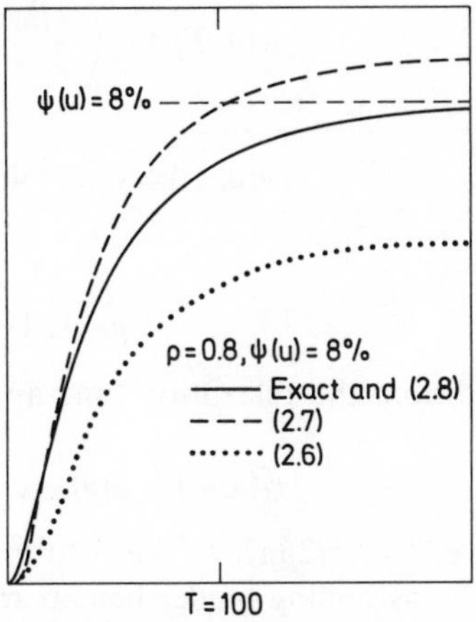

Fig. 2.2 Reproduced by permission of the Almqvist & Wiksell Periodical Company, Stockholm, from Asmussen (1984)

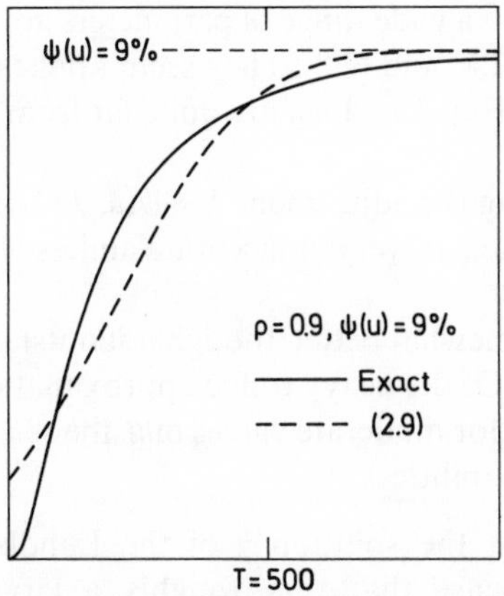

Fig. 2.3 Reproduced by permission of the Almqvist & Wiksell Periodical Company, Stockholm, from Asmussen (1984)

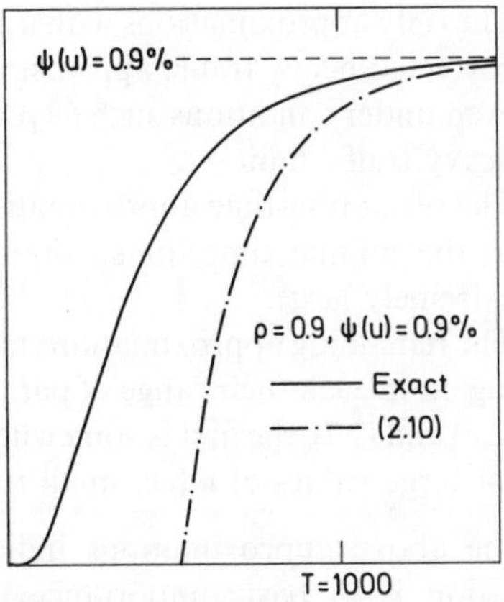

Fig. 2.4 Reproduced by permission of the Almqvist & Wiksell Periodical Company, Stockholm, from Asmussen (1984)

$$\psi(u,T) \cong G\left(\frac{T\beta_0 \mathbb{E}_0 U^2}{u^2} + \frac{b}{u}; -\frac{u\gamma}{2}; 1 + \frac{b}{u}\right), \tag{2.8}$$

$$\psi(u,\mathrm{T}) \cong C\mathrm{e}^{-\gamma u}\Phi\left(\frac{T - \lambda u}{\omega u^{1/2}}\right), \tag{2.9}$$

$$\lambda = \frac{1}{\beta_{\theta_L}\mathbb{E}_{\theta_L} U - 1}, \qquad \omega^2 = \lambda^3 \beta_{\theta_L}\mathbb{E}_{\theta_L} U^2.$$

In addition, the relaxation-time approximation XII.(2.3) is relevant here and yields

$$\psi(u,T) \cong \psi(u) - \exp\{T\kappa(\gamma_0) - \gamma_0 u\} c U_+^{(0)}(u), \tag{2.10}$$

where $c^{-1} = \gamma_0^2 (2\beta\pi \mathbb{E}_0 U^2 \, \mathbb{E}\mathrm{e}^{-\gamma_0 U})^{1/2}$ and $U_+^{(0)}$ is the renewal measure associated with the ascending ladder heights of the $\mathbb{P}_0$-process.

The symbol '$\cong$' in the above relations has a stringent meaning in the sense of a mathematical limit result. However, the limits are in general different. Thus (2.3), (2.4), (2.7) and (2.8) require $\theta_0 \uparrow 0$, $u\theta_0 \to \xi$; (2.5) and (2.9) require $u \to \infty$; and (2.10) is valid when $T \to \infty$. Therefore the range of parameters where the approximations are good can hardly be expected to be the same and numerical studies are required to judge the robustness. Some examples are given in Figs. 2.1–2.4, depicting $\psi(u,T)$ and approximations as functions of T for the $M/M/1$ case $\mathbb{P}(U > y) = \mathrm{e}^{-\delta y}$ with $\delta = 1$. In conjunction with more extensive investigations the following rough picture emerges:

1. The only approximations with a close fit in a wide range of parameters are the corrected heavy traffic approximations (2.4) and (2.8). They seem applicable even under conditions such as $\rho = 0.5$ in Fig. 2.1 which are quite far from the heavy traffic limit.
2. The relaxation-time approximation, giving the adjustment $\Delta = \psi(u,T) - \psi(u)$ to the infinite time (or steady-state) case, is very unaccurate unless T is extremely large.
3. The remaining approximations have a somewhat better, though not outstanding, fit in each their range of parameters. Of the heavy traffic approximations (2.6) and (2.7), the first is somewhat better for moderate values of u, the second for large values of u (i.e. small ruin probabilities).

The above approximations indicate that the solution γ of the Lundberg equation is of basic importance in the sense that very roughly a large γ corresponds to a sound business with small ruin probabilities. For this reason, γ is sometimes denoted in the literature as the *insolvency constant*, and when faced with a change in the company's policy, it may be of importance to study the resulting change in the insolvency constant. As an example, we shall study a reinsurance problem. Suppose the company has arranged for reinsurance such that of a single claim U, the company pays only $h(U) \leqslant U$ and the reinsurer the excess $U - h(U)$, the purpose from the company's point of view being to improve the safety by taking the top of the risk. For example, an obvious choice of h is

$h(u) = u \wedge M$, in which case one calls M the *retention limit*, and we shall show that in an appropriate setting such an arrangement is optimal in the sense of maximizing γ (the appropriate M will also be computed).

We shall assume that β, B are given and restrict attention to functions h with fixed mean claim size $\varepsilon = \mathbb{E}h(U)$. The safety loading is kept fixed in the sense that the premium received by the company (after paying the reinsurance premium) is $p = \beta\varepsilon(1+\eta)$. For a given h, we can then define the corresponding insolvency constant $\gamma^{(h)} > 0$ by

$$\kappa^{(h)}(\gamma^{(h)}) = \beta(\mathbb{E}\,\mathrm{e}^{\gamma^{(h)}h(U)} - 1) - p = 0. \tag{2.11}$$

Proposition 2.1. *For given $\beta, B, \varepsilon, \eta$ let the constant M be defined by $\mathbb{E}m(U) = \varepsilon$, where $m(x) = x \wedge M$. Then $\gamma^{(m)} \geqslant \gamma^{(h)}$ for all h with $\mathbb{E}h(U) = \varepsilon$.*

Proof. An easy graphical argument, using the shape of $\kappa^{(h)}$, shows that it is sufficient to show $\kappa^{(h)}(\gamma^{(m)}) \geqslant 0$. Since $\kappa^{(m)}(\gamma^{(m)}) = 0$, this will in turn follow from $\kappa^{(h)}(\alpha) \geqslant \kappa^{(m)}(\alpha)$ for all $\alpha > 0$, i.e. from

$$\mathbb{E}\,\mathrm{e}^{\alpha h(U)} \geqslant \mathbb{E}\,\mathrm{e}^{\alpha m(U)}, \qquad \alpha > 0. \tag{2.12}$$

Now $\mathrm{e}^z \geqslant 1 + z$ with $z = \alpha(h(y) - m(y))$ yields

$$\mathrm{e}^{\alpha h(y)} \geqslant \mathrm{e}^{\alpha m(y)} + \alpha\,\mathrm{e}^{\alpha m(y)}[h(y) - m(y)]. \tag{2.13}$$

Also

$$\mathrm{e}^{\alpha m(y)}[h(y) - m(y)] \geqslant \mathrm{e}^{\alpha M}[h(y) - m(y)] \tag{2.14}$$

[if $h(y) > m(y)$, then $m(y) = M$ because of $h(y) \leqslant y$; if $h(y) \leqslant m(y)$, then (2.14) follows from $m(y) \leqslant M$]. Hence

$$\mathrm{e}^{\alpha h(U)} \geqslant \mathrm{e}^{\alpha m(U)} + \alpha\,\mathrm{e}^{\alpha M}[h(U) - m(U)]$$

and (2.12) follows from $\mathbb{E}h(U) = \mathbb{E}m(U) = \varepsilon$. □

Notes

The numerical comparisons in Figs. 2.1–2.4 are from Asmussen (1984), where the formulae for the approximations are also discussed in more detail. A study in a somewhat similar vein is given in Embrechts *et al.* (1985) for the distribution of a compound Poisson process, i.e. up to a translation the distribution of the accumulated claims S_t which is another main objective of study in risk theory. The material of Proposition 2.1 the author learned from 1981 lectures by Howard Waters at the University of Copenhagen. Further discussion and references can be found in Van Wouwe *et al.* (1983).

3. COMPOUND POISSON DAMS WITH GENERAL RELEASE RULE

This model originates from problems of storage of water in dams or reservoirs. Water flows in, say from a river or several creeks, according to an input process $\{A_t\}_{t \geqslant 0}$ and is released at a rate $r(x)$ depending on the present content x of the dam. We let X_t be the content at time t and shall be interested in the ergodicity problem for the process $\{X_t\}_{t\geqslant 0}$. From a practical point of view, the stationary distribution π is of importance for assessing values of quantities like the proportion $\pi_0 = \pi(\{0\})$ of time the dam is empty and the average release rate $\int_0^\infty r(x)\pi(\mathrm{d}x)$. Thereby possibly also some guidelines for the choice of r are provided.

We shall assume that $\{A_t)_{t\geqslant 0}$ is a compound Poisson process with no drift term,

$$A_t = \sum_{n=1}^{N_t} U_n, \tag{3.1}$$

where $\{N_t\}_{t\geqslant 0}$ is a Poisson process with intensity β and $U_1, U_2, \ldots$ are independent identically distributed with $\mathbb{P}(U \leqslant x) = B(x)$ and independent of $\{N_t\}$ (here $U > 0$, i.e. $B(0) = 0$). In terms of water storage, this corresponds intuitively to the input to the dam being due mainly to rare large rainfalls. This assumption is acceptable for the dry climatic conditions for which the theory was initially developed, but certainly not always. Thus it would frequently be reasonable to add a drift term ct to (3.1), and also the effects of frequent small rainfalls may not be negligible which would lead to $\{A_t\}$ being a Lévy process with only positive jumps. These cases will, however, not be discussed here.

The dam is taken to be infinitely high, i.e. the state space for $\{X_t\}$ is $[0, \infty)$. The release rate being $r(x)$ at content x means that in between jumps, $\{X_t\}$ should satisfy the differential equation

$$\dot{x} = -r(x), \tag{3.2}$$

where $\dot{x}$ means left derivative. We shall assume that r is in $D(0, \infty)$ with

$$0 < \inf_{\varepsilon < x < \varepsilon^{-1}} r(x), \qquad \sup_{0 < x < \varepsilon^{-1}} r(x) < \infty \tag{3.3}$$

for any $\varepsilon \in (0, 1)$, and extend r to 0 by letting $r(0) = 0$ (in applications, $r(x)$ will typically be non-decreasing). It is then easily seen that (3.2) has for each $x_0 > 0$ a unique solution x_t starting at x_0. In fact, if we let

$$\theta(x; y) = \int_y^x \frac{1}{r(v)}\,\mathrm{d}v, \qquad x \geqslant y \tag{3.4}$$

then $(\mathrm{d}/\mathrm{d}t)\theta(x_0; x_t) = 1$ and $\theta(x_0; x_0) = 0$ yields $\theta(x_0; x_t) = t$. That is, x_t is the inverse function of $\theta(x_0; \cdot)$ or equivalently, $\theta(x; y)$ is the time required for x_t to pass from x to y. Note that in view of $r(0) = 0$, x_t gets absorbed at zero once zero is hit, which happens if and only if $\theta(x; 0) < \infty$ for $x > 0$.

The construction of the process is now obvious since we only have to start at say $X_0 = x$ and let the process move deterministically according to (3.2) until the first jump of A_t where X_t jumps the same amount. Then (3.2) governs the motion until the next jump and so on. These properties can be summarized in the so-called *storage equation*

$$X_t = x + A_t - \int_0^t r(X_s)\,\mathrm{d}s \tag{3.5}$$

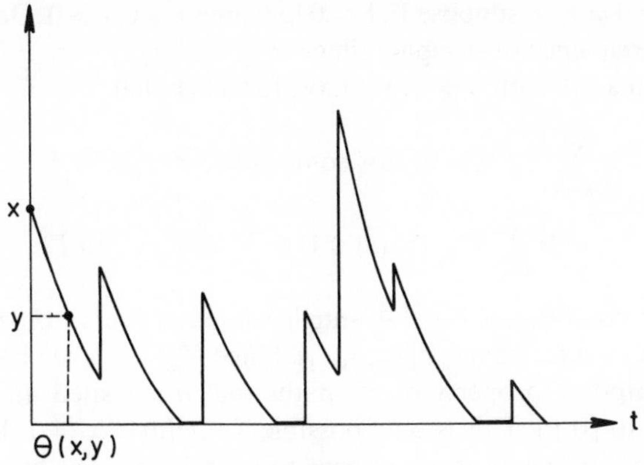

Fig. 3.1

and under the given set of assumptions, it is easy to see that there is a unique solution [(3.5) simply reflects that X_t has the same upwards jumps as A_t and moves according to (3.2), with $r(0)=0$, in between jumps]. It is also intuitively clear from the construction that $\{X_t\}$ is Markov, and this is readily checked as well as the strong Markov property, cf. Problem 3.1. An example of a sample path is in Fig. 3.1. Here $r(x)=1+x$ corresponding to $x_t=(x_0+1)e^{-t}-1$, and $\theta(x;0)<\infty$ because of $\underline{\lim}\, r(x)>0,\ x\to 0$. We note that if $r(x)=1$, then $\{X_t\}$ is simply the virtual waiting-time process of a $M/G/1$ queue, cf. Fig. III.1.4.

Preparing for the study of the ergodicity problem, let

$$\tau_+(u)=\inf\{t\geqslant 0: X_t>u\},\qquad \tau_-(u)=\inf\{t\geqslant 0: X_t\leqslant u\},$$
$$\tau(u)=\inf\{t\geqslant 0: X_t=u\}.$$

Lemma 3.1. *For any $u\in(0,\infty)$, $v<u$ and $T>0$, there is an $\varepsilon>0$ such that $\mathbb{P}_x(\tau_+(u)\leqslant T)\geqslant\varepsilon$ for all $x\leqslant v$.*

Proof. Define

$$m=\sup_{x\leqslant u} r(x),\qquad F=\{A_T>u+mT\}.$$

Then $\varepsilon=\mathbb{P}F>0$ since the distribution of A_T as compound Poisson has unbounded support. Furthermore, $\tau_+(u)\leqslant T$ on F, since otherwise $X_t\leqslant u$, $r(X_t)\leqslant m$ for all $t\leqslant T$, and the storage equation yields the contradiction

$$X_T>0+u+mT-\int_0^T r(X_t)\,dt\geqslant u. \qquad \square$$

Proposition 3.2. *The process is either transient in the sense that $\mathbb{P}_x(X_t\to\infty)=1$ for all x, or recurrent in the sense that for all $x\geqslant 0$, $u>0$ $\mathbb{P}_x(\tau(u)<\infty)=1$.*

Proof. Define $F=\{\underline{\lim}_{t\to\infty} X_t<v\}$, $v\geqslant 0$. If $\mathbb{P}_x F=0$ for all $x\geqslant 0$, $v>0$, it is clear that $\mathbb{P}_x(X_t \to \infty) = 1$, so suppose $\mathbb{P}_x F > 0$ for some $x \geqslant 0, v > 0$. On F, there exists an increasing sequence of stopping times $\{\sigma_k\}$ with $\sigma_{k+1} - \sigma_k \geqslant 1$ (say), $X_{\sigma_k} \leqslant v$. Then by Lemma 3.1 with $T = 1$, we have for $u > v$ that

$$\sum_{k=1}^{\infty} \mathbb{P}_x(X_t > u \quad \text{for some } t \in [\sigma_k, \sigma_{k+1}) | \mathscr{F}_{\sigma_k})$$
$$\geqslant \sum_{k=1}^{\infty} \mathbb{P}_{X_{\sigma_k}}(\tau_+(u) \leqslant 1) \geqslant \sum_{k=1}^{\infty} \varepsilon = \infty \quad \text{on } F.$$

Hence, by the conditional Borel–Cantelli lemma it occurs on F for infinitely many k that $X_t > u$ for some $t \in [\sigma_k, \sigma_{k+1})$. Since $X_{\sigma_{k+1}} \leqslant v$, we have thus by the downwards skip-free property of the paths that u is visited infinitely often in between visits to $[0, v]$. This is only possible if $\mathbb{P}_u(\tau(u) < \infty) = 1$. Hence, starting from u, there are infinitely many returns to u, and since any state $y > u$ can be reached from u, we must have $\mathbb{P}_y(\tau(u) < \infty) = 1$. To get $\mathbb{P}_0(\tau(u) < \infty) = 1$, just condition on the first state y entered at the jump away from zero. □

It follows easily from irreducibility properties of the process that in the recurrent case, $\mathbb{E}_u \tau(u)$ is either finite for all $u > 0$ or infinite for all $u > 0$. For obvious reasons, we refer to the two possibilities as *positive recurrence* and *null recurrence*, respectively. In either case, it follows from VI.3 that $\{X_t\}$ has a stationary measure v which is unique up to a constant, and which for any $u > 0$ may be written as $v = c_u v^{(u)}$ where

$$\int_0^{\infty} f(x) v^{(u)}(\mathrm{d}x) = \mathbb{E}_u \int_0^{\tau(u)} f(X_t) \mathrm{d}t. \tag{3.6}$$

Proposition 3.3. *For any $u > 0$, the function $z(t) = \mathbb{P}_u(X_t \leqslant u, \tau(u) > t)$ tends to zero exponentially fast. In particular, the stationary measure v is Radon ($v(A) < \infty$ for A* compact) *and in the null recurrent case $\mathbb{P}_x(X_t \leqslant u) \to 0$ for all $x, u \geqslant 0$. That is, $X_t \to \infty$ in $\mathbb{P}_x$-distribution.*

Proof. An inspection of the paths shows that $z(t) = \mathbb{P}_u(\tau_+(u) > t)$. Letting H_n be the conditional distribution of X_n given $\tau_+(u) > n$, Lemma 3.1 yields

$$\mathbb{P}_u(\tau_+(u) > n + 1) = \mathbb{P}(\tau_+(u) > n) \int_0^n \mathbb{P}_x(\tau_+(u) > 1) \mathrm{d}H_n(x)$$
$$\leqslant (1 - \varepsilon)\mathbb{P}(\tau_+(u) > n)$$

and the exponential decay of $z(t)$ follows. Clearly z is measurable, thus Lebesgue integrable. Therefore $v^{(u)}[0, u] = \int z < \infty$, the truth of which for all u shows that v is Radon. Since clearly the cycle length distribution is absolutely continuous with mean $\mathbb{E}_u \tau(u)$ which is infinite in the null recurrent case, VI.3.8 then yields $\mathbb{P}_x(X_t \leqslant u) \to 0$ for all x, u and $X_t \xrightarrow{\mathscr{D}} \infty$. □

Theorem 3.4. *The stationary measure v has an atom $v_0 = v(\{0\})$ at zero if and only if $\theta(x; 0) < \infty$ for some* (and then all) $x > 0$. *Also v is absolutely continuous* *on* $(0, \infty)$ *and there exists a version* g *of the density satisfying*

$$g(x) = \frac{1}{r(x)} \beta \int_0^x (1 - B(x-y))\nu(\mathrm{d}y) \tag{3.7}$$

$$= \frac{\beta}{r(x)} \left\{ \nu_0(1 - B(x)) + \int_0^x (1 - B(x-y))g(y)\mathrm{d}y \right\}. \tag{3.8}$$

Proof. We take $\nu = \nu^{(u)}$ for a while. Starting from u, we reach zero with positive probability if and only if $\theta(u;0) < \infty$, and then have a non-zero sojourn time. From this the statement concerning ν_0 is immediately apparent. For the absolute continuity, note that a particle moving at a speed at least δ spends at most time $\delta^{-1}|A|T$ in the set A within T units of time. Hence, if $A \subseteq (\varepsilon, \varepsilon^{-1})$ and $|A| = 0$, it follows by (3.3) that $\int_0^\infty I(X_t \in A)\mathrm{d}t = 0$, implying $\nu^{(u)}(A) = 0$ and absolute continuity.

For the proof of (3.7), we normalize ν by $\nu[0,1] = 1$. That is, $\nu = \nu^{(u)}/\nu^{(u)}[0,1]$ for all $u > 0$. Now consider the number K_u of upcrossings from $[0,u]$ to (u,∞) before time $\tau(u)$. The upcrossing rate is $\beta(1 - B(u-y))$ in state $y \leqslant u$, hence

$$\mathbb{E}_u K_u = \mathbb{E}_u \int_0^{\tau(u)} \beta(1 - B(u - X_t))I(X_t \leqslant u)\mathrm{d}t = \beta \int_0^u (1 - B(u-y))\nu^{(u)}(\mathrm{d}y).$$

But after an upcrossing, the process hits u when returning to $[0,u]$. Hence $\mathbb{P}_u(K_u = 1) = 1$ and

$$\frac{1}{\nu^{(u)}[0,1]} = \beta \int_0^u (1 - B(u-y))\nu(\mathrm{d}y).$$

Now for $v > u$ let $F_v = \{S_{\tau_+(u)} \geqslant v\}$,

$$\varphi(v) = \nu^{(u)}(u, v] = \mathbb{E}_u \int_0^{\tau(u)} I(u < X_t \leqslant v)\mathrm{d}t,$$

$$\varphi^*(v) = \mathbb{E}_u \left[\int_0^{\tau(u)} I(u < X_t \leqslant v)\mathrm{d}t \,|\, F_v \right].$$

Here the integral in the definition of $\varphi(v)$ is $(v-u)/r(u) + \mathrm{o}(v-u)$ for a.a. paths. To check that the expectation $\varphi(v)$ is so too, we condition upon the first jump after $\tau(v)$. Then for small v

$$\varphi^*(v) \leqslant \theta(v;u)\mathrm{e}^{-\beta\theta(v;u)} + \int_0^{\theta(v;u)} \{s + \varphi^*(v)\}\beta\,\mathrm{e}^{-\beta s}\mathrm{d}s$$

$$\leqslant \theta(v;u)\{1 + \beta\varphi^*(v) + \theta(v;u)\},$$

$$\varphi(v) \leqslant \varphi^*(v) \leqslant \frac{\theta(v;u) + \theta(v;u)^2}{1 - \beta\theta(v;u)}.$$

Also clearly $\mathbb{P}F_v \to 1$, $v \downarrow u$, and hence

$$\varphi(v) \geqslant \theta(v;u)\mathbb{P}F_v = \theta(v;u)\{1 - \mathrm{o}(1)\}.$$

Since $\theta(v;u)=(v-u)/r(u)+o(v-u)$, the claim concerning $\varphi(v)$ follows. Hence the right derivative of $\nu[0,v]$ at $v=u$ exists and equals

$$\frac{\varphi'(u)}{\nu^{(u)}[0,1]}=\frac{1}{r(u)}\int_0^u \beta(1-B(u-y))\nu(\mathrm{d}y). \tag{3.9}$$

But this being true for all $u>0$ implies that (3.9) is a version of the density, cf. Problem 3.3. □

In the positive recurrent case $\|\nu\|<\infty$, we can define the unique stationary distribution π by $\pi=\nu/\|\nu\|$ and have by general results on regenerative processes that $X_t\to\pi$ in $\mathbb{P}_x$-distribution for all x. In that case, the solution g to (3.7) and (3.8) exists and is integrable.

We shall next show that conversely the existence of an integrable solution to (3.7) and (3.8) implies positive recurrence. To this end, let Y_n be the content just before the $(n+1)$th jump. Then $\{Y_n\}$ is a Markov chain, and we have:

Lemma 3.5. (i) *Either* $\{Y_n\}$ *is transient in the sense that* $\mathbb{P}_x(Y_n\to\infty)=1$ *for all* x, *or* $\{Y_n\}$ *is recurrent in the sense that* $\mathbb{P}_x(Y_n\leqslant v$ infinitely often$)=1$ *for all* $x\geqslant 0$, $v>0$; (ii) *In the recurrent case,* $\{Y_n\}$ *is Harris recurrent*; (iii) *if a distribution* π *has the property that* (3.7) *and* (3.8) *hold for* $\pi=\nu$, *then* π *is stationary for* $\{Y_n\}$.

Proof. Here (i) is shown similarly as in Proposition 3.2. For (ii), let $R=[0,v]$ for some arbitrary $v>0$ and choose $0<c<d<f<\infty$ such that $\delta_1=\mathbb{P}(U\in(d,f])>0$. Now for $(a,b)\subseteq(c,d)$ and $x\in R$,

$$\begin{aligned}\mathbb{P}(Y_1\in(a,b)\mid Y_0=x)&\geqslant \mathbb{P}(x+U\in(d,f+v],\theta(x+U;b)<T<\theta(x+U;a))\\&\geqslant\delta_1\mathbb{P}(T>\theta(x+U;b))\mathbb{P}(T<\theta(x+U;a)\mid T>\theta(x+U;b))\\&\geqslant\delta_1\exp\{-\beta\theta(f+v;c)\}(1-\exp\{-\beta\theta(b;a)\})\end{aligned}$$

where T is the time between the first and second jump. But from (3.3), we can find $\delta_2>0$ such that $1-\exp\{-\beta\theta(b;a)\}\geqslant\delta_2(b-a)$ for $(a,b)\subseteq(c,d)$. Hence

$$\mathbb{P}(Y_1\in(a,b)\mid Y_0=x)\geqslant\delta_3(b-a),\qquad x\in R,$$

so that the minorization condition VI.(3.1) holds if we take λ to be the uniform distribution on (c,d). This proves (ii). For (iii), let $\mathbb{P}_{\pi-}$ refer to the initial conditions where the first jump occurs at time zero and $X_{0-}=Y_0$ has distribution π. Then X_0 has distribution $\pi*B$, and $\{Y_1\leqslant z\}$ will occur if either $X_0\leqslant z$, or $X_0=y>z$ and the next jump occurs after $\theta(y;z)$. Hence

$$\begin{aligned}\mathbb{P}_{\pi-}(Y_1\leqslant z)&=\pi*B(z)+\int_z^\infty \exp\{-\beta\theta(y;z)\}\pi*B(\mathrm{d}y)\\&=\int_0^\infty 1\wedge\exp\{-\beta\theta(y;z)\}\pi*B(\mathrm{d}y)\\&=1-\int_z^\infty\frac{\beta}{r(y)}\exp\{-\beta\theta(y;z)\}(1-\pi*B(y))\mathrm{d}y.\end{aligned} \tag{3.10}$$

Now clearly by (3.7),

$$1-\pi*B(y)=1-\pi(y)+\int_0^y(1-B(y-x))\pi(\mathrm{d}x)=1-\pi(y)+\frac{1}{\beta}g(y)r(y)$$

and hence (3.10) becomes

$$1-\int_z^\infty\frac{\beta}{r(y)}\exp\{-\beta\theta(y;z)\}(1-\pi(y))\mathrm{d}y-\int_z^\infty\exp\{-\beta\theta(y;z)\}g(y)\mathrm{d}y$$
$$=\pi(z)+\int_z^\infty\exp\{-\beta\theta(y;z)\}\pi(\mathrm{d}y)-\int_z^\infty\exp\{-\beta\theta(y;z)\}\pi(\mathrm{d}y)=\pi(z),$$

proving stationarity of π for $\{Y_n\}$. □

Theorem 3.6. *The process $\{X_t\}$ is positive recurrent if and only if there exists a probability measure $\pi(x)=\pi_0+\int_0^x g(y)\mathrm{d}y$ such that* (3.7) *and* (3.8) *hold for $v=\pi$. In that case, the solution to* (3.7) *and* (3.8) *is unique and the stationary distribution.*

Proof. The existence of a solution in the positive recurrent case follows from Theorem 3.4. If, conversely, π is a solution with $\|\pi\|=1$, then by Lemma 3.5(iii) π is a stationary distribution for $\{Y_n\}$, the existence of which implies first that $\{Y_n\}$ cannot be transient, cf. (i), and next by (ii) that $\{Y_n\}$ is positive recurrent and the solution to (3.7) and (3.8) is unique. It thus only remains to show that $\{X_t\}$ is indeed positive recurrent if π exists. But then by the PASTA property (V.3.3) the time-averages $\int_0^t I(X_s\leqslant u)\mathrm{d}s/t$ have non-zero limits, which excludes transience and (by Proposition 3.3 and Fatou's lemma) null recurrence. □

In the case $\theta(x;0)<\infty$, it is possible to give an alternative characterization of π. Define

$$K(x,y)=\beta(1-B(x-y))/r(x),\quad 0\leqslant y<x,\quad K_1=K,$$
$$K_{n+1}(x,y)=\int_y^x K(x,z)K_n(z,y)\mathrm{d}z=\int_y^x K_n(x,z)K(z,y)\mathrm{d}z.$$

Using $K(x,y)\leqslant\beta/r(x)$, it follows easily by induction that

$$K_{n+1}(x,y)\leqslant\beta^{n+1}\theta(x;y)^n/r(x)n!.$$

Hence $K^*=\sum_1^\infty K_n$ is well defined and finite, and we have:

Corollary 3.7. *If $\theta(x;0)<\infty$, $x>0$, then $\{X_t\}$ is positive recurrent if and only if*

$$\frac{1}{\pi_0}=1+\int_0^\infty K^*(x,0)\,\mathrm{d}x<\infty,\tag{3.11}$$

in which case $\pi(\{0\})=\pi_0$, $g(x)=\pi_0K^*(x,0)$, $x>0$.

Proof. We rewrite (3.8) as

$$g(x)=\pi(\{0\})K(x,0)+\int_0^x K(x,y)g(y)\mathrm{d}y.\tag{3.12}$$

Iterating $N-1$ times, this yields

$$g(x)=\pi(\{0\})\sum_{n=1}^{N}K_n(x,0)+\int_0^x K_N(x,y)g(y)\mathrm{d}y.$$

Letting $N\to\infty$ yields the desired conclusions in the positive recurrent case. If, conversely, (3.11) holds and we define π as indicated, the r.h.s of (3.12) becomes

$$\pi_0K(x,0)+\pi_0\int_0^x K(x,y)\sum_{n=1}^{\infty}K_n(y,0)\mathrm{d}y=\pi_0K^*(x,0)=g(x)$$

so that π is a probability measure satisfying (3.7). □

Also in the case $\theta(x;0)=\infty$, one can give a criterion for positive recurrence in terms of K^*:

Corollary 3.8. *The process is positive recurrent if and only if for some* (and then for all) $a>0$ $\int_a^\infty K^*(x,a)<\infty$.

Proof. Let $\{\tilde{X}_t\}$ correspond to $\tilde{r}(x)=r(x+a)$ and the same β, B. Then, in the obvious notation, $\tilde{\theta}(x;0)<\infty$ because of $\inf_{x+a\geqslant y\geqslant a}r(y)>0$ so that we may apply Corollary 3.7 to see that $\{\tilde{X}_t\}$ is positive recurrent if and only if $\tilde{K}^*(x,0)$ is integrable. But for $n=1$

$$\tilde{K}_n(x,y)=\frac{1}{r(x+a)}(1-B(x-y))=K_n(x+a,y+a)$$

and it follows easily by induction that this is also valid for $n=2,3,\ldots$. Hence $\int_0^\infty\tilde{K}^*(x,0)\mathrm{d}x$ and $\int_a^\infty K^*(x,a)\mathrm{d}x$ are equal, in particular finite at the same time, so that we have only to show that $\{X_t\}$ and $\{\tilde{X}_t\}$ are positive recurrent at the same time. But for $x>y>a$, $\mathbb{E}_x\tau(y)=\mathbb{E}_{x-a}\tilde{\tau}(y-a)$, and by irreducibility properties of the processes, it is easily seen that $\mathbb{E}_x\tau(y)<\infty$ exactly when $\{X_t\}$ is positive recurrent, and that $\mathbb{E}_{x-a}\tilde{\tau}(y-a)<\infty$ exactly when $\{\tilde{X}_t\}$ is positive recurrent. □

The above result concludes the treatment of the general theory, and we proceed in the next section to see how the derived formulae and criteria take a more explicit form in some particular cases.

Problems

3.1. Show that $\{X_t\}$ has the strong Markov property [hint: recall the proof of IV.1.5].

3.2. In a storage problem with $A_t=t/2$, $r(x)=1$, $x>0$, it seems reasonable to define $X_t=(X_0-t/2)^+$. Show that the storage equation is not satisfied.

3.3. Let F be a distribution on $[0,\infty)$ such that F has a density f on $(0,\infty)$, $F(x)=F(0)+\int_0^x f(y)\mathrm{d}y$, and that the right derivative $g(x)=\lim_{t\downarrow 0}\{F(x+t)-F(x)\}/t$ exists for all $x>0$. Show that $f(x)=g(x)$ a.e. [hint: suppose $F(0)=0$ and let u_n be the density of the uniform distribution on $(0,1/n)$. Show first that $f*u_n(x)\to g(x)$ for all $x>0$ and next that $f*u_n\to f$ in L_1 by considering first the case where f has a bounded continuous derivative].

3.4. Show by an example that it is possible that $\lim_{x\to\infty}\theta(x;y)<\infty$, and that there is then positive recurrence.

Notes

The results are contained in Brockwell *et al.* (1982), who also treat the Lévy case, and Harrison and Resnick (1976). However the proofs are somewhat different. In particular, the present imbedded Harris chain $\{Y_n\}$ replaces certain arguments based on semi-group tools, and the derivation of (3.7) and (3.8) is somewhat more direct (similar observations have been made by Jim Pitman, unpublished). It should be stressed that questions concerning the storage equation may be considerably more complicated for say the Lévy case (see also Problems 3.2). Some discussion and references are given by Brockwell *et al.* (1982).

4. SOME EXAMPLES

The model and notation is that of the preceding section. In the following let $\mu_B = \int_0^\infty x\,\mathrm{d}B(x) = \int_0^\infty (1 - B(x))\mathrm{d}x$.

Example 4.1. Constant release, say $r(x) = 1$. This case is already well known from the $M/G/1$ virtual waiting-time interpretation, but is treated here for the sake of illustration. Here $\theta(x;0) = x < \infty$ so that we may apply Corollary 3.7. Since $\int_0^\infty K(x,0)\mathrm{d}x = \beta\mu_B$, it is necessary for positive recurrence that $\mu_B < \infty$. If $\mu_B < \infty$, define $\rho = \beta\mu_B$, $B_0(\mathrm{d}x) = (1 - B(x))/\mu_B\,\mathrm{d}x$. Then $K(x,0) = \rho\,\mathrm{d}B_0(x)/\mathrm{d}x$ and it follows easily by induction that $K_n(x,0) = \rho^n\,\mathrm{d}B_0^{*n}(x)/\mathrm{d}x$. Hence $\pi_0^{-1} = \sum_0^\infty \rho^n$ is finite and we have positive recurrence if and only if $\rho < 1$, in which case the expression for π in Corollary 3.7 is immediately seen to coincide with the Pollaczek–Khintchine formula IX.(2.1). □

Example 4.2. Arbitrary release and exponential input, $B(\mathrm{d}x) = \delta\,\mathrm{e}^{-\delta x}\,\mathrm{d}x$. Here (3.8) becomes

$$g(x) = \frac{\beta}{r(x)}\left\{\pi_0\,\mathrm{e}^{-\delta x} + \int_0^x \mathrm{e}^{-\delta(x-y)}g(y)\mathrm{d}y\right\}$$

which, letting $\psi(x) = \int_0^x \mathrm{e}^{\delta y}g(y)\mathrm{d}y$, may be rewritten as

$$\psi'(x) = \frac{\beta}{r(x)}\{\pi_0 + \psi(x)\}.$$

This is a differential equation of standard type and any solution may be written in the form

$$\psi(x) = c\,\mathrm{e}^{\beta\theta(x;1)} - \pi_0 \tag{4.1}$$

so that

$$g(x) = \mathrm{e}^{-\delta x}\psi'(x) = \frac{c\beta}{r(x)}\exp\{\beta\theta(x;1) - \delta x\} \tag{4.2}$$

Thus the solution to (3.8) is unique up to a constant and we have positive recurrence if and only if

$$\alpha = \int_0^\infty \frac{\beta}{r(x)}\exp\{\beta\theta(x;1) - \delta x\}\mathrm{d}x < \infty.$$

It only remains to evaluate π_0 and c in the case $\alpha < \infty$. We have $\|\pi\| = 1$, which yields $\pi_0 + c\alpha = 1$. If $\theta(x;0) = \infty$, then $\pi_0 = 0$ and $c = \alpha^{-1}$. If $\theta(x;0) < \infty$, then $\psi(0) = 0$ and (4.1) with $x = 0$ yields $c = \pi_0 \exp\{\beta\theta(1;0)\}$,

$$\pi_0 = \frac{1}{1 + \alpha \exp\{\beta\theta(1;0)\}} = \frac{1}{1 + \displaystyle\int_0^\infty \frac{\beta}{r(x)} \exp\{\beta\theta(x;0) - \delta x\}\mathrm{d}x}.$$

Note that we may rewrite (4.2) as

$$g(x) = \frac{\pi_0 \beta}{r(x)} \exp\{\beta\theta(x;0) - \delta x\}. \qquad \square$$

Proposition 4.3. *Suppose $r(x)$ is non-decreasing. Then $\{X_t\}$ is positive recurrent if and only if*

$$\lim_{a\to\infty} \beta \int_0^\infty \frac{1 - B(x)}{r(x + a)} \mathrm{d}x < 1. \tag{4.3}$$

If $\mu_B < \infty$, (4.3) is equivalent to $\beta\mu_B < \lim_{x\to\infty} r(x)$.

Proof. The limit in (4.3) clearly exists and by the definition of $K = K_1$ is also the limit of $\int_a^\infty K(x,a)\mathrm{d}x$. Hence if (4.3) holds, there are a_0 and $\delta < 1$ such that $\int_a^\infty K(x,a)\mathrm{d}x \leqslant \delta$ for all $a \geqslant a_0$. Since

$$\begin{aligned}\int_a^\infty K_{n+1}(x,a)\mathrm{d}x &= \int_a^\infty \mathrm{d}x \int_a^x K_n(x,y)K(y,a)\mathrm{d}y \\ &= \int_a^\infty K(y,a)\mathrm{d}y \int_y^\infty K_n(x,y)\mathrm{d}x\end{aligned}$$

it follows by induction that $\int_a^\infty K_{n+1}(x,a)\mathrm{d}x \leqslant \delta^{n+1}$ for all n. Summing yields

$$\int_a^\infty K^*(x,a)\mathrm{d}x \leqslant \sum \delta^n < \infty, \quad a \geqslant a_0.$$

If conversely (4.3) fails, then $\int_a^\infty K(x,a)\mathrm{d}x \geqslant 1$ for all a and thus in the same way

$$\int_a^\infty K^*(x,a)\mathrm{d}x = \int_a^\infty \sum_{n=1}^\infty K_n(x,a)\mathrm{d}x \geqslant \sum_{n=1}^\infty 1 = \infty.$$

Reference to Corollary 3.7 completes the proof. $\square$

Example 4.4. Linear release rate, $r(x) = c + fx$. Since $r(x) \to \infty$ monotonically, positive recurrence is immediately obtained from Proposition 4.3 provided only that $\mu_B < \infty$. If $\mu_B = \infty$, we have

$$\beta \int_0^\infty \frac{1 - B(x)}{r(x + a)} = \frac{\beta}{f} \int_0^\infty \log\left(1 + \frac{fx}{c + fa}\right) B(\mathrm{d}x). \tag{4.4}$$

By elementary properties of the logarithm, this is finite if and only if $\int_0^\infty \log x$

$B(\mathrm{d}x) < \infty$. In that case, (4.4) tends to zero as $a \to \infty$ by monotone convergence and hence (4.3) is automatic. That is, we have positive recurrence if and only if $\int_0^\infty \log x B(\mathrm{d}x) < \infty$. Also clearly $\theta(1;0) < \infty$, i.e. $\pi_0 > 0$, if and only if $c > 0$. □

Notes

Again, the basic reference is Brockwell *et al.* (1982).

5. FURTHER DAM AND STORAGE MODELS

As for the compound Poisson insurance risk model, the dam model studied in Sections 3 and 4 may require variants and modifications in practical situations.

Example 5.1 (THE FINITE DAM). In practice, the capacity of the dam is apparently not infinite but the content cannot exceed some level, say v. If more input occurs at a jump than the dam can receive, the excess water simply flows over instantaneously (this corresponds formally to letting $r(x) = \infty$, $x > v$). An example of the sample paths of this process $\{X_t\}$ are in Fig. 5.1.

Concerning ergodicity questions, the type of arguments used already for the infinitely deep dam shows that any state $u \in (0, v]$ is recurrent with finite mean recurrence time. Therefore there always exists a limiting stationary distribution π, the characteristics of which can be evaluated in just the same way as for the infinite case. It follows that there is an atom π_0 at zero with $\pi_0 > 0$ if and only if $\theta(x;0) < \infty$, and that π is absolutely continuous on $(0, v)$ with density g satisfying

$$g(x) = \frac{\beta}{r(x)} \left\{ \pi_0 + \int_0^x (1 - B(x-y))g(y)\mathrm{d}y \right\}, \qquad 0 < x < v. \tag{5.1}$$

In addition, the uniqueness of the solution can be obtained just as for the infinite case.

It should be noted that instead of copying the proofs for the infinitely deep dam it is also possible to proceed by a direct comparison with such a process $\{\tilde{X}_t\}$

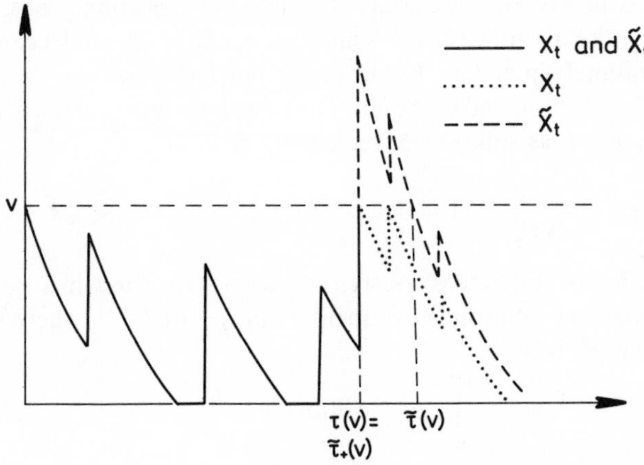

Fig. 5.1

having $\tilde{\beta} = \beta$, $\tilde{r}(x) = r(x)$ and $\tilde{B}(x) = B(x)$ for $x \leqslant v$. For $x > v$, $r(x)$ and $\tilde{B}(x)$ are chosen in a manner ensuring positive recurrence, (e.g. $\tilde{B}(x) = 1$, $r(x) = \beta\mu_{\tilde{B}} + \varepsilon$). Then obviously $\{\tilde{X}_t\}$ and $\{X_t\}$ develop in the same way in $[0, v]$ and starting from $x \leqslant v$, we may even realize the processes such that $\tilde{X}_t = X_t$ for $t \leqslant \tau(v) = \tilde{\tau}_+(v)$. If we take $x = v$, it is seen that the cycle interval $[0, \tilde{\tau}(v))$ for $\{\tilde{X}_t\}$ splits into the cycle interval $[0, \tau(v))$ for $\{X_t\}$ and an interval $[\tilde{\tau}_+(v))$, $\tilde{\tau}(v)$, where $\{\tilde{X}_t\}$ takes values only in (v, ∞). Thus the stationary measures defined by

$$\nu(A) = \mathbb{E}_v \int_0^{\tau(v)} I(X_t \in A)\mathrm{d}t, \qquad A \subseteq [0, v],$$

$$\tilde{\nu}(A) = \mathbb{E}_v \int_0^{\tilde{\tau}(v)} I(\tilde{X}_t \in A)\mathrm{d}t, \qquad A \subseteq [0, \infty)$$

satisfy $\nu(A) = \tilde{\nu}(A)$ for $A \subseteq [0, v]$. Normalizing so as to obtain stationary distributions $\pi = \nu/\|\nu\|$, $\tilde{\pi} = \tilde{\nu}/\|\tilde{\nu}\|$, it is seen that $\pi(A) = \tilde{\pi}(A)/\tilde{\pi}[0, v]$, $A \subseteq [0, v]$. That is, *the stationary distribution for the finite dam is simply obtained by conditioning the stationary infinite content to be in* $[0, v]$ (this is also clear by comparing (3.8) and (5.1)).

To exemplify the important characteristics of the finite dam, consider the average overflow in the steady state,

$$\beta \int_0^v \pi(\mathrm{d}x) \int_{v-x}^{\infty} (y - v + x)\mathrm{d}B(y). \tag{5.2}$$

□

Example 5.2 (the INFINITELY DEEP or BOTTOMLESS dam). Instead of approaching the finite dam via the infinitely high one, the suggestion has been made of using an approximation in terms of an infinitely deep or bottomless dam. This is reasonable in particular if overflow is a more predominant phenomenon than emptiness, as will be the case if the process in some sense has an upwards drift (say $r(x)$ constant and $\beta\mu_B > r$). The state space for the process $\{X_t\}$ now becomes $(-\infty, v]$ (equivalently, we may study the *deficit* or *depletion* process $\{v - X_t\}$. Without going into a rigorous treatment, it is seen that the model can be treated much as the infinitely high dam. For example, the derivation of (3.7) does not use the fact that $[0, x]$ is bounded to the left. Thus the density $g(x)$ of the content in the steady state is given as solution of the equation

$$g(x) = \frac{\beta}{r(x)} \int_{-\infty}^{x} (1 - B(x - y))g(y)\mathrm{d}y, \qquad -\infty < x \leqslant v. \tag{5.3}$$

A particularly important case is constant release $r(x) = r$, in which case it is seen by insertion that the solution is exponential, $g(x) = \eta\,\mathrm{e}^{\eta(x-v)}$, where $\eta > 0$ is the unique solution of

$$\int_0^{\infty} \mathrm{e}^{-\eta u}\,\mathrm{d}B(u) = 1 - \frac{\eta r}{\beta} \tag{5.4}$$

which exists when $\mu_B > r/\beta$. □

Example 5.3 (MORAN'S MODEL FOR THE DAM). This is a historically important discrete-time Markov chain model. The inputs $Y_0, Y_1, \ldots$ are assumed independent identically distributed and the release is constant, say m (if the content just before a release is $c < m$, only the amount c is released). We let v denote the capacity of the dam and X_n the content just before the nth input is received (or just after the $(n-1)$th release). Then the content just after the nth input is $(X_n + Y_n) \wedge v$ and thus

$$X_{n+1} = [(X_n + Y_n) \wedge v - m]_+. \tag{5.5}$$

Consider as an illustration the simplest case where $m = 1$ and the Y_n are $\mathbb{N}$-valued, $\mathbb{P}(Y = k) = q_k, k = 0, 1, 2, \ldots$. Then the state space is $0, 1, \ldots, v-1$ and with $r_k = q_{k+1} + q_{k+2} + \cdots$, the transition matrix is

$$P = \begin{pmatrix} q_0 + q_1 & q_2 & q_3 & \cdots & q_{v-1} & r_{v-1} \\ q_0 & q_1 & q_2 & \cdots & q_{v-2} & r_{v-2} \\ 0 & q_0 & q_1 & \cdots & q_{v-3} & r_{v-3} \\ & \vdots & & \ddots & & \vdots \\ 0 & 0 & 0 & & q_0 & r_0 \end{pmatrix}.$$

It is seen that we have irreducibility and hence ergodicity provided that $0 < q_0 < 1$, and the equation $\pi P = \pi$ yields

$$\pi_0 = (q_0 + q_1)\pi_0 + q_0\pi_1, \qquad \pi_1 = q_2\pi_0 + q_1\pi_1 + q_0\pi_2, \ldots$$

which is easily solved stepwise by letting $\pi_0 = 1$ (say), solving for $\pi_1, \ldots, \pi_{v-1}$ and finally normalizing so as to get $\pi_0 + \cdots + \pi_{v-1} = 1$. If we write $X_n = X_n^{(v)}$, $\pi = \pi^{(v)}$ to stress the dependence on v, it is also seen that $\pi_0^{(v)}, \ldots, \pi_{v-1}^{(v)}$ and $\pi_0^{(v+1)}, \ldots, \pi_{(v-1)}^{(v+1)}$ differ only by the normalizing constants. That is, $\pi^{(v)}$ is obtained by conditioning $\pi^{(v+1)}$ to $\{0, 1, \ldots, v-1\}$. This also follows simply by noting that since the chains are skip-free to the left, $\{X_n^{(v)}\}$ may be identified with the consecutive visits of $\{X_n^{(v+1)}\}$ to $\{0, 1, \ldots, v-1\}$, cf. I.3.8. Pushing the argument one step further, consider the Lindley process $X_{n+1}^{(\infty)} = [X_n^{(\infty)} + Y_n - 1]_+$. Then in the recurrent case $\mathbb{E}Y \leqslant 1$, $\{X_n^{(v)}\}$ is similarly the visits of $\{X_n^{(\infty)}\}$ to $\{0, 1, \ldots, v-1\}$ and in the ergodic case $\mathbb{E}Y < 1$, we obtain $\pi^{(v)}$ upon conditioning of $\pi^{(\infty)}$, the distribution of the maximum of a random walk with increments $Y_n - 1$ [of course $\pi^{(v)}$ also exists if $\mathbb{E}Y \geqslant 1$, but cannot then be identified in this way]. □

The above dam models are related to a number of further models for storage and inventories. Here in the discrete-time case, X_n becomes the stock level at time n, the release R_n corresponds to the amount being sold and the input Y_n to reordering of the material. However, here typically R_n is random, whereas the choice of Y_n is up to the policy of the firm. We shall give just one example.

Example 5.4 (THE (s, S) INVENTORY MODEL). Here $s < S$ and when the stock level falls below s, reordering is done to bring the level up to S (above s, no action is taken). Thus

$$X_n = \begin{cases} X_{n-1} - R_{n+1} & \text{if} \quad s \leqslant X_{n-1} < S \\ S \quad\;\; - R_{n+1} & \text{if} \quad 0 \leqslant X_{n-1} < s \end{cases}$$

(here X_n may be negative, i.e. the state space is $(-\infty, S)$). Suppose $R_1, R_2, \ldots$ are independent identically distributed with common distribution F. The stationary distribution π is easily obtained by observing that the epochs n with $X_{n-1} < s$ are regeneration points. Thus the zero-delayed case corresponds to $X_0 = S - R_1$, the cycle length is

$$C = \inf\{n \geqslant 1 : R_1 + R_2 + \cdots + R_n > S - s\}$$

and within a cycle the points visited in $[s, S)$ are of the form $S - y$ with $y \in (0, S - s)$ an epoch of a renewal process governed by F. Similarly, exactly one visit to $(-\infty, s)$ occurs at the point $s - z$, where z is the overshot $B(S-s)$ of $S - s$. With $U = \sum_0^\infty F * n$ and $H(z) = \mathbb{P}(B(S-s) \leqslant z)$, we therefore have

$$\pi(\mathrm{d}y) = \begin{cases} \dfrac{U(S - \mathrm{d}y)}{U(S-s)} & s \leqslant y < S, \\[2ex] \dfrac{H(s - \mathrm{d}y)}{U(S-s)} & -\infty < y < s. \end{cases} \qquad \square$$

Notes

Two of the classical books on storage processes are Arrow *et al.* (1958) and Moran (1959). More recent surveys can be found in Tijms (1972) and Prabhu (1980).

APPENDIX

Selected Background and Notation

A1. SOME NOTATION

Symbols like say A, η, etc. do not of course have the same meaning throughout the book and may interchangeably be used for real numbers, matrices, measures and so on. For queueing processes, some effort has been made to make the notation (introduced in III.1) reasonably consistent throughout the book. One inconvenience is that the associated random walk becomes $S_n = X_0 + \cdots + X_{n-1}$ and not $X_1 + \cdots + X_n$ as in Chapter VII. Of course, similar (hopefully minor) incidents occur at a number of other places. For example, in integrals with respect to the measure μ, we change freely between writing $\mathrm{d}\mu(x)$ or $\mu(\mathrm{d}x)$.

A basic principle for references within the book is to specify the chapter number only when it is not the current one. Thus, say, Proposition 1.3, formula (2.7) or Section 5 of Chapter IV is referred to as IV.1.3, IV.(2.7) and IV.5 respectively in all chapters other than IV where we write Proposition 1.3, (2.7) and Section 5. Ends of proofs or examples are marked by the symbol. □

Some main abbreviations are given in the following list (others occur locally):

LLN	law of large numbers
CLT	central limit theorem
LIL	law of the iterated logarithm
l.h.s.	left-hand side
r.h.s.	right-hand side
a.s.	almost surely
i.i.d.	independent identically distributed
i.o.	infinitely often
r.v.	random variable
t.v.	total variation
w.l.o.g.	without loss of generality
w.p.	with probability
w.r.t.	with respect to
d.R.i.	directly Riemann integrable
m.g.f.	moment generating function
c.g.f.	cumulant generating function
g.c.d.	greatest common divisor
supp.	support

The expression $\mathbb{E}[X; A]$ means $\mathbb{E}XI(A)$, where $I(A)$ is the indicator of A (if say $A = \{X > 0\}$, we write $\mathbb{E}[X; X > 0]$). By $X \overset{\mathcal{D}}{=} Y$ we mean equality in distribution and by $X_n \overset{\mathcal{D}}{\to} X$ convergence in distribution (weak convergence).

The typeface $\mathbb{P}$, $\mathbb{E}$ is used for probability and expectation. The standard sets are denoted as follows:

$\mathbb{N} = \{0, 1, 2, \ldots\}$	the natural numbers
$\mathbb{Z} = \{0, \pm 1, \pm 2, \ldots\}$	the integers
$\mathbb{Q} = \{p/q : p \in \mathbb{Z}, q \in \mathbb{N} \setminus \{0\}\}$	the rationals
$\mathbb{R} = (-\infty, \infty)$	the real numbers
$\mathbb{C} = \{x + iy : x, y \in \mathbb{R}\}$	the complex numbers

(no special notation like $\mathbb{R}_+$ is used for $[0, \infty)$ or $(0, \infty)$). Also, the set D of functions $\{x_t\}$ which are right-continuous ($x_t = \lim_{s \downarrow t} x_s$) and have left-hand limits $x_{t-} = \lim_{s \uparrow t} x_s$ is frequently encountered. If, say, t varies in $[0, 1]$ and x_t is E-valued, we may specify this by writing $D([0, 1], E)$. Most often D stands for $D[0, \infty) = D([0, \infty), \mathbb{R})$.

A2. POLISH SPACES AND WEAK CONVERGENCE

Polish spaces are of importance in probability theory by, on one hand, to provide a common framework comprising Euclidean space $\mathbb{R}^n$ and its nice subsets, discrete (finite or countable) sets and also some function spaces like D, and on the other, to possess many of the same regularity properties as $\mathbb{R}^n$ (e.g. in Polish spaces, Kolmogorov's consistency theorem holds and regular conditional distributions exist, see Neveu, 1965). Fundamental examples are:

(a) any locally compact space with a countable dense subset is Polish;
(b) if E is Polish, then $F \subseteq E$ is so if F is, say, closed or open (in fact, F is Polish if and only if F is a G_δ, i.e. of the form $\bigcap_0^\infty F_n$ with the F_n open);
(c) any countable product $E_0 \times E_1 \times \cdots$ of Polish spaces $E_0, E_1, \ldots$ is Polish;
(d) if E is Polish, then so are the sets $\mathcal{P}(E)$, $\mathcal{M}(E)$ of probability distributions on E equipped with the topology for weak convergence and (in the locally compact case) of non-negative Radon measures on E equipped with the topology for vague convergence, see below;
(e) if E is Polish, then so are function spaces like $D([0, 1], E)$ and $D([0, \infty), E)$ in the standard topology.

Now let E be Polish and $\mathscr{C}$ the set of bounded continuous functions $E \to \mathbb{R}$. The *Borel σ-algebra* $\mathscr{E}$ on E is defined as the σ-algebra generated by the open sets (or the $f \in \mathscr{C}$) and is used throughout for measure theory. The *topology for weak convergence* is the initial topology defined by the mapping $\mathbb{P} \to \int f \, \mathrm{d}\mathbb{P}$, $f \in \mathscr{C}$, i.e. the weakest topology making all these mappings continuous. If $\mathbb{P}_n \to \mathbb{P}$, then $\int f \, \mathrm{d}\mathbb{P}_n \to \int f \, \mathrm{d}\mathbb{P}$, not only for $f \in \mathscr{C}$ but also if f is bounded and measurable with $\mathbb{P}D_f = 0$, where D_f is the set of discontinuities for f. Occasionally we use *Prohorov's theorem*, stating that $\mathcal{P}_0 \subseteq \mathcal{P}$ is (weakly) relatively compact if and only if $\mathscr{P}_0$ is *tight*, i.e. if to each $\varepsilon > 0$ we can find a compact set $K \subseteq E$ with $\mathbb{P}K \geqslant 1 - \varepsilon$, $\mathbb{P} \in \mathscr{P}_0$. Note that in particular convergent sequences form relatively compact sets.

A closely related topology is the *topology for vague convergence* of Radon measures μ on a locally compact space E (Radon means that $\mu(K) < \infty$ when K is compact). This is defined as the initial topology defined by the set of mappings $\mu \to \int f \mathrm{d}\mu$ indexed by the continuous f with compact support.

Some standard references are Bauer (1981), Billingsley (1968), Parthasarathy (1967) and Pollard (1984). There are some modern trends to go into more general spaces like Lusin spaces, Souslin spaces or co-Souslin spaces. We shall not use this but refer to Dellacherie and Meyer (1978). For D-spaces see also Lindvall (1973) and Whitt (1980). For the present purposes, elementary treatments of weak convergence in $\mathbb{R}$ like that of Breiman (1968, Chap. 8) most often suffice. For some (certainly not all) purposes the case of a general E can be reduced to the compact and/or real case by noting that E is homomorphic to a subset (necessarily G_δ!) of $[0, 1]^{\mathbb{N}}$. This follows from the following simple lemma:

Lemma A2.1. *If E is Polish, then there exists a countable class $\mathscr{K}$ of continuous functions $f: E \to [0, 1]$ such that $x_n \to x$ in E if and only if $f(x_n) \to f(x)$ in $[0, 1]$ for all $f \in \mathscr{K}$.*

Proof. Take $y_1, y_2, \ldots$ as a countable dense subset, let d be some metric and let $f_{k,n}: E \to [0, 1]$ be continuous with $f_{k,n}(y) = 1$ for $d(y, y_k) \leqslant 1/n$, $f_{k,n}(y) = 0$ for $d(y, y_k) \geqslant 2/n$. Then $\mathscr{K} = \{f_{k,n}: k, n = 1, 2, \ldots\}$ is easily seen to have the desired property. □

A3. SAMPLE PATH PROPERTIES

The stochastic processes $\{X_t\}_{t \in T}$ encountered in this book have almost exclusively one-dimensional discrete ($T = \mathbb{N}$) or continuous ($T = [0, \infty)$) time parameter. Occasionally also the doubly infinite time case $T = \mathbb{Z}$ or $T = (-\infty, \infty)$ is encountered. The state space E is usually of an elementary type, discrete (finite or countable, e.g. $\mathbb{Z}$, $\mathbb{N}^p$, etc), a well-behaved subset like $[0, \infty)$, $(a, b]^p$, etc. of Euclidean space $\mathbb{R}^p$ or combinations like $[0, \infty) \times \{0, 1\}$, etc. In any case, it is more than sufficient to allow E to be a general Polish space which we then equip with the Borel σ-algebra $\mathscr{E}$. When we talk about a subset A of E, this is most often assumed to be in $\mathscr{E}$ without further notice.

The traditional definition of a stochastic process $\{X_t\}_{t \in T}$ with state space E means just an indexed set of measurable mappings from a probability space $(\Omega, \mathscr{F}, \mathbb{P})$ into a general measurable space $(E, \mathscr{E})$. In discrete time, this is quite sufficient, but difficulties arise in continuous time. This is due to the fact that in discrete time the relevants events such as

$$\{X_t > 0 \quad \text{for all} \quad t\}, \qquad \{X_t \to X\}, \qquad \left\{\max_{0 \leqslant t \leqslant T} X_t > u\right\}, \quad \text{etc.}$$

are virtually always measurable since they can be obtained from elementary measurable sets of the form $\{X_n \in A_n\}$, $A_n \in \mathscr{E}$, by countable combinations of elementary set operations like unions, intersections, differences, etc. In continuous time, this is not so, and in fact the examples above are not measurable events. A variety of suggestions (separability, joint measurability of $X_t(\omega)$ in (t, ω), etc.) to overcome such difficulties have been considered, but the standard point of view these days is to assume that for a.a. (or even all) $\omega \in \Omega$ the sample function $\{X_t(\omega)\}_{t \in T}$ belongs to a well-behaved space of functions, of which the standard choice is D. This is quite sufficient for the present purposes, and in fact, the sample paths of the processes under study exhibit most often even stronger regularity like being piecewise continuous.

Noteworthy properties of a D-function $\{x_t\}_{t \geqslant 0}$ are:

(a) $\{x_t\}_{t \geqslant 0}$ is given by the values on any dense countable set, say $\mathbb{Q}$ (this is elementary by right-continuity);
(b) $\{x_t\}_{t \geqslant 0}$ is bounded on compact intervals (this is easy by a compactness argument);
(c) $\{x_t\}_{t \geqslant 0}$ has at most countably many jumps. We shall prove this in a more general setting:

Proposition A3.1. *If x: $[0, \infty) \to E$ is right-continuous, then x is continuous except at a (at most) countable collection of points.*

Proof (adapted from Björnsson, 1984). In view of Lemma A2.1, we must show that $t \to f(x_t)$ has the desired property for each $f \in \mathscr{K}$. That is, it is sufficient to consider the case $E = [0, 1]$. For $t > 0$ let

$$y_t^{(1)} = \overline{\lim_{s \uparrow t}}\, x_s - \underline{\lim_{s \uparrow t}}\, x_s, \qquad y_t^{(2)} = \begin{cases} x_t - \lim\limits_{s \to t} x_s & \text{if} \quad y_t^{(1)} = 0 \\ 0 & \text{otherwise} \end{cases}.$$

Then if x is discontinuous at t, we have either $y_t^{(1)} > 0$ or $y_t^{(2)} > 0$, and it is sufficient to show that for any $\varepsilon > 0$ the sets $A^{(i)} = \{t : y_t^{(i)} > \varepsilon\}$, $i = 1, 2$, are at most countable. It is clear by right-continuity that

$$0 = \lim_{s \downarrow t} y_s^{(1)} = \lim_{s \downarrow t} y_s^{(2)} \quad \text{for any} \quad t \geqslant 0. \tag{A3.1}$$

In particular, $A^{(i)} \cap (0, \delta] = \varnothing$ for some $\delta > 0$ and hence

$$\tau^{(i)} = \sup\{\delta > 0 : A^{(i)} \cap (0, \delta] \text{ is at most countable}\}$$

is non-zero. But if $A^{(i)}$ was uncountable, we would have $\tau^{(i)} < \infty$, implying the existence of a sequence $\tau^{(i)}(n) \in A^{(i)} \cap (\tau^{(i)}, \infty)$ with $\tau^{(i)}(n) \downarrow \tau^{(i)}$. Then $\underline{\lim}_{s \downarrow \tau^{(i)}} y^{(i)}(s) \geqslant \varepsilon$, a contradiction. □

Now let $\{X_t\}_{t \geqslant 0}$ be a stochastic process with right-continuous paths and define

$$Y_{t,f}^{(i)}(\omega) = y_t^{(i)}(f(X_t(\omega))), \qquad f \in \mathscr{K},$$
$$C_u = \{\omega : X_t(\omega) \text{ is continuous at } t = u\} = \bigcap_{f \in \mathscr{K}} \{Y_{t,f}^{(i)} = 0, i = 1, 2\}.$$

It follows by right-continuity that $\underline{\lim}_{t\uparrow u} f(X_t)$ and $\overline{\lim}_{t\uparrow u} f(X_t)$ are both measurable. Hence $Y_{t,f}^{(i)}$ and C_u are so, and it makes sense to define u to be a *fixed discontinuity* of $\{X_t\}$ if $\mathbb{P}C_u < 1$. Now by (A3.1) and dominated convergence we have $\lim_{s\downarrow t} \mathbb{E}Y_{s,f}^{(i)} = 0$ for all t. Hence exactly as in Proposition A3.1 we may conclude that $\mathbb{E}Y_{t,f}^{(i)} = 0$ except for t in a at most countable set $N_f^{(i)}$. But when $t \notin N = \bigcup_{i,f} N_f^{(i)}$, we then have all $Y_{t,f}^{(i)} = 0$ a.s., implying $\mathbb{P}C_u = 1$. Hence;

Corollary A3.2. *A stochastic process with right-continuous paths has at most countably many fixed discontinuities.*

A4. POINT PROCESSES

A point process on E is in intuitive terms just a random collection of points in E. The simplest standard example is the Poisson process on $[0, \infty)$. In fact almost exclusively the point processes encountered in this book have $E = [0, \infty)$ or $E = \mathbb{R}$ and satisfy some further regularity conditions: (a) there are no multiple points; (b) the points do not accumulate. Processes of such types are easily brought in one-one correspondence with sequences $\{S_n\}$ of $[0, \infty)$-valued random variables (e.g. if $E = [0, \infty)$, we may just let S_n be the position of the nth point to the right of the origin) and thus no foundational difficulties arise. It may, however, frequently be revealing also to have the general abstract formulation in mind. One then requires E to be locally compact Polish and defines a point process on E to be a $\mathcal{N}$-valued random variable N, where $\mathcal{N}$ is the set of (Radon) counting measures μ, i.e. having the property $N(A) \in \mathbb{N}$ for all relatively compact $A \in \mathscr{E}$ (thus the connection to the setting above is $N(A) = \Sigma I(S_n \in A)$).

Of the many texts on point processes, we mention in particular Cox and Isham (1980), Kallenberg (1976) and Matthes *et al.* (1978). As mentioned above, the concept occurs in an elementary manner in the present book. One piece of terminology that is used occasionally is that of the *intensity measure* which is defined in the obvious way as the set function $A \to \mathbb{E}N(A)$ (thus for example for the (time homogeneous) Poisson process with intensity α, the intensity measure is Lebesgue measure scaled by α).

A5. STOCHASTICAL ORDERING

Let X, Y be real-valued random variables with distributions F, G. We then say that $X \leqslant Y$ in the sense of stochastical ordering if (a) $\mathbb{P}(X > t) \leqslant \mathbb{P}(Y > t)$ for all t, i.e. if $F \geqslant G$. Alternative formulations in the same situation are 'X is stochastically dominated by Y (or G)', 'F is stochastically smaller than G' and so on.

The area is surveyed in a broad setting in Stoyan (1983) and references given there. Some of the facts that are used here are the equivalence of the definition (a) to either of (b) $\mathbb{E}f(X) \leqslant \mathbb{E}f(Y)$ for all increasing $f: \mathbb{R} \to \mathbb{R}$, or (c) there exists random variables X', Y' with $X \overset{\mathscr{D}}{=} X', Y \overset{\mathscr{D}}{=} Y', X' \leqslant Y'$ a.s.

A6. TOTAL VARIATION CONVERGENCE

Let $(E, \mathscr{E})$ be a measurable space and ν a signed measure on $(E, \mathscr{E})$. Then the *total variation* (t.v.) of ν is defined as $\|\nu\| = \sup_{A\in\mathscr{E}}|\nu(A)|$. If $\nu \geqslant 0$, then $\|\nu\| = \nu(E)$. However, the main case for our applications is $\nu = \mathbb{P}' - \mathbb{P}''$ with $\mathbb{P}', \mathbb{P}''$ probabilities. Then $2\|\nu\| = \nu(E_+) + \nu(E_-)$, where $E = E_+ \cup E_-$ is the Jordan–Hahn decomposition of E with respect to ν. We say that $\nu_n \to \nu$ in t.v. if $\|\nu_n - \nu\| \to 0$, i.e. if $\nu_n(A) \to \nu(A)$ uniformly in $A \in \mathscr{E}$, which in turn is easily seen to be equivalent to $\int f\,\mathrm{d}\nu_n \to \int f\,\mathrm{d}\nu$ uniformly in the measurable f with $\|f\|_\infty \leqslant 1$. Similarly, $X_n \to X$ in t.v. means that $\pi_n \to \pi$ in t.v. where $\pi_n(A) = \mathbb{P}(X_n \in A)$, $\pi(A) = \mathbb{P}(X \in A)$.

It is immediately apparent that t.v. convergence entails weak convergence. One important example of t.v. convergence is provided by *Scheffe's theorem* (Billingsley, 1968, p. 224) which states that if ν_n, ν are probabilities with densities f_n, f with respect to μ, and $f_n(x) \to f(x)$ for μ a.a. x, then $\nu_n \to \nu$. This means in particular that for a discrete E, the notions of weak convergence and t.v. convergence coincide. In fact, if μ is counting measure on E, then $\nu_n \to \nu$ and E being discrete implies that

$$f_n(x) = \nu_n(\{x\}) \to \nu(\{x\}) = f(x) \quad \text{for all} \quad x.$$

Note that in some of the literature, the definition of $\|\nu\|$ differs by a factor of 2. Thus, for example, in the coupling inequality VI.(2.1) the r.h.s. is frequently encountered as $2\mathbb{P}(T > t)$ rather than $\mathbb{P}(T > t)$.

A7. TRANSFORMS

Transforms of a distribution F are denoted by $\hat{F}$ and may mean either characteristic function (ch. *f.*) $\hat{F}(s) = \int_{-\infty}^{\infty} \mathrm{e}^{isx}\,\mathrm{d}F(x)$, Laplace transform $\hat{F}(s) = \int_{-\infty}^{\infty} \mathrm{e}^{-sx}\,\mathrm{d}F(x)$, moment generating function (m.g.f.) $\hat{F}(s) = \int_{-\infty}^{\infty} \mathrm{e}^{sx}\,\mathrm{d}F(x)$, or, if F is concentrated on $\mathbb{Z}$ with point probabilities $\{f_n\}$, generating function $\hat{F}(s) = \hat{f}(s) = \sum_{-\infty}^{\infty} s^n f_n$. The cumulant generating function (c.f.f.) is $\log \hat{F}(s)$, where $\hat{F}$ is the m.g.f.

In the text, we use without further reference a number of standard facts such as that F is uniquely determined by $\hat{F}$, that $\widehat{F*G} = \hat{F}\hat{G}$, that moments can be expressed in terms of derivatives of $\hat{F}$ and so on.

A8. STOPPING TIMES AND WALD'S IDENTITY

Let $T = \mathbb{N}$ or $T = [0, \infty)$, and let $\{\mathscr{F}_t\}_{t\in T}$ be a family of σ-fields which is *increasing*, i.e. $\mathscr{F}_s \subseteq \mathscr{F}_t$, $s, t \in T$, $s \leqslant t$. A random time $\tau \in T \cup \{\infty\}$ is then a *stopping time with respect to* $\{\mathscr{F}_t\}_{t\in T}$ if

$$\{\tau \leqslant t\} \in \mathscr{F}_t \quad \text{for all} \quad t \in T. \tag{A8.1}$$

The *stopping time σ-field* $\mathscr{F}_\tau$ (sometimes called *pre-τ-field*) is then defined as the collection of all sets $A \in \mathscr{F}_\infty = \sigma(\bigcup_{t\in T}\mathscr{F}_t)$ such that $A \cap \{\tau \leqslant t\}$ belongs to $\mathscr{F}_t$ for all $t \in T$.

In applications, it is convenient to note that measurability is an automatic result of (A8.1) and need not be checked separately, and also that for $T=\mathbb{N}$ (A8.1) is equivalent to $\{\tau=n\}\in\mathscr{F}_n$ for all $n\in\mathbb{N}$ and $A\in\mathscr{F}_\sigma$ to $A\cap\{\tau=n\}\in\mathscr{F}_n$ for all $n\in\mathbb{N}$. The following result is standard and easy to prove:

Proposition A8.1. *Let τ be as stopping time. Then* (a) *τ is $\mathscr{F}_\tau$-measurable;* (b) *if $\{X_t\}_{t\in T}$ is a stochastic process such that X_t is $\mathscr{F}_t$-measurable for each $t\in T$, and that the paths are right-continuous for $T=[0,\infty)$ then $X_\tau I(\tau<\infty)$ is $\mathscr{F}_\tau$-measurable;* (c) *if σ is an additional $\{\mathscr{F}_t\}$-stopping time, then so are $\sigma\wedge\tau, \sigma\vee\tau, \sigma+\tau$. If $\sigma\leqslant\tau$, then $\mathscr{F}_\sigma\leqslant\mathscr{F}_\tau$.*

Part (a) of the following result is referred to as *Wald's identity* (sometimes called also *Wald's lemma*), and part (b) as *Wald's second moment identity*. For a proof, see e.g. Neveu (1972, pp. 83–85) and for higher order versions, Chow *et al.* (1965).

Proposition A8.2. *Let τ be an a.s. finite stopping time with respect to $\{\mathscr{F}_n\}_{n\in\mathbb{N}}$. Further, let $X_1, X_2, \ldots$ be independent identically distributed random variables such that for any n X_n is $\mathscr{F}_n$-measurable and $X_{n+1}, X_{n+2}, \ldots$ are independent of $\mathscr{F}_n$, and write $S_n=X_1+\cdots+X_n$, $\mu=\mathbb{E}X_1$. Then:*

(a) *if either $\mathbb{E}|X_1|<\infty$, $\mathbb{E}\tau<\infty$ or $X_1\geqslant 0$, then $\mathbb{E}S_\tau=\mu\mathbb{E}\tau$;*
(b) *if $\sigma^2=\mathbb{V}\mathrm{ar}\,X_1<\infty$ and $\mathbb{E}\tau<\infty$, then $\mathbb{E}(S_\tau-\tau\mu)^2=\sigma^2\mathbb{E}\tau$.*

Alternatively, one might formulate the condition by $X_1, X_2, \ldots$ being independent identically distributed and $\{\tau\leqslant n\}$ being independent of $X_{n+1}, X_{n+2}, \ldots$ for any n. This means that τ is a *randomized stopping time*, see Pitman and Speed (1973). Further examples of this notion occur in VI.3 and VIII.3.

A9. DISCRETE SKELETONS

Limit theory for continuous-time stochastic processes ($T=[0,\infty)$), being concerned with the question of existence of limits of functions like $f(t)=\mathbb{P}(X_t\in A)$, can sometimes be reduced to the discrete case $T=\mathbb{N}$ by means of the study of discrete skeletons $\{X_{n\delta}\}_{n\in\mathbb{N}}$. For example, elementary topology yields:

Proposition A9.1. *If $f:[0,\infty)\to\mathbb{R}$ is uniformly continuous and $\lambda(\delta)=\lim_{n\to\infty}f(n\delta)$ exists for each $\delta>0$, then $\lambda(\delta)=\lambda$ does not depend on δ, and furthermore $f(t)\to\lambda$ as $t\to\infty$ continuously.*

It is frequently much easier to show that $f(t)=\mathbb{P}(X_t\in A)$ is just continuous rather than uniformly continuous. In fact, this is sufficient:

Proposition A9.2. *The conclusion of Proposition A9.1 holds true if f is continuous and $\lambda(\delta)=\lim_{n\to\infty}f(n\delta)$ exists for each $\delta>0$.*

This result is known in the literature as the *Croft–Kingman lemma.* The proof (Kingman, 1963) is again real topology, but much less elementary as for Proposition A9.1.

References

Allen, A. O. (1978) *Probability, Statistics and Queueing Theory with Computer Applications.* Academic Press, New York, San Francisco, London.

Arjas, E., Nummelin, E., and Tweedie, R. L. (1978) Uniform limit theorems for non-singular renewal and Markov renewal processes. *J. Appl. Probab.*, **15**, 112–125.

Arjas, E., and Speed, T. P. (1973) Symmetric Wiener–Hopf factorisations in Markov additive processes. *Z. Wahrscheinlichkeitsth. verw. Geb.*, **26**, 105–118.

Arndt, K. (1984) On the distribution of the supremum of a random walk on a Markov chain. In: *Limit Theorems and Related Problems* (A. A. Borovkov, ed.), pp. 253–267. Optimizations Software, New York.

Arrow, K., Karlin, S., and Scarf, H. (1958) *Studies in the Mathematical Theory of Inventory and Production.* Stanford University Press, Stanford.

Asmussen, S. (1981) Equilibrium properties of the $M/G/1$ queue. *Z. Wahrscheinlichkeitsth. verw. Geb.*, **58**, 267–281.

Asmussen, S. (1982) Conditioned limit theorems relating a random walk to its associate, with applications to risk reserve processes and the $GI/G/1$ queue. *Adv. Appl. Probab.*, **14**, 143–170.

Asmussen, S. (1984) Approximations for the probability of ruin within finite time. *Scand. Act. J.*, **1984**, 31–57; **1985**, 64.

Asmussen, S. (1985) Conjugate processes and the simulation of ruin problems. *Stoch. Proc. Appl.*, **20**, 213–229.

Asmussen, S., and Hering, H. (1977) Some modified branching diffusion models. *Math. Biosc.*, **35**, 281–299.

Asmussen, S., and Hering, H. (1983) *Branching Processes.* Birkhäuser, Basel, Boston, Stuttgart.

Asmussen, S., and Johansen, H. (1986) Über eine Stetigkeitsfrage betreffend das Bedienungssystem $GI/GI/s$. *Elektron. Inf. Kyb.*, **22**, 565–570.

Asmussen, S., and Thorisson, H. (1987) A Markov chain approach to periodic queues. *J. Appl. Probab.*

Athreya, K. B., McDonald, D., and Ney, P. (1978) Limit theorems for semi-Markov processes and renewal theory for Markov chains. *Ann. Probab.*, **6**, 788–797.

Athreya, K. B., and Ney, P. (1978a) A new approach to the limit theory of recurrent Markov chains. *Trans. Amer. Math. Soc.*, **245**, 493–501.

Athreya, K. B., and Ney, P. (1978b) A Markov process approach to systems of renewal equations with applications to branching processes. In: *Branching Processes* (A. Joffe and P. Ney, eds.) Advances in Probability and Related Topics **5**. Marcel Dekker, New York, Basel.

Athreya, K. B., and Ney, P. (1982) A renewal approach to the Perron–Frobenius theory of non-negative kernels on general state spaces. *Math. Z.*, **179**, 507—529.

Bahadur, R. R., and Rao, R. R. (1960) On deviations of the sample mean. *Ann. Math. Statist.*, **31**, 1015–1027.
von Bahr, B. (1974) Ruin probabilities expressed in terms of ladder height distributions. *Scand. Act. J.*, **1974**, 190–204.
Bauer, H. (1981) *Probability Theory and Elements of Measure Theory*. Academic Press, London, New York, Toronto, Sydney, San Francisco.
Barndorff-Nielsen, O. (1978) *Information and Exponential Families in Statistical Theory*. Wiley, New York.
Barndorff-Nielsen, O., and Cox, D. R. (1979) Edgeworth and saddlepoint approximations with statistical applications. *J. Roy Statist. Soc.*, **B41**, 279–312.
Beard, R. E., Pentikäinen, T., and Pesonen, E. (1984) *Risk Theory* (3rd edn.). Chapman and Hall, London, New York.
Berbee, H. C. P. (1979) *Random Walks with Stationary Increments and Renewal Theory*. Mathematical Centre Tracts **112**. Mathematisch Centrum, Amsterdam.
Berndtsson, B., and Jagers, P. (1979) Exponential growth of a branching process usually implies stable age distribution. *J. Appl. Probab.*, **16**, 651–656.
Bhattacharya, R. N., and Rao, R. R. (1976) *Normal Approximations and Asymptotic Expansions*. Wiley, New York, London, Sydney.
Bickel, P. J., and Yahav, J. A. (1965) Renewal theory in the plane *Ann. Math. Statist.*, **36**, 946–955.
Billingsley, P. (1968) *Convergence of Probability Measures*. Wiley, New York, London, Sydney, Toronto.
Björnsson, O. J. (1984) Notes on right- (left-) continuous functions. *Preprint 1984*, No. **4**. Institute of Mathematical Statistics, University of Copenhagen.
Blomqvist, N. (1974) A simple derivation of the $GI/G/1$ waiting time in heavy traffic. *Scand. J. Statist.*, **1**, 41–48.
Borovkov, A. A. (1976) *Stochastic Processes in Queueing Theory*. Springer-Verlag, New York, Heidelberg, Berlin.
Borovkov, A. A. (1984) *Asymptotic Methods in Queueing Theory*. Wiley, Chichester, New York, Brisbane, Toronto, Singapore.
Brandt, A., and Lisek, B. (1983) On the continuity of $G/GI/m$ queues. *Math. Operationsforsch. Statist.*, Ser. Statist., **12**, 577–587.
Braun, H. (1978) Stochastic stable population theory in continuous time. *Scand. Act. J.*, **1978**, 185–203.
Breiman, L. (1968) *Probability*. Addison-Wesley, Reading, Mass.
Bremaud, P. (1981) *Point Processes and Queues*. Springer-Verlag, New York, Heidelberg, Berlin.
Brockwell, P. J., Resnick, S. I., and Tweedie, R. L. (1982) Storage processes with general release rule and additive inputs. *Adv. Appl. Probab.*, **14**, 392–433.
Brown, M., and Ross, S. M. (1972) Asymptotic properties of cumulative processes. *SIAM J. Appl. Math.*, **22**, 93–105.
Bühlmann, H. (1970) *Mathematical Methods in Risk Theory*. Springer-Verlag, Heidelberg.
Carlsson, H. (1983) Remainder term estimates of the renewal function. *Ann. Probab.*, **11**, 143–157.
Carlsson, H., and Wainger, S. (1984) On the multi-dimensional renewal theorem. *J. Math. Anal. Appl.*, **100**, 316–322.
Charlot, F., Ghidouche, M., and Hamami, M. (1978) Irréducibilité et recurrence au sens de Harris des 'Temps d'attente' des files $GI/G/q$. *Z. Wahrscheinlichkeitsth. verw. Geb.*, **43**, 187–203.
Chover, J., and Ney, P. (1968) The non-linear renewal equation. *J. D'Analyse Math.*, **21**, 381–413.
Chow, Y. S., Hsiung, C., and Lai, T. (1979) Extended renewal theory and moment convergence in Anscombe's theorem. *Ann. Probab.*, **7**, 304–318.

Chow, Y. S., Robbins, H., and Teicher, H. (1965) Moments of randomly stopped sums. *Ann. Math. Statist.*, **36**, 789–799.

Chung, K. L. (1967) *Markov Chains with Stationary Transition Probabilities* (2nd ed.). Springer-Verlag, New York, Heidelberg, Berlin.

Chung, K. L. (1974) *A Course in Probability Theory* (2nd edn.). Academic Press, New York, San Francisco, London.

Chung, K. L. (1982) *Lectures from Markov Processes to Brownian Motion.* Springer-Verlag, New York, Heidelberg, Berlin.

Çinlar, E. (1975) *Introduction to Stochastic Processes.* Prentice-Hall, Englewood Cliffs, N. J.

Coffman, E. G. Jr., Muntz, R. R., and Trotter, H. (1970) Waiting time distributions for processor-sharing systems. *J. Ass. Comp. Mach.*, **17**, 123–130.

Cohen, J. W. (1975) The Wiener–Hopf technique in applied probability. In: *Perspectives in Probability and Statistics* (J. Gani, ed.), 145–156. Academic Press, London.

Cohen, J. W. (1976) *On Regenerative Processes in Queueing Theory.* Lecture Notes in Economics and Mathematical Systems, **121**. Springer-Verlag, New York, Heidelberg, Berlin.

Cohen, J. W. (1982) *The Single Server Queue* (2nd edn.). North-Holland, Amsterdam.

Cohen, J. W. (1984) On processor sharing and random service. *J. Appl. Probab.*, **21**, 937.

Cooper, R. B. (1981) *Introduction to Queueing Theory* (2nd edn.). North-Holland, New York.

Cox, D. R. (1962) *Renewal Theory.* Methuen, London.

Cox, D. R., and Isham, V. (1980) *Point Processes.* Chapman and Hall, London.

Cox, D. R., and Smith, W. L. (1961) *Queues.* Methuen, London.

Crane, M. A., and Lemoine, A. J. (1977) *An Introduction to the Regenerative Method for Simulation Analysis.* Lecture Notes in Control and Information Sciences, **4**. Springer-Verlag, New York, Heidelberg, Berlin.

Cumani, A. (1982) On the canonical representation of homogeneous Markov processes modeling failure-time distributions. *Microelectron. Reliab.*, **22**, 583–602.

Daley, D. J. (1976) Queueing output processes. *Adv. Appl. Probab.*, **8**, 395–415.

Dellacherie, C., and Meyer, P. -A. (1978) *Probabilities and Potential.* North-Holland, Amsterdam.

Disney, R. L., and König, D. (1985) Queueing networks: a survey of their random processes. *SIAM Review*, **27**, 335–403.

van Doorn, E. (1980) *Stochastic Monotonicity and Queueing Applications of Birth–Death Processes.* Lecture Notes in Statistics, **4**. Springer-Verlag, New York, Heidelberg, Berlin.

Embrechts, P., Jensen, J. L., Maejima, M., and Teugels, J. L. (1985) Approximations for compound Poisson and Polya processes. *Adv. Appl. Probab.*, **17**, 623–637.

Erdös, P., Feller, W., and Pollard, H. (1949) A property of power series with positive coefficients. *Bull. Amer. Math. Soc.*, **55**, 201–204.

Erickson, K. B. (1970) Strong renewal theorems with infinite mean. *Trans. Amer. Math. Soc.*, **151**, 263–291.

Feller, W. (1966) *An Introduction to Probability Theory and its Applications.* Vol. 1 (3rd edn.). Wiley, New York.

Feller, W. (1971) *An Introduction to Probability Theory and its Applications.* Vol. 2 (2nd edn.). Wiley, New York.

Fienberg, S. E. (1974) Stochastic models for single neuron firing trains. *Biometrics*, **30**, 399–427.

Foss, S. G. (1980) Approximation of multichannel queueing systems. *Sib. Math. J.*, **21**. 851–857.

Franken, P., König, D., Arndt, U., and Schmidt, V. (1981) *Queues and Point Processes.* Akademie-Verlag, Berlin.

Freedman, D. (1971) *Markov Chains.* Holden-Day, San Francisco, Cambridge.

Fristedt, B. (1974) Sample functions of stochastic processes with stationary, independent increments. In: *Advances in Probability and Related Topics*, Vol. **3** (P. Ney and S. Port eds.), pp. 241–396. Marcel Dekker, New York.

Gantmacher, F. R. (1959) *The Theory of Matrices.* Chelsea, New York.

Garsia, A., and Lamperti, J. (1962/63) A discrete renewal theorem with infinite mean. *Comment. Math. Helv.*, **37**, 221–234.

Gaver, D. P., and Lehoczky, J. P. (1982) Channels that cooperatively service a data stream and voice messages. *IEEE Trans. Com.*, **30**, 1153–1161.

Gerber, H. U. (1979) *An Introduction to the Mathematical Risk Theory.* S. S. Huebner Foundation Monographs, University of Pennsylvania.

Gnedenko, B. V., and Kolmogorov, A. N. (1954) *Limit Distributions for Sums of Independent Random Variables.* Addison-Wesley, Reading, Mass.

Gnedenko, B. W., and König, D. (1983/84) *Handbuch der Bedienungstheorie*, Vols. I–II. Akademie-Verlag, Berlin.

Gnedenko, B., and Kovalenko, I. N. (1968) *An Introduction to Queueing Theory.* Israel Program for Scientific Translations, Jerusalem.

Grandell, J. (1977) A class of approximations of ruin probabilities. *Scand. Act. J., Suppl.*, **1977**, 37–52.

Grandell, J. (1978) A remark on 'A class of approximations of ruin probabilities'. *Scand. Act. J.*, **1978**, 77–78.

Griffeath, D. (1978) Coupling methods for Markov for processes. In: *Advances in Mathematics Supplementary Series*, Vol. 2. *Studies in Probability and Ergodic Theory.*

Grimmett, G., and Stirzaker, D. (1982) *Probability and Random Processes.* Clarendon Press, Oxford.

Gross, D., and Harris, C. M. (1974) *Fundamentals of Queueing Theory.* Wiley, New York.

Grübel, R. (1983) Functions of probability measures: rates of convergence in the renewal theorem. *Z. Wahrscheinlichkeitsth. verw. Geb.*, **64**, 341–357.

Gut, A. (1986) *Stopped Random Walks. Limit Theory and Applications.* Book Manuscript, University of Uppsala.

Harris, T. E. (1963) *The Theory of Branching Processes.* Springer-Verlag, Berlin, Göttingen, Heidelberg.

Harrison, J. M. (1985) *Brownian Motion and Stochastic Flow Systems.* Wiley, Chichester, New York, Brisbane, Toronto, Singapore.

Harrison, J. M., and Lemoine, A. J. (1976) On the virtual and actual waiting time of a $GI/G/1$ queue. *J. Appl. Probab.*, **13**, 833–836.

Harrison, J. M., and Resnick, S. I. (1976) The stationary distribution and first exit probabilities of a storage process with general release rule. *Math. Opns. Res.*, **1**, 347–358.

Heathcote, C. R. (1967) Complete exponential convergence and related topics. *J. Appl. Probab.*, **4**, 1–40.

Heathcote, C. R., and Winer, P. (1969) An approximation for the moments of waiting times. *Opns. Res.*, **17**, 175–186.

Henningsen, I., and Liestøl, K. (1983) A model of neurons with pacemaker behaviour receiving strong synaptic input. *IEEE Trans.*, **SMC-13**, 720–727.

Heyman, P., and Sobel, M. J. (1982) *Stochastic Models in Operations Research.* McGraw-Hill, New York.

Hillier, F. S., and Yu, O. S. (1981) *Queueing Tables and Graphs.* North-Holland, New York, Oxford.

Hordijk, A., and Tijms, H. C. (1976) A simple proof of the equivalence of the limiting distribution of the continuous-time and the embedded process of the queue size in the $M/G/1$ queue. *Statist. Neerlandica*, **36**, 97–100.

Iglehart, D. L., and Shedler, G. S. (1979) Regenerative simulation of response times in networks of queues. *Lecture Notes in Control and Information Sciences* **26**. Springer-Verlag, New York.

Iversen, V. B. (1981) *An Introduction to Teletraffic Theory* (in Danish). Department of Mathematical Statistics and Operations Research, Technical University of Denmark, 511 pp.

Jagers, P. (1975) *Branching Processes with Biological Applications.* Wiley, London, New York, Sydney, Toronto.

Jagers, P., and Nerman, O. (1984) The growth and composition of branching populations. *Adv. Appl. Probab.*, **16**, 221–259.

Johnson, N. L., and Kotz, S. (1969) *Distributions in Statistics. Continuous Univariate Distributions*, Vol. 1. Houghton Mifflin, Boston.

Kallenberg, O. (1976) *Random Measures.* Akademie-Verlag, Berlin.

Karlin, S., and Taylor, H. G. (1975) *A First Course in Stochastic Processes* (2nd edn.). Academic Press, New York, San Francisco, London.

Keiding, N., and Hoem, J. (1976) Stochastic stable population theory with continuous time. *Scand. Act. J.*, **1976**, 150–175.

Keilson, J. (1965) *Green's Function Methods in Probability Theory.* Griffin, London.

Keilson, J. (1979) *Markov Chain Models—Rarity and Exponentiality.* Springer-Verlag, New York, Heidelberg, Berlin.

Keilson, J., and Wishart, D. M. G. (1965) Boundary problems for additive processes defined on a finite Markov chain. *Proc. Camb. Philos. Soc.*, **61**, 173–190.

Kelly, F. P. (1979) *Reversibility and Stochastic Networks.* Wiley, Chichester, New York, Brisbane, Toronto.

Kelly, F. P. (1983) Invariant measures and the Q-matrix. In: *Probability, Statistics and Analysis* (J. F. C. Kingman and G. E. H. Reuter eds.), pp. 143–160. London Math. Soc. Lect. Not. Ser. **79**. Cambridge University Press, Cambridge.

Kemeny, J. G., Snell, J. L., and Knapp, A. W. (1976) *Denumerable Markov Chains* (2nd edn.). Springer-Verlag, New York, Heidelberg, Berlin.

Kemperman, J. H. B. (1961) *The Passage Problem for a Markov Chain.* University of Chicago Press, Chicago.

Kendall, D. G. (1951) Some problems in the theory of queues. *J. Roy. Statist. Soc.*, **B13**, 151–173.

Kendall, D. G. (1953) Stochastic processes occurring in the theory of queues and their analysis by means of the imbedded Markov chain. *Ann. Math. Statist.*, **24**, 338–354.

Kesten, H. (1974) Renewal theory for Markov chains. *Ann. Probab.*, **3**, 355–387.

Khintchine, A. Y. (1960) *Mathematical Methods in the Theory of Queueing.* Griffin, London.

Kiefer, J., and Wolfowitz, J. (1955) On the theory of queues with many servers. *Trans. Amer. Math. Soc.*, **78**, 1–18.

Kiefer, J., and Wolfowitz, J. (1956) On the characteristics of the general queueing process with applications to random walks. *Ann. Math. Statist.*, **27**, 147–161.

Kingman, J. F. C. (1963) Ergodic properties of continuous time Markov processes and their discrete skeletons. *Proc. London Math. Soc.*, **13**, 593–604.

Kingman, J. F. C. (1972) *Regenerative Phenomena.* Wiley, New York.

Kleinrock, L. (1975) *Queueing Systems.* Volume 1: *Theory.* Wiley, New York, London, Sydney, Toronto.

Kleinrock, L. (1976) *Queueing Systems.* Volume 2: *Computer Applications.* Wiley, New York, London, Sydney.

Köllerström J. (1981) A second-order heavy traffic approximation for the queue $GI/G/1$. *Adv. Appl. Probab.*, **13**, 167–185.

Kotz, S., Johnson, N. L., and Read, C. B. (1982/–) *Encyclopedia of Statistical Sciences.* Wiley, New York, Chichester, Brisbane, Toronto, Singapore.

Le Gall, P. (1962) *Les Systémes avec ou sans Attente et les Processus Stochastiques.* Dunod, Paris.

Lemoine, A. J. (1974) On two stationary distributions for the stable $GI/G/1$ queue. *J. Appl. Probab.*, **11**, 849–852.

Lemoine, A. J. (1976) On random walks and stable $GI/G/1$ queues. *Math. Opns. Res.*, **1**, 159–164.

Lindvall, T. (1973) Weak convergence of probability measures and random functions in the function space $D[0, \infty)$. *J. Appl. Probab.*, **10**, 109–121.

Lindvall, T. (1977) A probabilistic proof of Blackwell's renewal theorem. *Ann Probab.*, **5**, 482–485.

Lindvall, T. (1982) On coupling of continuous time renewal processes. *J. Appl. Probab.*, **19**, 82–89.

Liptser, R. S., and Shiryaev, A. N. (1977) *Statistics of Random Processes*, Vol. I. Springer-Verlag, New York, Heidelberg, Berlin.

Lisek, B. (1982) A method for solving a class of recursive stochastic equations. *Z. Wahrscheinlichkeitsth. verw. Geb.*, **60**, 151–161.

Lorden, G. (1970) On excess over the boundary. *Ann. Math. Statist.*, **41**, 520–527.

Loynes, R. M. (1962) The stability of a queue with non-independent interarrival and service times. *Proc. Camb. Philos. Soc.*, **58**, 497–520.

Lucantoni, D. M., and Ramaswami, V. (1985) Efficient algorithms for solving the non-linear matrix equations arising in phase type queues. *Stochastic Models*, **1**, 29–51.

McDonald, D. (1975) Renewal theorem and Markov chains. *Ann. Inst. Henri Poincaré*, **XI**, 187–197.

Maisonneuve, B. (1974) Systèmes régénératifs. *Astérisque*, **15**. Soc. Math. France, Paris.

Marshall, K. T. (1968) Some relationships between the distributions of waiting time, idle time and interoutput time in the $GI/G/1$ queue. *SIAM J. Appl. Math.*, **16**, 324–327.

Matthes, K., Kerstan, J., and Mecke, J. (1978) *Infinitely Divisible Point Processes*. Wiley, New York.

Mertens, J. -F., Samuel-Cahn, E., and Zamir, S. (1978) Necessary and sufficient conditions for recurrence and transience of Markov chains in terms of inequalities. *J. Appl. Probab.*, **15**, 848–851.

Miller, D. R. (1972) Existence of limits in regenerative processes. *Ann. Math. Statist.*, **43**, 1275–1282.

Miller, D. R. (1974) Limit theorems for path-functionals of regenerative processes. *Stoch. Proc. Appl.*, **2**, 141–161.

Minh, D. L., and Sorli, R. M. (1983) Simulating the $GI/G/1$ queue in heavy traffic. *Opns. Res.*, **31**, 966–971.

Miyazawa, M. (1977) Time and customer processes in queues with stationary inputs. *J. Appl. Probab.*, **14**, 349–357.

Moran, P. A. P. (1959) *The Theory of Storage*. Methuen, London.

Morse, P. M. (1958) *Queues, Inventories and Maintenance*. Wiley, New York.

Neuts, M. F. (1981) *Matrix–Geometric Solutions in Stochastic Models*. Johns Hopkins University Press, Baltimore, London.

Neuts, M. F. (1986) The caudal characteristic curve of queues. *Adv. Appl. Probab.*, **16**, 221–254.

Neveu, J. (1965) *Mathematical Foundations of the Calculus of Probability*. Holden-Day, San Francisco, London, Amsterdam.

Neveu, J. (1972) *Martingales à Temps Discret*. Masson, Paris.

Newell, G. F. (1982) *Applications of Queueing Theory* (2nd edn.). Chapman and Hall, London, New York.

Nummelin, E. (1978) A splitting technique for Harris recurrent markov chains. *Z. Wahrscheinlichkeitsth. verw. Geb.*, **43**, 309–318.

Nummelin, E. (1984) *General Irreducible Markov Chains and Non-Negative Operators*. Cambridge University Press, Cambridge.

Olver, F. W. J. (1965) Bessel functions of integer order. *Handbook of Mathematical Functions* (*M. Abramowitz and I. A. Stegun, eds.*), pp. 355–434. Dover, New York.

Orey, S. (1971) *Lecture Notes on Limit Theorems for Markov Chain Transition Probabilities*. Van Nostrand Reinhold, London.

Ott, T. J. (1984) The sojourn time in the $M/G/1$ queue with processor sharing. *J. Appl. Probab.*, **21**, 360–378.

Ovuworie, G. C. (1980) Multi-channel queues: a survey and bibliography. *Int. Stat. Rev.*, **48**, 49–71.

Parthasarathy, K. R. (1967) *Probability Measures on Metric Spaces.* Academic Press, New York.

Pitman, J. W. (1974) Uniform rates of convergence for Markov chain transition probabilities. *Z. Wahrscheinlichkeitsth. verw. Geb.*, **29**, 193–227.

Pitman, J. and Speed, T. P. (1973) A note on random times. *Stoch. Proc. Appl.*, **1**, 369–374.

Pollard, D. (1984) *Convergence of Stochastic Processes.* Springer-Verlag, New York, Berlin, Heidelberg, Tokyo.

Pollard, J. H. (1973) *Mathematical Models for the Growth of Human Populations.* Cambridge University Press, Cambridge, London, New York, Melbourne.

Prabhu, N. U. (1965) *Queues and Inventories.* Wiley, New York, London, Sydney.

Prabhu, N. U. (1974) Wiener–Hopf techniques in queueing theory. In: *Mathematical Methods in Queueing Theory* (A. B. Clarke ed.). Lecture Notes in Economics and Mathematical Systems **98**, 81–90. Springer-Verlag, Berlin.

Prabhu, N. U. (1980) *Stochastic Storage Processes. Queues, Insurance Risk and Dams.* Springer-Verlag, New York, Heidelberg, Berlin.

Ramalhoto, M. F., Amaral, J. A., and Cochito, M. T. (1983) A survey of J. Little's formula. *Int. Stat. Rev.*, **51**, 255–278.

Ramaswami, V. (1984) The sojourn time in the $GI/M/1$ queue with processor sharing. *J. Appl. Probab.*, **21**, 437–442.

Revuz, D. (1975) *Markov Chains.* North-Holland, Amsterdam.

Rolski, T. (1981) *Stationary Random Processes Associated with Point Processes.* Lecture Notes in Statistics, **5**. Springer-Verlag, New York, Heidelberg, Berlin.

Saaty, T. L. (1961) *Elements of Queueing Theory.* McGraw-Hill, New York, Toronto, London.

Schaefer, H. H. (1966) *Topological Vector Spaces.* Macmillan, New York, London.

Schassberger, R. (1973) *Warteschlangen.* Springer-Verlag, Vienna, New york.

Schassberger, R. (1984) A new approach to the $M/G/1$ processor sharing queue. *Adv. Appl. Probab.*, **16**, 202–213.

Seal, H. L. (1969) *The Stochastic Theory of a Risk Business.* Wiley, New York.

Seal, H. L. (1978) *Survival Probabilities.* Wiley, New york.

Seelen, L. P., Tijms, H. C., and Van Hoorn, M. H. (1985) *Tables for Multi-Server Queues.* North-Holland, Amsterdam.

Segerdahl, C. -O. (1955) When does ruin occur in the collective theory of risk? *Skand. Aktuar Tidsskr.*, **1955**, 22–36.

Seneta, E. (1981) *Non-Negative Matrices and Markov Chains* (2nd edn.). Springer-Verlag, New York, Heidelberg, Berlin.

Siegmund, D. (1969) The variance of one-sided stopping rules. *Ann. Math. Statist.*, **40**, 1074–1077.

Siegmund, D. (1975) The time until ruin in collective risk theory. *Mitteil. Verein Schweiz. Versich. Math.*, **75**, 157–166.

Siegmund, D. (1976) Importance sampling in the Monte Carlo study of sequential tests. *Ann. Statist.*, **4**, 673–684.

Siegmund, D. (1979) Corrected diffusion approximations in certain random walk problems. *Adv. Appl. Probab.*, **11**, 701–719.

Siegmund, D. (1985) *Sequential Analysis.* Springer-Verlag, New York, Berlin, Heidelberg, Tokyo.

Skorokhod, A. V. (1965) *Studies in the Theory of Random Processes.* Krieger, Huntington.

de Smit, J. H. A. (1973) On the many server queue with exponential service times. *Adv. Appl. Probab.*, **5**, 170–182.

Smith, W. L. (1955) Regenerative stochastic processes. *Proc. Roy. Soc.*, Ser. A, **232**, 6–31.

Stam, A. J. (1969/71) Renewal theory in r dimensions. *Compos. Math.*, **21**, 383–399; *ibid.*, **23**, 1–13.

Stidham, S. (1972) Regenerative processes in the theory of queues, with applications to the alternating-priority queue. *Adv. Appl. Probab.*, **4**, 542–577.

Stone, C. (1966) On absolutely continuous distributions and renewal theory. *Ann. Math. Statist.*, **37**, 271–275.

Stoyan, D. (1983) *Comparison Methods for Queues and Other Stochastic Models* (D. J. Daley ed.). Wiley, Chichester, New York, Brisbane, Toronto, Singapore.

Syski, R. (1960) *Introduction to Congestion Theory in Telephone Systems.* Oliver and Boyd, London.

Takacs, L. (1962) *Introduction to the Theory of Queues.* Oxford University Press, New York.

Teugels, J. L. (1977) On the rate of convergence of a compound Poisson process. *Bull. Soc. Math. Belgique*, **39**, 205–216.

Teugels, J. L. (1982) Estimation of ruin probabilities. *Insurance: Mathematics and Economics*, **1**, 163–175.

Thorisson, H. (1983) The coupling of regenerative processes. *Adv. Appl. Probab.*, **15**, 531–561.

Tijms, H. C. (1972) *Analysis of the* (*s*, *S*) *Inventory Model.* Mathematical Centre Tracts, **40**. Mathematisch Centrum, Amsterdam.

Tweedie, R. L. (1982) Operator-geometric stationary distributions for Markov chains, with applications to queueing models. *Adv. Appl. Probab.*, **14**, 368–391.

Tweedie, R. L. (1983) Criteria for rates of convergence of Markov chains, with applications to queueing and storage models. In: *Probability, Statistics and Analysis* (J. F. C. Kingman and G. E. H. Reuter eds.), pp. 260–276. London Math. Soc. Lect. Not., Ser. **79**. Cambridge University Press, Cambridge.

Van Wouve, M., De Vylder, F., and Goovaerts, M. (1983) The influence of reinsurance limits on infinite time ruin probabilities. In: *Premium Calculation in Insurance* (F. De Vylder, M. Goovaerts, J. Haezendonck eds.). Reidel, Dordrecht, Boston, Lancaster.

Veraverbeke, N., and Teugels, J. L. (1975/76) The exponential rate of convergence of the distribution of the maximum of a random walk. *J. Appl. Probab.*, **12**, 279–288; *ibid.*, **13**, 733–740

Watson, G. N. (1958) *A Treatise on the Theory of Bessel Functions.* Cambridge University Press, Cambridge.

Whitt, W. (1972) Embedded renewal processes in the $GI/GI/s$ queue. *J. Appl. Probab.*, **9**, 650–658.

Whitt, W. (1974) Heavy traffic limit theory for queues: a survey. In: *Mathematical Methods in Queueing Theory* (A. B. Clarke ed.). Lecture Notes in Economics and Mathematical Systems 98, 307–350. Springer-Verlag, Berlin.

Whitt, W. (1980) Some useful functions for functional limit theorems. *Math. Opns. Res.*, **5**, 67–81.

Williams, D. (1979) *Diffusions, Markov Processes and Martingales.* Wiley, Chichester, New York, Brisbane, Toronto.

Wolff, R. W. (1977) An upper bound for multi-channel queues. *J. Appl. Probab.*, **14**, 884–888.

Wolff, R. W. (1982) Poisson Arrivals See Time Averages. *Opns. Res.*, **30**, 223–231.

Wolff, R. W. (1984) Conditions for finite ladder height and delay moments. *Opns. Res.*, **32**, 909–916.

Wolff, R. L. (1987) Upper bounds on work in system for multi-channel queues. *J. Appl. Prob.*, in press.

Woodroofe, M. (1982) *Nonlinear Renewal Theory in Sequential Analysis.* SIAM, Philadelphia.

Index

Applied Probability and Statistics (Continued)

JUDGE, HILL, GRIFFITHS, LÜTKEPOHL and LEE • Introduction to the Theory and Practice of Econometrics

JUDGE, GRIFFITHS, HILL, LÜTKEPOHL and LEE • The Theory and Practice of Econometrics, *Second Edition*

KALBFLEISCH and PRENTICE • The Statistical Analysis of Failure Time Data

KISH • Survey Sampling

KUH, NEESE, and HOLLINGER • Structural Sensitivity in Econometric Models

KEENEY and RAIFFA • Decisions with Multiple Objectives

LAWLESS • Statistical Models and Methods for Lifetime Data

LEAMER • Specification Searches: Ad Hoc Inference with Nonexperimental Data

LEBART, MORINEAU, and WARWICK • Multivariate Descriptive Statistical Analysis: Correspondence Analysis and Related Techniques for Large Matrices

LINHART and ZUCCHINI • Model Selection

LITTLE and RUBIN • Statistical Analysis with Missing Data

McNEIL • Interactive Data Analysis

MAINDONALD • Statistical Computation

MANN, SCHAFER and SINGPURWALLA • Methods for Statistical Analysis of Reliability and Life Data

MARTZ and WALLER • Bayesian Reliability Analysis

MIKÉ and STANLEY • Statistics in Medical Research: Methods and Issues with Applications in Cancer Research

MILLER • Beyond ANOVA, Basics of Applied Statistics

MILLER • Survival Analysis

MILLER, EFRON, BROWN, and MOSES • Biostatistics Casebook

MONTGOMERY and PECK • Introduction to Linear Regression Analysis

NELSON • Applied Life Data Analysis

OSBORNE • Finite Algorithms in Optimization and Data Analysis

OTNES and ENOCHSON • Applied Time Series Analysis: Volume I, Basic Techniques

OTNES and ENOCHSON • Digital Time Series Analysis

PANKRATZ • Forecasting with Univariate Box-Jenkins Models: Concepts and Cases

PIELOU • Interpretation of Ecological Data: A Primer on Classification and Ordination

PLATEK, RAO, SARNDAL and SINGH • Small Area Statistics: An International Symposium

POLLOCK • The Algebra of Econometrics

PRENTER • Splines and Variational Methods

RAO and MITRA • Generalized Inverse of Matrices and Its Applications

RÉNYI • A Diary on Information Theory

RIPLEY • Spatial Statistics

RIPLEY • Stochastic Simulation

RUBIN • Multiple Imputation for Nonresponse in Surveys

RUBINSTEIN • Monte Carlo Optimization, Simulation, and Sensitivity of Queueing Networks

SCHUSS • Theory and Applications of Stochastic Differential Equations

SEAL • Survival Probabilities: The Goal of Risk Theory

SEARLE • Linear Models

SEARLE • Matrix Algebra Useful for Statistics

SPRINGER • The Algebra of Random Variables

STEUER • Multiple Criteria Optimization

STOYAN • Comparison Methods for Queues and Other Stochastic Models

TIJMS • Stochastic Modeling and Analysis: A Computational Approach

TITTERINGTON, SMITH, and MAKOV • Statistical Analysis of Finite Mixture Distributions

UPTON • The Analysis of Cross-Tabulated Data

UPTON and FINGLETON • Spatial Data Analysis by Example, Volume I: Point Pattern and Quantitative Data

(*continued from front*)